Internet of Things

Technology, Communications and Computing

Series Editors

Giancarlo Fortino, Rende (CS), Italy

Antonio Liotta, School of Computing, Edinburgh Napier University, Edinburgh, UK

The series Internet of Things - Technologies, Communications and Computing publishes new developments and advances in the various areas of the different facets of the Internet of Things. The intent is to cover technology (smart devices, wireless sensors, systems), communications (networks and protocols) and computing (theory, middleware and applications) of the Internet of Things, as embedded in the fields of engineering, computer science, life sciences, as well as the methodologies behind them. The series contains monographs, lecture notes and edited volumes in the Internet of Things research and development area, spanning the areas of wireless sensor networks, autonomic networking, network protocol, agent-based computing, artificial intelligence, self organizing systems, multi-sensor data fusion, smart objects, and hybrid intelligent systems.

Indexing: *Internet of Things* is covered by Scopus and Ei-Compendex **

Ajay Prasad · Thipendra P. Singh ·
Samidha Dwivedi Sharma
Editors

Communication Technologies and Security Challenges in IoT

Present and Future

Springer

Editors
Ajay Prasad
School of Computer Science
University of Petroleum and Energy Studies
Dehradun, India

Thipendra P. Singh
School of Computer Science Engineering
and Technology
Bennett University
Greater Noida, Uttar Pradesh, India

Samidha Dwivedi Sharma
Department of Information Technology
Saudi Electronic University
Abha, Saudi Arabia

ISSN 2199-1073 ISSN 2199-1081 (electronic)
Internet of Things
ISBN 978-981-97-0051-6 ISBN 978-981-97-0052-3 (eBook)
https://doi.org/10.1007/978-981-97-0052-3

This Springer imprint is published by the registered company Springer Nature Singapore Pte Ltd.
The registered company address is: 152 Beach Road, #21-01/04 Gateway East, Singapore 189721, Singapore

Paper in this product is recyclable.

Preface

IoT is becoming part of our day-to-day activities. There are lots of areas where IoT has made its mark. There are many areas where IoT solutions are being or will be explored. The IoT is being adopted, incorporated, and envisioned in multiple areas of human life. It is also an area that is fast evolving. Also, many new communication technologies [8] are taking shape specifically suited for IoT in terms of problem solutions or domains. Major application domains include health care, industrial IoT, smart vehicles, agriculture, homes [6], cities, sustainability, etc. Different application domains will have a different set of challenges [1] in terms of implementation, mass expectations, mass acceptance, change management, privacy, security, etc. Challenges in terms of implementation will involve technology-related challenges. The available technology around IoT solutions has constraints and needs careful study before incorporation. The domains of IoT solutions as discussed in [5] desire high scalability, reliability, and availability. Above all, since the IoT adds to the comfort zones of people and organizations it is always going to be vulnerable to various attacks. There are challenges about sensor technologies, communication systems, communication protocols, edge computing [4], fog computing [2], etc. Both the technology and application areas in IoT involve several challenges that need to be pointed out. The book will present a collection of various challenges corresponding to both IoT application domains and IoT technology domains. The book will also present best practices that need to be followed while carrying out designing and developing IoT solutions. Every aspect of applications and corresponding technology will be covered through a set of case studies from several sources and industrial inputs. The book will be useful for IoT solution designers and developers to get initialization guidance and will help them channelize their approaches. Communication technologies play a vital role in IoT, and hence must be fast to adapt to the evolving security threats. It is obvious that apart from ensuring basic requirements of IoT solutions, security challenges of IoT and communication technologies and protocols are vital aspects of which a developer must be aware. The book is also designed to present the overall communication technologies and protocols used in IoT like focusing on the architecture and threat perseverance of each. The book also presents the new/future technological additions and changes coming up in communication systems like blockchain in IoT [3] and its possible security aspects. The book will also cover security aspects in communication mechanisms in domain-specific IoT solutions about healthcare, smart cities, smart homes, smart vehicles, etc. The overall objective of the book is to assist IoT developers to have a good insight into available and upcoming communication technologies so that they can employ the best possible practices while designing and developing IoT solutions.

Organization of the Book

The book is organized in a fashion where the Communication methodologies and their security challenges are discussed in the first ten chapters. The range of discussion starts from various communication modes and models towards the quantum era. The later nine chapters focus on a few application areas of IoT ranging from smart logistics, Smart Homes to healthcare. Here the case studies and architectures are presented with suggestive approaches addressing several security issues. The last six chapters present the advancements in IoT communication each with a different view and related security aspects. The book is organized into twenty-six chapters. A brief description of each of the chapters follows:

Chapter "Communication Technologies and Security Challenges in IoT: An Introduction" sets the preamble of entire topic and gives a comprehensive outlook on the topic.

Chapter "IoT Communication Models and Modes of Connectivity" presents various standard ways in which sensor data can be communicated which are hereby referred to as "communications models" and various means that are used to establish connections between different entities of IoT which are hereby referred to as "modes of communication". The technologies pertaining to these are also presented.

Chapter "Challenges and Key Issues in IoT Privacy and Security" addresses the challenges of IoT security and suggests methods to improve its security so that more benefits can be realized by the growing number of connected devices and services.

Chapter "Security Challenges in IOT" presents another way of looking into IoT Security Challenges. It discusses various IoT security challenges, IoT security architectures, IoT security solution trust zones and boundaries, potential risks of IoT devices, notable cases of IoT security breaches, solutions to IoT security breaches, strategies for securing IoT data, and best IoT security practices.

Yet another way to look into IoT Challenges is to perceive the fact that IoT platforms are rapidly changing. Chapter "Survey of IoT Security: Application Areas, Threats and Solutions" presents different technologies currently in use and those still in the development stage. The enhancement of IoT security is the focus of current research into four distinct technologies: edge computing, fog computing, blockchain, and machine learning.

People can use SaaS, PaaS, IaaS, and other benefits of cloud computing even though many cloud computing providers, like Google, Microsoft, IBM, and Amazon, are moving toward adopting cloud technology, which is causing a noticeable increase in the utilization of various cloud services. There are still numerous difficulties that users must overcome. Chapter "Security and Privacy Issues in Cloud and IoT Technology and Their Countermeasures" brings the Cloud into picture and relates the security challenges thus.

Chapter "Security Threats in IOT and Their Prevention" looks at IoT as layered architecture and discusses various security issues deeply in layer-wise fashion. It describes different kinds of security issues pertaining to each layer as a reference point for the beginners in IoT and discusses the means and methods to deal with these security threats.

In the context of IoT, Chapter "IPV6: Strengths and Limitations" focuses on the strengths and weaknesses of IPV6. It does give a good background of IPv6 and its transitioning intricacies.

Chapter "A Survey: Internet of Things (IoTs) Technologies, Embedded Systems and Sensors" brings in an elaborate study of various embedded systems such as Raspberry Pi, ZYNQ, Jetson, Altera Cyclone II, Beagle Bone, Odroid, STM32, and PSoC, were investigated, along with IoT communication protocols including IPv6, 6LowPAN, ZigBee, Bluetooth, Wi-Fi, LoRaWAN, Z-Wave, NFC, and Sigfox. Furthermore, a comparative analysis of the sensors utilized in these systems and their respective application areas has been provided.

Though the IP-based communications are more prevalent since the internet, the non-IP communications are taking charge in largely low-powered IoT setups. Chapter "Non-IP Based Communication in IoT: Handling Security Challenges" discusses IoT non-IP communication security issues. It outlines threats and attacks that can compromise the integrity and confidentiality of non-IP-based communication technologies and their inherent risks and vulnerabilities. Real-world case studies demonstrate non-IP-based IoT communication and analyze the security concerns and solutions in each scenario.

Bringing in the inevitable. Chapter "Security Challenges to IOT and Cloud-Based Systems in the Era of Quantum Attacks" discusses the threats to internetworked devices and their mitigation in the era of quantum computers.

Chapter "Overview of Internet of Things-Based Smart Logistics Systems" outlines essential technologies and effects of IoT-based smart logistics research and implementations, as well as their industry and regional distributions, are shown through a bibliometric analysis of articles published between 2008 and 2019.

Chapter "Fault Sensor Detection and Authentication Mechanism for Improving Quality of Services in Smart Homes" presents a fault sensor detection and authentication mechanism for improving the quality of services in smart homes. This chapter presents a fault identification mechanism of various sensors that help to resolve the fault and get the proper and enough information for various services. Further, the chapter validates the data source which helps to improve the data integrity.

Chapter "Security in IOT-Enabled Smart Agriculture Systems" discusses criteria for security based on a proposed architecture for smart agriculture that is irrespective of any core technologies that could be deployed. The authors confer the technology of IoT-based smart ecosystem for agriculture by estimating their framework, and their claim. In addition, it also presents opportunities and trends of IoT claims for precise agriculture and specifies the open challenges and issues of IoT in agriculture.

Chapter "Comprehensive Study of Intelligent Transference System in a Contemporary IoT Commuting Environment" discusses resourceful message communication amid the automobiles and RSUs in IoT setup for the Intelligent Transference System.

Chapter "Redefining Urban Development: Technology and Automation in India" explores the significance of technology and automation in transforming urban development in India. It highlights the key areas where technology and automation can revolutionize urban planning, infrastructure, governance, and sustainability. Additionally, the challenges and opportunities associated with implementing technology-driven solutions are discussed, along with examples of successful initiatives in Indian cities.

Chapter "Trust-Free Homes: The Zero-Trust Paradigm in a Smart Home Setting" is aimed to serve as a guide to implementing a Zero-Trust network model in Smart Home systems.

Chapter "Security Challenges of IoT-Enabled Vehicular Communications and Their Countermeasures" provides a comprehensive overview of the state of the art in this area and identifies challenges and solutions for securing current and future IoT-enabled vehicular communications.

Chapter "Industrial IoT Security Infrastructures and Threats" discusses essential ways to mitigate cyber-attacks on Industrial IoT systems and is presented with a comprehensive review and security strategy that includes risk assessment, threat modeling, and vulnerability testing as well as the proposed countermeasure models keeping IoT infrastructure in context.

Chapter "Industrial IOT: Security Threats and Counter Measures" comprehensively examines the security threats faced by industrial IoT systems and presents effective countermeasures to mitigate these risks. By exploring the unique characteristics of IIoT environments, such as interconnectedness, heterogeneity, and resource constraints, this study identifies potential vulnerabilities that can be exploited by adversaries.

Chapter "Internet-of-Things Enabled Smart Health Monitoring System Using AutoAI: A Graphical Tool of IBM Watson Studio" presents the use of internet-of-things (IoT) enabled smart health monitoring systems to improve healthcare service quality using a method including IBM Watson studio and AUTO AI services to produce a real-time warning and precise predictions of patient health.

Chapter "Phase Noise Performance of MIMO—GFDM Systems for Millimeter Wave 5G Technology" provides an understanding of phase noise effects and their implications for MIMO-GFDM systems, thus enhancing the development of efficient communication systems for millimeter wave 5G networks.

Chapter "Credit Card Fraud Detection Using Deep Learning for Internet-of-Things Enabled Smart Digital Financial System" discusses cases of a few machine learning-based and deep learning-based methods for Internet-of-Things-enabled smart digital financial systems and the associated difficulties.

Chapter "Development of a Secure and Transparent Blockchain for Electricity Bill Management in Smart Cities Using Enhanced Proof of Energy Consumption" presents a case of a blockchain-based model for electricity use in smart cities to address the issues of maintaining transparency and protecting consumers' privacy.

Chapter "A Comparative Study of Threat Detection for IoT Devices Using Machine Learning Techniques" brings forth a study of threat Detection for IoT Devices Using Machine Learning Techniques and examines how machine learning is applied to the IoT defense.

Chapter "Exhaustive Theoretical Study of Practical Free Space Optical Cooperative Relaying Technology: New Trends in IoT Communication" discusses a new trend of communication in IoT, i.e., "Optical Cooperative Relaying Technology". It brings insights into the key components of Optical Wireless transmission and to highlight various factors that need to be included in mixed radio/ Optical Wireless relaying systems to quantify the impact of its practical problems.

Dehradun, India — Ajay Prasad
Greater Noida, India — Thipendra P. Singh
Abha, Saudi Arabia — Samidha Dwivedi Sharma

References

1. Abiodun, O.I., Abiodun, E.O., Alawida, M. et al. A Review on the Security of the Internet of Things: Challenges and Solutions. Wireless Pers Commun 119, 2603–2637 (2021). https://doi.org/10.1007/s11277-021-08348-9
2. Adel, A. Utilizing technologies of fog computing in educational IoT systems: privacy, security, and agility perspective. J Big Data 7, 99 (2020). https://doi.org/10.1186/s40537-020-00372-z
3. Alfa, A.A., Alhassan, J.K., Olaniyi, O.M. et al. Blockchain technology in IoT systems: current trends, methodology, problems, applications, and future directions. J Reliable Intell Environ 7, 115–143 (2021). https://doi.org/10.1007/s40860-020-00116-z
4. Jazaeri, S.S., Jabbehdari, S., Asghari, P. et al. Edge computing in SDN-IoT networks: a systematic review of issues, challenges and solutions. Cluster Comput 24, 3187–3228 (2021). https://doi.org/10.1007/s10586-021-03311-6
5. Kaur, J., Jaskaran, Sindhwani, N., Anand, R., Pandey, D. (2023). Implementation of IoT in Various Domains. In: Sindhwani, N., Anand, R., Niranjanamurthy, M., Chander Verma, D., Valentina, E.B. (eds) IoT Based Smart Applications. EAI/Springer Innovations in Communication and Computing. Springer, Cham. https://doi.org/10.1007/978-3-031-04524-0_10
6. Touqeer, H., Zaman, S., Amin, R. et al. Smart home security: challenges, issues and solutions at different IoT layers. J Supercomput 77, 14053–14089 (2021). https://doi.org/10.1007/s11227-021-03825-1
7. Waverley Team and Iryna Hladun, TOP 12 TECHNOLOGY TRENDS IN IOT TO WATCH FOR IN 2023, June 21, 2023, https://waverleysoftware.com/blog/top-iot-tech-trends/

Acknowledgements

Editing a book is harder than we thought. None of this would have been possible without so many persons behind us. And this is next to impossible to enlist them here. The editors want to thank all those who were directly or indirectly involved in the project.

The editors want to thank their families, colleagues, and all well-wishers who helped them in achieving this milestone. We also want to put on record our appreciation for our respective Organizations **University of Petroleum and Energy Studies, Bennett University**, and **Saudi Electronic University** and their respective management to provide such a wonderful opportunity to let us grow academically.

Although the entire period of editing the book was filled with many ups and downs, our time with colleagues at office was worth of priceless. Many a time we felt low but people around us kept going. Our heartfelt thanks to you all.

Above and over, we feel thankful to the almighty to shower all His blessings upon us to complete this project.

Dr. Ajay Prasad
Dr. Thipendra P. Singh
Dr. Samidha Dwivedi Sharma

Contents

Communication Technologies and Security Challenges in IoT: An Introduction

Ambrish Kumar, Ajay Prasad, and Thipendra P. Singh

Abstract The Internet of Things (IoT) has become integral to our daily lives. IoT is tightly governed by the principles of communication and the technologies around it. Thus an in-depth discussion on Communication Technologies and Security Challenges in IoT is required for a professional to design and develop IoT applications. The core aspects that need elaborate discussion are being brought forward in this chapter. It is identified here that three major aspects of IoT Security are vital and require extensive discussion. They are, Communication methodologies and their security challenges, Application areas and approaches addressing several security issues, and the advancements/New trends in IoT communication and related security aspects.

Introduction

Kevin Ashton first used the term "Internet of Things" (IoT) in 1999, and the International Telecommunication Union (ITU) formally adopted it in 2005. The interconnectivity of IoT devices is also referred to as a "Network of Things" (NoT) [1]. Every NoT has a distinct function and goal, and together they make up the overall Internet of Things [2]. A global network of interconnected physical and virtual objects that are embedded with sensors, software, and network connectivity that allows these objects to gather and exchange data is what the ITU defines as the Internet of Things. These gadgets can be anything from basic household items to highly advanced industrial tools. Establishing independent connections between actual objects and applications

A. Kumar · T. P. Singh
School of Computer Science Engineering and Technology, Bennett University, Greater Noida, India
e-mail: ambrish.kumar@bennett.edu.in

T. P. Singh
e-mail: thipendra@gmail.com

A. Prasad (✉)
School of Computer Science UPES, Dehradun, India
e-mail: aprasad@ddn.upes.ac.in

A. Prasad et al. (eds.), *Communication Technologies and Security Challenges in IoT*, Internet of Things, https://doi.org/10.1007/978-981-97-0052-3_1

to facilitate information transfer and intelligent decision-making is the core concept of the Internet of Things. IoT makes it possible to link real-world activities to digital ones, which spurs the creation of a wide range of services and applications that are advantageous to professionals, enterprises, and individuals alike. The Internet of Things (IoT) is growing quickly; projections indicate that there will be more than 75 billion connected devices globally by 2025. Technological developments, falling costs, and the growing need for connected devices and services are the main drivers of this growth.

IoT System Model

The IoT is having a transformative impact on various industries, including healthcare, manufacturing, logistics, and smart cities. It enables new levels of automation, efficiency, and personalization. A generic IoT system model can be represented by three key layers [3]. Perception Layer, Transportation Layer, and Application Layer. Each of these layers summarized in Fig. 1 has its technologies that bring issues and some possible security weaknesses.

- **Perception Layer (PL)**: The Perception layer is responsible for collecting the data from the physical objects using sensors, actuators, and other embedded devices.
- **Transportation Layer (TL)**: The transportation layer is mainly responsible for transmitting the data between the perception layer(PL) and the application layer (AL). The data is typically transmitted over wireless networks, for example, Wi-Fi, Bluetooth, and cellular networks.
- **Application Layer (AL)**: In addition to offering services to consumers, the application layer is in charge of handling and assessing the data it receives from the

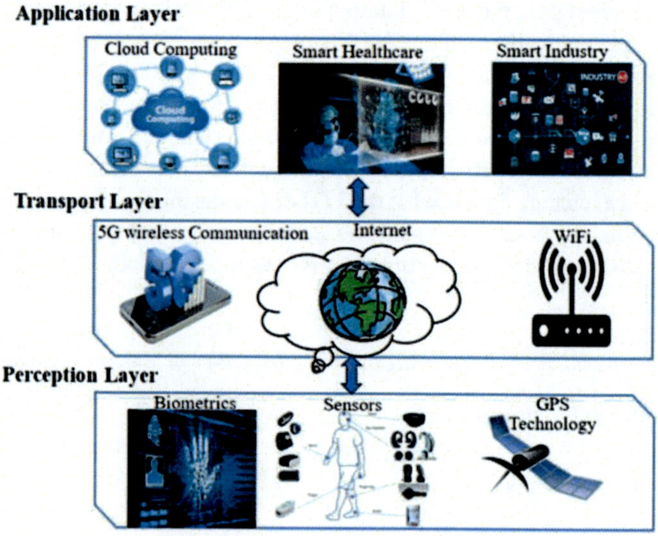

Fig. 1 IoT system model

layer of perception. This layer typically consists of software applications that can control the overall IoT system.

IoT Communication Models and Technology Requirements

The best communication technology for a particular IoT application will depend on several factors, including the range required, and the power consumption. The most commonly used communication technologies in IoT are as under:

- **IoT (Internet of Things) Sensors and Models of Communication Systems** [4]: Wireless sensor networks (WSNs) are used to connect and manage large numbers of low-power devices. These networks typically use short-range communication protocols such as ZigBee, Thread, and Bluetooth Low Energy (BLE). The networks are often used in industrial and environmental monitoring applications such as monitoring the temperature, humidity, and vibration of equipment in a factory or the air quality in a city.
- **Low-Power Communication Systems** [5]: Low-power communications are used to connect devices over long distances with low power consumption. Networks typically use protocols such as LoRaWAN, NB-IoT, and Sigfox. These networks are often used in smart cities and smart agriculture applications, for example, to monitor the level of water in a reservoir or the temperature of crops in a field.
- **Long-Range Communication Systems** [6]: Cellular networks are used to provide high-speed, reliable connectivity for long-range communication to IoT devices. Cellular networks typically use protocols such as LTE-M and LTE Cat-1. These networks are often used in connected car and wearable device applications, for example, to track the location of a vehicle or to monitor the health of a person wearing a fitness tracker.
- **Non-Internet Protocol (NON-IP) based Communication Systems** [7]: Cellphones and additional gadgets are frequently connected to Internet of Things (IoT) devices via Bluetooth connectivity, a short-range communication technology. IoT devices are frequently connected to smartphones and other devices via Bluetooth. It can be used to connect a smartwatch to a smartphone or a smart thermostat to a smartphone app.
- **Wireless Fidelity (*Wi-Fi*)**: IoT devices are frequently connected to home networks and other Wi-Fi networks via Wi-Fi, a medium-range communication protocol. IoT devices are frequently connected to home networks and other Wi-Fi networks via Wi-Fi. Additionally, it can be used to link a smart doorbell to a Wi-Fi-enabled doorbell chime or a smart speaker to a home network.
- **Ethernet**: IoT devices are frequently connected to wired networks via Ethernet, a high-speed communication protocol. IoT device connections to wired networks are frequently made using it. Additionally, it can be used to link a smart TV or security camera to a wired network.

IoT Sensors and Models of Communication

Sensors are key components of the IoT system for gathering information or data from physical objects and converting the data into digital signals for processing and analysis. These sensors can be extremely important in Internet of Things applications since they allow for automation, data-driven decision making, and continuous surveillance. Communication models in IoT systems define the way data is exchanged between devices, networks, and applications. These models play a significant role in ensuring efficient and reliable data transmission, especially in large-scale IoT deployments.

IoT Sensors: According to the Measurement Capabilities, IoT Sensors can be Defined as Follows

- **Environment Sensors**: These sensors measure environmental parameters such as temperature, humidity, pressure, light, and air quality. They are widely used in weather monitoring, smart homes, and industrial control systems.
- **Motion Sensors**: These sensors detect motion, including acceleration, tilt, and rotation. They are used in security systems, wearables, and robotics.
- **Proximity Sensors**: These sensors detect the presence of nearby objects. They are used in collision avoidance systems, proximity sensing for user interaction, and gesture recognition.
- **Chemical Sensors**: These sensors detect the presence and concentration of specific chemicals. They are used in environmental monitoring, industrial process control, and healthcare applications.
- **Biological Sensors**: These sensors detect biological phenomena such as heart rate, blood pressure, and blood glucose levels. They are used in wearable health monitoring devices and medical diagnostics.

Communication Models

Common Communication Models for IoT Applications Are as Follows [8].

- **Client–Server Model**: In this model, IoT devices act as clients, requesting data or services from servers. The server responds to client requests and provides the requested information or services. This model is widely used in web-based IoT applications.
- **Publish-Subscribe Model**: In this model, devices publish data to a central broker or messaging system. Subscribers interested in the data can subscribe to the topic and receive updates whenever new data is published. This model is well-suited for real-time data dissemination and event-driven applications.
- **Peer-to-Peer Model**: In this model, devices communicate directly with each other without the need for a central server or broker. This model is often used in mesh networks and applications that require decentralized control and data sharing.
- **Hybrid Models**: Combinations of the above models can be used to create more complex and flexible communication architectures for specific IoT applications (Fig. 3).

Communication Methodologies and Their Security Challenges

Many IoT devices are vulnerable to attack due to weak or default passwords, inadequate authorization mechanisms, insecure communication protocols, lack of regular software updates and security patches, limited resources, and the heterogeneous nature of the IoT landscape. An overview of security challenges in IoT systems that needs to be tackled is presented in Fig. 2. IoT device growth presents a number of security problems that should be carefully considered. Vulnerabilities in IoT devices due to their limited computing capacity can make them susceptible to viruses and hacking attempts, exacerbated by the absence of regular security updates. Privacy and data integrity are paramount, particularly in sensitive domains like healthcare and smart cities, where securing the confidentiality and integrity of data is imperative. Additionally, the scalability of IoT ecosystems, often comprising millions of devices, poses significant challenges to conventional centralized security methods. Ensuring that only authorised users can interact with IoT devices requires robust access control and authentication procedures. This is because poor or nonexistent authentication can result in unauthorised access and data breaches. This study examines the various security issues that arise in the Internet of Things environment and addresses appropriate risk mitigation techniques. Therefore, security is a critical concern for IoT systems, as they collect and transmit a vast amount of sensitive data. By understanding the security challenges at each layer of the IoT system, organizations can take steps to mitigate these risks and protect their data and systems. The security challenges at different layers of the IoT system are described in Table 1.

Fig. 2 An overview of security challenges in IoT systems

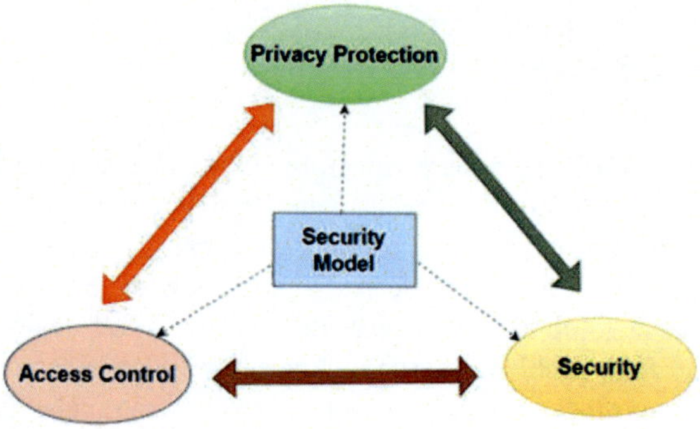

Fig. 3 Security model of the IoT system

Table 1 Threats in IoT system model

Layer	Main threats
Prediction layer	Data leakage
	DoS attacks
	Malicious code injection
Transportation layer	Routing attacks
	DoS attacks
	Data transmit attacks
Application layer	Physical attacks
	Impersonation
	DoS attacks
	Routing attacks (e.g. in WSN, RSN)
	Data transmit attacks (e.g. in WSN, RSN)

Challenges at the Perception Layer

- **Data Tampering at Physical Object**: An Internet of Things system's data obtained through a sensor node may contain inaccurate or deceptive information. Furthermore, a physical injection of malicious malware could grant an attacker access to the IoT system.
- **Impersonation**: During the authentication process, an attacker may use a fictitious identity to launch harmful attacks against an IoT system.
- **Denial of Service (DoS)**: An attacker uses the legitimate node's limited processing power to prevent it from accepting new requests for service.
- **Routing attacks**: The routing pathway of the network is changed by a malicious intermediary node during the data collection and forwarding operation.
- **Data transit attacks**: These happen when an attacker compromises the integrity and confidentiality of information being transported throughout the data processing process..

Security Challenges at the Transportation Layer

- **Interception**: Internet of Things (IoT) gadgets may be employed to track adherence to medications, keep an eye on the vitals of patients, and offer distant treatment.
- **Manufacturing**: IoT devices can be used to optimize production processes, monitor equipment health, and track inventory levels.
- **Logistics**: IoT devices can be used to track shipments in real-time, optimize delivery routes, and reduce costs.
- **Smart Cities**: IoT devices can be used to monitor traffic congestion, optimize energy consumption, and improve city services.

Security Challenges at the Application Layer

- **Unauthorized Access**: A malicious actor can access the application server or cloud infrastructure that can steal or modify the system configurations in order to disturb the IoT system.
- **Vulnerabilities in Application Software**: An adversarial node may utilise the application software to obtain private information or take over the Internet of Things.
- **Data Breaches**: A malicious node can expose the sensitive information about the users of the IoT system itself.

Communication Protocols in IoT Systems

The Internet of Things (IoT) depends on protocols for communication to facilitate data flow and device engagement [8]. By defining the guidelines and formats for data transfer, these protocols guarantee interoperability and compatibility across various internet of things networks and devices.

- **Internet Protocol Version 6 (IPv6) protocols**

  IPv6 is the successor to IPv4, the current protocol that underpins the Internet. IPv6 has been in development since the early 1990s and was officially standardized in 1998. However, it has only recently begun to be widely adopted. Some key IPv6 protocols can be described as follows.
- **Control Message Protocol version 6 (CMPv6)**: CMPv6 is responsible for sending control messages between IPv6 devices, such as error reports and diagnostic information.
- **Neighbor Discovery Protocol (NDP)**: NDP protocol is used to discover and maintain neighbor relationships between IPv6 devices on a local network.
- **Stateless Address Autoconfiguration (SLAAC) Protocol**: SLAAC allows IPv6 devices to automatically configure their own IP addresses without requiring manual configuration or DHCP servers.
- **Dynamic Host Configuration Protocol version 6 (DHCPv6)**: It is an additional sophisticated method for configuring IPv6 devices with an IP address alongside additional network-related settings.
- **Path MTU Discovery (PMTUD) Protocol**: In order to prevent message fragmentation and increase transmitting effectiveness, the PMTUD protocol is used to calculate the maximal transmitting unit (MTU) size over a network path.

- **Multicast Listener Discovery (MLD) Protocol**: In order to control multicast group membership and guarantee that multicast packets are only sent to interested devices, the MLD standard is utilised.
- **Secure Neighbor Discovery (SND) Protocol**: This protocol provides cryptographic authentication for neighbor discovery messages to prevent attackers from impersonating other devices and disrupting network communications.
- **Authentication Header (AH) Protocol**: The AH protocol provides optional authentication and integrity protection for IPv6 packets.
- **Encapsulating Security Payload (ESP) Protocol**: It can provide optional encryption and authentication for IPv6 packets.

Internet Protocol (IP)-Based Protocols

Internet protocol (IP)-based is the fundamental aspect of the IoT system. It is an underlying protocol suite that enables devices to exchange data between objects over the internet [9]. In IoT, IP-based communication is very important as it can manage and connect a vast network of devices by enabling them to share information. IP-based communication is a cornerstone of the IoT, providing a standardized, scalable, and secure framework for connecting and managing a vast network of devices. Its integration with existing infrastructure, support for diverse network technologies, and ability to accommodate future protocols make it an essential enabler for the continued growth and development of the IoT. The key IP-based communication protocols given in Table 2 can be described as follows.

- **Transmission Control Protocol (TCP)**: The connection-oriented protocol TCP offers dependable and systematic transfer of information across apps. In order to ensure the confidentiality of information and avoid loss or fraud, it connects both ends, divides the data into segments, then reconstructs these at the point of reception.

Table 2 Communication protocols in IoT systems

IPv6 protocol	IP based protocol	Non-IP based protocol	RFID protocols	NFC protocols
CMPv6	TCP	LoRaWAN	EPC Class 1 Generation 2 (ISO 18000–3)	TNEP
NDP	UDP	Sigfox	Mifare (ISO 14443)	DOP
SLAAC	IP	Zigbee	EPCglobal Gen 2	BCP
DHCPv6	ICMP	BLE	UHF RFID	LAC
PMTUD	ARP	6LoWPAN	HF RFID	SNEP
MLD	DHCP		LF RFID	SNEP +
SND				
AH				
ESP				

- **User Datagram Protocol (UDP)**: UDP is a connectionless protocol that provides best-effort data delivery between applications. It sends data packets without establishing a prior connection, making it faster and less overhead-intensive than TCP. UDP is suitable for applications where data delivery is less critical, such as streaming audio or video.
- **Internet Protocol (IP)**: The network layer protocol known as IP is responsible for addressing and directing information packets over the Internet. It gives devices and networks distinct IP addresses, allowing routing algorithms to choose the most efficient route for information packets to take in order to get to their intended location.
- **Internet Control Message Protocol (ICMP)**: ICMP is a network layer protocol used for sending control messages between devices. It provides error reporting, diagnostic information, and network management capabilities, enabling troubleshooting and monitoring of network connectivity and performance.
- **Address Resolution Protocol (ARP)**: ARP is a protocol at the network layer that maps IP addresses to MAC addresses. By doing so, devices on the same local network are able to communicate directly with one another by learning the physical MAC address of a device linked to an IP address that is particular.
- **Dynamic Host Configuration Protocol (DHCP)**: A protocol at the network layer called DHCP is used to give IP addresses and other network configuration parameters to connected devices dynamically. It lowers administrative cost and streamlines network administration, especially in large networks with changing configurations of devices.

Protocols not Based on IP

The IoT applications can benefit greatly from non-IP-based connectivity [7]. These protocols are appropriate for limited IoT contexts due to their low power consumption, low latency, and immunity to interference, in contrast to IP-based communications. The following is a description of the non-IP-based protocols listed in Table 2.

- **Wide Area Network for Long Range (LoRaWAN)**: Using chirp spread spectrum technology, LoRaWAN is a low-power wide-area network (LPWAN) protocol that enables low-power, long-range communication. It works especially well with Internet of Things devices that need to transmit data only sometimes, like asset tracking devices and sensors for environmental monitoring.
- **Sigfox**: This other LPWAN protocol makes use of ultra-narrowband technology to provide low-power, long-range communication. While operating at a higher frequency than LoRaWAN, it is comparable and allows for quicker data throughput. IoT devices like smart parking systems and smart metres that need to transmit data sometimes are a good fit for Sigfox.
- **Zigbee**: Zigbee is a low-power wireless personal area network (WPAN) protocol that uses the unlicensed 2.4 GHz frequency. It is frequently utilised in enclosed spaces, such smart homes and buildings, for short-range communication between

IoT devices. Mesh networking is made possible via Zigbee, which allows devices to relay data for greater coverage.

- **Bluetooth Low Energy (BLE)**: BLE is a low-power Bluetooth wireless protocol variation created especially for short-range, low-power communication. In order to facilitate data sharing and configuration, it is frequently used to couple Internet of Things devices with smartphones or tablets. Beacon technology, which enables proximity-based location tracking applications, is another application that uses BLE.
- IPv6 over Low-Power Wireless Personal Area Networks (6LoWPAN): Zigbee and 802.15.4 are examples of low-power wireless networks that can support IPv6 communication thanks to the 6LoWPAN protocol. It changes IPv6 packets into more manageable, smaller datagrams that can be sent over these low-power networks. IoT devices can join bigger IoT ecosystems and connect to the Internet thanks to 6LoWPAN.

Protocols for radio frequency identification (RFID)

RFID [10, 11] technology can wirelessly identify and monitor objects, it has become a critical enabler for the Internet of Things. RFID readers enable real-time tracking and monitoring of a variety of assets by the attachment of RFID tags, which are embedded with unique IDs. These guidelines consist of.

The most popular RFID protocol is EPC Class 1 Generation 2 (ISO 18000–3), which is especially useful for supply chain management and asset monitoring applications. It can read at a distance and operates in the ultra-high frequency (UHF) band.

- Mifare (ISO 14443): Applications for contactless payments and proximity cards frequently employ this protocol. It provides a good mix of performance and range while operating in the high frequency (HF) band.

Another well-liked protocol for contactless and proximity card applications is iCOde (ISO 15693). It provides a performance and security combination while operating in the HF band.

- EPCglobal Gen 2: Developed by EPCglobal to improve interoperability and compatibility among RFID systems, this is a standardised version of ISO 18000–3. It specifies extra requirements for reader communication, tag identification, and data encoding.
- RFID systems that operate in the UHF band, usually at frequencies between 860 and 960 MHz, are referred to as Ultra High Frequency (UHF) systems. Because UHF RFID tags can read over a long distance, they are useful for inventory management and asset tracking applications.
- High Frequency (HF) RFID: This describes RFID systems that use frequencies in the HF band, usually 3.56 MHz to 13.56 MHz. Due to its ability to balance performance and range, HF RFID tags are a good fit for contactless payment applications, access control systems, and proximity cards.

- RFID systems that operate in the LF region, usually at frequencies between 30 and 500 kHz, are referred to as low-frequency (LF) systems. Although LF RFID tags have a limited reading range, they are less prone to interference and can pass through metal and water.

Protocols for Near Field Communication (NFC)

NFC [12] is a 13.56 MHz short-range wireless communication technology which enables data sharing between two devices when they are near together, usually less than 10 cm. NFC is commonly utilised in mobile phones, contactless payment cards, and other devices for peer-to-peer data transfer, access control, and secure transactions. Numerous NFC protocols are optimised for Internet of Things applications, providing improved functionality and features catered to the particular requirements of IoT networks and devices. These guidelines consist of.

- Tag NDEF Exchange Protocol (TNEP): NFC-enabled devices can communicate with one other by exchanging NDEF (Near Field Communication Data Exchange Format) messages thanks to this lightweight protocol. NDEF messages are appropriate for a variety of Internet of Things applications since they can contain multiple data kinds, including text, URLs, and contacts.
- Device Orientation Protocol (DOP): NFC devices can detect their relative positions and orientations by exchanging orientation information using the standardised method provided by DOP. Applications like augmented reality and indoor navigation that depend on exact placement need this information.
- Battery Check Protocol (BCP): BCP enables power management and guarantees that devices have enough power for communication by enabling NFC devices to inquire about the battery level of other NFC devices.
- Logical Access Control (LAC) Protocol: NFC devices can transmit sensitive data, including encryption keys and authentication credentials, via a secure channel with LAC. Applications that need higher security, such financial transactions and access control systems, depend on this protocol.
- Simple NDEF Exchange Protocol (SNEP) Protocol: This protocol is designed to make it easier for NFC devices to exchange NDEF messages with one another. It is especially helpful for low-power Internet of Things sensors and tags, as well as other devices with constrained memory or processing capabilities.
- Secure NDEF Exchange Protocol (SNEP+): SNEP+ uses Secure Sockets Layer (SSL) encryption to improve the security of NDEF message exchange. Applications requiring the highest level of security, like sensitive data sharing and financial transactions, are advised to use this protocol.

Securing IoT Applications

The network of physically connected objects that communicate and share data with one another over the Internet is known as the Internet of Things (IoT). IoT applications offer creative answers to a variety of problems across a broad range of sectors and use cases. Figure 4 displays a few typical IoT applications from various industries.

The domains listed in Fig. 4 are just a few examples, and the applications of IoT continue to expand as technology advances. The potential of gadgets together and exchange data in real time opens up numerous possibilities for improving efficiency, reducing costs, and creating new services across various industries.

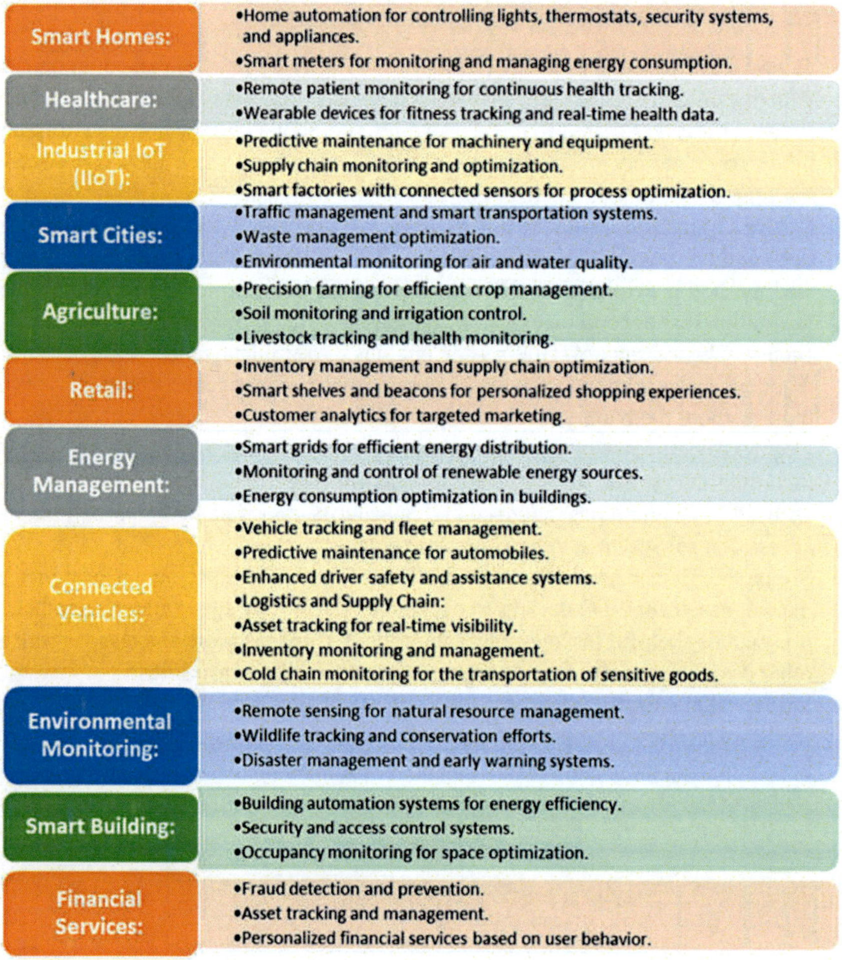

Fig. 4 IoT application domains

Ensuring the security of IoT applications is paramount in safeguarding the interconnected network of devices that form the Internet of Things (IoT) [13]. With the proliferation of connected gadgets across various industries, the complexity of securing these applications has grown exponentially. Robust security measures begin with strong device authentication, utilizing unique identifiers and secure key management protocols. Secure communication channels, employing encryption methods such as TLS or DTLS, are important to safegaurd data in transit. Regular firmware updates and patch management are critical for addressing vulnerabilities and maintaining the resilience of IoT devices. Network security, both through segmentation and the deployment of firewalls and intrusion detection systems, helps mitigate the impact of potential breaches. Physical security measures, including tamper-evident packaging and access controls, add a layer of protection. Privacy considerations, user education, and compliance with regulatory standards contribute to a holistic approach to IoT application security. Ongoing security testing, incident response planning, and vendor security assessments are integral components of a comprehensive strategy to identify and address potential threats in this dynamic and evolving landscape.

When studying IoT (Internet of Things) application security, there are several factors that need to be considered. Here are some key areas to focus on:

Device Security

Put robust permission and authentication processes in place to manage who has access to IoT networks and devices. To limit access to authorised users exclusively, utilise restricted access lists, encryption, and distinctive passwords. There are a number of factors to consider when thinking about IoT device safety [14]. Following are a few important things to think about:

- Secure Hardware Design: Ensuring that the hardware components of IoT devices are designed with security in mind. This includes using secure elements for storing cryptographic keys, implementing tamper-resistant hardware, and protecting against physical attacks.
- Secure Software Development: Following secure coding practices during the software development process for IoT devices. This includes minimizing vulnerabilities such as buffer overflows, input validation issues, and insecure configurations. Secure coding practices should also address authentication, encryption, and access control mechanisms.
- Authentication and Authorization: Implementing best authentication mechanisms to verify the identity of IoT devices and users accessing the devices. This may involve using unique device credentials, digital certificates, or secure tokens. Additionally, implementing proper authorization mechanisms to control device access and permissions is crucial.
- Encryption and Data Protection: Encrypting sensitive data both at rest and in transmission. This includes using strong encryption algorithms to safegaurd information from unauthorized access or tampering. Encryption should cover communication between devices, as well as data stored on the device and transmitted to the cloud or other systems.

- Over-the-Air (OTA) Updates: Implementing a safe and robust mechanism for delivering OTA updates to IoT devices. This includes verifying the authenticity and integrity of updates, ensuring secure communication channels for delivering updates, and providing rollback mechanisms in case of failed updates.
- Physical Security Measures: Protecting the physical integrity of IoT devices to prevent unauthorized access, tampering, or theft. This may involve using tamper-evident seals, secure enclosures, and physical access controls.
- Secure Communication Protocols: Using secure communication protocols, such as HTTPS, MQTT with TLS, or CoAP with DTLS, to establish encrypted and authenticated communication between devices and other components in the IoT ecosystem. This helps protect data privacy and integrity.
- Device Management and Monitoring: Implementing robust device management and monitoring capabilities to detect and respond to security events. This includes monitoring device behavior, detecting anomalies, and implementing mechanisms for remote device management, such as revoking device access or disabling compromised devices.
- Privacy Considerations: Considering privacy implications when designing IoT devices. Minimizing the collection and use of personal data, implementing privacy-by-design principles, and complying with relevant privacy regulations are important aspects of IoT device security.
- Security Testing and Vulnerability Management: Conduct regular security testing, including penetration testing and vulnerability assessments, to identify and remediate security vulnerabilities in IoT devices. Establishing a process for managing and patching vulnerabilities is crucial to maintaining the security of IoT devices over their lifecycle.

It's important to note that IoT device security requires a holistic approach, considering both hardware and software components, as well as the broader ecosystem in which the devices operate.

Data Security

Encrypt data at rest and in transit to protect it from unauthorized interception or tampering. Use secure protocols like HTTPS, TLS/SSL, and IPSec for data transmission. Implement data encryption algorithms like AES, RSA, or ECC to safeguard sensitive data. Data security in IoT (Internet of Things) applications is a critical aspect that involves protecting the confidentiality, integrity, and availability of data generated and exchanged by connected devices. Several key considerations contribute to robust data security in IoT applications [15]:

- Encryption: Utilize strong encryption algorithms to protect data both in transit and at rest. This prevents unauthorized access to sensitive information as it travels between devices and the cloud.
- Authentication and Authorization: Implement secure authentication mechanisms to ensure that only authorized devices and users can access IoT data. Proper authorization controls define what actions and data each device or user can access.

- Secure APIs: Use secure Application Programming Interfaces (APIs) for communication between devices and backend systems. API security involves employing authentication tokens, encrypted connections, and proper access controls.
- Data Integrity: Implement mechanisms to ensure the integrity of IoT data. This involves verifying that data has not been tampered with during transmission or storage, ensuring its accuracy and reliability.
- Role-Based Access Control (RBAC): Enforce role-based access controls to limit access to sensitive data based on the roles and responsibilities of users and devices. This helps prevent unauthorized users from gaining access to critical information.
- Secure Device Onboarding and Lifecycle Management: Securely onboard devices to the IoT network by using secure provisioning methods. Implement proper lifecycle management to handle device activation, deactivation, and updates securely.
- Data Minimization: Adopt a principle of data minimization, where only the necessary data is collected and stored. This reduces the potential impact in case of a security breach.
- Secure Cloud Services: If data is stored in the cloud, ensure that cloud services are secured with strong authentication, access controls, and encryption. Regularly monitor and audit cloud infrastructure for security vulnerabilities.
- IoT Gateway Security: Secure IoT gateways that mediate communication between devices and the cloud. Apply security measures such as firewalls, intrusion detection systems, and regular security updates.
- Privacy by Design: Integrate privacy considerations into the design and development of IoT applications. This involves minimizing the collection of personally identifiable information (PII) and ensuring that privacy controls are implemented.
- Security Analytics and Monitoring: Implement security analytics and monitoring tools to detect unusual patterns or anomalies in data traffic. Prompt identification of security incidents allows for swift response and mitigation.
- Regulatory Compliance: Adhere to relevant data protection and privacy regulations. Understand the legal requirements for handling IoT data and ensure compliance with standards such as GDPR, HIPAA, or industry-specific regulations.

A comprehensive approach to data security in IoT applications involves a combination of technological measures, secure development practices, and ongoing monitoring and adaptation to emerging threats. As the IoT landscape continues to evolve, maintaining a proactive stance on data security is essential to building and maintaining trust in connected systems.

Other application security measures

- User Awareness and Training: Educate users about IoT security risks and best practices. Train users on strong password management, phishing awareness, and secure device handling procedures.
- Vulnerability Management: Regularly scan IoT devices and networks for vulnerabilities and misconfigurations. Use vulnerability scanning tools and penetration testing to identify and remediate security weaknesses.

- Incident Response: Establish an incident response plan to effectively handle security incidents. Define clear roles, responsibilities, and communication protocols for incident response.
- Compliance: Adhere to relevant industry regulations and data privacy laws. Implement data governance practices to ensure compliance with data protection requirements.
- Continuous Improvement: Continuously review and update IoT security practices to adapt to evolving threats and technologies. Stay informed about emerging security trends and vulnerabilities, and implement appropriate countermeasures.

New Trends in IoT Communication and Related Security Aspects

IoT security and communication technologies have a bright future ahead of them, with a number of exciting new innovations planned. The creation of new, more secure communication protocols is one important trend. One new area of study is post-quantum cryptography, which is creating new cryptographic algorithms that are immune to assault by quantum computers. In the future, these algorithms might be applied to secure Internet of Things communication protocols. The creation of fresh security solutions created especially for the Internet of Things is another important trend. Moreover, new intrusion detection and prevention systems that are more adept at recognising and obstructing malicious traffic can be created using artificial intelligence (AI) and machine learning (ML) technology. The Internet of Things is being more and more linked with other technologies, such edge computing and 5G, which opens up new possibilities for processing data and facilitating safe communication. Furthermore, Edge computing can handle data from Internet of Things devices closer to the point of collection, which can lower latency and boost security. Future IoT communication technologies may be secured by promising new technologies like homomorphic encryption, zero-trust security, and post-quantum cryptography. New security solutions that are being developed expressly for the Internet of Things include safe device management solutions, blockchain-based security solutions, and AI and ML-based intrusion detection and prevention systems. In the Internet of Things, edge computing, 5G, and other cutting-edge technologies are opening up new possibilities for data processing and secure communication. IoT security and connectivity technologies have a bright future ahead of them. An increasingly dependable and safe Internet of Things is being created via the creation of new integrated technologies, security solutions, and communication protocols.

Because it allows businesses and organisations to make well-informed decisions based on data acquired from IoT sensors, forecasting is a crucial component of the Internet of Things. IoT forecasting can offer useful insights into future behaviour by evaluating past data and seeing patterns and trends. This allows for proactive planning and resource allocation. Furthermore, protecting linked devices, networks, and the data they produce requires addressing risks associated with the Internet of Things.

Potential security hazards are growing along with the prevalence and sophistication of IoT devices. A multi-layered strategy that addresses device security, network security, data security, and user behaviour is required to effectively counter these threats.

New IoT communication trends and changing security considerations follow the rapid advancement of technology. The following are some noteworthy developments in IoT communication and their security implications:

- 5G Connectivity: By enabling low-latency apps and offering quicker and more dependable connectivity, the rollout of 5G networks improves Internet of Things communication. New security issues brought about by 5G include possible weaknesses in the network design. To meet these issues, security protocols—such as secure edge computing and 5G network slicing—must be modified [16].
- Edge Computing: Edge computing [17] lowers latency and improves real-time data processing by bringing processing power closer to Internet of Things devices. It becomes imperative to secure edge devices, which involves putting strong access controls, encryption, and intrusion detection technologies in place at the edge. Closer to the source, edge security aids in the protection of sensitive data.
- Blockchain Integration: Research is being done to improve the security and reliability of data transfers and Internet of Things transactions using blockchain technology [18]. Blockchain technology can offer a decentralised, impenetrable ledger, boosting the security of Internet of Things transactions and data integrity. However, there are difficulties with scalability and interoperability when adopting blockchain.
- AI and Machine Learning in Security: In IoT ecosystems, anomaly detection, threat analysis, and predictive security are achieved through the application of AI and ML [19]. By spotting patterns suggestive of possible security risks, AI and ML can increase the effectiveness of security monitoring. Nonetheless, there is increasing worry about protecting AI models from hostile assaults.
- Software-Defined Networking (SDN): SDN facilitates more adaptable and dynamic network administration, resulting in effective device communication. SDN must be properly secured to thwart potential network configuration manipulation and unauthorised access. Regular monitoring, encryption, and authentication should all be part of your security protocol.
- Zero Trust Security Model: Using a zero trust security model means that by default, no entity—internal or external to the network—is trusted [20]. Strict access limits, ongoing authentication, and careful monitoring are all part of zero trust, which lowers the attack surface and strengthens security posture overall.
- IoT Device Identity Management: To avoid unwanted access, there is a greater emphasis on controlling and safeguarding the identities of IoT devices. Secure onboarding, lifecycle management, and making sure that only authorised and authenticated devices are able to access the network are all necessary for the implementation of strong device identity management.

- Quantum-Safe Cryptography: Researching quantum-safe cryptographic solutions is a result of anticipating how quantum computing may affect existing cryptographic techniques [21]. To keep IoT communications secure as quantum computing develops, encryption techniques must switch to ones that can withstand quantum attacks.
- Green IoT and sustainability: There is a growing focus on IoT devices and communication protocols that are sustainable and low-energy [22]. Energy-efficient gadgets and low-power communication protocols may be used in sustainable Internet of things solutions. In order to prevent sacrificing security in the sake of sustainability, security concerns ought to be incorporated into the design of such systems.
- Regulatory Compliance: Frameworks addressing IoT security and privacy are receiving more attention from regulators. To maintain compliance, organisations need to stay up to date on the latest legislation and standards. This covers laws pertaining to data protection, standards unique to a given industry, and recommendations for Internet of Things security.

Building and maintaining a robust and secure IoT ecosystem requires keeping up with these changes and resolving the related security issues. IoT technologies are dynamic, thus security needs to be proactive and flexible.

Conclusion

In the framework of the present and future, the chapter provides an overview of the three main points of view on communication technologies and security challenges in the Internet of Things. The areas are: issues related to IoT application domains; challenges related to communication technologies; and security challenges in both current and future IoT contexts. Understanding IoT communication types and modalities of connectivity, as well as the main privacy and security concerns and problems they present, as well as the threats and solutions found in the IoT application areas, is crucial. Cloud and Internet of Things security and privacy issues, IOT security dangers and how to prevent them, IPv6's advantages and disadvantages, embedded systems, sensors in the Internet of Things, and security issues, Managing Security Issues with IP-Based and Non-IP-Based Internet of Things Communication The application of blockchain technology in the Internet of Things, threat detection for IoT devices using machine learning techniques, and many other new trends in IoT communication are among the security challenges facing IoT and cloud-based systems in the era of quantum attacks. Other challenges include security issues in various IoT application domains and current and upcoming developments and their impact on IoT, such as 5G systems.

References

1. Voas, J. M. (2016). *Networks of 'things': National institute of standards and technology*. Gaithersburg, MD, NIST SP, pp 800–183. https://doi.org/10.6028/NIST.SP.800-183
2. Garrity, J. (2015). *Harnessing the Internet of Things for global development*. Rochester, NY. https://doi.org/10.2139/ssrn.2588129
3. Mashal, I., Alsaryrah, O., Chung, T.-Y., Yang, C.-Z., Kuo, W.-H., & Agrawal, D. P. (2015). Choices for interaction with things on Internet and underlying issues. *Ad Hoc Networks, 28*, 68–90. https://doi.org/10.1016/j.adhoc.2014.12.006
4. Tayeb, S., Latifi, S., & Kim, Y. (2017). A survey on IoT communication and computation frameworks: An industrial perspective. In *2017 IEEE 7th Annual Computing and Communication Workshop and Conference (CCWC)* (pp. 1–6). https://doi.org/10.1109/CCWC.2017.7868354
5. Durand, T. G., Visagie, L., & Booysen, M. J. (2019). Evaluation of next-generation low-power communication technology to replace GSM in IoT-applications. *IET Communications, 13*(16), 2533–2540. https://doi.org/10.1049/iet-com.2019.0168
6. Centenaro, M., Vangelista, L., Zanella, A., & Zorzi, M. (2016). Long-range communications in unlicensed bands: The rising stars in the IoT and smart city scenarios. *IEEE Wireless Communications, 23*(5), 60–67. https://doi.org/10.1109/MWC.2016.7721743
7. Giuliano, R., Mazzenga, F., Neri, A., & Vegni, A. M. (2014). Security access protocols in IoT networks with heterogenous non-IP terminals. In *2014 IEEE International Conference on Distributed Computing in Sensor Systems* (pp. 257–262). https://doi.org/10.1109/DCOSS.2014.50
8. Kulkarni, S., & Kulkarni, S., Communication models in Internet of Things: A survey *3*(11).
9. Heer, T., Garcia-Morchon, O., Hummen, R., Keoh, S. L., Kumar, S. S., & Wehrle, K. (2011). Security challenges in the IP-based Internet of Things. *Wireless Personal Communications, 61*(3), 527–542. https://doi.org/10.1007/s11277-011-0385-5
10. Ranky, P. G. (2006). An introduction to radio frequency identification (RFID) methods and solutions. *Assembly Automation, 26*(1), 28–33. https://doi.org/10.1108/01445150610645639
11. Karygiannis, A. T., Eydt, B., Barber, G., Bunn, L., & Phillips, T. (2007). Guidelines for securing radio frequency identification (RFID) systems. National institute of standards and technology. Gaithersburg, MD, NIST SP, pp 800–98. https://doi.org/10.6028/NIST.SP.800-98
12. Al-Ofeishat, H. A. (2012). Near field communication (NFC)
13. Fernandez, E. B., Washizaki, H., Yoshioka, N., & Okubo, T. (2021). The design of secure IoT applications using patterns: State of the art and directions for research. *Internet Things, 15*, 100408. https://doi.org/10.1016/j.iot.2021.100408
14. Kim, H. -J., Chang, H. -S., Suh, J. -J., & Shon, T. (2016) A study on device security in IoT convergence. In *2016 International Conference on Industrial Engineering, Management Science and Application (ICIMSA)* (pp. 1–4). https://doi.org/10.1109/ICIMSA.2016.7503989
15. "Data security and privacy in the Internet of Things (IoT) Environment | SpringerLink." Accessed: Nov. 21, 2023. https://link.springer.com/chapter/https://doi.org/10.1007/978-3-319-33124-9_11
16. Li, S., Xu, L. D., & Zhao, S. (2018). 5G Internet of Things: A survey. *Journal of Industrial Information Integration, 10*, 1–9. https://doi.org/10.1016/j.jii.2018.01.005
17. Cao, K., Liu, Y., Meng, G., & Sun, Q. (2020). An overview on edge computing research. *IEEE Access, 8*, 85714–85728. https://doi.org/10.1109/ACCESS.2020.2991734
18. Viriyasitavat, W., Xu, L. D., Bi, Z., & Hoonsopon, D. (2019). Blockchain Technology for Applications in Internet of Things—Mapping From System Design Perspective. *IEEE Internet of Things Journal, 6*(5), 8155–8168. https://doi.org/10.1109/JIOT.2019.2925825
19. Chen, J., & Ran, X. (2019). Deep learning with edge computing: A review. *Proceedings of the IEEE, 107*(8), 1655–1674. https://doi.org/10.1109/JPROC.2019.2921977
20. He, Y., Huang, D., Chen, L., Ni, Y., & Ma, X. (2022). A survey on zero trust architecture: Challenges and future trends. *Wireless Communications and Mobile Computing, 2022*, e6476274. https://doi.org/10.1155/2022/6476274

21. Cavaliere, F., Mattsson, J., & Smeets, B. (2020). The security implications of quantum cryptography and quantum computing. *Network Security, 2020*(9), 9–15. https://doi.org/10.1016/S1353-4858(20)30105-7
22. Almalki, F. A., et al. (2023). Green IoT for eco-friendly and sustainable smart cities: future directions and opportunities. *Mobile Networks and Applications, 28*(1), 178–202. https://doi.org/10.1007/s11036-021-01790-w

IoT Communication Models and Modes of Connectivity

Ajay Prasad, Prachi Kapoor, and Thipendra P. Singh

Abstract The IoT is a rapidly growing field that enables communication with various devices and objects. It allows for the exchange of information within physical objects and virtual applications, regardless of location or time. The fundamental idea behind IoT is to establish autonomous connections between real objects and applications, enabling the transfer of information and enabling intelligent decision-making. IoT enables the connection of physical activities to virtual ones, leading to the development of numerous services and applications that can benefit individuals, professionals, and businesses. Many gadgets and devices, such as smartphones, laptops, and wearable devices, are now equipped with sensors that gather data and transmit it to the internet. This interconnected network of devices with embedded sensors opens up new possibilities for innovative services and applications. This chapter presents various standard ways in which sensor data can be communicated which are hereby referred to as "communications models" and various means that are used to establish connections between different entities of IoT which are hereby referred to as "modes of communication". The technologies pertaining to these are also presented.

Keywords IoT · Communication models

A. Prasad (✉)
School of Computer Science UPES, Dehradun, India
e-mail: aprasad@ddn.upes.ac.in

P. Kapoor
UPES, Dehradun, India

T. P. Singh
School of Computer Science Engineering and Technology, Bennett University, Greater Noida, India
e-mail: thipendra@gmail.com

A. Prasad et al. (eds.), *Communication Technologies and Security Challenges in IoT*, Internet of Things, https://doi.org/10.1007/978-981-97-0052-3_2

Introduction

We are moving toward a world where everyone and everything can communicate and be connected thanks to the rapid growth of IoT [1]. The IOT is a type of technology that makes communication between any two parties, regardless of location or time (IoT). In the coming future Machine-to-machine can get support from IoT, which can be the Internet of tomorrow. Establishing autonomous, self-governing connections that are reliable and enable the transfer of information between real, tangible objects and real applications is the basic thought behind the Internet of Things [2].

IoT makes it possible to connect actual physical activities to virtual ones. There are many gadgets connected to the Web nowadays, and the number of these gadgets is growing rapidly. PDAs, laptops, smartphones, palmtops, and many other types of portable embedded devices are examples of these types of gadgets. Nowadays, many of these smartphones have a variety of sensors built into them that allow them to sense, calculate, and send the gathered data to cyberspace. Using the interconnection of gadgets and devices where sensors are embedded can also lead to the development of numerous services and apps that can be extremely helpful on a personal, professional, and financial level. An advanced system called the Internet of Things (IoT) combines intelligent things with sensors, processing units that can be found on the cloud, and objects with action-invoking systems, network connectivity, and decision-making systems [3]. Figure 1 reflects the span of things, systems, devices, etc. that are connected via the Internet.

In the past several years, the internet has rapidly expanded to connect millions and millions of things all over the globe. These objects range in size, processing power, and other characteristics, and they may serve a wide range of applications. Consequently, it is necessary to combine the current Internet with the Internet of Devices, a wiser version of the Internet of the future. IoT enables connections between

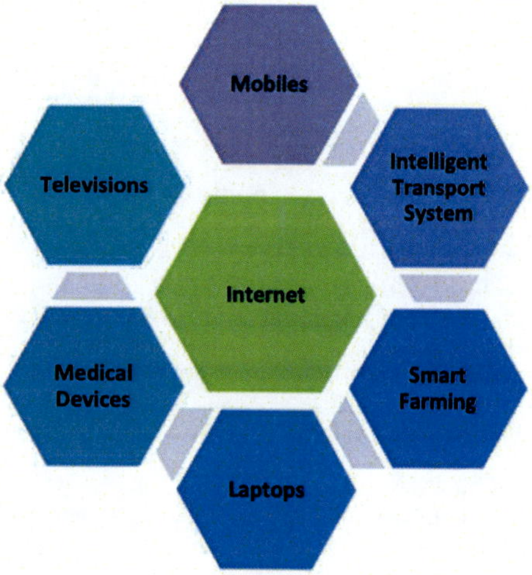

Fig. 1 Generic IOT

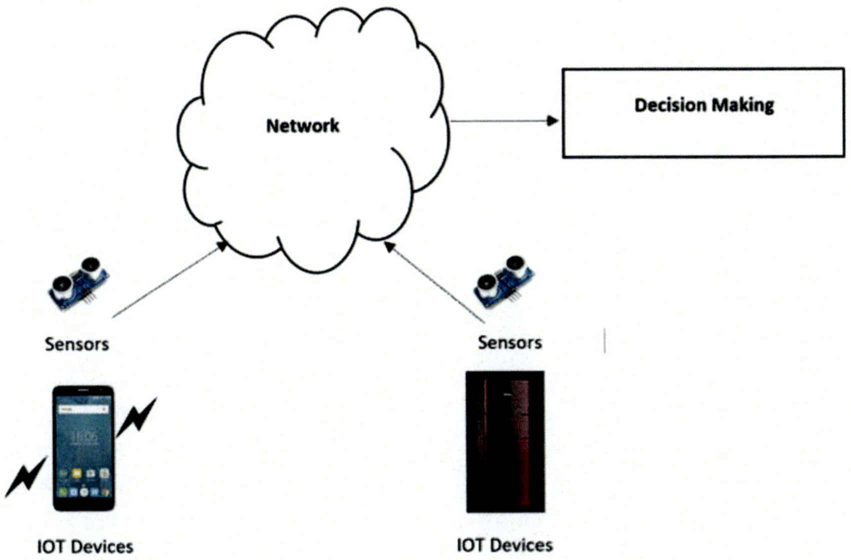

Fig. 2 Basic communication in IoT

actual, tangible objects while also giving them the intelligence to process information unique to each thing. Also, this knowledge is put to good use in independent judgment calls [2, 4].

The storage and processing capability of the gadgets rose dramatically as technology developed, which led to a reduction in their size. These sophisticated gadgets use a variety of actuators and sensors. These gadgets can interact because they can connect to the Internet. This then creates a fresh set of opportunities [5]. The basic Internet of Things network that is used to deploy numerous services and applications is shown in the Fig. 2. These items can be linked to other items that provide the same service type.

Basic communication of IOT involves sensing a thing, recognizing it, and expressing information specific to it. Information may comprise sensed data, such as motion, humidity, and temperature. Depending on the various types of sensors smart applications can be developed and implemented. The obtained thing-specific data is then processed by a smart device processing unit. An action that needs to be taken is decided upon following the processed information. The IoT system or application can offer a wide range of services and can have a mechanism for informing the controller or user about the status of the current scenario of the system. Decisions can be framed based on the information/status gathered [6].

Modes of Communications in IoT

The architectures and standards that are utilised to allow communication between different IoT systems and devices are referred to as communication modes in the Internet of Things (IoT). Here are three IoT communication models that are often used.

Device-To-Device (D2D) Communication

In this mode, IoT devices directly communicate with each other without the need for a centralized system. D2D communication is typically used for localized interactions and data sharing between nearby devices. It allows devices to exchange information, collaborate, and make real-time decisions without relying on cloud or internet connectivity.

Device-To-Cloud (D2C) Communication

D2C communication involves IoT devices transmitting data to a cloud-based platform or server. Devices collect data from sensors or other sources and send it to the cloud for storage, processing, and analysis. This mode enables centralized management, remote monitoring, and data aggregation across a large number of IoT devices. It is commonly used for applications such as remote monitoring, data analytics, and system control.

Device-To-Gateway (D2G) Communication

In this mode, IoT devices communicate with an intermediate device called a gateway. The gateway acts as an intermediary between the devices and the cloud, performing tasks such as data aggregation, protocol translation, and security enforcement. D2G communication is useful in scenarios where devices use different communication protocols or when there is a need for local data processing before transmitting it to the cloud.

These communication modes play a crucial role in establishing connectivity, enabling interoperability, and facilitating data exchange in IoT ecosystems. The choice of the communication mode depends on factors such as the specific use case, network infrastructure, scalability requirements, and latency constraints.

Models of Communication in IoT

In each case, the protocols used might be varied. However, the sensor data that is brought onto the cloud may be consumed for various purposes by different endpoints. Apart from traditional client server and peer-to-peer ways of communication, the data in IoT are generally communicated in further following four communication models in IoT, which are discussed here. They are:

1. Request and Response Model
2. Publisher-Subscriber Model
3. Push–Pull Model
4. Exclusive Pair.

In a P2P model, IoT gadgets interact directly with each other without relying on a central server. This model is useful in decentralized IoT environments and can be employed when direct device-to-device communication is required. The client–server model is a conventional interaction model where IoT devices act as clients, and the central system acts as the server. Devices request data or services from the server, and the server acknowledge with the necessary information. This model is commonly used in IoT applications where centralized control and data processing are preferred.

Request-Response (RR) Model

The Request-Response (RR) Model is a common communication pattern used in various computing systems, particularly in distributed systems and client–server architectures like the IoT. It defines a simple way for two entities (client and server) to interact with each other by sending and receiving messages in a sequential manner. This model enables communication between different components in a system or across a network.

Here's How the Request-Response Model Works

Client sends a request: The process begins with the client (often called the requester) sending a request message to the server (the responder). The request can be a specific command, an inquiry, or a data packet that the client wants the server to process or respond to.

The server processes the request: Upon receiving the request, the server processes the information contained within it. The server's logic determines how to handle the request and generates a response based on the request's content.

The server sends a response: The server replies to the client with a response message after handling the request. The answer includes the outcome of the requested operation, the data the client requested, or any additional pertinent information.

The client receives the response: The client, upon receiving the response from the server, processes the information contained within the response message. It may then take further action based on the received data, or the interaction between the client and server may be considered complete.

End of interaction: With the exchange of the response, the current interaction between the client and server is concluded. If the client needs to perform additional tasks, it may initiate another request, and the process repeats.

The Request-Response The model is frequently used in a variety of situations, such as online communication using HTTP (Hypertext Transfer Protocol), in which the client makes an HTTP request to the server, which then returns an HTTP response providing the desired web page or data. It's also used in remote procedure calls (RPC) and various application protocols in networking.

Advantages of the Request-Response Model

Simplicity: The RR model is straightforward to implement and understand, making it a popular choice for communication in various applications.

Statelessness: Each request-response interaction is independent of previous interactions, making it easier to scale and handle multiple clients.

Clear flow control: The sequential nature of the model ensures that there is a clear flow of communication between the client and server.

However, the Request-Response Model may not be the best fit for all scenarios, especially when real-time bidirectional communication or long-lived connections are required. In such cases, other communication patterns like the Publish-Subscribe model or WebSockets may be more suitable (Fig. 3).

Publisher-Subscriber (PS) Model

The Publisher-Subscriber (PS) Model is a messaging pattern used in distributed systems and event-driven architectures like the IoT. Also known as the Publish-Subscribe pattern, it facilitates communication between different components of a system in a more loosely coupled manner compared to the Request-Response Model. In this model, components do not directly communicate with each other but rather interact through an intermediary known as the message broker or event bus.

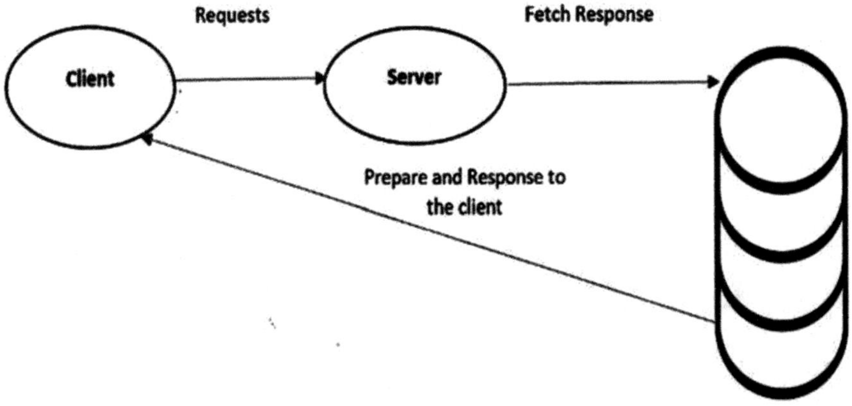

Fig. 3 Request and response model

Here's How the Publisher-Subscriber Model Works

Publishers: Publishers are components or entities that generate and send messages or events. These events could be updates, notifications, or any other relevant information that subscribers might be interested in. Publishers do not need to know the identities of the subscribers, and their responsibility is solely to produce events and send them to the message broker.

Message Broker/Event Bus: The message broker (also known as the event bus) acts as an intermediary or hub in the system. It receives events from publishers and then distributes these events to interested subscribers. The message broker can be a dedicated server or service that handles the event distribution efficiently.

Subscribers: Subscribers are components or entities that have registered an interest in certain types of events. They subscribe to specific event types or topics they want to receive updates about. When a publisher generates an event and communicate it to the message broker, the broker ensures that the event is delivered to all relevant subscribers.

Event Distribution: When a publisher sends an event to the message broker, the broker efficiently disseminates the event to all subscribers who have expressed interest in that particular event type. Each subscriber receives the event and processes it independently based on its requirements.

Advantages of the Publisher-Subscriber Model

Loose coupling: The PS model promotes loose coupling between components since publishers and subscribers do not need to be aware of each other's existence. This allows for greater flexibility and scalability in the system.

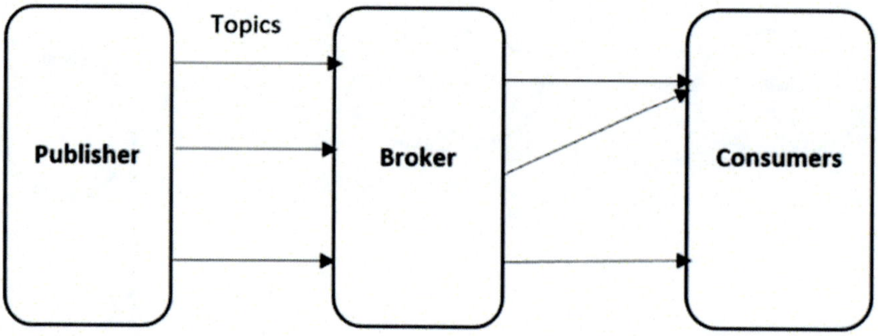

Fig. 4 Publisher-subscriber model

Event-driven: The PS model is inherently event-driven, making it suitable for scenarios where components need to respond to changes or updates in real-time.

Scalability: The use of a message broker allows for efficient event distribution, making it easier to scale the system as the number of components and events grows.

Decoupled architecture: Components can be added or removed without affecting the functionality of other components, as long as they adhere to the agreed-upon event format.

The Publisher-Subscriber Model is commonly used in various systems, including messaging systems, real-time applications, event-driven microservices, and IoT (Internet of Things) applications. Examples of message brokers that support this pattern include Apache Kafka, RabbitMQ, and Amazon SNS (Simple Notification Service) (Fig. 4).

Push–Pull (PP) Model

The push–pull paradigm is made up of data queues, publishers, and consumers. After publishing a message or piece of data, publishers push it onto the queue. The consumers who are present on the opposite side pull the data out of the queue. The queue serves like a buffer in which the message get stored, the major difference is noted in the rate of data push and pull.

Queues can be used to divide the messages sent among the producer and the consumer. In cases when the rate at which data producers push it and consumers pull it varies, queues also act as an additional layer.

In the relation of the Internet of Things (IoT), the Push–Pull (PP) Model refers to a data communication pattern used to exchange information between IoT devices and cloud-based or centralized platforms. This model allows IoT devices to share data with the cloud (or other services) and enables the cloud to push commands or updates back to the IoT devices when necessary.

Here's How the Push–Pull Model Works in IoT

Data Push from IoT Devices: IoT devices, equipped with sensors or actuators, collect data from their environment. These devices periodically send the collected data to a cloud-based platform or server without waiting for a specific request from the cloud. This data push can be triggered based on predefined intervals or specific events.

Cloud-based Data Storage and Processing: The cloud-based platform receives the data sent by the IoT devices. It stores this data in a database or processes it to generate insights, perform analytics, or trigger certain actions based on predefined rules and algorithms. The cloud platform acts as a centralized repository for IoT data, enabling data storage and advanced data processing capabilities.

Data Pull or Commands from Cloud to IoT Devices: In the Push–Pull Model, the cloud platform can also push commands or updates back to specific IoT devices when needed. This can be done to reconfigure device settings, trigger actions, or provide real-time updates to the IoT devices. Instead of relying solely on data pushed from the IoT devices, the cloud can initiate communication with individual devices as required.

Reactive Communication: The Push–Pull Model allows for more reactive communication between IoT devices and the cloud platform. IoT devices don't need to be continuously active or polling the cloud for updates, which can help conserve power and resources, especially in low-power IoT devices.

The Push–Pull Model in IoT facilitates bidirectional communication between IoT devices and the cloud platform, enabling real-time data exchange and control. It is especially useful for applications that require timely updates, remote monitoring, and control of IoT devices. Examples of such applications include smart home automation, industrial IoT (IIoT) systems, remote asset management, and environmental monitoring. By using this model, IoT systems can efficiently manage and respond to data from multiple distributed devices in a scalable and responsive manner.

Advantages of the Push–Pull Model in IoT

Real-time Responsiveness: The Push aspect of the model allows critical data to be transmitted in real-time, enabling rapid response to time-sensitive events or emergencies.

Reduced Latency: By using the Push aspect for immediate data delivery and the Pull aspect for less time-sensitive data, the PP Model minimizes data transmission latency and ensures efficient resource utilization.

Efficient Data Collection: The Pull aspect of the model allows the central server to gather non-critical data from multiple IoT devices efficiently, reducing unnecessary data transmission.

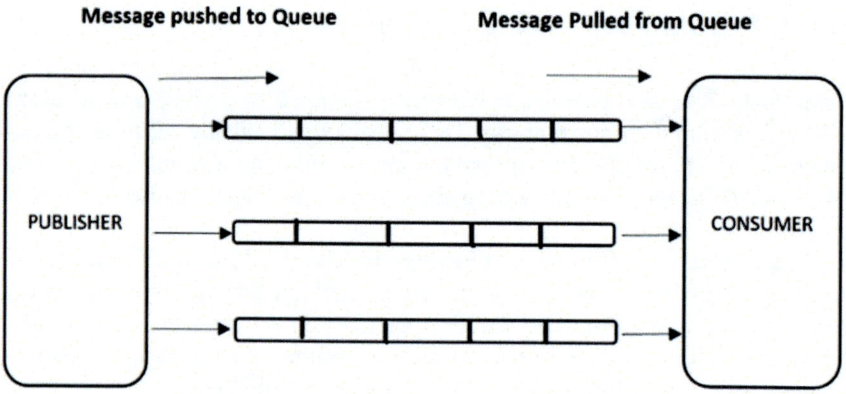

Fig. 5 Push and pull (PP) model

Scalability: The PP Model can be easily scaled to accommodate a large number of IoT devices, as it allows the central server to efficiently manage data retrieval from multiple sources.

Overall, the Push–Pull Model is a versatile approach to IoT data communication, providing a balance between real-time data delivery and efficient data collection, making it well-suited for various IoT applications and use cases (Fig. 5).

Exclusive Pair (EP) Model

A bi-directional method called Exclusive Pair provides full-duplex client and server interaction. The connection between the two is always open and continuous unless the customer submits an inquiry to break it. The server keeps track of every communication that has been established. Since this is a state-full relationship architecture, the host computer has knowledge of each active connections. The entire WebSocket-based communication API is built on this principle.

The "exclusive-pair communication model" could refer to a specific type of communication pattern or protocol used in IoT systems where devices are paired exclusively with one another for communication. In this model, two IoT devices establish a one-to-one connection, and they exclusively communicate with each other without involving other devices or a centralized server.

Here's How the Exclusive-Pair Communication Model Might Work

Device Pairing: Two IoT devices are initially paired with each other. The pairing process could be initiated by a user, by exchanging unique identification codes, or through a secure authentication process.

Communication Establishment: Once the devices are paired, they establish a direct communication link between each other. This link could be a wireless connection (e.g., Bluetooth, Zigbee, or Wi-Fi Direct) or a wired connection (e.g., Ethernet or USB).

Exclusive Communication: After successful pairing and communication establishment, the paired devices can exchange data, commands, or messages exclusively with each other. They do not share this information with other devices on the network or through a central server.

Limited Scope: The exclusive-pair communication model restricts communication to only the paired devices. As a result, the range of communication is limited to the physical proximity of the paired devices.

Potential Use Cases

The exclusive-pair communication model might be employed in specific IoT applications where limited and direct communication between two devices is desired. Some possible use cases include:

IoT Device Setup: During the initial setup or configuration of an IoT device, it may establish an exclusive pair with a smartphone, tablet, or computer to facilitate the setup process.

Secure Data Transfer: In scenarios where security is paramount, two IoT devices can form an exclusive pair to ensure that sensitive data is only exchanged between trusted devices.

Collaborative Actions: Two IoT devices with complementary functions can pair exclusively to achieve collaborative actions or combined functionality (Fig. 6).

Communication Technologies Used in Internet of Things

The Internet of Things (IoT) encompasses a wide variety of devices and applications, each with specific requirements for communication. Various communication technologies are used in IoT to connect devices, transfer data, and enable seamless interactions. Here are some of the key communication technologies commonly used in IoT.

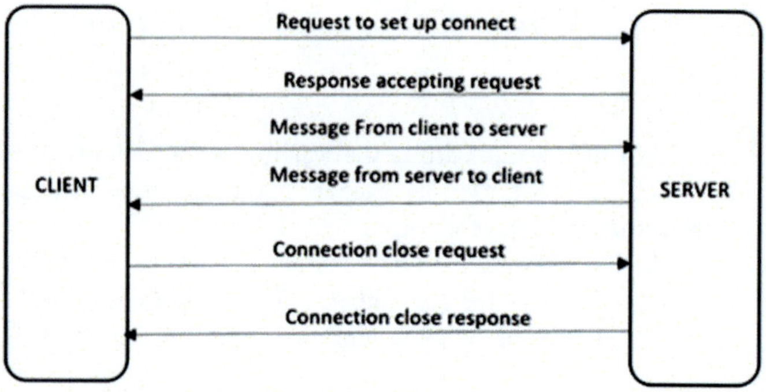

Fig. 6 Exclusive pairing

Wi-Fi (Wireless Fidelity): Wi-Fi is one of the most widely used communication technologies in IoT due to its widespread availability and high data transfer rates. IoT devices equipped with Wi-Fi modules can connect to existing Wi-Fi networks in homes, offices, or public spaces, enabling them to exchange data with the cloud and other devices on the network.

Bluetooth: Bluetooth is a short-range wireless communication technology that is commonly used for connecting IoT devices in close proximity. Bluetooth Low Energy (BLE) is particularly popular for IoT applications due to its low power consumption, making it suitable for battery-powered devices like wearables, smart sensors, and beacons [6].

Zigbee: Zigbee is a low-power, low-data-rate wireless communication protocol designed for IoT applications that require mesh networking capabilities. It is commonly used in smart home and industrial automation scenarios where devices need to communicate with one another in a self-organizing network [7].

Z-Wave: Z-Wave is another wireless communication protocol optimized for low-power IoT devices and is primarily used in smart home applications. It operates on a mesh network, allowing devices to extend the network's reach by acting as repeaters [8].

LoRa (Long Range): LoRa is a long-range, low-power wireless technology designed for IoT applications that require long-range communication, such as smart city solutions and agricultural monitoring. It operates in unlicensed frequency bands and is capable of covering large areas with minimal power consumption.

NB-IoT (Narrowband IoT): NB-IoT is a cellular communication technology specifically designed for IoT devices that require low data rates and long battery life. It operates on existing cellular networks and provides better coverage in hard-to-reach areas, making it suitable for IoT deployments in remote locations.

LTE-M (LTE for Machines): LTE-M is another cellular communication technology that offers higher data rates and lower latency than NB-IoT, making it suitable for more data-intensive IoT applications like asset tracking and video surveillance.

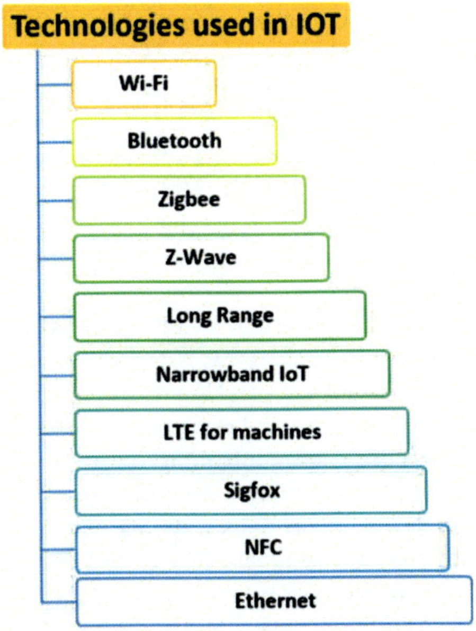

Fig. 7 IoT technologies

Sigfox: Sigfox is a low-power, long-range communication technology that operates on a global public network dedicated to IoT. It is ideal for low-bandwidth applications, such as environmental monitoring and asset tracking [9].

NFC (Near Field Communication): NFC is a short-range communication technology commonly used in IoT for contactless data exchange between devices, such as smartphones and NFC tags. It is often used for access control, payment systems, and smart packaging.

Ethernet: For wired connections, Ethernet is commonly used in industrial IoT applications and certain smart home setups where devices are connected via Ethernet cables for reliable and high-speed data transfer.

Each of these communication technologies has its strengths and weaknesses, making them suitable for different IoT use cases and environments. The choice of communication technology depends on factors such as range, data rate, power consumption, coverage area, and deployment cost (Fig. 7).

Current Advances in IoT Communication

5G Integration: The rollout of 5G networks has the potential to significantly impact IoT communication. With higher data rates, lower latency, and greater capacity, 5G enables more efficient and real-time communication between IoT devices. This facilitates the growth of advanced IoT applications, such as augmented reality, virtual reality, and autonomous vehicles.

Edge Computing: Edge computing has emerged as a critical solution for reducing latency and bandwidth usage in IoT communication. By processing data closer to the source (at the edge of the network), edge computing enables faster decision-making and more efficient use of network resources [10].

LPWAN (Low Power Wide Area Network): LPWAN technologies, such as NB-IoT, LTE-M, and LoRaWAN, continue to gain traction in IoT deployments. These technologies offer long-range communication with low power consumption, making them well-suited for applications that require extended battery life and wide area coverage.

MQTT (Message Queuing Telemetry Transport): MQTT is a lightweight messaging protocol that is becoming increasingly popular for IoT communication due to its efficiency and low overhead. It is well-suited for scenarios where bandwidth and power consumption are critical considerations [11].

Secure Communication: As the number of IoT devices increases, the need for secure communication becomes more pronounced. Advancements in cryptographic protocols, secure bootstrapping, and authentication mechanisms help ensure the integrity and confidentiality of data exchanged between IoT devices and the cloud.

Satellite Communication: Satellite-based IoT communication solutions have seen advancements, enabling connectivity in remote and challenging environments. This is particularly valuable for applications like asset tracking, environmental monitoring, and precision agriculture [12].

AI-Driven Communication: Artificial intelligence (AI) and machine learning are being integrated into IoT communication to optimize network performance, predict connectivity issues, and dynamically manage IoT device connections for better efficiency and reliability [13].

Mesh Networking: Mesh networking in IoT is gaining attention as it offers enhanced connectivity and redundancy. IoT devices can form self-organizing networks where each device acts as a relay to extend coverage and improve communication robustness.

IPv6 Adoption: With the proliferation of IoT devices, the adoption of IPv6 has become crucial to accommodate the large number of connected devices and enable direct communication between them [14].

Standardization Efforts: Various standardization bodies, such as the Industrial Internet Consortium (IIC), the Open Connectivity Foundation (OCF), and the Thread Group, are actively working on defining communication protocols and interoperability standards for IoT devices.

Conclusion

The Internet of Things (IoT) is a rapidly evolving field that enables seamless communication between physical devices and virtual applications, regardless of location or time. The fundamental idea behind IoT is to establish autonomous connections between real objects and applications, enabling the transfer of information and intelligent decision-making. This interconnected network of devices, equipped with sensors, has opened up new possibilities for innovative services and applications across various sectors.

The IoT communication landscape is diverse, offering several communication models and modes of connectivity to suit different use cases and requirements. The communication models, such as the Request-Response, Publisher-Subscriber, Push–Pull, and Exclusive Pair, enable devices to exchange data and commands efficiently. These models are chosen based on factors like real-time responsiveness, scalability, and resource efficiency.

To support IoT communication, a variety of communication technologies are utilized, each catering to specific needs. Technologies like Wi-Fi, Bluetooth, Zigbee, LoRa, NB-IoT, and more play crucial roles in connecting IoT devices and facilitating seamless data transfer. Advances in 5G integration, edge computing, LPWAN, secure communication, and satellite communication further enhance the capabilities of IoT deployments.

As the IoT ecosystem continues to grow, security and standardization efforts are paramount. Technologies like AI-driven communication and mesh networking help optimize network performance and enhance reliability. Adopting IPv6 becomes essential to accommodate the ever-increasing number of connected devices.

In summary, the IoT communication landscape is dynamic and continually evolving, enabling the realization of various smart applications and services that benefit individuals, industries, and society as a whole. As researchers and innovators continue to explore new frontiers, IoT's potential to revolutionize how we interact with technology and the world around us becomes increasingly apparent. Ensuring secure and standardized IoT communication will be crucial for realizing the full potential of this transformative technology.

References

1. Lin, J., Yu, W., Zhang, N., Yang, X., Zhang, H., & Zhao, W. (2017). A survey on Internet of Things: Architecture, enabling technologies, security and privacy, and applications. *IEEE Internet of Things Journal, 4*(5).
2. Khan, R., Khan, S. U., Zaheer, R., & Khan, S. (2012). Future internet: The Internet of Things architecture, possible applications and key challenges. In *10th International Conference on Frontiers of Information Technology (FIT)*.
3. Siegel, J. E., Kumar, S., & Sarma, S. E. (2018). The future Internet of Things: Secure, efficient, and model-based. *IEEE Internet of Things Journal, 5*(4).
4. Yang, Y., Wu, L., Yin, G., Li, L., & Zhao, H. (2017). A survey on security and privacy issues in Internet-of Things. *IEEE Internet of Things Journal, 4*(5).
5. Padma, B., & Erukala, S. B. (2023). End-to-end communication protocol in IoT-enabled ZigBee network: Investigation and performance analysis. *Internet of Things, 22*, 100796. ISSN 2542-6605.
6. Stankovic, J. A. (2014). Research directions for the Internet of Things. *IEEE Internet of Things Journal, 1*(1).

7. Qays, M. O., Ahmad, I., Abu-Siada, A., Hossain, M. L., & Yasmin, F. (2023). Key communication technologies, applications, protocols and future guides for IoT-assisted smart grid systems: A review. *Energy Reports, 9*, 2440–2452. ISSN 2352-4847.
8. Bang, A. O., Rao, U. P., Visconti, A., Brighente, A., & Conti, M. (2022). An IoT inventory before deployment: A survey on IoT protocols, communication technologies, vulnerabilities, attacks, and future research directions. *Computers & Security, 123*, 102914. ISSN 0167-4048.
9. Mekki, K., Bajic, E., Chaxel, F., & Meyer, F. (2019). A comparative study of LPWAN technologies for large-scale IoT deployment. *ICT Express, 5*(1), 1–7. ISSN 2405-9595.
10. Laroui, M., Nour, B., Moungla, H., Cherif, M. A., Afifi, H., & Guizani, M. (2021). Edge and fog computing for IoT: A survey on current research activities and future directions. *Computer Communications, 180*, 210–231. ISSN 0140-3664.
11. Alatram, A., Sikos, L. F., Johnstone, M., Szewczyk, P., & Kang, J. J. (2023). DoS/DDoS-MQTT-IoT: A dataset for evaluating intrusions in IoT networks using the MQTT protocol. *Computer Networks, 231*, 109809. ISSN 1389-1286.
12. Ye, N., Yu, J., Wang, A., & Zhang, R. (2022). Help from space: Grant-free massive access for satellite-based IoT in the 6G era. *Digital Communications and Networks, 8*(2), 215–224. ISSN 2352-8648.
13. Chahed, H., Usman, M., Chatterjee, A., Bayram, F., Chaudhary, R., Brunstrom, A., Taheri, J., Ahmed, B. S., & Kassler, A. (2023). AIDA—A holistic AI-driven networking and processing framework for industrial IoT applications. *Internet of Things, 22*, 100805. ISSN 2542-6605.
14. Atlam, H. F., Hemdan, E. E., Alenezi, A., Alassafi, M. O., & Wills, G. B. (2020). Internet of Things forensics: A review. *Internet of Things, 11*, 100220. ISSN 2542-6605.
15. Li, J., Silva, B. N., Diyan, M., Cao, Z., & Han, K. (2018). A clustering based routing algorithm in IoT aware wireless mesh networks. *Sustainable Cities and Society, 40*, 657–666. ISSN 2210-6707.

Challenges and Key Issues in IoT Privacy and Security

Atul B. Kathole, Vinod V. Kimbahune, Sonali D. Patil, Avinash P. Jadhav, and Kapil N. Vhatkar

Abstract "Internet of Things" (IoT) refers to a system that enables the interconnection of many different types of things and intelligent devices. In order to function, devices connected to the Internet of Things must be able to sense their surroundings, process that information, and then transmit the results. The new Internet of Things (IoT) technology has many applications today, including in the domestic sphere, the medical field, the media, the natural world, the industrial sector, the construction industry, and the provision of essential services like water and electricity. Internet of Things (IoT) technologies differentiate themselves from more conventional computer platforms like desktops, laptops, and mobile phones through the use of embedded devices. Because of the potential for real and virtual worlds to merge and the exchange of sensitive sensor data, security for IoT devices is becoming increasingly crucial. The Internet of Things (IoT) faces a variety of challenges that must be surmounted. They include faulty device upgrades, weak security standards, user illiteracy, and the infamous active device monitoring. Our work aims to address the challenges of IoT security and improve its security so that more benefits can be realized by the growing number of connected devices and services.

Keywords Cloud computing · Edge computing · Internet of Things · Security regulations · Cloud computing

A. B. Kathole (✉) · K. N. Vhatkar
Computer Engineering, Dr. D. Y. Patil Institute of Technology, Pune, India
e-mail: atul.kathole1910@gmail.com

V. V. Kimbahune
Artificial Intelligence and Data Science, Dr. D. Y. Patil Institute of Technology, Pune, India

S. D. Patil
Information Technology Department, Pimpri Chinchwad College of Engineering, Pune 411044, India

A. P. Jadhav
Computer Science and Engineering, DRGIT&R, Amravati, India

A. Prasad et al. (eds.), *Communication Technologies and Security Challenges in IoT*, Internet of Things, https://doi.org/10.1007/978-981-97-0052-3_3

Introduction

The "Internet of Things" (IoT) refers to the interconnection of everyday objects and technologies through some kind of network infrastructure. The rapid ascent of the Internet of Things can be traced in part to the proliferation of related technologies in fields as diverse as transportation, education, and economic development. Hyperconnectivity was pioneered by IoT, allowing for rapid communication between individuals and businesses in different regions of the world. In 1999, Kevin Ashton coined the term "Internet of Things" to promote the idea of Radio Frequency Identification (RFID), which uses embedded sensors and actuators. But the suggestion itself dates back to that decade. At the time, some people used terms like "embedded Internet" or "pervasive computer" to describe it. Ashton suggested the IoT to enhance operations all across the supply chain. Nevertheless, IoT's skyrocketing popularity since the summer of 2010 is largely attributable to the various advantages of the technology. Over the next five years, the Chinese government plans to pour a lot of money into the Internet of Things (IoT). The number of connected devices is up to 26.66 billion right now [1]. The market took off in a big way in 2011, when items like smart energy metres, wearable electronics, and home automation were widely available. The exponential growth of IoT has benefited businesses in many ways, especially in the fields of market research and strategic planning. The IoT's contribution to people's quality of life has expanded to include automation services. However, such unchecked growth gives rise to new fears for people's safety and security.

Sensitive data stored in IoT systems are more susceptible to cyberattacks and harmful apps when users are careless, don't change passwords often, and don't update software. A security event like a data leak becomes more likely when such behavior occurs. Due to its lack of comprehensive security procedures and guidelines, the Internet of Things (IoT) is frequently considered as a prime target for cyberattacks. While the IoT has numerous protections in place to ward off cyberattacks, not much is known about them [2]. As a result, consumers were unable to employ safeguards designed to deter online fraudsters. Hackers have been creating new sorts of malware since the eve of 2008 in order to break into IoT devices. Many different methods were employed by phishers to induce people to give personal information [3]. Therefore, widespread attacks compromise the safety of business and personal computers and gadgets on a regular basis. With an accurate image of the dangers, manufacturers of protective gear and cyber security professionals can create an impenetrable defense.

The Internet of Things has seen widespread adoption in the business sector and other sectors of industry [4]. These apps provide an advantage for these companies in the market. Smart gadgets are becoming increasingly common, and with them comes a slew of new privacy and security concerns for organisations as customers' information is shared and integrated. Professionals are required to deal with these security risks and create all-encompassing security policies and processes to safeguard company assets and guarantee service dependability. Connected home kitchen appliances with Internet of Things (IoT) capabilities, for example, could provide hackers with a backdoor to acquire private and/or confidential commercial data or even to control and interrupt corporate activity.

The field of technology is constantly evolving, with new discoveries and innovations being made every day. Take the state-of-the-art 5G network as an illustration. The systems and apps that run in the background of the Internet of Things will rely heavily on 5G. As a result of its short wavelength, more base stations must be installed in order to provide the same coverage as with other wireless technologies, but its high frequency and bandwidth have piqued the interest of researchers concerned about privacy and security. Dangers, such as mobile base stations, are now present in the new layout. Understanding the potential threats to security and the steps that might be taken to lessen them is essential [5].

The purpose of this work is to provide readers with a comprehensive introduction to the Internet of Things, including its possible advantages, disadvantages and uses. Further, either the deployment and evaluation of existing schemes or the creation of new ones will necessitate a framework to investigate and define the best security practices. Our findings inform our suggestions for closing these security loopholes and reducing these hazards [6]. In order to better enforce policies and educate end-users, entities, and stakeholders participating in IoT on how to put in stronger security and privacy measures, this study will help regulatory bodies.

We used a narrative review to investigate the evolution of IoT systems, the issues they've faced, and the solutions they've applied. The extremely wide and global IoT idea was investigated for potential threats to users' online safety and privacy. The studies in this work were carefully designed to test the model described below. Potential security dangers and threats, however, are not highlighted, nor are the characteristics, benefits, and problems of the numerous IoT applications (smart health, smart cities, supply chain, transportation, and so on). There are countless situations like this in the published literature. The goal of this effort was to offer context, illustrate the feasibility of the notion, and open the door to future research.

Related Work

The security and privacy problems that afflict the Internet of Things have received a lot of attention as of late. Concerns concerning the safety of IoT devices have been the subject of a great deal of research and writing.

The developers of [5] claim that many attacks, including jamming and spoofing, have compromised the security of users' data. It's possible that there are choices available to help individuals better secure their IoT gadgets. IoT devices and the networks they rely on are vulnerable to a variety of new privacy threats, as detailed in [6]. It can be challenging to keep tabs on the security of Internet of Things (IoT) devices in business and government environments. To check for privacy issues and lessen the likelihood of a breach, businesses should implement scanning and monitoring solutions for all IoT devices. Network traffic analyzers and interceptors help find and investigate possible security holes.

Several reports and services have been written about the current state of IoT security [7]. The problems encountered by various Internet of Things devices and their defenses have been exacerbated by a number of services. The availability of multiple platforms that can confirm this security protocol, as well as the availability of various simulation tools and modelers, can all contribute to the creation of the one-of-a-kind IoT security protocol. Since the advent of powerful simulation tools and modelers, research into IoT security has moved quickly. In the event that Internet of Things gadgets failed, serious issues would arise.

Despite the many benefits it brings, the authors in [8] suggest that it is vital to think about the risks involved with the Internet of Things. Most people are concerned about their personal information being compromised or exposed in the online world. Both of these factors are posing serious challenges for several organizations, both private and public. The proliferation of high-profile cyberattacks has demonstrated the vulnerability of Internet of Things (IoT) technologies. This is because the interconnection of networks in the Internet of Things necessitates novel security measures, such as allowing access from the anonymous and untrusted Internet. It is essential, however, to emphasize the principles and key concepts of the IoT Cyber Security Framework while putting the IoT security system into reality. According to [9], one of the most important things to do is to terminate a contract that involves gadgets that support several kinds of network protocols for exchanging data. Protocol variations prevent the implementation of separate service contracts, which are essential to the cyber security architecture of any Internet of Things. He demonstrated that the reliability of the IoT framework can be reduced by taking even minimal precautions against cybersecurity threats to the IoT. The authors in [9] also shown how scalability is crucial to the development of a cybersecurity framework for the Internet of Things. The Internet of Things ecosystem, so the experts say, must be robust enough to handle a billion Internet and cybersecurity issues. The publication also showed how testability is crucial in the Internet of Things cybersecurity context for reducing risk. Integration testing, component testing, system testing, and compliance testing are all part of this category.

Low-end system-specific remedies to safety and personal difficulties are proposed by Yang et al. [7] in their survey. IoT security in terms of networks, devices, and systems is discussed briefly by a number of authors [2, 3]. Weber and Gopi and Rao's reviews [8, 9] break down the limitations of IoT devices into four categories: First, we need to find ways to make batteries last longer; second, lightweight processing; third, categorizing security assaults; and fourth, access control techniques and architecture. We also talk about the presentation, network, transport, and application layers of the IoT stack.

Security and privacy are discussed in Weber's survey, and a security paradigm for IoT-based devices is provided by researchers [8]. The Internet of Things (IoT) is gaining traction in many regions because it makes use of other cutting-edge technologies already widely used in the international distribution network. The vast majority of people are getting used to using this technology. The low-end devices can talk to one another and share information and data from their many sensors. The main issue with IoT devices is the threat they present to users' safety and privacy. Keeping such a massive amount of information organised to ensure reliable and safe machine processing is difficult. These IoT can pose a risk to people's safety, confidentiality, and privacy. The authors of this study address the increasing urgency of closing the gap between automated surveillance by IoT-based devices and the legally protected rights of individuals who are unaware of the dangers to their privacy posed by these technologies through the adoption of appropriate regulations and technicalities. In [10], authors Aleisa and Renaud explain the issues of IoT privacy, including its principles, potential threats, and suggested solutions.

Tewari and Gupta provided a second comprehensive review of vulnerabilities in IoT gadgets. The layered architecture of IoT devices is analyzed, and new security issues are revealed. They provided strategies and instruments for investigating the Internet of Things, with a special emphasis on the challenges of heterogeneous integration across layers [11]. In their 2019 study, Noor and Hassan assess and contrast numerous research across a wide range of factors, such as modeling tools, methods, Internet of Things device security, and privacy. Authentication, encryption, trust management, and other approaches to safeguarding the Internet of Things (IoT) are among those explored [12].

The authors of [10], available in the same location, discuss some contemporary approaches to IoT cybersecurity. While the company admits that it is not in their best financial interest to constantly supply high-quality solutions, they have taken some basic security precautions. Therefore, it is highly unlikely that enterprises will develop a workable solution to the cybersecurity issues posed by the Internet of Things.

According to the authors in [11], mobile and cyber-physical systems are now ubiquitous, present in everything from factory control systems and modern autos to life-supporting networks. Constantly, these systems generate, process, and exchange a wealth of valuable data. Due to the significance people place on privacy and confidentiality, cyber-attacks on the Internet of Things system might cause physical harm and disturbance in people's daily lives. The security of our personal information and the prevalence of online dangers are two of the most urgent problems we face today. The increasing sophistication and potential damage of cyberattacks pose new risks to interconnected industrial IoT devices. Solutions to security and privacy concerns may be found in generic security frameworks for industrial IoT systems. The safety of the planned operations cannot be guaranteed by the current state of IoT security.

Because of this, it's crucial that the numerous security issues in the IoT be researched and analyzed. Therefore, research into diverse IoT security is getting the necessary impetus [12] with the aid of various simulation tools and several computing platforms.

In addition, a paper is provided that details concerns raised by experts in the field on the lack of privacy and security in IoT gadgets. Experts on the Internet of Things (IoT) have attempted to understand security and privacy concerns and have proposed new security methods for efficient SPMs [13]. Almost any connected device comes with significant risks and attack possibilities.

The intruder's reasons for acting violently could be anything from curiosity to vengeance. The two most common threats to society are natural disasters and terrorist acts. Even if data is encrypted, physical damage to devices could render them useless. Furthermore, a lot of individuals have worked hard to guarantee the security of Internet of Things devices against attacks and threats from humans.

IoT Security Concerns

The various sectors of interest raise unique challenges for securing the Internet of Things.

DOS

A denial-of-service (DOS) assault is an attempt to overwhelm a system with so much traffic that neither humans nor legitimate software can use it. It's the standard and most reliable method of assault. It is good knowledge that an attacker can use a flooding attack to exhaust a system's resources including memory, CPU, and bandwidth [18]. When he engages with us, he either completely disables the system or significantly slows down its ability to meet our needs.

Pirates can launch this assault in a number of ways, such as sending spam over the network or delivering unwanted packets. This prevents actual users from making use of the service's capabilities. An attack on the rewind, paragraph [14].

Attacks

Replay attacks are among the oldest forms of assault on communication networks and, in particular, on authentication and key-exchanging protocols. The pirate has the option of recording all legal traffic or only a selection of it. With the victim's permission, an attacker on a public network can listen in on a conversation, record it, and then transmit it on to either another participant in the conversation or a completely different target. Since a replay attack permits the unintended storing and subsequent transmission of previously sent data, it is seen as a security risk in IoT networks. The goal of this type of attack is to deceive the target into doing some kind of action without their knowledge or consent [13]. For instance, a smart home's temperature could be monitored by a sensor connected to the HVAC system, with the data then being sent to the home's central processing unit. With these inputs, the system can determine if it needs to turn on the air conditioning to keep the facility at a comfortable temperature. However, if the temperature sensor is stolen, the intruder can collect data throughout the day and deliver it afterward. Because of this, the air conditioner will not function normally.

The timestamp, the nonce, and the response challenge are the three main tactics employed by existing systems to resist a replay attack. The first is a technique for ensuring the message has been freshly received, which can help avoid replay attacks. However, maintaining accurate time synchronization among IoT devices is difficult. The second mechanism is called a nonce and it is a sequence of random numbers. However, the node's RAM is insufficient to hold the list of received nonces, therefore this approach has limitations [15]. The challenge-response method is presented last. The aim is to check that both of you are problem solvers. However, this approach is only useful if both parties are already privy to the same secret.

Password

Cybercriminals have devised a wide array of methods for obtaining passwords since they are used in so many different authentication protocols and are so crucial to the authentication process. Therefore, the most prevalent kind of attack is guessing a user's password. This assault, in particular, can be carried out in both online and offline settings [16]. A third party eavesdrops on a discussion between two parties during authentication in order to steal private information. In order to crack authentication, a hacker must test every possible combination of credentials.

Attacking with a Spoofing Attempt

When an untrusted source creates a faked parameter, it is called a spoofing attack because the network's security has been breached. The target servers are tricked into thinking the attacker is authentic. As a result, the government decides to put the pirate in a position of trust. For example, a hacker in the field of smart healthcare could try to use a victim's phone number to bypass an authentication server. Once he's been authenticated, he'll be able to make a request for the victim's sensor and access sensitive medical data [17].

Potential Threats to Personal Information from the Internet of Things

IoT had several positive effects on its users but has also caused some problems. Experts and academics have identified cyber security and privacy risks as two of the industry's most pressing problems. These two factors pose a significant challenge for numerous businesses and government agencies. The proliferation of high-profile cyberattacks has demonstrated the vulnerability of Internet of Things (IoT) technologies. The interconnected nature of IoT networks makes them susceptible to attack because they may be accessed from anywhere on the anonymous and untrusted Internet [13].

Concerns over data privacy and security make many users wary of IoT devices. After a security breach has already occurred, it can be difficult for users to grasp the entire extent of the damage, which might include the loss of vital information. Users' privacy has been exposed on a regular basis due to security lapses, and as a result, users' tolerance for insufficient protection has decreased. Recent studies on the safety and privacy of consumer-grade IoT devices have come up short. Many vulnerable areas exist in today's automobile electronics.

Protection

The Internet of Things (IoT) poses special security concerns due to the ways in which it differs from traditional PCs and other types of computing devices [14]. Many IoT gadgets are designed with massive deployment in mind. The use of sensors is a perfect example of this. Most devices used in an IoT deployment will be the same or very similar to one another. Since they are so similar, any vulnerability in their security would have a disproportionately negative effect on all of them. In addition, many groups have published guides on how to conduct risk assessments. This breakthrough demonstrates an unprecedented degree of communication among IoT gadgets. It's also evident that many of these devices can connect to one another and communicate with one another automatically, albeit in a disorganized fashion. They require analysis of available tools, techniques, and approaches for protecting the Internet of Things (IoT).

Although the problem of security in the IT sector is not new, the proliferation of the Internet of Things (IoT) has presented new barriers. As Internet of Things (IoT) gadgets and services become more pervasive and unobtrusive in people's daily lives, it is crucial that they inspire confidence that they will be secure. Inadequately secured Internet of Things (IoT) devices and services are a major entry point for cyber-attacks and the exposure of user data.

Because of how interconnected they are, IoT devices can have far-reaching consequences if they are not properly safeguarded and secured. The simple cause of this behavior is the use of so many comparable IoT devices. Because certain IoT devices are designed to be physically attached to one another, their users and creators have an added obligation to ensure the safety of their local networks and the wider Internet. An effective and acceptable response to the problems requires a currently existing unified strategy in the IoT [14].

For example, authentication remains a significant challenge when it comes to ensuring the safety of many IoT-based services, despite the proliferation of security measures. The authentication technique can only prevent one specific kind of attack, like a DoS or a replay attack. Information security is one of the most crucial weak points in IoT authentication due to the growth of applications that offer a security risk due to the inherent plurality of data collection in the IoT environment. As an example, consider the rise of contactless credit cards. Since this sort of card permits the reading of card numbers and names, it is possible for hackers to exploit the card number and the cardholder's identity to make purchases without the authentication of IoT.

Common in the IoT is the practice of "man in the middle" assaults, in which a third party takes control of a communication channel to hide their true identities. Man-in-the-middle attacks are effective against financial institutions because the attacker does not have to know the genuine identity of the victim [13].

Privacy

Users' opinions on the IoT's value hinge on how well it protects their personal information. Concerns about privacy and the possible hazards that come with it may be significant impediments to the wide adoption of the Internet of Things. It's crucial to keep in mind that users have a right to privacy and that their data will be secured when engaging with the Internet of Things, the connected device, and the services linked with it. There is a lot of work being done to make sure that the privacy issues raised by the Internet of Things, such as the increase in spying and tracking, are rethought. Since there are now more places where data collecting and dissemination could take place in the Internet of Things, the widespread presence of intelligently integrated artifacts creates privacy concerns. Understanding this problem is vital because of the ease with which private information may be accessible from anywhere in the world because of the ubiquitous availability of the Internet [15].

Interoperability

It is common knowledge that a fragmented ecosystem of proprietary IoT technology implementation reduces value for consumers. Even if full interoperability is not always achievable between goods, users may be hesitant to acquire those with limited personalization possibilities and the possibility of dealer lock-in. Poorly planned IoT gadgets may have a negative impact on shared network resources [16].

Cryptography is another fundamental part since it has been used for decades to fortify many different types of software against security flaws [9]. There are too many different kinds of attacks for any single security program to be effective. Thus, numerous precautions are required to forestall attacks on the IoT authentication procedure.

Adding more complex security features to products is the greatest approach to stopping attacks. Users can avoid this security by opting for products that already have built-in safeguards. Some have suggested that cybersecurity frameworks be used to safeguard IoT [17]. Attempts to protect the information stored on Internet of Things devices face additional challenges due to factors like:

IoT manufacturers typically only update their products once every three months with new security patches. System software and bug fixes both get regular updates [18]. Because of this, bad actors will have plenty of chances to bypass safeguards and steal sensitive data.

Support workers can remotely upgrade software and fix operating system faults if they have access to the passwords stored in the integrated storage of IoT devices. This vulnerability could be exploited by hackers to get unauthorized access to devices.

Businesses and consumers alike make heavy use of IoT devices' automation capabilities for data collection and the streamlining of administrative operations. Without restricting access to potentially harmful URLs, a system is vulnerable to attack.

Connectivity from afar: The network protocols utilized by IoT devices for remote access include Wi-Fi, ZigBee, and Z-Wave, to name a few. Rarely are the kinds of restrictions that could be used to stop cybercriminals actually disclosed. Therefore, hackers might easily establish an adversarial connection via these remote access protocols.

There are a plethora of third-party programs available on the web, allowing firms to perform an array of functions. However, verifying the authenticity of these submissions was a major challenge. Anyone using or installing such a program runs the risk of malevolent actors corrupting the embedded database.

Most Internet of Things apps don't make use of authentication mechanisms, making networks susceptible to intrusion. This makes people vulnerable to burglars, who represent a threat if they get in. Weak All IoT producers implement unique device identifiers as a means of inventory management. However, not all manufacturers use a stringent data protection policy. As a result, monitoring suspicious behavior on the internet has grown challenging.

Internet of Things Authentication Methods

Due to the pervasive nature of the IoT, it is vital that users' private information is protected from any unauthorized access. In addition, unauthorized individuals must be prevented from gaining access to the devices. Therefore, the user must be verified as legitimate before access may be granted. Therefore, there are numerous approaches to authenticating a user's identity.

However, the most common kind of authentication makes use of the secret distribution of a key or password in advance. Therefore, in this section, we take a look at how authentication is strengthened in an IoT environment.

Using a One-Time Password (OUPA)

One-time passwords (OTPs) are passwords that can only be used once in a given authentication process and are generated dynamically each time. In order to ensure the security of communication in an IoT environment, many different OTP authentication techniques have been created.

Time synchronization, hash algorithms (MD5, SHA1, and SHA256), and RSA cryptography are only some of the foundations on which these protocols rest. All of these strategies are predicated on Lamport's OTP algorithm. Unfortunately, vulnerabilities in these protocols can be exploited by attacks like those stated. A safe OTP method based on bilinear maps was presented, however, by Lee and Kim in 2013.

Complex calculations are still needed, though. The problem was used by Shivraj et al. to present a secure OTP system for the Internet of Things. Lamport's One-Time Pad and the concepts of lightweight identity-based elliptic curve encryption form the basis of the proposed protocol.

Mutual Authentication Using ECC

Most IoT gadgets don't have enough processing power. Communication in real time between sensors, actuators, objects, and nodes is also essential. Therefore, it is critical to provide an IoT-friendly authentication solution that doesn't drag everything down. As a result, Azrour et al. developed a secure authentication method for the IoT. Elliptic curve cryptography (ECC) is used in this protocol, which has been proven to be more secure than the standard RSA approach. Furthermore, in [19], further authentication algorithms based on ECC are presented. Elliptic curve cryptography is the ideal technique of encryption in low-resource situations, such as those with limited memory or processing power, due to its higher efficiency and security.

Using a Username and Password for Authentication

Using a user's identification information to verify their identity helps reduce security risks. The user's identification is checked before granting access or revoking it. In this context, "user ID" refers to any piece of information that can be used to uniquely identify a single user as opposed to all users.

Using this technique, several protocols are proposed for use in an IoT environment. The server-client architecture, however, sees the most use of this authentication method. Therefore, an IoT environment requires a server to safely store user credentials.

However, as will be seen below, there are issues with the ID-based authentication technique.

How do the user's records ever get to the server in the first place? Is there enough protection on the server to thwart insider attacks and lost verifiers? The second issue is that patrons can forget their credentials. The consequent authentication cannot be performed by them. Sensitive information should not be stored on a computer, tablet, or smartphone in this situation, regardless of whether or not the device is linked to a network. User ID transfer via a public network presents an odd additional challenge. A cryptographic hash function or method should be used in this scenario.

Certificate-Based Authentication

Rather than relying on user IDs and passwords, an alternative method was proposed. Certificate-based authentication is a mechanism used to confirm a user's claimed identity. Certificate-based authentication has been adopted by other apps. For example, Hiltgen et al. suggested a new certificate-based authentication method for verifying a user's identification in a financial application. The concept has also been applied to the realm of the Internet of Things. Certificate-based authentication is more secure than other methods, but it can't always be used because it demands a lot of processing power that isn't always present in IoT devices. As a result, Internet of Things devices cannot use this approach.

Blockchain

Blockchain is a novel data structure with several applications. This database is unlike others because of the unconventional method it uses to store data. Blockchains keep track of data using blocks linked together through networks. In recent years, a few authors have made use of this state-of-the-art technology to develop authentication protocols for the Internet of Things.

Conclusion

Through the IoT, communities are opening up previously inaccessible lines of communication with one another. As a result, it makes it easier to have effective conversations. By allowing for the creation of smart systems, like those used in smart homes, smart farms, and more, it is also improving people's quality of life. Despite its usefulness in many areas, bad actors are always looking for new methods to use it to break into the Internet of Things and steal sensitive data from users who don't realize they've been targeted. This highlights the importance of developing measures to protect IoT infrastructure. Therefore, sensitive information is protected.

Internet of Things (IoT) system security and privacy have quickly risen to the forefront of important issues. Concerns about privacy and safety range in severity. Not all assaults have the same impact. Not all assaults come from without; sometimes they come from within. Although there are many different kinds of attacks, they all have the same damaging effect and get increasingly dangerous as they progress.

The goal of this study was to survey the field of Internet of Things (IoT) security and privacy in its entirety. Security and privacy issues in the Internet of Things were also broken down into their component parts. Possible security risks, their underlying mechanisms, and how to deal with them were also detailed. A taxonomy of attacks, rationale for their occurrence, and preventative measures were also supplied by the survey. Recent research has examined the privacy and security issues plaguing IoT systems.

References

1. Khvoynitskaya, S. (2020). The history and future of the Internet of Things. https://www.itransition.com/, https://www.itransition.com/blog/iot-history. Accessed 25 March 2020.
2. Conti, M., Dehghantanha, A., Franke, K., & Watson, S. (2018). Internet of Things security and forensics: Challenges and opportunities. *Future Generation Computer Systems, 78*, 544–546.
3. Monther, A. A., & Tawalbeh, L. (2020). Security techniques for intelligent spam sensing and anomaly detection in online social platforms. *International Journal of Electrical and Computer Engineering, 10*, 2088–8708.
4. Makhdoom, I., Abolhasan, M., Lipman, J., Liu, R. P., & Ni, W. (2018). Anatomy of threats to the Internet of things. *IEEE Communications Surveys & Tutorials, 21*, 1636–1675.
5. Meng, Y., Zhang, W., Zhu, H., Shen, X. S. (2018). Securing consumer IoT in the smart home: Architecture, challenges, and countermeasures. *IEEE Wireless Communications, 25*, 53–59.
6. Siby, S., Maiti, R. R., Tippenhauer, N. O. (2017). Iotscanner: Detecting privacy threats in IoT neighborhoods. In *Proceedings of the 3rd ACM International Workshop on IoT Privacy, Trust, and Security, Abu Dhabi United Arab Emirates, 2 April 2017* (pp. 23–30).
7. Hassan, W. H. (2019). Current research on Internet of Things (IoT) security: A survey. *Computer Networks, 148*, 283–294.
8. Leloglu, E. (2016). A review of security concerns in Internet of Things. *Journal of Computer and Communications, 5*, 121–136.
9. Liu, X., Zhao, M., Li, S., Zhang, F., & Trappe, W. (2017). A security framework for the internet of things in the future internet architecture. *Future Internet, 9*, 27.
10. Ali, S., Bosche, A., & Ford, F. (2018). *Cybersecurity is the key to unlocking demand in the Internet of Things*. Boston, MA, USA: Bain and Company.
11. Sadeghi, A. -R., Wachsmann, C., & Waidner, M. Security and privacy challenges in industrial internet of things. In *Proceedings of the 2015 52nd ACM/EDAC/IEEE Design Automation Conference (DAC), San Francisco, CA, USA, 8–12 June 2015* (pp. 1–6).
12. Izzat, A., Chuck, E., & Lo'ai, T. (2020). *The NICE cyber security framework, cyber security management*. Springer: Basel, Switzerland. ISBN 978-3-030-41987-5.
13. Kathole, A. B., Katti, J., Lonare, S., & Dharmale, G. (2023). Identify and classify pests in the agricultural sector using metaheuristics deep learning approach. Franklin Open, 100024. https://doi.org/10.1016/j.fraope.2023.100024
14. Kumbhare, S, Kathole, A. B., & Shinde, S. (2023). Federated learning aided breast cancer detection with intelligent Heuristic-based deep learning framework. *Biomedical Signal Processing and Control, 86*, Part A, 105080. https://doi.org/10.1016/j.bspc.2023.105080

15. Nagaraj, S., Kathole, A. B., Arya, L., Tyagi, N., Goyal, S. B., Rajawat, A. S., Raboaca, M. S., Mihaltan, T. C., Verma, C., & Suciu, G. (2023). Improved secure encryption with energy optimization using random permutation pseudo algorithm based on internet of thing in wireless sensor networks. Energy efficiency in wireless networks, MDPI.
16. Kathole, A. B., Katti, J., Dhabliya, D., Deshpande, V., Rajawat, A. S., Goyal, S. B., Raboaca, M. S., Mihaltan, T. C., Verma, C., & Suciu, G. (2022). Energy-aware UAV based on blockchain model using IOE application in 6g network-driven cybertwin. Energy efficiency in wireless networks. MDPI.
17. https://www.mdpi.com/1996-1073/15/21/8304
18. Tawalbeh, L. A., Somani, T. F. (2016). More secure Internet of Things using robust encryption algorithms against side-channel attacks. In *Proceedings of the 2016 IEEE/ACS 13th International Conference of Computer Systems and Applications (AICCSA), 29 November–2 December 2016* (pp. 1–6). Agadir, Morocco.
19. Dalipi, F., Yayilgan, S. Y. (2016). Security and privacy considerations for IoT application on smart grids: Survey and research challenges. In *Proceedings of the in Future Internet of Things and Cloud Workshops (FiCloudW), 22–24 August 2016* (pp. 63–68). Vienna, Austria.
20. Bugeja, J., Jacobsson, A., & Davidsson, P. (2016). On privacy and security challenges in smart connected homes. In *Proceedings of the European Intelligence and Security Informatics Conference (EISIC)* (pp. 172–175), Uppsala, Sweden, 17–19 August.
21. Culbert, D. (2020). Personal data breaches and securing IoT deviceshttps://betanews.com/2019/08/13/securing-iot-devices/. Accessed 15 Sept 2019.
22. Gemalto. (2020). Securing the IoT-building trust in IoT devices and data. https://www.gemalto.com/: https://www.gemalto.com/iot/iot-security. Accessed 17 Feb 2020.

Security Challenges in IOT

Kingsley Igulu, Barilemena Johnson, Agbeb Nornu Stephen, and Tarandeep Kaur Bhatia

Abstract The concept of the Internet of Things (IoT) revolves around the exchange of information among low-power embedded devices, linked to the internet, in order to enable seamless communication. The IoT has a profound impact on various aspects of modern life, from mobile devices and sensors that keep track of the surrounding environment to smart industrial gadgets. While the Internet of Things offers numerous advantages, it also presents security and privacy concerns. The information transmitted through the IoT includes sensitive data such as banking information, geographic data, environmental data, medical information, and other personal information. Hence, it is crucial to acknowledge the security challenges posed by the IoT and address them appropriately. This chapter presents comprehensive insights into the security challenges associated with the Internet of Things, while considering the vast scope of the topic and existing literature. It discusses various IoT security challenges, IoT security architectures, IoT security solution trust zones and boundaries, potential risks of IoT devices, notable cases of IoT security breaches, solutions to IoT security breaches, strategies for securing IoT data and best IoT security practices.

Keywords Internet of Things (IoT) · Communication · Security · Embedded devices · Security architectures

B. Johnson
Department of Computer Science, Ignatius Ajuru University of Education, Port Harcourt, Nigeria

K. Igulu (✉) · A. N. Stephen
Department of Computer Science, Ken Saro-Wiwa Polytechnic, Rivers State, Bori, Nigeria
e-mail: igulukt@gmail.com

T. K. Bhatia
School of Computer Science, UPES, Dehradun, India
e-mail: drtarandeepkaurbhatia@gmail.com

A. Prasad et al. (eds.), *Communication Technologies and Security Challenges in IoT*, Internet of Things, https://doi.org/10.1007/978-981-97-0052-3_4

Introduction

The IoT has the potency to completely alter the way human engage with the physical world from our homes and workplaces to our vehicles and cities. But with this revolution comes a host of security challenges, from data privacy and hacking to cyber-attacks and malware. As the IoT revolution progresses, it is essential that we understand the risks and develop effective strategies to protect ourselves [13].

The IoT is a rapidly evolving field that is transforming industries such as agriculture, healthcare transportation, and manufacturing. It is projected that by 2030, there will be 50 billion IoT devices, with an associated economy worth of $11 trillion [53]. Nevertheless, the growth of IoT devices raises various security concerns that must be addressed to guarantee that the benefits of IoT do not outweigh the hazards.

IoT devices are vulnerable to attacks from hackers who can exploit their weaknesses to gain unauthorized access to data, manipulate systems, and cause disruptions. These security challenges are compounded by the fact that IoT devices are often constructed with small memory, restricted processing power, and battery life, which limits the implementation of robust security protocols. Furthermore, many IoT devices are deployed in remote and harsh environments where maintenance and security updates are challenging to implement [78].

In this chapter, we shall look at the security challenges facing the IoT revolution and what can be done to address them. We shall explore the potential risks of IoT technology, the need for robust security measures, and the steps that can be taken to protect ourselves and our data. With the right approach, we can ensure that the IoT revolution is a safe and secure one.

IOT Security Architecture

The term "IoT security architecture" refers to the planning and execution of protective measures for the IoT devices, networks, and applications from cyber threats. An IoT security architecture typically involves several layers of security controls to ensure the confidentiality, availability, and integrity of transmitted data between IoT devices and systems. Here are some key components of an IoT security architecture [13, 15, 57, 73]:

- **Device security**: This layer involves securing IoT devices at the hardware and firmware levels, which includes secure boot, encryption, and device authentication. In addition to this, it involves making sure that the devices in question have the most recent security patches and firmware updates installed.
- **Network security**: This layer involves securing communication between devices in IoT and networks. This includes authorization and authentication, secure data transmission, encryption, and intrusion detection.

- **Cloud security**: This layer involves securing the cloud infrastructure and services that store and process IoT data. This includes data encryption, access control, and backup and recovery mechanisms.
- **Application security**: This layer involves securing the IoT applications that run on top of the infrastructure. This includes authentication, authorization, input validation, and encryption of sensitive data.
- **Data security**: This layer involves securing the data transmitted and stored by IoT devices and systems. This includes data encryption, access control, and data loss prevention mechanisms.
- **Physical security**: This layer involves securing the physical environment where IoT devices and systems are deployed. This includes access control, surveillance, and monitoring mechanisms.

Overall, an effective IoT security architecture should be designed to cater for the peculiar security challenges eventuated by IoT devices and networks, such as the massive scale of devices, the variety of communication protocols, and the diversity of application domains.

Components of IoT Security Architecture

The solution developed during the procedure of designing an architecture for the IoT would have many different parts. Each of these parts serves a different purpose and provides essential building blocks that may be put together in any way to create an IoT solution. Any of these parts can be used in the design and construction of an IoT system.

- **Devices**: The sensors and other parts of the Internet of Things are connected to specific devices. Devices may also be linked to additional parts. The "things" in "Internet of Things" come from the devices themselves.
- **Field Gateway(s)**: This is a software or hardware component that acts as an intermediary between the endpoint devices and the cloud. By acting as a hub for both cloud-based and on-premises components, they can be leveraged to improve security. Message or protocol translation, aggregation of events and messages in communication, and other system-level services are only a few of the many uses for field gateways.
- **Edge Devices**: One subset of Field Gateways are the Edge Gateways (also known as edge devices). It's superior because it allows Cloud functions to be executed locally, in close proximity to other devices, hence minimizing delays in the feedback loop of real-time processing. Edge gateways can be outfitted to perform tasks typically performed in the cloud, such as event stream processing and machine learning.

- **Cloud Gateway**: These are equivalent to the popular Field Gateways; the main difference is that they are hosted in the cloud rather than on a company's own servers. By being in the cloud rather than on a local device, Cloud Gateways can perform the same duties as Field Gateways.
- **Services**: The Services section of an IoT system's backend encompasses everything else, such as REST APIs, databases, and so on. Depending on their role in the overall IoT solution architecture, these services can be setup in the cloud, on-premises, or hybridized.

Boundaries and Trust Zones for IoT Security Solutions

Every Internet of Things solution is assembled from a wide variety of individual components. The many components of an Internet of Things solution will each be compartmentalized into their own distinct Trust Zones and Boundaries as part of the system's overarching security architecture. These diverse zones and borders enable both physical and software-based isolation layers to separate the various system components in order to ensure their security [13].

As depicted in Fig. 1, security measures are utilized to partition the various components of an Internet of Things (IoT) solution, isolating each component from the others in a more secure manner. This is done to give Trust Zones and Boundaries with segmented protection.

When developing any kind of Internet of Things Security Architecture, the following trust zones are the most important ones to bear in mind:

i. Local Zone
ii. Device Zone
iii. Cloud Gateway Zone
iv. Remote User Zone
v. Field Gateway Zone
vi. Gateway and Services Zone.

Trust Zone boundaries should be separate. This isolates each zone, protecting it. As you move closer to the cloud, adding security to each successively higher zone verifies the communications between the zone below it and the local zone or on-premises components.

Except for the Device Zone and the Field Gateway Zone, most zone boundaries are depicted in Fig. 1. Field Gateway(s) will receive event data from the Device Zone devices via direct communication. If your Internet of Things (IoT) solution calls for or benefits from a higher level of security, you may want to consider extending this paradigm with a Trust Boundary. The boundaries and trust zones of your Internet of Things architecture are something you design.

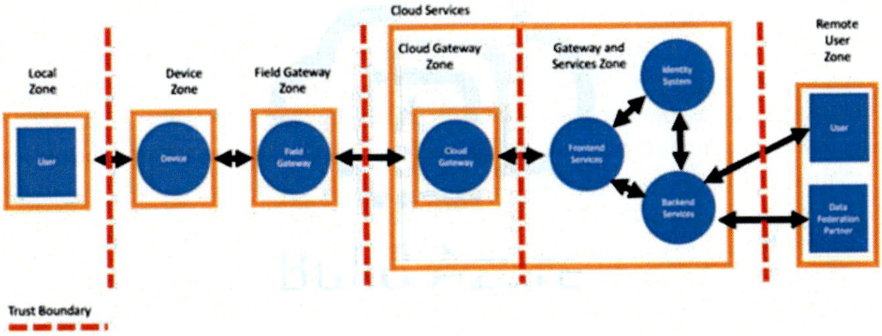

Fig. 1 IoT trust zones and boundaries [13]

Security Challenges Facing the IoT Revolution- an Overview

The IoT has the potential to completely alter the way we engage with the physical world, from our homes and workplaces to our vehicles and cities [20, 44, 56]. But with this revolution comes a host of security challenges, from data privacy and hacking to cyber-attacks and malware. As the IoT revolution progresses, it is essential that we understand the risks and develop effective strategies to protect ourselves. In this overview, we'll look at the security challenges facing the IoT revolution and what can be done to address them [78].

We shall explore the potential risks of IoT technology, the need for robust security measures, and the steps that can be taken to protect ourselves and our data. With the right approach, we can ensure that the IoT revolution is a safe and secure one.

Security Challenges of the IoT Revolution

The security challenges posed by the IoT revolution are numerous. Increases in the number of Internet-connected devices have also increased the potential for malicious actors to exploit weaknesses in these systems. IoT devices are often vulnerable to a range of security challenges which are listed and discussed below according to [60, 63].

Vulnerabilities in Devices

Typically, the processing power, memory, and battery life of IoT devices are all on the low end. This often leads to vulnerabilities that can be exploited by attackers to gain access to sensitive data or take control of the device. Manufacturers must ensure that IoT devices are designed with security in mind, including secure communication protocols, firmware updates, and secure data storage. In the context of IoT security, device security is considered to be a crucial aspect according to the findings of several researchers [83, 84, 87]. Yu et al. [84] have discussed the presence of known vulnerable devices that can be exploited for DDoS attacks due to issues like hardcoded administrative login credentials and open DNS resolvers. Airehrour et al. [3] have reported a 2012 case where TRENDNET IP cameras streamed live footage without the need for a password. Patton et al. [58] have conducted a thorough investigation of 35,737 different IoT devices, and the results show that the majority of devices were openly accessible over the internet without any identification requirements.

IoT devices have various potential vulnerabilities that can be exploited by attackers. These vulnerabilities can be classified into three categories: physical, software, and network.

Physical Vulnerabilities

Physical vulnerabilities are related to physical access to IoT devices. IoT devices can be physically accessed by attackers to manipulate the device, extract sensitive data or install malware. For example, attackers can physically tamper with the device to extract passwords or install malware.

Software Vulnerabilities

Software vulnerabilities are related to the software running on the IoT devices. IoT devices are often designed with limited resources, which means the software may not be fully secure. Additionally, IoT devices may be running outdated or unpatched software, which can leave them vulnerable to attacks. For example, attackers can exploit a software vulnerability to remotely control the device.

Network Vulnerabilities

The network connections of IoT devices present security risks. Since many IoT devices are online, they are susceptible to cybercriminal activity. Additionally, IoT devices are often connected to other devices, which can be exploited to gain access to the IoT device. For example, attackers can exploit a network vulnerability to intercept data transmitted between the device and the cloud server.

Inadequate Identification, Authentication and Authorization

Many As per industry experts, one of the major security challenges in the Internet of Things (IoT) is the lack of authentication and authorization mechanisms in many IoT devices. Attackers can easily exploit this vulnerability to gain unauthorized access to the device and compromise its security. To combat this issue, manufacturers need to implement robust authentication mechanisms such as multi-factor authentication and role-based access control to restrict access to only authorized users.

The identification and authentication of devices are complex tasks in IoT due to the vast number of devices connected. Authentication is crucial to ensure the integrity of a device's data stream. Additionally, authorization serves as a crucial access control measure to prevent hacking attempts. A study by [56] indicates that current security mechanisms lack access control and privacy protection. Therefore, it is imperative to implement authorization servers in server-based protocols to ensure access control in IoT.

Lack of Encryption

Encryption is critical to protect data transmitted between IoT devices and servers. However, many IoT devices do not have built-in encryption capabilities or use weak encryption algorithms. This leaves data vulnerable to interception and manipulation by attackers. It is essential to implement strong encryption mechanisms to protect data transmitted between devices and servers.

Environment Constraints

The environment presents a significant problem for loT security. Hossain et al. [35] enumerate them. The computing power, memory, and battery life of devices are cited as the first hardware restrictions. Therefore, operations that are computationally complex and memory expensive are not a good fit for the IoT. Next, they focus on software restrictions. Thin network stacks are a feature of loT devices' operating systems, which may prevent them from being remotely reprogrammed. This restricts the development of security modules and the distribution of security updates for certain platforms. Finally, they talk about the limits imposed by the network. The security design is complicated by the networks' mobility, scale, and heterogeneity. The computational and network limits are agreed upon by [11, 19] as limitations to loT security.

Legislative Issues

As Weber [78] pointed out, the loT business was mostly self-regulated at the time, but new legal frameworks will be required to protect consumers' privacy. Weber claimed that such rules might not be sufficient to guarantee true anonymity or safety. Weber argued that a global IoT necessitates a global regulatory framework. In the wake of increasing global concerns over security, it is imperative that the IoT industry is regulated by a comprehensive security framework. As emphasized by Weber [79, 71] the IoT poses significant challenges to national security, company secrets, and personal privacy, and therefore, requires legal provisions to ensure its safe and secure growth.

Enforcement Mechanisms

The enforcement mechanisms of the Internet of Things are either flawed or nonexistent, as stated in [83, 84]. Because IoT devices have limited resources and vary widely in their characteristics, traditional security measures like antivirus software are not available. Moreover, unlike regular networked devices, IoT devices do not have automated software updates. Instead, they rely on firmware updates, which need to be carried out separately by each manufacturer and for each device. Another issue is that most network security approaches depend on strong static perimeter defenses, such as firewalls, which become ineffective when insecure IoT devices are integrated deeply within the network. Kumar et al. [44] also expressed concerns regarding the absence of security updates for the Internet of Things.

Privacy Concerns

The privacy aspect of the IoT is a significant concern due to the increased risks of security threats like eavesdropping, unauthorized access, data manipulation, data forgery, and remote tampering with devices. Researchers such as Ram Mohan Rao et al. (2018), [65] have highlighted this issue. Data forgery, which involves unauthorized modification of device data by external parties, is another potential security risk, as pointed out by [49]. The data collected in IoT, such as personal information like names, addresses, and insurance policy numbers, is often sensitive, as emphasized by [3], and transmitting it to cloud environments introduces additional challenges. "A vast number of the services and apps offered by the internet of things (IoT) give sensitive and personal information that is accessible and can be exploited by an adversary" [49]. Sensitive information that is not encrypted may be accessed by unauthorized persons.

IoT devices often collect and transmit personal data, such as location, health, and biometric information. This data must be protected from unauthorized access and use. Manufacturers must implement appropriate data privacy policies and ensure that personal data is collected and stored securely.

Data privacy is another significant concern in IoT. The vast volumes of data produced by Internet of Things (IoT) devices can encompass confidential and private information pertaining to individuals and organizations. This data can include location data, personal health information, and financial data. The following are some data privacy challenges in IoT:

- Collection and Storage of Data: IoT devices collect and store large amounts of data. This data can be vulnerable to theft or misuse.
- Data Ownership: Who exactly owns the information collected by IoT gadgets is often not made apparent. This can create challenges when it comes to data privacy and security.
- Consent: It is often unclear whether individuals have given consent for their data to be collected and used. This can create challenges for organizations when it comes to compliance with data privacy regulations.

Overall, data privacy in IoT requires a combination of technical and organizational measures to ensure that data is protected at every stage of the data lifecycle.

Distributed Denial-Of-Service (DDoS) Attacks

IoT devices are often connected to the internet through home networks, which are typically not designed to handle large-scale DDoS attacks. Attackers can use compromised IoT devices to launch DDoS attacks, which can cripple servers and websites. Manufacturers must ensure that IoT devices are designed with built-in security mechanisms that can detect and prevent DDoS attacks.

Overall, the security challenges in IoT are complex and require a holistic approach that involves manufacturers, users, and regulators. Manufacturers must build security into their devices from the ground up, while users must be educated on how to use IoT devices safely and responsibly. Regulators must also develop policies and standards to ensure that IoT devices meet minimum security requirements [17, 30, 85].

Cross-Device Dependencies

According to Yu et al. [84], the interconnectedness of the Internet of Things creates extra opportunities for cyberattacks. They describe a scenario in which an attacker disables an air conditioner, raising the room temperature, which in turn triggers another system to open the windows, posing a threat to the physical security of the facility. It is not unusual to see devices with many connections. Yu et al. [84], offers a few examples: There are 188 cross-device policies available through the NEST Protect home security system, 227 through the Wemo Plugin, and 63 through the Scout Alarm.

Data Theft

Theft of data or information is the illegal acquisition or compromise of private or sensitive data kept in computer systems, networks, or electronic media belonging to a company. As the world becomes increasingly interconnected and reliant on technology, data theft in IoT is of increasing concern. With the increasing use of IoT devices, the risk of data theft is also increasing. Cybercriminals are constantly finding new ways to exploit vulnerabilities in IoT devices, and organizations must be vigilant in protecting their sensitive and valuable data.

One of the main reasons for data theft in IoT is the lack of security measures in place. Many IoT devices do not have adequate security protocols, making them an easy target for hackers. Assaults including man-in-the-middle attacks, denial-of-service attacks, and ransomware can easily infiltrate these devices. In addition, IoT devices are often connected to a network that is also vulnerable to attack. This means that if one device is compromised, it can provide a gateway for hackers to access other devices on the network.

The substantial quantity of data collected and transmitted by IoT devices contributes to the risk of data theft. This data encompasses sensitive information, including personal identification details, financial data, and health information. If this data is accessed by unauthorized individuals, it can be exploited for identity theft, financial fraud, or other nefarious intentions. In addition, the sheer volume of data being collected can make it difficult for organizations to monitor and control access to this data.

The majority of data breaches occur because hackers stole sensitive data for either identity theft or sale on the dark web. If a hacker or threat actor manages to get unauthorized access to sensitive information, they may change, erase, or even completely block the owners from viewing the data.

Keep in mind that when discussing data theft, the terms data leak and data breach are typically used interchangeably. Data leaks and breaches are two different things. Intentional cyber attacks are what we call a "data breach." On the other side, a data leak occurs when private information is inadvertently shared online or when hard drives or other storage devices are lost or stolen, allowing threat actors to gain access to the data without any additional effort on their part.

Data Tampering

Tampering refers to the act of making unauthorized changes to or deletions from a resource. An application that can be accessed via the internet and a web browser on a device is referred to as a web application. When discussing web applications, "data tampering" refers to any incident in which an unauthorized user (such as a hacker) gains access to a system and then tampers with the data stored inside. Indirect tampering can also be accomplished by a hacker or other malicious user using a script exploit. This means that the hacker would get the script to execute by disguising it as a user input from a page or as a web link [40, 82].

Tampering presents a major threat to the security of the Internet of Things. Files stored in cloud-based apps, which are used by millions of companies throughout the world, can be edited and updated with this tool. In the late 1980s, tampering was first used as a method to sabotage data or plant a harmful or destructive program with the intention of deleting data. Since that time, it has developed further and improved over the course of the years. In the year 2000, hackers were able to fabricate data and falsify information in order to trick the consumers of a web application. Each year brings new and better technology that makes it easier for bad actors to hack web applications. This is due to the fact that these technologies provide simple tools and application programs that make it easier to tamper with data or manipulate data within computer systems [13, 28, 73, 87].

Utilizing a firewall and the security features offered by Windows to restrict access to critical files, directories, and other resources is a first line of defense against the alteration of data. Additionally, the web application ought to function with the least amount of privileges possible. In order to prevent script vulnerabilities from being exploited, it is important to never put any faith in the information that originates from a user or even a database. When acquiring information from untrusted sources, appropriate and safe precautions should be taken to ensure that the material does not contain any executable code that may be used for malevolent purposes.

[76] provides a categorization of these security challenges or requirement as given in Fig. 2.

Potential Risks of IoT Devices

IoT devices also pose various risks that can have serious consequences. These risks can be classified into three categories: privacy, financial, and physical [53, 53].

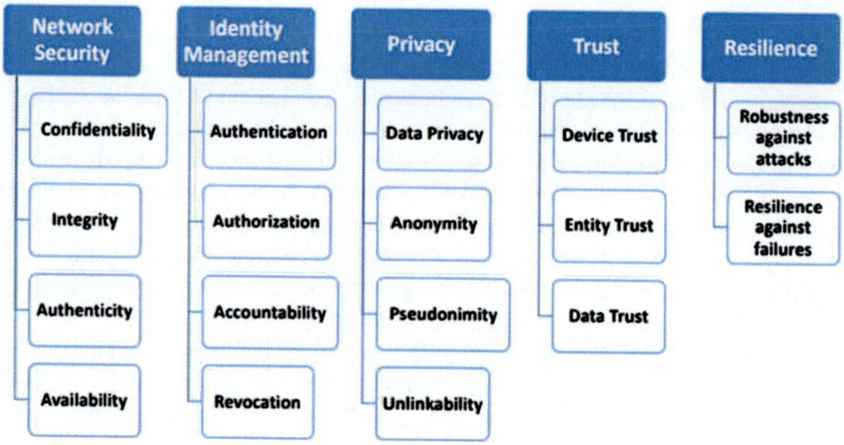

Fig. 2 Security requirement classification according to [76]

Privacy Risks

Privacy risks are related to the protection of personal information of the user. IoT devices collect and transmit sensitive personal data, such as health data or location data. If this data is intercepted, it can be used for malicious purposes. For example, attackers can steal personal data and use it for identity theft or blackmail.

Financial Risks

Financial risks are related to the financial loss of the user. IoT devices can be used for financial transactions, such as online shopping or banking. If an IoT device is compromised, attackers can steal financial information and use it to make unauthorized purchases or transfers.

Notable Examples of IoT Security Breaches

Several high-profile IoT security breaches have occurred in recent years, highlighting the need for better security measures. Some notable examples include:

- **Mirai Botnet**: In 2016, the Mirai botnet infected more than 600,000 IoT devices, including cameras, routers, and digital video recorders. In order to disrupt DNS services, the botnet was utilized to perform a massive, distributed denial of service (DDoS) attack [5, 86].

- **Jeep Hack**: In 2015, a group of hackers demonstrated how they could take control of a Jeep Cherokee remotely. The hackers were able to manipulate the car's air conditioning, radio, and even its brakes [6, 24, 54].
- **St. Jude Medical Hack**: In 2017, a security researcher discovered that it was possible to hack into St. Jude Medical's pacemakers and defibrillators. The vulnerability could have been exploited to remotely control the devices or to drain their batteries [10, 26, 29].
- **Smart Home Device Hacks**: Many smart home gadgets, including smart locks and security cameras, were discovered to be hackable in 2020. Because of this, hackers may be able to enter people's houses or spy on them [55, 57, 73] (Shilpa Sharma 2021).
- **Target breach**: Using a flaw in a contracted HVAC (Heating, Ventilation, and Air Conditioning) system, hackers were able to breach Target's payment system in 2013.The hackers' success in stealing credit card information for millions of Target consumers highlights the dangers of interconnected Internet of Things (IoT) devices and vital infrastructure [50, 59, 66].
- **Baby monitor hack**: In 2015, a family in Houston, Texas, discovered that their baby monitor had been hacked when they heard a stranger's voice coming from the device. The baby monitor had been connected to the internet without proper security measures, allowing the attacker to access the device and monitor the family's activities [23, 32, 33].
- Stuxnet worm: In 2010, Iran's nuclear program was attacked using the Stuxnet worm, which specifically targeted the country's industrial control systems. The centrifuges used in uranium enrichment were controlled by programmable logic controllers (PLCs), which the worm exploited, causing them to malfunction and destroying the program [38].

These examples demonstrate the importance of IoT security, and the potential risks associated with insecure IoT devices and systems. It is essential to implement proper security measures, including access control, network segmentation, and behavioral analytics, to protect IoT devices and data from potential security breaches.

Physical Risks

Physical risks are related to the safety of the user. IoT devices can be used to control physical systems, such as door locks or medical equipment. If an IoT device is compromised, attackers can control the physical system and cause harm to the user. For example, attackers can remotely unlock the door of a house and break into it.

Solutions to IoT Security Challenges

Even though Internet of Things comes with tremendous security issues when compared to traditional systems, numerous researchers have also provided solutions for the challenges that Internet of Things presents to security. In the parts that follow, we'll talk about some of the ways in which the problems with IoT security could be fixed. Vasilomanolakis et al. [76] provide a taxonomy of such approaches.

Trust Management

According to both Yu et al. [84] and Hossain et al. [35], trust management is crucial to the IoT. Having a system in place to manage trust can help individuals deal with the risks and uncertainties of the Internet of Things. Trust involves guaranteeing someone's safety and keeping their business private. According to Bekkali et al. [11], Ferraris et al. [20] and Tsunoda et al. [74], trust is a key component of the IoT. They claim that consumers' experiences with the Internet of Things' various features are crucial to their faith in the network as a whole. Users need agency over their service experiences and transparency into their interactions with systems. They also think that establishing a reliable governance system can boost people's faith in IoT.

Trust management in the IoT is multidimensional as explored in [42]. New decentralized trust models, trust mechanisms for cloud computing, and the creation of trust-based apps at the node level are the three focus areas for IoT trust research. They argue that trust evaluations must be automated and unbiased.

Authentication

Zhang and Green [85] provide several different authentication models for use with the internet of things. Some examples of these models are the gateway model, the trust chain model, the security token model, and the global trust tree model of authentication. Every model comes with both positives and negatives specific to it. Mahmoud et al. [47] discusses authentication mechanisms as well. They describe a method that only requires one cipher and one time to use it, and utilizes a request-reply technique.

Though very trivial but strong passwords and multi-factor authentication should be adopted.

Strong Passwords: Strong passwords are essential for IoT devices. Passwords should be long and complex, and different for each device.

Multi-Factor Authentication (MFA): With MFA, users are required to submit not one but two different types of authentication before getting access to an IoT device.

In addition to authentication, the following are examples of various forms of access control that should be implemented in in IoT deployments:

- Role-based access control: Access to IoT resources and data should be restricted based on the user's role or job function. This can help to limit access to sensitive data or critical systems and reduce the risk of unauthorized access or data breaches.
- Device authentication: IoT devices should be authenticated before they are allowed to connect to the network or access IoT resources. This can be achieved through digital certificates, unique identifiers, or other forms of device authentication.
- Network segmentation: IoT networks can be segmented to limit the scope of a potential security breach and reduce the attack surface of IoT devices and systems. Access to each network segment can be restricted based on the user's role or job function.
- Audit logging: All access to IoT devices, applications, and data should be logged and monitored. This can help to detect and prevent unauthorized access or data breaches and provide a record of all access activity.

Privacy Solutions

Data privacy is also an important consideration when implementing IoT security measures. Companies need to take precautions to safeguard their data and the privacy of their customers. Encryption, firewalls, and other forms of access control are all part of this strategy. Companies must also keep their systems patched and updated to fix any security holes that may have appeared.

Tsunoda et al. [74] present several potential answers to the problems with privacy. One of these principles is called "privacy by design," and it stipulates that users should be provided with the opportunity to take control of their own data. Transparency is yet another guiding principle. In the context of loT, transparency refers to the idea that users should be aware of which entities are handling their data, as well as how and when those companies are using the data. The management of data is the third recommendation that they provide. Taking this step requires determining who oversees keeping the secrets. Data management policies and enforcement mechanisms should address an extensive array of usage cases. To deal with the issue of managing IoT data in cloud environments, [31] propose a method called User-driven Privacy Enforcement for Cloud-based Services on the Internet of Things (UPECSI). With UPECSI, users may manage their sensitive information before it is even sent to the cloud.

Users also need to be educated on the security concerns associated with the IoT revolution, and this is something that businesses should take care of. Giving users access to the resources they need to keep their data safe and conducting routine reviews of security policies and procedures are both essential. Additionally, businesses should make testing and monitoring of their systems a regular practice.

Finally, businesses need to test and monitor their systems frequently for problems. Automated tools for incident detection and response, as well as routine security audits, fall under this category. To further guarantee the safety of their data in the event of a security breach, businesses must also have a thorough backup and recovery plan in place.

Encryption can help in privacy management by securing data that is transmitted between IoT devices and servers, as well as data that is stored on the devices themselves. End-to-end encryption ensures that only the intended recipients can read the data.

Manufacturers of IoT devices should develop privacy policies that are open, simple to comprehend, and compliant with industry standards.

Policy Enforcement

An approach to the security of loT devices that is based on software is presented in [84]. Their answer is a security architecture that is made up of miniature security functions that are denoted with the symbol mboxes. A centralized loTSec controller is included in the architecture. This controller keeps an eye on the surrounding environment and generates a global picture to facilitate the enforcement of policies across multiple devices. From this point of view, administrators have the ability to configure and instantiate new mboxes, as well as their routing methods.

Fault Tolerance

In order for Internet of Things systems to be reliable, [74] list a number of requirements. There are three necessities for fault tolerance in the loT to be realized. First and foremost, security needs to be turned on by default for all devices. The second essential requirement is to provide all IoT devices with the capability to continuously monitor the network and its services without restrictions. Moreover, it is crucial for every device to possess self-defense mechanisms against potential threats and network failures. These components should be capable of promptly responding to service attacks and efficiently restoring them to their normal functioning state.

Secure Communication

According to [44], the protocol stack of IoT devices will seek to mimic that of conventional Internet hosts in order to provide the groundwork for an enlarged internet. They argue that this allows the Internet of Things to take advantage of many existing security measures. In addition, secure communication protocols for the IoT are explored in [56]. They look into both asymmetric key-based security solutions and symmetric pre-distributed-key-based security solutions.

Secure communication protocols can reduce the network vulnerabilities of IoT devices. In today's digital era, security has become a top priority for almost every business. With the expanding use of Internet of Things devices, guaranteeing the security of transmitted data has become a major challenge. To limit the hazards connected with IoT devices, it is vital to employ secure communication protocols that encrypt the transferred data, such as TLS or SSL. In order to strengthen the overall security of IoT devices, it is essential to prioritize the deployment of these protocols.

Secure Routing

In the realm of IoT security, secure routing protocols play a crucial role. These protocols ensure that sensitive data transmitted across networks remains secure and protected from potential attacks. There are several examples of such protocols, including the trust-aware secure routing framework (TSFR), the secure multi-hop routing protocol (SMRP), the group-based trust management scheme (GTMS), the two-way acknowledgment-based trust (2-ACKT) protocol, and the collaborative lightweight trust-based routing protocol [3]. As IoT continues to evolve, the need for robust security measures will only increase, making secure routing protocols a top priority for organizations and businesses alike.

DDoS Protection

A Leaming Automata (LA) has been proposed as a countermeasure to Distributed Denial of Service assaults in IoT networks [62, 85]. The LA would make an intelligent determination based on the environment regarding the packet sampling rate. The DDoS protection component works by constantly monitoring the requests received by each device during the detection phase. If the device detects a pre-set maximum capacity being surpassed, it immediately issues a DDoS warning to the neighboring nodes. This warning notifies the neighboring nodes about the threat and allows them to take necessary precautions to mitigate it. In essence, the DDoS protection component is a crucial feature that ensures the security and safety of IoT devices. With its robust monitoring capabilities and preemptive warning system, it plays a vital role in safeguarding the Internet of Things from potential cyber threats. As the use of IoT devices continues to expand, it is essential to prioritize their security to ensure their safe and successful integration into our daily lives. As soon as the warning is sent, the devices will begin to conduct an IP address scan in an effort to identify the attacker. After the attacker has been located, the remaining nodes will be informed of this threat and will immediately discard any packets that originate from the attacker's IP address [17, 22, 30, 69, 85].

[80] present a method based on this principle for detecting and averting DDoS attacks in IoT networks. Taking a copy of the sink node's configuration is another option (a node that collects data from sensors). This new node would function as a redundant channel, performing a portion of the sink node's duties. It has been calculated that this strategy is financially viable [62, 85].

Spam Prevention

It is imperative to have measures in place to prevent spam and unwarranted dissemination of material. Using digital signatures to verify data in 2D barcodes is a popular approach described by [67]. This technique entails encoding the original content, digitally signed content, and the public key of the individual who made the barcode within the barcode itself. Once a user scans the barcode into their computer, they can verify the author's identity by visiting the URL generated by the code. Additionally, an application could be used to verify the contents of the QR code and its position in the certificate chain [67].

With the use of digital signatures, the security of IoT devices can be significantly enhanced. It provides an effective way of ensuring that the content being accessed is authentic and has not been tampered with. As such, this solution is gaining popularity among businesses and organizations that rely on the security of IoT devices to conduct their operations.

Secure Software Development

Secure software development can reduce the software vulnerabilities of IoT devices. IoT devices must be designed with security in mind from the beginning. This includes regular software updates and patches to fix vulnerabilities [22, 72].

LoT Architectures

Several other IoT architectures are presented in the literature [76]. The objective of an IoT architecture is to link physical devices with their digital counterparts that generate services and the like. IoT architectures include Open source cloud solutions for the Internet of Things (OpenioT), the IoT at Work (IoT@Work), and the Internet of Things architecture (IoT-A) [76].

Regulatory Solutions

The Internet of Things (IoT) is a rapidly evolving concept that has caught the attention of governments around the world. As Weber [78] explains, the European government has been proactive in its efforts to regulate the IoT. In 2007, the European Union recognized the IoT as a legitimate concept and in 2009, a 14-point strategic action plan was developed. However, as the technology continued to advance, concerns about data protection and security emerged.

In 2012, it became evident that there were significant differences of opinion between users and industry stakeholders regarding data protection in the context of the IoT. The European Commission recognized the need for clear, comprehensive rules governing the use of IoT devices, and in 2013, the task of developing such rules was assigned to the European firm RAND [14, 48, 68].

Since then, the issue of IoT security has remained at the forefront of public debate, with governments and industry leaders working together to find solutions that will safeguard personal information and protect against cyber-attacks [16, 48, 70, 77, 83].

However, things are less cut and unambiguous in the United States. Most discussions happen inside the many federal departments that focus on different aspects of the Internet of Things. In 2013, the Federal Trade Commission (FTC) initially asked for public feedback on the privacy and security of the Internet of Things. More than 60% of the 27 responses were negative to regulatory intervention. The FTC hosted a session on the IoT later that year. The study's authors concluded that regulation should be based on whether or not businesses expect to make money solely from selling IoT devices, or whether or not they expect to make money from selling user data as well [48, 70].

IoT (Internet of Things) devices and the data they collect and transmit are only as safe and private as the security standards and regulations that govern them. The healthcare, transportation, and energy sectors all rely heavily on IoT devices, making them prime targets for destructive cyberattacks [48].

Security standards and regulations set guidelines and requirements for manufacturers and service providers to follow when designing, developing, and deploying IoT devices. They provide a framework for assessing and managing security risks, implementing security controls, and ensuring that the devices comply with industry best practices.

Some of the key security standards and regulations in IoT include:

ISO/IEC 27,001: Information security management systems should adhere to this standard since it has international credibility. Protecting private information, such as that gathered by Internet of Things gadgets, is made easier with this methodical approach [14, 75].

NIST Cybersecurity Framework: The National Institute of Standards and Technology developed the NIST framework as a set of standards, recommendations, and best practices to aid in the control of cyber risks. It helps organizations better prepare for, respond to, and recover from cyberattacks [48, 77].

GDPR: The General Data Protection Regulation is a comprehensive privacy law that must be followed by all European Union (EU) countries. It lays up stringent guidelines for the collection, processing, and storage of sensitive personal information. If an IoT device gathers or processes data about EU residents, it must do so in accordance with the General Data Protection Regulation (GDPR) [2, 8, 43, 52, 68, 70].

HIPAA: The federal government of the United States enacted the Health Insurance Portability and Accountability Act to protect the privacy of medical records. Medical IoT devices must be HIPAA-compliant so that sensitive patient information is protected [2, 25].

IEEE 802.1X: This network access control standard provides a way of authentication for IoT devices that connect to a network. It aids in preventing unwanted access and guarantees that only authorized devices have network access [9, 41].

The basis for protecting the security and privacy of IoT devices and data is provided by security standards and legislation. Compliance with these standards and regulations is essential for protecting critical infrastructure and maintaining trust in the IoT ecosystem.

Given the complex nature of the IoT, ensuring security and data protection will require ongoing collaboration between stakeholders and a commitment to implementing best practices and standards. It is only through such efforts that the vast potential of the Internet of Things can be realized in a manner that is safe, secure, and sustainable over the long term. It determined that "soft law," which incorporates rules, oversight, and ethical character while yet guaranteeing industry independence, is the optimal regulatory approach for the Internet of Things.

Physical Security

Locking cabinets or rooms in which IoT devices are stored can help prevent illegal physical access to these devices. Physical security on the IoT refers to the precautions taken to prevent the loss, compromise, or tampering of IoT hardware and infrastructure. Protecting IoT devices and infrastructure against unauthorized physical access and tampering is a crucial part of overall IoT security [7, 82].

Here are some ways physical security can be implemented in IoT [12, 81, 82]:

- Device placement: IoT devices should be placed in secure locations where they are less likely to be accessed by unauthorized individuals. For example, devices can be placed in locked rooms or cabinets, or secured to walls or floors.
- Access control: Access to Internet of Things devices must be controlled to only authorized individuals. This is possible using physical keys, access cards, or biometric authentication.
- Video surveillance: Video cameras can be used to monitor the physical environment where IoT devices are deployed. This can help to deter theft or vandalism and provide evidence in case of a security incident.

- Tamper-evident seals: Tamper-evident seals can be used to indicate if a device has been tampered with or opened without authorization. This can help to detect and prevent physical attacks on IoT devices.
- Environmental monitoring: Environmental sensors can be used to monitor the physical environment where IoT devices are deployed. This can help to detect changes in temperature, humidity, or air quality that may indicate a security threat.

Physical security is an important aspect of IoT security that should not be overlooked. By employing physical security measures, corporations may prevent attackers from physically accessing or tampering with Internet of Things (IoT) devices and systems, as well as secure the confidentiality, integrity, and availability of IoT data.

Network Segmentation

Network segmentation can help prevent a compromised IoT device from spreading malware or other threats to other devices on the same network. Network segmentation in IoT involves dividing an IoT network into smaller, more secure subnetworks or segments. This can help to improve security by limiting the scope or spread of a potential security breach and reducing the attack surface of IoT devices and systems [12, 36].

Here are some ways network segmentation can be used in IoT [16, 36, 37]:

- Device segmentation: IoT devices can be grouped into smaller subnetworks based on device type, location, or function. This can assist in limiting the spread of a potential security breach and lower the likelihood of unauthorized access to sensitive data or essential systems.
- Application segmentation: IoT applications can be separated into smaller subnetworks based on their function or sensitivity. This can help to protect critical applications and data from unauthorized access or tampering.
- User segmentation: Users can be assigned to specific subnetworks based on their role or access requirements. This can help to limit the exposure of sensitive data or systems to unauthorized users.
- Data segmentation: IoT data can be separated into smaller subnetworks based on its sensitivity or importance. This can help to protect critical data from unauthorized access or tampering.

Overall, network segmentation in IoT can help to improve security by limiting the scope of a potential security breach, reducing the attack surface of IoT devices and systems, and providing more granular control over access to sensitive data or critical systems. However, network segmentation can also increase complexity and management overhead, so it should be implemented carefully and with consideration of the overall security posture of the IoT network.

Data Anonymization

IoT devices can anonymize data by removing personally identifiable information from it. This safeguards users' privacy while granting businesses access to valuable data for research and insight. Data privacy protection relies on the removal of PII, or personally identifiable information. Data anonymization is a crucial method for ensuring the privacy of consumers whose information is being gathered and processed by Internet of Things (IoT) devices and systems [27, 45].

Here are some ways data anonymization can be implemented in IoT:

- Pseudonymization: Pseudonymization involves replacing identifiable data with pseudonyms or unique identifiers to protect the identity of individuals. This can be accomplished by substituting personal data such as addresses, names, and mobile numbers with random or unique IDs.
- Data aggregation: Data aggregation involves combining data from multiple sources to create a larger dataset while removing personally identifiable information. As a result, it becomes more challenging to identify particular persons from the data, which might aid in protecting individuals' privacy.
- Data masking: Data masking entails changing or removing portions of data that could be used to identify specific individuals. This can include masking or deleting sensitive data such as credit card numbers, ssn, or medical information.
- Differential privacy: Using noise or randomization, differential privacy makes it harder to determine the identities of specific people in a dataset. This method can assist safeguard personal information while still facilitating research.

In general, data anonymization is a crucial method for shielding the privacy of individuals whose information is being processed by Internet of Things (IoT) gadgets and infrastructure. Organizations can adhere to data protection requirements such as the General Data Protection Regulation and the California Consumer Privacy Act as well as preserve the privacy of their customers and employees by using data anonymization procedures.

Behavioral Analytics

Behavioral analytics can help identify unusual behavior on IoT devices that may indicate a security threat. It involves using data analytics techniques to monitor and analyze the behavior of IoT devices and users. It helps to identify patterns and anomalies in device behavior, which can indicate potential security threats or performance issues [18, 21].

Here are some ways behavioral analytics can be used in IoT [4, 34, 45]:

- Anomaly detection: Behavioral analytics can be used to detect anomalies in device behavior that may indicate a security breach or malfunction. By analyzing patterns of behavior, such as communication patterns or usage patterns, behavioral analytics can identify deviations from normal behavior and trigger alerts.

- Predictive maintenance: Behavioral analytics can be used to identify patterns in device behavior that may indicate potential maintenance issues. By analyzing device performance data, such as sensor readings or error logs, behavioral analytics can predict when a device may require maintenance or repair.
- User behavior analysis: User activity can be analyzed using behavioral analytics to discover potential security threats. By analyzing user activity, such as logins or data access patterns, behavioral analytics can detect anomalies in user behavior that may indicate a security breach.
- Fraud detection: Behavioral analytics can be used to detect fraud in IoT systems. By analyzing patterns in device behavior or user activity, behavioral analytics can detect fraudulent activities, such as fake sensor readings or unauthorized data access.
- Performance optimization: Behavioral analytics can be used to optimize device performance. By analyzing usage patterns and device behavior, behavioral analytics can identify areas where device performance can be improved and make recommendations for optimization.

Overall, behavioral analytics in IoT can help to improve security, performance, and reliability of IoT devices and systems by providing insights into device behavior and identifying potential issues before they become major problems.

Preventing Hacking and Malware Attacks

In today's digital age, it is of the utmost importance to ensure the security of organizational systems. With the proliferation of IoT devices, the threat of hacking and malware assaults has grown. Organizations must take proactive steps to protect their systems from such malicious activities. Strong security measures such as the use of complex passwords, two-factor authentication, and access control measures must be adopted [39, 46, 64].

Moreover, it is essential to regularly update and patch the systems to address any new vulnerabilities that may arise. The consequences of a security breach can be disastrous, leading to data loss, financial loss, and reputational damage. Therefore, organizations must make security a top priority and invest in the necessary resources and technologies to safeguard their systems [1, 55, 61].

Given the evolving nature of cyber threats, it is crucial to stay vigilant and keep up with the latest security trends. Organizations must conduct regular security audits to identify potential weaknesses and take corrective measures. By adopting a proactive security strategy, firms may reduce the likelihood of hacking and malware assaults, thereby protecting the safety and confidentiality of their data [53].

Firms also need to do their part to make sure their users understand the security dangers associated with the IoTs. This involves assessing security policies and procedures on a regular basis and giving users the resources, they need to keep their data safe. In addition, organizations should guarantee that their systems are routinely tested and monitored for any problems [53].

Lastly, companies should ensure that their systems are routinely tested and monitored for any problems. This includes the use of automated tools to detect and respond to security incidents, as well as the implementation of regular security audits. Additionally, organizations should ensure that they have a comprehensive backup and recovery plan in place to ensure that their data is safe in the event of a security breach [53].

The Need for Robust IoT Security Measures

Given the potential security risks posed by the IoT revolution, it is essential that organizations take steps to protect themselves. To protect data and ensure the security of IoT devices, strict security measures must be implemented. Access control techniques, encryption, and firewalls are all part of this effort to keep out intruders. In addition, businesses must keep their systems patched and updated to close security holes when they are discovered [51, 53].

Companies also need to do their part to make sure their customers understand the security dangers associated with the Internet of Things. Giving users access to security-related resources and conducting routine reviews of security policies and procedures are both essential. Finally, businesses should make testing and monitoring their systems a routine practice.

Conclusion

To provide a safe and secure environment, the IoT revolution brings numerous and sophisticated security concerns that must be addressed. Organizations must take proactive steps to protect their data and equipment. This involves implementing robust security measures like strong passwords, two-factor authentication, and access controls. Regularly updating and patching systems is also crucial to address emerging vulnerabilities.

Furthermore, educating users about IoT security risks is essential. Providing users with necessary tools and resources to protect their data and conducting regular reviews of security policies and procedures are important steps. Continuous testing and monitoring of systems are vital to identify and address potential issues. Automated tools can aid in detecting and responding to security incidents, while regular security audits ensure ongoing evaluation.

Moreover, organizations should establish comprehensive backup and recovery plans to ensure data safety in case of a security breach. By addressing these challenges through the appropriate approach, organizations can navigate the complexities of the IoT revolution and establish a secure environment for its advancement.

References

1. Aazam, M., Huh, E. N., Zeadally, S., Harras, K. A., AbuKhousa, E., Mohamed, N., Al-Jaroodi, J., Access, O. P., Fung, C., Kadiyala, K., Jalali, F., Charlotte, U. N. C., Jue, J. P., Acm, I., Brogi, A., Forti, S., Ibrahim, A., Agarwal, S., Yadav, S., ... Keyvanpour, M. R. (2017). Fog computing—Glimps of upcoming research area (#4). *Future Generation Computer Systems, 5*(2).
2. Abdulghani, H. A., Nijdam, N. A., Collen, A., & Konstantas, D. (2019). A study on security and privacy guidelines, countermeasures, threats: IoT data at rest perspective. *Symmetry, 11*(6). https://doi.org/10.3390/sym11060774
3. Airehrour, D., Gutierrez, J., & Ray, S. K. (2016). Secure routing for internet of things: A survey. *Journal of Network and Computer Applications, 66*. https://doi.org/10.1016/j.jnca.2016.03.006
4. Andión, J., Navarro, J. M., López, G., Álvarez-Campana, M., & Dueñas, J. C. (2018). Smart behavioral analytics over a low-cost IoT Wi-Fi tracking real deployment. *Wireless Communications and Mobile Computing, 2018*. https://doi.org/10.1155/2018/3136471
5. Antonakakis, M., April, T., Bailey, M., Bernhard, M., Bursztein, E., Cochran, J., ... & Zhou, Y. (2017). Understanding the Mirai botnet. In *26th USENIX security symposium (USENIX Security 17)* (pp. 1093–1110). https://www.usenix.org/conference/usenixsecurity17/technical-sessions/presentation/antonakakis
6. Ashford, W. (2015). Jeep hack raises questions over responsibility for security (cover story). Computer Weekly, August.
7. Attkan, A., & Ranga, V. (2022). Cyber-physical security for IoT networks: A comprehensive review on traditional, blockchain and artificial intelligence based key-security. *Complex and Intelligent Systems, 8*(4). https://doi.org/10.1007/s40747-022-00667-z
8. Badii, C., Bellini, P., Difino, A., & Nesi, P. (2020). Smart city IoT platform respecting GDPR privacy and security aspects. *IEEE Access, 8*. https://doi.org/10.1109/ACCESS.2020.2968741
9. Baird, M., Ng, B., & Seah, W. (2017). WiFi network access control for IoT connectivity with software defined networking. In *Proceedings of the 8th ACM Multimedia Systems Conference, MMSys 2017*. https://doi.org/10.1145/3083187.3084021
10. Baranchuk, A., Refaat, M. M., Patton, K. K., Chung, M. K., Krishnan, K., Kutyifa, V., Upadhyay, G., Fisher, J. D., & Lakkireddy, D. R. (2018). Cybersecurity for cardiac implantable electronic devices: What should you know? Journal of the American College of Cardiology, 71(11). https://doi.org/10.1016/j.jacc.2018.01.023
11. Bekkali, A. El, Boulmalf, M., Essaaidi, M., & Mezzour, G. (2019). Securing the Internet of Things (IoT). In *Proceedings—2018 International Conference on Wireless Networks and Mobile Communications, WINCOM 2018*. https://doi.org/10.1109/WINCOM.2018.8629652
12. Chang, C. W., & Hung, W. H. (2021). Strengthening existing internet of things system security: Case study of improved security structure in smart health. *Sensors and Materials, 33*(42). https://doi.org/10.18494/SAM.2021.3163
13. Chris Pietschmann. (2022). IoT security architecture: Trust zones and boundaries. https://build5nines.com/iot-security-architecture-trust-zones-and-boundaries/
14. Danielis, P., Beckmann, M., & Skodzik, J. (2020). An ISO-compliant test procedure for technical risk analyses of IoT systems based on STRIDE. In *Proceedings—2020 IEEE 44th Annual Computers, Software, and Applications Conference, COMPSAC 2020*. https://doi.org/10.1109/COMPSAC48688.2020.0-203
15. Desai, N., Kulkarni, P., Joy, L., & Raut, P. (2020). IoT based post crash assistance system. In *2020 4th International Conference on Trends in Electronics and Informatics (ICOEI) (48184)* (pp. 734–739). https://doi.org/10.1109/ICOEI48184.2020.9142894

16. Dhar, S., & Bose, I. (2021). Securing IoT devices using zero trust and blockchain. *Journal of Organizational Computing and Electronic Commerce, 31*(1). https://doi.org/10.1080/10919392.2020.1831870
17. Elejla, O. E., Anbar, M., & Belaton, B. (2017). ICMPv6-based DoS and DDoS attacks and defense mechanisms: Review. *IETE Technical Review (Institution of Electronics and Telecommunication Engineers, India), 34*(4). https://doi.org/10.1080/02564602.2016.1192964
18. Empl, P., & Pernul, G. (2023). Digital-twin-based security analytics for the Internet of Things. *Information (Switzerland), 14*(2). https://doi.org/10.3390/info14020095
19. Ferrando, R., & Stacey, P. (2017). Classification of device behaviour in Internet of Things infrastructures. *ACM International Conference Proceeding Series*. https://doi.org/10.1145/3109761.3109791
20. Ferraris, D., Bastos, D., Fernandez-Gago, C., El-Moussa, F., & Lopez, J. (2019). An analysis of trust in smart home devices. WISA Workshop 2019.
21. Fotopoulou, E., Zafeiropoulos, A., Terroso-Sáenz, F., Şimşek, U., González-Vidal, A., Tsiolis, G., Gouvas, P., Liapis, P., Fensel, A., & Skarmeta, A. (2017). Providing personalized energy management and awareness services for energy efficiency in smart buildings. *Sensors (Switzerland), 17*(9). https://doi.org/10.3390/s17092054
22. Galeano-Brajones, J., Carmona-Murillo, J., Valenzuela-Valdés, J. F., & Luna-Valero, F. (2020). Detection and mitigation of DoS and DDoS attacks in iot-based stateful SDN: An experimental approach. Sensors (Switzerland), 20(3). https://doi.org/10.3390/s20030816
23. Gonsalves, A. (2013). Baby monitor hack highlights manufacturers' security shortfalls. Network World.
24. Greenberg, A. (2015). After Jeep Hack, Chrysler Recalls 1.4M Vehicles for Bug Fix. Wired.
25. Griggs, K. N., Ossipova, O., Kohlios, C. P., Baccarini, A. N., Howson, E. A., & Hayajneh, T. (2018). Healthcare blockchain system using smart contracts for secure automated remote patient monitoring. *Journal of Medical Systems, 42*(7). https://doi.org/10.1007/s10916-018-0982-x
26. Groeneveld, S. A., Jongejan, N., de Winter, B. J. S. A. A. F., & Fiolet, A. T. L. (2019). [Hacking into a pacemaker; risks of smart healthcare devices]. Nederlands Tijdschrift Voor Geneeskunde, 163.
27. HaddadPajouh, H., Khayami, R., Dehghantanha, A., Choo, K. K. R., & Parizi, R. M. (2020). AI4SAFE-IoT: An AI-powered secure architecture for edge layer of Internet of things. *Neural Computing and Applications, 32*(20). https://doi.org/10.1007/s00521-020-04772-3
28. Hassan, S., Mwangi, E., & Kihato, P. K. (2022). IoT based monitoring system for epileptic patients. *Heliyon, 8*(6). https://doi.org/10.1016/j.heliyon.2022.e09618
29. Hay Newman, L. (2018). A new pacemaker hack puts malware directly on the device. WIRED.
30. Hendriks, L., De Oliveira Schmidt, R., Van Rijswijk-Deij, R., & Pras, A. (2017). On the potential of IPv6 open resolvers for DDoS attacks. Lecture Notes in Computer Science (Including Subseries Lecture Notes in Artificial Intelligence and Lecture Notes in Bioinformatics), 10176 LNCS. https://doi.org/10.1007/978-3-319-54328-4_2
31. Henze, M., Hermerschmidt, L., Kerpen, D., Häußling, R., Rumpe, B., & Wehrle, K. (2016). A comprehensive approach to privacy in the cloud-based Internet of Things. *Future Generation Computer Systems, 56*, 701–718. ISSN 0167–739X. https://doi.org/10.1016/j.future.2015.09.016
32. Hill, K. (2013a). How a creep hacked a baby monitor to say lewd things to a 2-year-old. Forbes.Com.
33. Hill, K. (2013b). "Baby Monitor Hack" could happen to 40,000 Other Foscam Users. Forbes.
34. Hong-tan, L., Cui-hua, K., Muthu, B. A., & Sivaparthipan, C. B. (2021). Big data and ambient intelligence in IoT-based wireless student health monitoring system. *Aggression and Violent Behavior*. https://doi.org/10.1016/j.avb.2021.101601
35. Hossain, M. M., Fotouhi, M., & Hasan, R. (2015). Towards an analysis of security issues, challenges, and open problems in the Internet of Things. In *Proceedings—2015 IEEE World Congress on Services, SERVICES 2015*. https://doi.org/10.1109/SERVICES.2015.12

36. Huang, C., Zong, Y., Chen, J., Liu, W., Lloret, J., & Mukherjee, M. (2021). A deep segmentation network of stent structs based on IoT for interventional cardiovascular diagnosis. *IEEE Wireless Communications, 28*(3). https://doi.org/10.1109/MWC.001.2000407
37. Kamal, S., Shende, V. G., Swaroopa, K., Bindhu Madhavi, P., Akram, P. S., Pant, K., Patil, S. D., & Sahile, K. (2022). FCN network-based weed and crop segmentation for IoT-aided agriculture applications. *Wireless Communications and Mobile Computing, 2022.* https://doi.org/10.1155/2022/2770706
38. Karnouskos, S. (2011, November). Stuxnet worm impact on industrial cyber-physical system security. In *IECON 2011-37th Annual Conference of the IEEE Industrial Electronics Society* (pp. 4490–4494). Melbourne, VIC, Australia. https://doi.org/10.1109/IECON.2011.6120048.
39. Katal, A., Sharma, K., & Sethi, V. (2021). IoT based safety system: LPG/CNG detection and alert. *International Conference on Intelligent Technologies (CONIT), 2021,* 1–6. https://doi.org/10.1109/CONIT51480.2021.9498285
40. Khalil, U., Malik, O. A., Uddin, M., & Chen, C. L. (2022). A comparative analysis on blockchain versus centralized authentication architectures for IoT-enabled smart devices in smart cities: A comprehensive review, recent advances, and future research directions. *Sensors, 22*(14). https://doi.org/10.3390/s22145168
41. Kim, K. W., Han, Y. H., & Min, S. G. (2017). An authentication and key management mechanism for resource constrained devices in IEEE 802.11-based IoT access networks. *Sensors (Switzerland), 17*(10). https://doi.org/10.3390/s17102170
42. Køien, G., & Abomhara, M. (2014). Security and privacy in the Internet of Things: Current status and open issues. https://doi.org/10.1109/PRISMS.2014.6970594.
43. Kounoudes, A. D., & Kapitsaki, G. M. (2020). A mapping of IoT user-centric privacy preserving approaches to the GDPR. *Internet of Things (Netherlands), 11.* https://doi.org/10.1016/j.iot.2020.100179
44. Kumar, S. A., Vealey, T., & Srivastava, H. (2016). Security in internet of things: Challenges, solutions and future directions. In *Proceedings of the Annual Hawaii International Conference on System Sciences, 2016-March.* https://doi.org/10.1109/HICSS.2016.714
45. Li, W., Chai, Y., Khan, F., Jan, S. R. U., Verma, S., Menon, V. G., Kavita, & Li, X. (2021). A comprehensive survey on machine learning-based big data analytics for IoT-enabled smart healthcare system. *Mobile Networks and Applications, 26*(1). https://doi.org/10.1007/s11036-020-01700-6
46. Mahali, M. I., Marpanaji, E., Dewanto, S. A., Wulandari, B., Rochayati, U., & Hasanah, N. (2018). Smart traffic light based on IoT and mBaaS using high priority vehicles method. In *2018 5th International Conference on Electrical Engineering, Computer Science and Informatics (EECSI)*, pp. 703–707. https://doi.org/10.1109/EECSI.2018.8752694
47. Mahmoud, R., Yousuf, T., Aloul, F., & Zualkernan, I. (2016). Internet of Things (IoT) security: Current status, challenges and prospective measures. In *2015 10th International Conference for Internet Technology and Secured Transactions, ICITST 2015.* https://doi.org/10.1109/ICITST.2015.7412116
48. Malatji, M., Marnewick, A. L., & Von Solms, S. (2022). Cybersecurity capabilities for critical infrastructure resilience. *Information and Computer Security, 30*(2). https://doi.org/10.1108/ICS-06-2021-0091
49. Malina, L., Hajny, J., Fujdiak, R., & Hosek, J. (2016). On perspective of security and privacy-preserving solutions in the Internet of Things. *Computer Networks, 102*, 83–95, ISSN 1389–1286. https://doi.org/10.1016/j.comnet.2016.03.011
50. Manworren, N., Letwat, J., & Daily, O. (2016). Why you should care about the target data breach. *Business Horizons, 59.* https://doi.org/10.1016/j.bushor.2016.01.002
51. Marjani, M., Nasaruddin, F., Gani, A., Karim, A., Hashem, I. A. T., Siddiqa, A., & Yaqoob, I. (2017). Big IoT data analytics: Architecture, opportunities, and open research challenges. *IEEE Access, 5.* https://doi.org/10.1109/ACCESS.2017.2689040
52. Merzeh, H. R. J., Kara, M., Aydın, M. A., & Balık, H. H. (2022). Gdpr compliance IoT authentication model for smart home environment. *Intelligent Automation and Soft Computing, 31*(3). https://doi.org/10.32604/IASC.2022.021297

53. Mohanty, S., Cormican, K., & Dhanapathi, C. (2021). Analysis of critical success factors to mitigate privacy risks in IoT Devices. *Procedia Computer Science, 196*.https://doi.org/10.1016/j.procs.2021.12.005
54. Narayanan, S. N., Khanna, K., Panigrahi, B. K., & Joshi, A. (2018). Security in smart cyber-physical systems: A case study on smart grids and smart cars. *Smart Cities Cybersecurity and Privacy*. https://doi.org/10.1016/B978-0-12-815032-0.00011-1
55. Nath, M. P., Priyadarshini, S. B. B., & Mishra, D. (2021). A comprehensive study on security in IoT and resolving security threats using machine learning (ML). Lecture Notes in Networks and Systems, 202 LNNS. https://doi.org/10.1007/978-981-16-0695-3_51
56. Nguyen, K. T., Laurent, M., & Oualha, N. (2015). Survey on secure communication protocols for the Internet of Things. *Ad Hoc Networks, 32*. https://doi.org/10.1016/j.adhoc.2015.01.006
57. Oyshi, M. T., Bonny, M. Z., Saha, S., & Tumpa, Z. N. (2021). IoT security issues and possible solution using blockchain technology. In *Lecture notes in networks and systems* (vol. 127). https://doi.org/10.1007/978-981-15-4218-3_12
58. Patton, M., Gross, E., Chinn, R., Forbis, S., Walker, L., & Chen, H. (2014). Uninvited connections: A study of vulnerable devices on the internet of things (IoT). In *Proceedings—2014 IEEE Joint Intelligence and Security Informatics Conference, JISIC 2014*. https://doi.org/10.1109/JISIC.2014.43
59. Plachkinova, M., & Maurer, C. (2018). Security breach at target. *Journal of Information Systems Education, 29*.
60. Porras, J., Pänkäläinen, J., Knutas, A., & Khakurel, J. (2018). Security in the Internet of Things—A systematic mapping study. In *Proceedings of the Annual Hawaii International Conference on System Sciences, 2018-January*. https://doi.org/10.24251/hicss.2018.473
61. Puli, S., & Chaparala, S. C. (2022). An efficient swift routing model with node trust identity factor (SRM-NTIF) to perform secure data transmission among IoT gadgets. *Revue d'Intelligence Artificielle, 36*(1). https://doi.org/10.18280/ria.360104
62. Pundir, S., Wazid, M., Singh, D. P., Das, A. K., Rodrigues, J. J. P. C., & Park, Y. (2020). Intrusion detection protocols in wireless sensor networks integrated to Internet of Things deployment: Survey and future challenges. *IEEE Access, 8*. https://doi.org/10.1109/ACCESS.2019.2962829
63. Qaim, W. Bin, Ometov, A., Molinaro, A., Lener, I., Campolo, C., Lohan, E. S., & Nurmi, J. (2020). Towards energy efficiency in the internet of wearable things: A systematic review. *IEEE Access, 8*. https://doi.org/10.1109/ACCESS.2020.3025270
64. Raj, J. V., & Sarath, T. V. (2019). An IoT based real-time stress detection system for fire-fighters. *International Conference on Intelligent Computing and Control Systems (ICCS), 2019*, 354–360. https://doi.org/10.1109/ICCS45141.2019.9065866
77. Rao, P. M., & Deebak, B. D. (2023). Security and privacy issues in smart cities/industries: Technologies applications and challenges. *Journal of Ambient Intelligence and Humanized Computing, 14*(8), 10517–10553. https://doi.org/10.1007/s12652-022-03707-1
66. Radichel, T. (2014). Case study: Critical controls that could have prevented target breach. *SANS Gold GIAC Certification*. Retrieved from https://www.sans.org/readingroom/whitepapers/casestudies/case-studycritical-controls-prevented-target-breach-35412
67. Razzak, F. (2012). Spamming the Internet of Things: A possibility and its probable solution. *Procedia Computer Science, 10*, 658–665. http://dx.doi.org/10.1016/j.procs.2012.06.084
68. Rhahla, M., Abdellatif, T., Attia, R., & Berrayana, W. (2019). A GDPR controller for IoT systems: application to e-Health. In *Proceedings—2019 IEEE 28th International Conference on Enabling Technologies: Infrastructure for Collaborative Enterprises, WETICE 2019*. https://doi.org/10.1109/WETICE.2019.00044
69. Saad, R. M. A., Anbar, M., Manickam, S., & Alomari, E. (2016). An intelligent ICMPv6 DDoS flooding-attack detection framework (V6IIDS) using back-propagation neural network. *IETE Technical Review (Institution of Electronics and Telecommunication Engineers, India), 33*(3). https://doi.org/10.1080/02564602.2015.1098576
70. Seo, J., Kim, K., Park, M., Park, M., & Lee, K. (2018). An analysis of economic impact on IoT industry under GDPR. Mobile Information Systems, 2018. https://doi.org/10.1155/2018/6792028

71. Suo, H., Wan, J., Zou, C., & Liu, J. (2012). Security in the internet of things: A review. In 2012 *International Conference on Computer Science and Electronics Engineering* (pp. 648–651). Hangzhou, China. https://doi.org/10.1109/ICCSEE.2012.373
72. Tashtoush, Y. M., Darweesh, D. A., Husari, G., Darwish, O. A., Darwish, Y., Issa, L. B., & Ashqar, H. I. (2022). Agile approaches for cybersecurity systems, IoT and intelligent transportation. *IEEE Access, 10*. https://doi.org/10.1109/ACCESS.2021.3136861
73. Tayag, M. I., Napalit, F., & Napalit, A. (2020). IoT security: Penetration testing of white-label cloud-based IoT camera compromising personal data privacy. *International Journal of Computer Science and Information Technology, 12*(5). https://doi.org/10.5121/ijcsit.2020.12503
74. Tsunoda, H., Roman-Castro, R., Lopez, J., & Keeni, G. M. (2018). Feasibility of societal model for securing Internet of Things. *KSII Transactions on Internet and Information Systems, 12*(8). https://doi.org/10.3837/tiis.2018.08.003
75. Ueda, T., & Ikeda, Y. (2018). Socio-economics and educational case study with cost-effective IOT campus by the use of wearable, tablet, cloud and open E-learning services. In *Proceedings of the 2017 ITU Kaleidoscope Academic Conference: Challenges for a Data-Driven Society, ITU K 2017, 2018-January*. https://doi.org/10.23919/ITU-WT.2017.8246994
76. Vasilomanolakis, E., Daubert, J., Luthra, M., Gazis, E., Wiesmaier, A., & Kikiras, P. (2015, March). On the security and privacy of Internet of Things architectures and systems.https://doi.org/10.1109/SIOT.2015.9
77. Webb, J., & Hume, D. (2018). Campus IoT collaboration and governance using the NIST cybersecurity framework. In *IET Conference Publications, 2018(CP740)*. https://doi.org/10.1049/cp.2018.0025
78. Weber, R. H. (2010). Internet of Things—New security and privacy challenges. *Computer Law and Security Review, 26*(1). https://doi.org/10.1016/j.clsr.2009.11.008
79. Weber, R. H. (2015). Internet of things: Privacy issues revisited. *Computer Law & Security Review, 31*(5), 618–627, ISSN 0267–3649. https://doi.org/10.1016/j.clsr.2015.07.002
80. Xu, R., Baracaldo, N., & Joshi, J. (2021). Privacy-preserving machine learning: Methods, challenges and directions. *ArXiv, abs/2108.04417*.
81. Yang, X., Shu, L., Liu, Y., Hancke, G. P., Ferrag, M. A., & Huang, K. (2022). Physical security and safety of IoT equipment: A survey of recent advances and opportunities. *IEEE Transactions on Industrial Informatics, 18*(7). https://doi.org/10.1109/TII.2022.3141408
82. Yao, X., Farha, F., Li, R., Psychoula, I., Chen, L., & Ning, H. (2021). Security and privacy issues of physical objects in the IoT: Challenges and opportunities. *Digital Communications and Networks, 7*(3). https://doi.org/10.1016/j.dcan.2020.09.001
83. Yu, T., Sekar, V., Seshan, S., Agarwal, Y., & Xu, C. (2015a). Handling a trillion (unfixable) flaws on a billion devices. https://doi.org/10.1145/2834050.2834095
84. Yu, T., Sekar, V., Seshan, S., Agarwal, Y., & Xu, C. (2015b). Handling a trillion (unfixable) flaws on a billion devices: Rethinking network security for the Internet-of-Things. In *Proceedings of the 14th ACM Workshop on Hot Topics in Networks, HotNets-XIV 2015*. https://doi.org/10.1145/2834050.2834095
85. Zhang, C., & Green, R. (2015). Communication security in internet of thing: Preventive measure and avoid DDoS attack over IoT network. *Simulation Series, 47*(3).
86. Zhang, X., Upton, O., Beebe, N. L., & Choo, K. K. R. (2020). IoT botnet forensics: A comprehensive digital forensic case study on Mirai botnet servers. *Forensic Science International: Digital Investigation, 32*, 300926, ISSN 2666–2817. https://doi.org/10.1016/j.fsidi.2020.300926
87. Zhang, Z.-K., Cho, M., & Shieh, S. (2015). Emerging security threats and countermeasures in IoT. In *ASIACCS 2015—Proceedings of the 10th ACM Symposium on Information, Computer and Communications Security*, 1–6. https://doi.org/10.1145/2714576.2737091
88. Zhou, Z., Zhang, W., Li, S., & Yu, N. (2019). Potential risk of IoT device supporting IR remote control. *Computer Networks, 148*. https://doi.org/10.1016/j.comnet.2018.11.014

Survey of IoT Security: Application Areas, Threats and Solutions

Galiveeti Poornima, Y. Sudha, and R. Pallavi

Abstract The current communication era is commonly called the Internet of Things (IoT). The IoT facilitates data generation, reception, and transmission among tangible entities. A multitude of IoT applications prioritize the automation of diverse processes to endow non-living entities with autonomous functionality. Thanks to current and future IoT applications, user comfort, effectiveness, and level of automation will probably increase. However, such a world needs high-security, private, authentic, and attack-recovery mechanisms. Implementing the necessary changes to the building blocks of IoT applications is essential to achieving secure end-to-end IoT environments. This chapter will comprehensively examine the obstacles related to security and the conceivable hazards presented by applications of the IoT. Following a discussion of the potential risks associated with the IoT, a look will be taken at the many different technologies currently in use and those still in the development stage. The enhancement of IoT security is the focus of current research into four distinct technologies: Edge computing, Fog computing, Blockchain, fog computing, Edge computing, and machine learning.

Keywords IoT · Threats · IoT layers · Vulnerabilities · IoT security

Introduction

The rate at which the devices and infrastructure around us are becoming Internet-enabled is accelerating. A Gartner study estimate that there will be 80 trillion connected devices in use across the globe by the year 2020. They are expected to increase to 25 billion by 2024 [1]. The developments for the IoT are becoming increasingly popular around the world. China, West Europe, and North America are the leading nations that are driving this trend [1]. 27 billion machine-to-machine (M2M) connections are projected between 2016 and 2024 [1]. With these numbers increasing, the IoT is now recognized as among the most encouraging emerging economies that

G. Poornima (✉) · Y. Sudha · R. Pallavi
School of CSE and IS, Presidency University, Bangalore, India
e-mail: galiveetipoornima@presidencyuniversity.in

A. Prasad et al. (eds.), *Communication Technologies and Security Challenges in IoT*, Internet of Things, https://doi.org/10.1007/978-981-97-0052-3_5

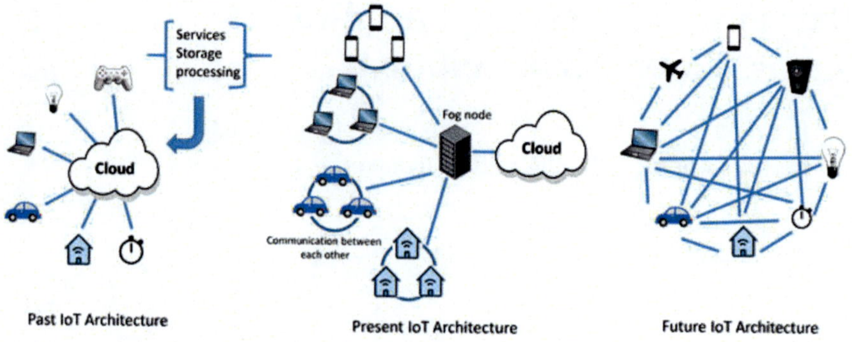

Fig. 1 IoT's future architecture

could support the growth of the digital economy. According to Muhammad et al. [2], it is projected that the IoT sector will yield a revenue of approximately $4 trillion by 2025, which is a significant increase from the nearly $100 billion generated in 2018. Machine-to-machine (M2M) connections have a diverse array of applications, such as in the domains of intelligent environments, smart cities, smart agriculture, smart retail, and smart grids [3]. Figure 1 demonstrates the IoT architectures of earlier times, current, and future. Devices are anticipated to interact with one another and the Internet in addition to being connected to other devices today and in the future. SIoT (Social IoT) is emerging in addition to connected devices and objects. SIoT will make it possible for social network users to connect and share devices over the Internet [4].

The extensive IoT applications have raised apprehensions about security and privacy. Emerging IoT applications will only be able to keep up with demands for them and may lose their potential with a trustworthy and interoperable IoT ecosystem. The online world, mobile networks, and WSNs have security issues, but the IoT has its own, including privacy, authentication, management, and information storage. Table 1 summarizes the factors that make securing an IoT environment significantly more complex than securing conventional information technology (IT) devices. IoT applications are a haven for numerous cyber threats as a result of these flaws and vulnerabilities. Numerous safety and privacy violations have been reported concerning commonly utilized IoT applications. The Mirai 2016 fourth quarter attack infected 2.5 million Internet-connected devices and launched a DDoS attack [5]. After Mirai, Hajime and Reaper were major botnet attacks on IoT devices [5]. Insecure and low-powered IoT devices allow hackers to access user data in home and corporate networks. IoT is expanding beyond objects. The implantation of IoT devices to monitor organ health has been successful many times [6]. Adversaries can use such devices to track a person or falsify data. If compromised, such devices could be hazardous.

The growth of the IoT has also benefited the cyber-physical systems (CPS) field. When physical changes are detected in physical objects in the environment, CPS takes appropriate action. CPSs contain critical infrastructure, such as transportation and

Table 1 IT versus IOT security

Widespread IT security	IoT security
Resources are abundant in IT devices	IoT devices need security measures
It uses powerful devices	IoT devices have software and hardware limitations
Complex algorithms are used for security and low capabilities	Algorithms should be lightweight
Homogeneous technology provides high security	Heterogeneous IoT data increases the attack surface

power grids, so security flaws in these systems have serious consequences. However, CPS security issues have unique characteristics of their own.

Any IoT ecosystem has four essential layers. Sensors and actuators in the first layer detect data and perform various functions. Second-layer data are transmitted over a communication network. The middleware bridges the networks and the application layers in most developed IoT applications. Intelligent grids, innovative grid applications, transportation, manufacturing, and other related areas are all included in the IoT applications end-to-end layer four is designated explicitly for. Security issues arise at each layer. These layers are connected by many gateways that allow data to flow. There are also security risks specific to these gateways.

Critical IoT Security Applications

Most deployed IoT apps require security. IoT applications are spreading across most industries. While some of these IoT applications can benefit from existing networking technologies, others require more rigorous security support. In this section, we discuss several security-sensitive Internet-of-Things applications.

Smart Cities

Gharaibeh et al. [7] smart claim that smart cities use cutting-edge communications and computer technology to enhance the standard of life for their inhabitants. This includes intelligent utilities, intelligent homes, and intelligent traffic and disaster management. Governments around the world are offering incentives to make cities smarter [8]. Intelligent apps are meant to improve citizens' lives, but they risk their privacy. Smart card services may compromise card data and buying habits. Smart mobility apps can reveal the location of the user. Parents can track their kids with apps. Hacking such apps could put children at risk.

Smart Environment

IoT applications in smart environments include forest fire detection, snow level monitoring in high- altitude regions, landslide prevention, early earthquake recognition, pollution monitoring, and more. Inhabitants of these areas, both human and animal, have a lot riding on the success of IoT initiatives. Information from IoT applications will also be shared with these government bodies. Any IoT application area where there are security breaches can have serious effects. The occurrence of false positives and negatives in IoT applications can result in severe consequences. The government and businesses will lose money if the application falsely detects earthquakes. If the earthquake prediction application fails to forecast seismic activity accurately, there is a risk of human casualties and property damage. Therefore, intelligent environment applications must exhibit high precision and prevent potential security breaches or data manipulation.

Intelligent Grid and Metering Systems

Smart metering encompasses evaluation, tracking, and administration applications. Intelligent meters are commonly employed in smart grids to monitor electricity usage. Electricity theft can also be prevented with smart meters [9]. Smart meters can monitor and measure the quantity of fluids contained within vessels, such as reservoirs and storage tanks. Intelligent meters are employed to monitor and optimize the acquisition of solar energy by dynamically adjusting the orientation of photovoltaic cells, thereby augmenting the efficacy of solar power facilities. IoT-based intelligent metering applications can measure product weight and water pressure in water distribution networks. Unlike analog meters, intelligent metering systems are vulnerable to environmental and cyber hazards. The Advanced Metering Infrastructure (AMI), also known as smart meters, is engineered to exceed the mere task of monitoring energy consumption. With the help of the Home Area Network (HAN), smart meters and other electrical appliances in the home can be linked together to form an efficient and effective smart grid. This equipment collects data that can be used for load and expense management. Malicious actors or consumers infiltrating communication systems may result in monetary losses for service providers and customers, as Namboodiri et al. [10] reported.

Security and Emergencies

In addition, a number of IoT applications are being used in crucial security and emergency situations. It has uses like allowing only authorized personnel access to areas that are restricted. Identifying hazardous gas emissions near industrial or chemical plants is another domain utilization. Notifications may be dispatched in the event of elevated radiation levels. Measurements of radiation levels can be conducted in proximity to both nuclear power plants and cellular base stations. Numerous edifices contain systems that accommodate or retain confidential data. Security applications may be employed to protect personal information and valuable assets. For example, identifying various liquids through the IoT applications can prevent corrosion and structural failure in fragile structures. The occurrence of security breaches within these applications may result in various adverse consequences. For example, malevolent actors may attempt to gain entry to restricted areas by exploiting a vulnerability in an application. In addition, false radiation alarms can have negative consequences in the short and long run. Children are especially vulnerable to the harmful effects of radiation exposure, especially if they are repeatedly exposed.

Smart Retail

The retail sector makes extensive use of IoT applications. Many instruments have been devised to monitor the moisture and temperature levels of commodities during transportation throughout the supply chain. The IoT is utilized for efficient restocking procedures, including monitoring warehouse inventory. Based on a customer's preferences, routines, allergies to particular ingredients, etc., various smart shopping applications are currently being created to help them. Additionally, systems have been developed that use augmented reality to give offline retailers the same shopping experience as online retailers. There have been security concerns with the integration and use of various IoT applications by retail businesses. The enterprises above encompass Sony, Apple, Home Depot, and JP Morgan Chase, as stated in the source [11]. Bad actors can try to increase sales by spreading false information about products through IoT apps by altering data about the conditions under which they are stored. Attackers may steal customer contact details, email addresses, banking and credit card information, and other confidential data if safety precautions aren't put in place in smart retail, costing both consumers and retailers money.

Smart Agriculture and Animal Farming

Tracking soil moisture levels, controlling humidity and temperature, and mainly irrigating in arid regions are all part of the larger category of "smart agriculture," including other methods. High yields can be achieved and farmers can save money by utilizing such cutting-edge features in agriculture. The management of both humidity and temperature during the cultivation of vegetable and grain crops can prevent the proliferation of fungal and other microbial contaminants. Optimizing crop and vegetable yield and quality may be facilitated by implementing weather control measures. Applications for the IoT can track farm animals' activities and health by strapping sensors to them, much like crop monitoring. If these applications are breached, there may be crop damage from intruders as well as theft of farm animals.

Automation in the Home

Home automation is a frequently employed and prevalent IoT application. The subject matter concerns software applications specifically developed to remotely manage electrical appliances with the aim of conserving energy, as well as security systems that are set up on doors and windows to prevent any unauthorized access. Individuals are incentivized to economize on resources and finances by utilizing monitoring mechanisms that oversee their energy and water consumption. According to the authors of the research paper titled "Improving Home Security Using Logic-Based Security Algorithms" [12], logic-based security algorithms can be employed to enhance home security. By contrasting user actions in essential areas of the residence with typical user conduct, intrusions are detected. Notwithstanding, malevolent actors may gain unauthorized access to IoT devices within a residential abode and endeavor to cause harm to its users. For instance, since the setting up of different home automation systems, home thefts have increased significantly [12]. Historically, antagonists have endeavored to evaluate the character and quantity of online data transmission to and from the intelligent domicile to gauge the inhabitants' whereabouts and conduct.

Threats to IoT Applications

The architecture of an IoT application comprises four distinct layers, namely:

- Detection Layer,
- Network Layer,
- Middleware Layer, and
- Application Layer.

A variety of security risks and problems are introduced by the use of various technologies in each one of these sections of an IoT application. The illustration depicted in Fig. 2 showcases a range of technologies, devices, and software programs utilized across the four discrete tiers. These four layers of IoT applications are covered by this section's discussion of various security risks. The illustration presented in Fig. 3 portrays potential attacks that may occur across the four layers above.

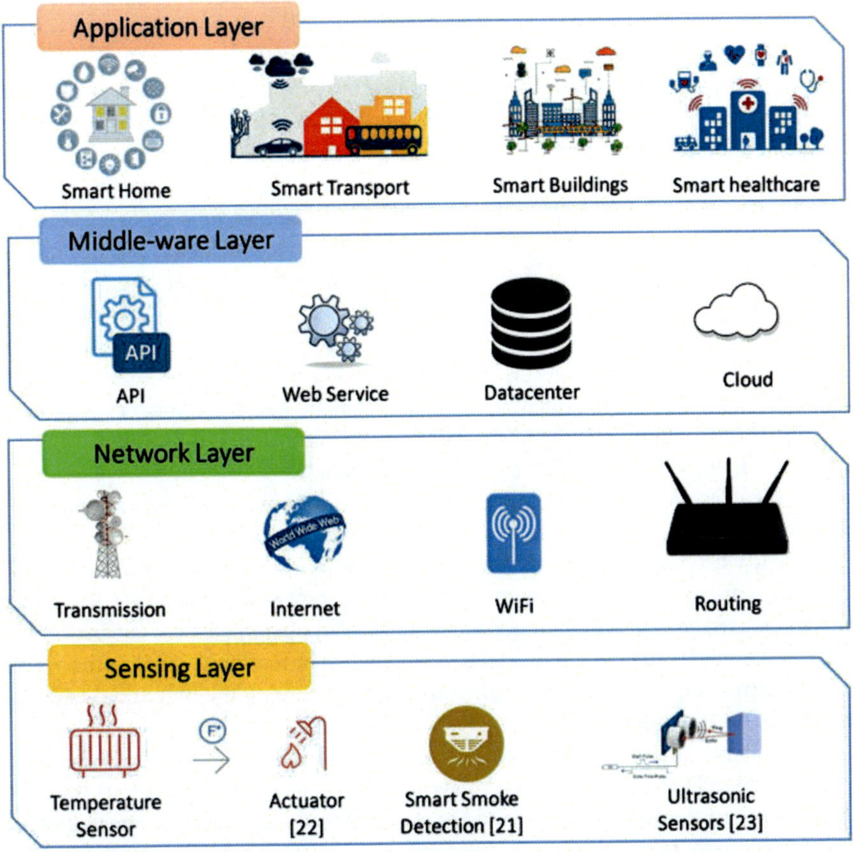

Fig. 2 Layers in IoT system

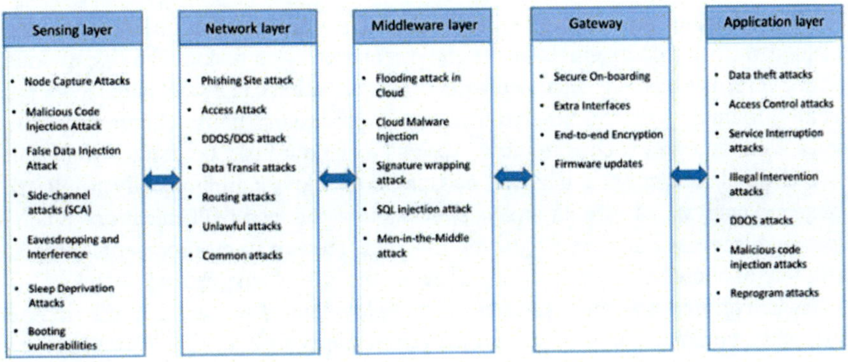

Fig. 3 Types of attack on IoT

Problems with Security at the Sensing Layer

Physical events in the surrounding environment are detected by sensors, as noted in the literature [13, 14]. Actuators utilize the data from identified objects to execute a specific operation on the natural surroundings. Various sensors, such as ultrasonic sensors, cameras, smoke detectors, temperature, and humidity sensors, are available for detecting and measuring different types of data. Various types of sensors, including mechatronic, electrical, electronic, and chemical sensors, can detect and measure physical phenomena. Various IoT applications utilize sensing layer technologies such as RFID, GPS, WSNs, and RSNs. The detection layer contains the major security risks listed below.

Node Capturing: Applications that run on the IoT use many nodes that consume little power, such as indicators and actuators. There is cause for concern regarding the vulnerable nature of these node locations to various attacks launched by the opponent. In an IoT system, malevolent actors can attempt to kidnap or substitute a node with a harmful one. The attacker maintains control over the newly added node, even though it appears fully integrated within the system.

Attack with Malicious Code Injection: The attacker's method of attack entails inserting malicious code directly into the node's memory. It is typical for IoT nodes to receive software or firmware updates over the air, opening a door for hackers to introduce malicious code. Male- factors may endeavor to obtain entry to the whole IoT infrastructure by employing malicious code or coercing the individual nodes to execute undesired operations.

False Data Injection Attack: A hacker's node acquisition could facilitate injecting fabricated data into the IoT system. The production of erroneous data and the inability of the IoT application to function correctly are both outcomes that are feasible under these circumstances. In addition, a malicious actor can use the method described above to launch a distributed denial- of-service attack (also known as a DDoS attack).

Side Channel Attacks: The potential for private information leakage exists through various side-channel attacks and direct attacks on nodes. The risk of confidential information being accessed without authorization can be attributed to various factors, including the micro-architectures of processing units, electromagnetism emanation, and power consumption. Various techniques can be utilized to execute side-channel attacks, such as power consumption analysis, timing analysis, electromagnetic analysis, electricity analysis, or photons-induced fault injection. When it comes to implementing cryptographic modules, modern microprocessors integrate a variety of countermeasures to help alleviate the risk of side-channel attacks.

Evasdropping and Interference: The deployment of numerous nodes in open spaces is a common characteristic of IoT applications, as Liao et al. (2018) noted in their study on eavesdropping [15]. Consequently, individuals who engage in eavesdropping can take advantage of these IoT applications. The security of data transmission and authentication procedures is susceptible to interception and eavesdropping by malicious actors.

Step Deprivation Attacks: These assaults target low-power, edge devices on the IoT to drain their batteries. All of the connected devices using the IoT service will eventually crash as a result. Furthermore, the malicious code can cause edge devices to operate in an infinite loop or artificially augment their power consumption.

Attacks by booting: Several attacks can be launched against the edge equipment while booting up. This results from the internal security mechanisms being inactive at the time. During the restart of node devices, malicious actors can exploit the vulnerability present in the system and launch an attack. Securing the boot process in edge devices is crucial because they frequently have low power consumption and occasionally experience sleep–wake cycles.

Network-Level Security Concerns

The sensing layer collects data, which is then sent to the network layer and, ultimately, the computational unit. The following are the most common types of threats discovered in the network layer.

Pishing Site Attack: Phishing attacks can target many IoT devices with minimal exertion from the attacker. As per the assailants, the assault is anticipated to be efficacious against a minimum of a few apparatus. Web surfers coming into contact with phishing sites when visiting websites. In a cyberattack, wherein a user's username, password, and account information are stolen, the security of an IoT environment may be put at risk. Phishing site attacks are very likely to happen on IoT networks.

Access Attack: Access attack (APT) is a type of advanced persistent threat. The assault above enables an evil entity or external agent to access the network of interconnected devices, commonly known as the IoT. The criminal can remain undetected in the system for extended periods. This form of the target aims to extract crucial information or data from the network rather than cause harm to it. The susceptibility of IoT applications to attacks is heightened due to the continuous exchange and reception of significant data, as noted by Li et al. [16].

DDoS /DoS Attack: In the context of information technology security, the offender will start a DDoS attack by bombarding the victim's servers with many requests to which they were not invited. Consequently, the services offered to real users are halted, and the server under attack becomes non-functional. A DDoS attack is characterized by the inundation of a target server with traffic from multiple sources, as executed by the attacker. Even though these attacks are not exclusive to IoT applications, IoT networks' intricate and diverse nature renders the layer vulnerable. Furthermore, numerous IoT devices in IoT applications possess inadequate configurations, creating weak entry points for malevolent actors to initiate DDoS attacks against the targeted servers. Section I of the present text discusses the attack by the Mirai botnet, which exploited a vulnerability in IoT devices that were not correctly configured. The attack consisted of a continuous stream of requests that ultimately blocked multiple servers [17].

Data Transit Attacks: IoT applications facilitate exchange and storing of essential data. Unfortunately, due to the substantial value of data, it is consistently targeted by malicious actors such as hackers and other adversaries. Even though transporting information from one location to another increases its vulnerability to cyberattacks, data security risks are present whether it is stored locally or in the cloud. The cloud, sensors, actuators, and other IoT components all exchange a lot of data with one another. Due to the use of various connection technologies in these data movements, IoT applications are vulnerable to security breaches resulting in unauthorized access to sensitive data.

Attacks on the Routing: During these attacks, malicious nodes within an application running on a connected device may attempt to change the data's routing paths while it is in transit. In the form of a routing attack known as a "sinkhole attack," an adversary will pose as the owner of a fictitious shortest routing path and try to persuade other nodes to direct their traffic through it. Wormhole attacks are another method that can compromise a system's safety, particularly when combined with sinkhole assaults. Warm holes connect two nodes that enable rapid packet transfers outside the band. They are also known as "warm bridges." An adversary attempting to overcome the basic safety protocols of an IoT application may create a "warm hole" between a device connected to the Internet and a node being attacked.

Security Problems in the Middleware Layer

When used in the IoT, middleware bridges the application and network layers. This leads to the formation of a layer of abstraction positioned between the layers above. In addition, Bandyopadhyay and Sen pointed out in their study [18] that middleware has an opportunity to provide robust storage and computing functionalities. APIs, or application programming interfaces, are made available by the layer in question. These APIs were explicitly designed to fulfill the requirements of the application layer. The middleware layer comprises various components such as machine learning algorithms, queuing systems, data that persists stores, and brokers. When it comes to ensuring the dependability and resiliency of applications for the IoT, the middleware layer is essential. However, it can be com- promised by several different types of malicious attacks. The compromised middleware facilitates the attacks mentioned above to hijack the entirety of the IoT application. Additionally, security concerns about databases and cloud storage significantly impact the middleware layer. The subsequent section will discuss various assaults that may be perpetrated toward the middleware layer.

Man-in-the-Middle Attack: To implement the publish-subscribe model and facilitate information exchange between clients and subscribers, the MQTT protocol makes use of the MQTT broker, which performs the role of a proxy. This model is used for establishing connections between clients and subscribers. Since the publishing and subscribing clients are not linked, messages can be sent without knowing which one will ultimately be received. An intruder can secretly take over all communication by acting as a broker and dictating terms to both parties.

SQL Injection Attack: SQL Injection (SQLi) is a threat not only to servers but also to middleware. Here, the attacker can compromise a program by inserting malicious SQL statements [19, 20]. After that, hackers can steal sensitive information from users and even alter their profiles in databases [21]. The Open Web Application Security Project (OWASP) named SQL injection the most critical vulnerability facing websites today in their OWASP Top 10 2018 document.[1]

Signature Wrapping Attack: XML signatures are employed in the middle layer to authenticate web services [22]. An attacker using a signature-wrapping attack can run programs or change intercepted messages by taking advantage of Simple Object Access Protocol (SOAP) vulnerabilities.

Cloud Malware Injection: Using cloud malware injection, a hacker could take over, plant a virtual machine, or inject malicious code. The malicious actor masquerades as a legitimate service to launch a malicious service module or VM. The attacker can then gather sensitive data that can be changed by viewing the victim's service requests.

Flooding Attack in Cloud: The assailant can see what services the victim has requested and gain access to private information that can be altered. Attackers repeatedly ask a service for resources from the cloud in order to deplete them. Such assaults significantly impact cloud systems because of the increased load on cloud servers.

Security Issues at Gateways

In order to connect various devices, people, things, and cloud services, a gateway is a substantial layer that is essential. Gateways help with IoT gadget hardware and software. Gates also translates protocols and encrypts IoT data. Today's IoT systems are heterogeneous, use TCP/IP stacks, and use LoraWan, ZigBee, Z-Wave, and numerous gateways as the intermediaries. IoT gateway security issues are covered in more detail below.

Secure On-boarding: Security of encryption keys is an ongoing concern for any IoT system that adds new sensors or devices. Gateways transmit authentication credentials for new devices and management services. When a user is onboarded to a gateway, there is a heightened risk of eavesdropping and attacks known as man-in-the-middle, making encryption keys vulnerable to theft.

Extra Interfaces: It is essential to minimize the attack surface when installing IoT devices [23]. Manufacturers of IoT gateways should only include the most fundamental interfaces and protocols in their products. In addition, limiting the capabilities of end users is necessary to prevent unauthorized identification or data breaches.

End-to-End Encryption: Confidentiality of data must be safeguarded through proper end-to- end application layer security [24]. The intended recipient should be

[1] https://www.acunetix.com/blog/%20articles/owasp-top-10-2017/).

the only one to decrypt the messages using the software. The Zigbee and Zwave protocols support encryption, but because the gateways have to initially decrypt the messages and then re-encrypt them to convert the data from one protocol to another, end-to-end encryption is not possible. This gateway-level decryption raises the possibility of data breaches.

Firmware Updates: Most IoT gadgets don't have the user interface or processing power necessary for downloading and installing firmware updates because they have constrained resources. The typical method of downloading and installing firmware updates is through gateways. IoT applications are safeguarded against data theft by implementing diverse techniques and protocols, including but not limited to information encryption, data in a state of isolation, user and network authorization, and privacy management. To perform secure firmware updates, keeping track of the firmware's previous and current iterations and verifying the signatures' legitimacy is essential.

Aspects of Security at the Application Layer

The application layer is in charge of communicating with and serving end users directly. The layer in question exhibits various applications of the IoT, such as smart grids, cities, homes, and meters. Distinct security considerations are exclusive to this particular layer. For example, the absence of data theft and privacy concerns in other layers makes them unique to this layer. This layer's security concerns vary widely depending on the application. In numerous IoT applications, there exists a stratum of software that is positioned between the "network" and "application" layers, which is commonly referred to as the "application support layer" or "middleware layer." Allocation of resources, computational intelligence, and other business services are all aided by the support layer. This article will review some of the most critical security concerns the application layer must address.

Data Thefts: Numerous sensitive data are used in IoT applications. As a result of the extensive data movement involved in IoT applications, it is even more susceptible to hackers than data at rest. Due to the significant data transfer involved in IoT applications, the susceptibility of such data to hacking is even greater than that of data at rest. The large amount of data transfer in IoT applications makes it even more hackable than data sitting idle. If users perceive an increased risk of data theft when using IoT applications, they will be less likely to provide personal information when registering.

Access Control Attacks: Access control allows access, so only legitimate individuals can interpret the information or account. Access control attacks are crucial in IoT applications because, once accessibility becomes compromised, every aspect of the IoT application is vulnerable.

Service Interruption Attacks: The terms "DDoS attacks" and "illegal interruption attacks" have also been used to describe these attacks in earlier writing. There have been multiple cyber assaults targeting IoT applications. These types of attacks impede the ability of authorized users to utilize the functionalities of IoT applications by intentionally overwhelming the servers or network.

Malicious Code Injection Attacks: Such attacks artificially overload the servers or network, preventing authorized users from using the capabilities of IoT applications. If a system exhibits susceptibility to malicious scripts and deceitful tactics, coupled with insufficient code verification measures, it serves as a potential hacker's primary point of ingress. Cross-site scripting (XSS) is a malicious technique attackers employ to inject malicious scripts into a trustworthy website. In the event of a successful cross-site scripting (XSS) attack, an IoT account may be compromised, potentially resulting in the cessation of IoT system operations.

Sniffing Attacks: Sniffer applications allow attackers to monitor internet traffic in IoT applications. Insufficient security measures may result in an attacker's unauthorized access to sensitive user data [25].

Reprogram Attacks: If the programming of an IoT device lacks adequate security measures, malicious actors may attempt to reprogram the device remotely. As a result, the IoT network may be taken over [26].

IoT Security Through the Use of Blockchain

The IoT and blockchain technologies are considered vital and are anticipated to impact the data and communication technology industries significantly. The primary objectives of these two technologies are to raise user levels of convenience, visibility, openness, and confidence. The utilization of sensors in IoT devices facilitates the provision of real-time data. At the same time, incorporating blockchain technology, characterized by its distributed, decentralized, and shared ledger, is imperative for ensuring data security [27].

The fundamental concept underpinning the blockchain is a distributed ledger, alternatively called replicated log files. The temporal information of the data is recorded through time stamping and organized chronologically on the blockchain. The ledger's entries are securely connected to the preceding one through cryptographic hash keys.

Miners work only to earn rewards; they are not personally invested in the transactions. The miners need to find out who is behind the transactions. Additionally, there is intense rivalry among the many miners processing an identical batch of transactions to be the very first to add them to the blockchain. These distinguishing qualities of the blockchain—robust, tamper-proof, distributed, and open—make it unique [28]. Numerous frameworks and platforms are being developed in both the academic and commercial sectors to support the creation and upkeep of blockchain. These platforms consist of Ripple, Ethereum, and Hyperledger Fabric [29].

Blockchain for IoT: Benefits

Applications for the IoT can benefit significantly from blockchain technology. Table 2 summarizes some specific IoT security challenges and possible blockchain-based solutions. IoT applications face several security issues. Here are a few significant advantages of incorporating blockchain technology into IoT applications.

Blockchain's decentralized nature secures data: The distributed blockchain design can reduce the risk of a single point of failure, which plagues many cloud-based IoT applications. The storage of data generated by devices on the blockchain is a secure and effortless process, irrespective of the distance between the devices [38].

Blockchain can hold IoT data: Applications use various interconnected devices for the IoT. These devices are also connected to and controlled by other devices. This configuration has an additional connection to the cloud, enabling the execution of IoT applications from any location. In addition to the data movement, blockchain promises to store and protect data from misuse. Blockchain can be a suitable data storage and transmission solution no matter what layer an IoT application uses.

Table 2 IoT Problems and a Blockchain-Based Potential Fix

Challenges in IoT	Specification	Potential use of blockchain technology
IoT devices and privacy	IoT devices can disclose confidential user data	Permissioned Blockchain, which can secure IoT devices, is a potential solution to this problem [30–32]
Price and volume	Appropriate measures must be taken to effectively handle the rapid proliferation of IoT device	The implementation of blockchain technology is being employed as a means to progress towards a decentralized system. Rather than utilizing centralized servers to communicate, the devices can establish direct connections and share with peers [3, 33, 34]
WSN node jamming	Man in the Middle Attack	Data security in cloud computing
IoT devices' flawed architecture	Single points of failure in every component affect the net- work and the entire machine	Due to blockchain, the legitimacy of devices is confirmed. Furthermore, the data undergoes cryptographic validation to guarantee that solely the initial sender can transmit it [35]
Modification of data	The data extracted from IoT devices is often misused after being manipulated	The blockchain has interlocked devices. The system rejects any device updates to data [36, 37]

Security against information leakage and spoofing attacks: Introducing a malicious node into the IoT network and making it look like a legitimate member of the original network is required to carry out a spoofing attack on an IoT application. A competitor can quickly observe, capture, or introduce data into the network by spoofing. An excellent deterrent to these assaults is the blockchain. Because each valid user or device is recorded on the blockchain, they can promptly verify and validate their identity without intermediaries such as central brokers as certification entities.[2] Due to their low power consumption, IoT devices risk losing data. Under certain circumstances, data loss may transpire on the sender's and receiver's end due to external environmental factors. The implementation of blockchain technology has the potential to mitigate such losses as it possesses an immutable nature, whereby once a block is appended to the chain, it cannot be altered or deleted [39].

Secure data storage via a cryptographically-verified secret hash: Blockchain technology enables storing the 256-bit data hash key while the actual data remains unalterable and unmodifiable. The primary data may be associated with a hash key, while the original information can be stored in a cloud-based system. In the event of any modifications to the data, the corresponding hash value will undergo alteration as well. At present, the data is safeguarded and protected from unauthorized access. Since the blockchain only stores the results of the hash algorithm, the size of the chain is independent of the size of the data being stored. Access to data stored in the cloud is restricted to authorized parties designated to use it and can only be achieved through encrypted data. Utilizing blockchain technology reduces the likelihood of storing corrupt data from devices, as each data set undergoes a comprehensive verification process by a network of miners.

Decommissioning of all Cloud-based Centralized Servers: Integrating blockchain technology can bolster the security of IoT systems by eliminating the need for centralized cloud servers and transitioning towards a decentralized peer-to-peer network. Centralized cloud servers are the primary target of most individuals who steal data. Instead, data distribution among all network nodes can be achieved using a blockchain and a cryptographic hash function.

Resource-constrained blockchain proxy-based architecture: While blockchain does offer a number of security benefits for a distributed environment, the limited resources of the IoT pose a unique challenge. IoT devices cannot store large ledgers due to their severe resource constraints. Several attempts have been made to simplify how to use blockchain in IoT. One feasible method for allowing Internet of Things devices to utilize blockchain technology is a proxy-based architecture. To protect data and other network assets, proxy servers can be set up in a network's infrastructure. The end-user can obtain the encrypted resources of the proxy servers for downloading [40].

Unauthorized access prevented by blockchain: Numerous IoT applications entail recur- rent communication among various nodes. Blockchains use public and private keys to facilitate communication, which helps to ensure that only the intended node or party can gain access to the data. This is done by ensuring that only the

---

[2] https://venturebeat.com/business/how-blockchain-can-change-the-future-of-iot/.

intended node or party has the private key. Because the data has been encrypted with keys, even if a third party is able to obtain access to it, they will still require assistance in deciphering the information contained within it. The data structure known as blockchain was developed specifically to address the numerous safety concerns posed by IoT applications.

Utilizing FoG Computing for IoT Security

The IoT and cloud computing are different technological advancements that offer many applications. The IoT has facilitated the availability of diverse, intelligent devices and programs to end-users. Like- wise, utilizing cloud technology presents a highly effective approach to data management and storage, which can be accessed from any geographical location, and is widely adopted by numerous enterprises. The Internet infrastructure is under great stress due to the IoT's production of an unprecedented volume of data. Integrating IoT and cloud technologies has created a novel epoch of prospects and complexities in data handling, storage, management, and security optimization. Combining the IoT and cloud has been an attempt by business and research organizations to solve some technology issues. Commercial enterprises and academic research institutions have explored the integration of the IoT and cloud technology as a potential solution to various technological challenges. The advantages of doing this make it necessary to be done, but they only partially solve all IoT problems. In 2012, Cisco introduced the concept of fog computing. Fog computing is an augmentative technology to cloud computing rather than a replacement.

Design of FoG Computing

The primary aim of fog computing is to facilitate data processing from IoT devices at a local level, thereby improving management. This requires the implementation of a multi-tiered architecture. Ni et al. [41] reported that the system under investigation employs the Fog-Device and Fog-Cloud- Device frameworks. The initial framework consists of the device and the fog layer, whereas the sub- sequent framework includes the device, fog, and cloud layers. Their aptitudes dictate the disposition of strata for both information retention and computational capabilities. Inter-layer communication employs wireless technologies such as WiFi and Bluetooth and wired technologies like optical fiber and Ethernet. Integrating fog nodes in the Fog-Device framework facilitates the delivery of various services to end-users, thereby preventing the necessity for cloud servers. According to Sehgal [42], the Fog- Cloud-Device framework delegates the responsibility of complex decision-making to the cloud, while the fog layer handles less complicated decisions. The diagram depicting the structure of the Fog-Cloud Device framework is presented in Fig. 4. The research conducted by Sarkar et al. [43] aimed to evaluate the effectiveness of the fog computing model in comparison to the conventional cloud computing architecture in terms of service latency and energy utilization. The authors' primary focus was on the architectural aspects of fog computing. As per the findings of Varghese et al. [44], the adoption of fog computing leads to a noteworthy reduction of 20% in the average response time of users, along with a significant decrease of 90% in data traffic between the cloud and network edge, as compared to a model that solely relies on cloud computing. This highlights the feasibility of fog computing as a promising approach for enhancing the performance of network systems. The academic literature discussed in reference [45] provides a comprehensive explanation and assessment of fog computing while also com- paring it to related concepts such as "mobile edge computing" (MEC) and "mobile cloud computing" (MCC). In the publication referenced as [45], the authors have presented various applications that can be utilized in fog computing. These include augmented reality (AR), content delivery networks, caching, and big data mobile analytics.

Fig. 4 Architecture of FoG computing

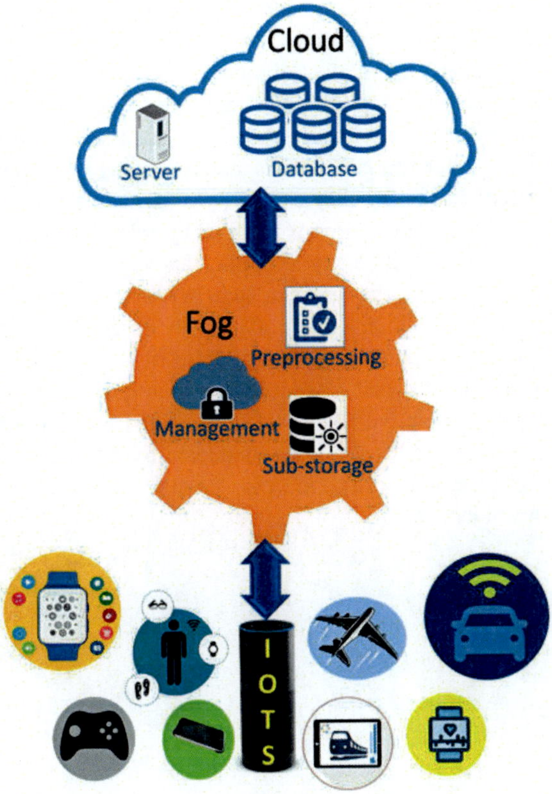

Benefits of FoG Compared to Cloud

Every day, IoT devices produce enormous amounts of data. Regrettably, transmitting this data to the cloud for instantaneous analysis is unfeasible. Thus, the concept of fog computing was developed. The diagram depicted in Fig. 4 elucidates the placement and operation of the fog stratum in an IoT framework. The concept of fog computing pertains to the extension of cloud computing and its associated services beyond the network's periphery. Fog computing, an infrastructure for decentralized data analysis, can efficiently and expeditiously maintain and analyze time-sensitive data. The main objectives of this initiative are to enhance the overall efficacy of IoT applications, reduce cloud-based data storage, prevent data breaches, and enhance security measures. The reduced latency observed in fog computing can be attributed to the proximity of the fog layer to the devices compared to cloud computing. The cloud storage system retains only essential and carefully selected information for permanent storage. Fog computing finds utility in various domains, such as software-defined networks, intelligent retail, intelligent traffic management, smart homes, precision agriculture, and healthcare. The handling and evaluation of the extensive quantity of data generated by IoT devices would result in considerable costs and necessitate a significant duration if executed in the cloud. According to a credible source,[3] fog computing alleviates the strain on network bandwidth and minimizes the frequency of bidirectional communication between IoT devices and the cloud.

Fog nodes collect the data, and according to a reliable source (https://www.thesecuritybuddy.com/), they can analyze 40% of the data in a fog architecture. The IoT device's response time is reduced by transferring information to the core network. A fog node can be anything that can connect to a network, process data, and store data, like a security camera, switch, or router. Fog nodes can be deployed in various locations, including but not limited to vehicles or factory floors, provided that a network connection is established. Data can be sent to a fog node, an aggregation node, or the cloud, depending on how quickly it needs to be sent. Fog nodes use cryptographic calculations to ensure IoT applications can communicate securely. According to the research conducted by Alrawais et al. [46], it has been observed that basic sensors and IoT devices may need more internal resources to accomplish this objective.

[3] https://www.techtarget.com/iotagenda/.

Overcoming IoT Security Threats with the Help of FoG Computing

Man-in-the-middle Attack: Cloud or IoT systems and end users are separated by a fog computing security layer. The fog layer can detect abnormal activity and shield IoT systems from potential threats and attacks. It can also help the system from being penetrated by such action.

Data Transit Attacks: Compared with IoT devices, the storage and management of data exhibits superior performance when executed on secure fog nodes. Furthermore, compared to data storage on end-user devices, data security can be enhanced by storing them on fog nodes. Moreover, the utilization of fog nodes serves to strengthen the availability of user data.

Evasdropping: The implementation of fog nodes facilitates communication exclusively between the end user and the fog node, thereby circumventing the need to transmit data across the entire network. As a result, reducing network traffic significantly decreases the probability of an adversary attempting to eavesdrop.

Resource-Constraint Issues: Perpetrators exploit the restricted resources in IoT devices. The perpetrators endeavor to compromise the edge devices to access the system. FoG nodes can support edge devices while simultaneously ensuring protection against possible security breaches. A nearby fog node can effectively execute intricate security measures for protection purposes.

Incident Response Service: Fog nodes can be programmed in a manner that enables them to offer prompt incident response services. If encountering questionable data or requests, fog nodes can communicate with the IoT system or the end users. The utilization of fog computing enables the detection of malware and prompt resolution of issues during transit. Attaining a comprehensive cessation of system operations to resolve malware occurrences in vital applications is impracticable. Fog nodes can aid in tackling the concerns mentioned above, contingent upon the system's sustained operability.

Man-in-the-middle Attack: Cloud or IoT systems and end users are separated by a fog computing security layer. The fog layer can detect abnormal activity and shield IoT systems from potential threats and attacks. It can also help the system from being penetrated by such action.

Machine Learning-Based IoT Security

Machine learning (ML) has garnered considerable attention in recent years. The application of machine learning is presently being employed in the IoT security domain and in the progression of diverse other disciplines. The utilization of machine learning (ML) has the potential to serve as an efficacious measure in protecting IoT devices from cyberattacks. This is due to the unique approach to defense that it provides, which differs from conventional methods.

Threats to Security Are Addressed by ML-Based Solutions

This discourse aims to elucidate the security hazards and their corresponding solutions being addressed by Machine Learning (ML).

DoS Attack: The issue of Denial-of-Service (DoS) attacks, whether directed toward IoT devices or emanating from them, is of considerable importance. Implementing a multi-layer perceptron (MLP), protocol has been proposed as a potential strategy for mitigating DoS attacks, as dis- cussed in the literature [47]. The authors of reference [48] utilized a multi-layer perceptron (MLP) to augment the security of wireless networks. Implementing machine learning methodologies has the potential to augment the security of IoT devices that are susceptible to Denial of Service (DoS) attacks. Furthermore, implementing these techniques has the potential to enhance the accuracy of the process of deduction.

Evasdropping: Malicious actors can intercept and monitor data transmissions, thereby engaging in eavesdropping. Machine learning (ML) techniques, specifically non-parametric Bayesian techniques and Q-learning-based offloading strategy, have been proposed as potential defense mechanisms against these types of attacks, as documented in references [49] and [50]. Machine learning techniques, such as Q-learning and Dyna-Q, can enhance the security of devices by mitigating the risk of eavesdropping. Xiao et al. [51] empirically assessed said schemes through experimentation and reinforcement learning. The details of their evaluation can be found in their cited work.

Spoofing: The techniques of Q-learning, Dyna-Q, Support Vector Machines (SVM), Deep Neural Network (DNN) model, Incremental Aggregated Gradient (IAG), and Distributed FrankWolfe (dFW), have been proposed as potential defenses against spoofer attacks, as reported in the literature [51–55]. The utilization of these techniques results in a reduction of the mean error rate and false alarm rate while simultaneously enhancing the precision of detection and classification.

Breach of Privacy: The gathering of individual data, encompassing health records, geographic coordinates, and images, poses a potential risk to the confidentiality of the user. To mitigate the risk of unauthorized disclosure of confidential data, it is advisable to utilize privacy-preserving scientific computations (PPSC) as expounded upon in the academic publication by Yan et al. [56]. Li et al. [57] have proposed the Commodity Integrity Detection Algorithm (CIDA) as a prospective resolution to establish trust in IoT applications. The algorithm utilized in this study is founded on the Chinese Remainder Theorem (CRT), as documented in the publication by Li et al. [57].

Digital Fingerprinting: Digital fingerprinting is a recently developed and up-and-coming technique for safeguarding IoT systems and fostering users' confidence in applications. Fingerprints are commonly employed to access smartphones, authorize financial transactions, grant entry to vehicles and residences, and perform various other functions. The utilization of digital fingerprinting as a biometric identification technique is gaining popularity owing to its cost-effectiveness, reliability, widespread acceptance, and superior security features.[4] Digital fingerprinting has several advantages; The effective deployment of this technology within the IoT framework is confronted with various obstacles, including but not limited to fingerprint categorization, image refinement, and characteristic correlation. The following discourse delves into a selection of machine learning algorithms devised to provide innovative remedies for the challenges above.

Support Vector Machine (SVM): The Support Vector Machine (SVM) is a machine learning algorithm for diverse applications such as principal component analysis, speaker identification, text categorization, regression, and linear and nonlinear classifications. The aim is to optimize the separation between the decision boundaries and the training patterns. The scholarly work referenced as [58] thoroughly examines the utilization of Support Vector Machines (SVM) in digital fingerprinting. Moreover, the authors have conducted a comparative analysis between the model above and other established frameworks. The generation of the feature vector is based on the pixel values of the fingerprint, which is then employed in the training of the Support Vector Machine (SVM). Through the examination of diverse fingerprint patterns, the process of identifying a fingerprint is accomplished by utilizing established methodologies.

Artificial neural networks (ANNs): ANNs represent a prevalent machine learning algorithm. The concept above offers numerous advantages, such as the ability to generalize, adapt to changing circumstances, and exhibit fault tolerance. The article referenced as [59] presents a proposed framework for utilizing artificial neural networks (ANN) to recognize digital fingerprints. The back propagation algorithm of Artificial Neural Networks (ANN) is utilized for training the network by feeding it with digital values of diverse fingerprint characteristics such as minutiae, ridge endings, and bifurcations as input. The verification of fingerprints is conducted by referencing the experiential values stored in the database.

IoT Security Using Edge Computing

See Fig. 5.

Edge Computing for IoT Security and Improvement

The solutions that edge computing provides or can provide to counteract these security threats are covered in the following section.

Losses of data: The concept of edge computing pertains to the localized processing and storage of data on a computer or network near the data source. The

[4] haaattps://waaaww.iotevolutionworld.com%2fiot%2farticles%2f438690-biometric-authentication-future-iot-security-solutions.htm.

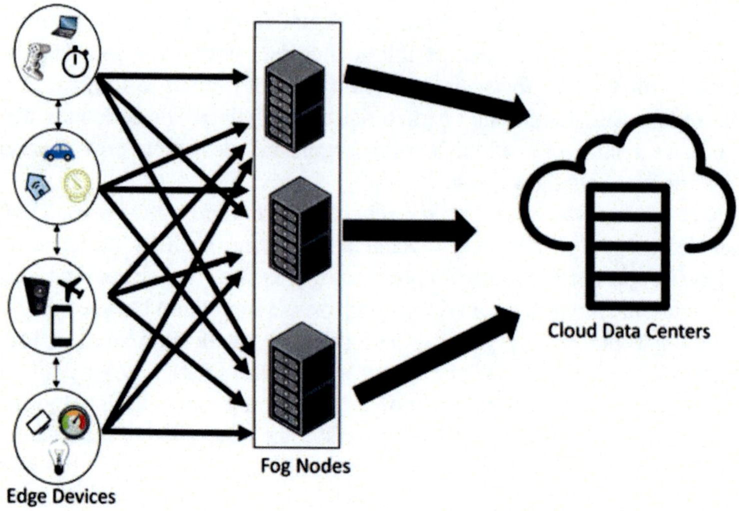

Fig. 5 Architecture of edge computing

data transfer from its origin to the machine is not taking place. Preventing data transmission effectively mitigates the potential hazards of data breaches and theft. Premsankar et al. [60] have noted in their research that transferring data from equipment to the fog layer in fog computing may present potential benefits to adversaries.

Issues related to compliance with data regulations: Several nations have adopted rigorous regulatory policies, including the General Data Protection Regulation (GDPR) implemented by the European Union, to inhibit the transfer of data beyond their respective territorial confines. Corporations can use edge computing to maintain data sovereignty by constraining it within their territories.[5]

Safety Issues: A heightened emphasis accompanies expanding cyber-physical systems on security and safety as critical considerations. The occurrence of physical safety concerns could be linked to any delay, no matter how small, in the promptness of responses. For example, in the event of an impending collision, airbags must be promptly deployed based on the signals received by a vehicle's sensors. The potential for injuries or fatalities may increase if sensor technology solely relies on transmitting all data to the cloud and awaiting a response before initiating any action. Edge computing has the potential to enhance the capabilities of surveillance cameras by enabling them to independently analyze anomalies and transmit condensed and suspicious data to data centers, thereby reducing response times.

Issues With Bandwidth: The IoT applications produce a significant amount of data at a high velocity. However, a significant portion of the available data still needs to be processed and processes limited value. In addition to the challenges associated with ensuring secure data transfer, transmitting all data to cloud-based storage incurs

[5] https://www.techtarget.com/iotagenda/feature/6-significant-issues-that-edge-computing-in-IoT-solves.

significant bandwidth expenses. Guo et al. [61] assert that the utilization of edge computing facilitates the implementation of a significant proportion of cleaning of data and aggregation at the edge nodes. Therefore, it is advisable to transmit only concise data to the cloud.

Conclusion

Multiple security risks have been identified across different layers of an IoT application. The concerns about the application, network, middleware, and sensing layers have been deliberated upon. The discourse has also encompassed contemporary and prospective security menaces and remedies in the IoT realm, including blockchain, fog computing, edge computing, and machine learning. In addition, conversations have occurred regarding various outstanding matters and challenges arising from the proposed resolution. Finally, the latest advancements in security for the IoT have been deliberated alongside prospective avenues for future research. The survey is anticipated to function as a valuable resource for enhancing security in future IoT applications.

References

1. Kandaswamy, R., Furlonger, D. (2018). Blockchain-based transformation: A gartner trend insight report. Gartner IT Glossary.
2. Muhammad, S., Umar, M. M., Khan, S., Alrajeh, N. A., & Mohammed, E. A. (2023). Honesty-based social technique to enhance cooperation in social Internet of Things. *Applied Sciences, 13*(5), 2778.
3. Fernández-Caramés, T. M., & Fraga-Lamas, P. (2018). A review on the use of blockchain for the internet of things. *IEEE Access, 6*, 32979–33001.
4. Frustaci, M., Pace, P., Aloi, G., & Fortino, G. (2017). Evaluating critical security issues of the IoT world: Present and future challenges. *IEEE Internet of Things Journal, 5*(4), 2483–2495.
5. Premkumar, M. S., Manikannan Kaliyaperumal, M., Abirami, A., Bhuvaneswari, M. S., Sahu, D. N., & Kasireddy, I. (2022). Intelligent IoT communication system based on block chain. *Annals of Forest Research, 65*(1), 3888–3900.
6. Yang, G., Jiang, M., Ouyang, W., Ji, G., Xie, H., Rahmani, A. M., Liljeberg, P., & Tenhunen, H. (2017). IoT-based remote pain monitoring system: From device to cloud platform. *IEEE Journal of Biomedical and Health Informatics, 22*(6), 1711–1719.
7. Gharaibeh, A., Salahuddin, M. A., Hussini, S. J., Khreishah, A., Khalil, I., Guizani, M., & Al- Fuqaha, A. (2017). Smart cities: A survey on data management, security, and enabling technologies. *IEEE Communications Surveys & Tutorials, 19*(4), 2456–2501.
8. Eckhoff, D., & Wagner, I. (2017). Privacy in the smart city—Applications, technologies, challenges, and solutions. *IEEE Communications Surveys & Tutorials, 20*(1), 489–516.
9. Xia, X., Xiao, Y., & Liang, W. (2018). Absi: An adaptive binary splitting algorithm for malicious meter inspection in smart grid. *IEEE Transactions on Information Forensics and Security, 14*(2), 445–458.
10. Namboodiri, V., Aravinthan, V., Mohapatra, S. N., Karimi, B., & Jewell, W. (2013). Toward a secure wireless-based home area network for metering in smart grids. *IEEE Systems Journal, 8*(2), 509–520.

11. Dlamini, N. N., & Johnston, K. (2016). The use, benefits and challenges of using the Internet of Things (IoT) in retail businesses: A literature review. In *2016 International Conference on Advances in Computing and Communication Engineering (ICACCE)*. IEEE (pp. 430–436).
12. Jose, A. C., & Malekian, R. (2017). Improving smart home security: Integrating logical sensing into smart home. *IEEE Sensors Journal, 17*(13), 4269–4286.
13. Mondal, S., Mukherjee, S., Banerjee, S. (2022). Machine learning based malicious node detection in IoT environment. In *Advanced Techniques for IoT Applications: Proceedings of EAIT 2020*. Springer (pp. 316–326).
14. Hassija, V., Chamola, V., Saxena, V., Jain, D., Goyal, P., & Sikdar, B. (2019). A survey on IoT security: Application areas, security threats, and solution architectures. *IEEE Access, 7*, 82721–82743.
15. Liao, C. -H., Shuai, H. -H., & Wang, L. -C. (2018). Eavesdropping prevention for heterogeneous internet of things systems. In *2018 15th IEEE Annual Consumer Communications & Networking Conference (CCNC)*. IEEE (pp. 1–2).
16. Li, C., & Chen, C. (2011). A multi-stage control method application in the fight against phishing attacks. In *Proceeding of the 26th Computer Security Academic Communication Across the Country* (p. 145).
17. Kolias, C., Kambourakis, G., Stavrou, A., & Voas, J. (2017). DDoS in the IoT: Mirai and other botnets. *Computer, 50*(7), 80–84.
18. Bandyopadhyay, S., Sengupta, M., Maiti, S., & Dutta, S. (2011). A survey of middleware for internet of things. In *Recent Trends in Wireless and Mobile Networks: Third International Conferences, WiMo 2011 and CoNeCo 2011* (pp. 288–296). Ankara, Turkey, June 26–28, 2011. Proceedings. Springer.
19. Zhang, Q., & Wang, X. (2009). SQL injections through back-end of RFID system. In *2009 International Symposium on Computer Network and Multimedia Technology* (pp. 1–4). IEEE.
20. Dorai, R., & Kannan, V. (2011). Sql injection-database attack revolution and prevention. *International Journal of Commercial Law & Technology, 6*, 224.
21. Razzaque, M. A., Milojevic-Jevric, M., Palade, A., & Clarke, S. (2015). Middleware for Internet of Things: A survey. *IEEE Internet of Things Journal, 3*(1), 70–95.
22. Kumar, J., Rajendran, B., Bindhumadhava, B., & Babu, N. S. C. (2017). Xml wrapping attack mitigation using positional token. In *2017 International Conference on Public Key Infrastructure and its Applications (PKIA)* (pp. 36–42). IEEE.
23. Stanciu, A., Balan, T. -C., Gerigan, C., & Zamfir, S. (2017). Securing the IoT gateway based on the hardware implementation of a multi pattern search algorithm. In *2017 International Conference on Optimization of Electrical and Electronic Equipment (OPTIM) & 2017 International Aegean Conference on Electrical Machines and Power Electronics (ACEMP)* (pp. 1001–1006). IEEE.
24. Cha, S. -C., Chen, J. -F., Su, C., & Yeh, K. -H. (2018). A blockchain connected gateway for BLE-based devices in the Internet of Things. *IEEE Access, 6*, 24639–24649.
25. Swamy, S. N., Jadhav, D., & Kulkarni, N. (2017). "Security threats in the application layer in IoT applications. In *2017 International Conference on i-SMAC (IoT in Social, Mobile, Analytics and Cloud) (i-SMAC)* (pp. 477–480). IEEE.
26. Abdul-Ghani, H. A., Konstantas, D., & Mahyoub, M. (2018). A comprehensive IoT attacks survey based on a building-blocked reference model. *International Journal of Advanced Computer Science and Applications, 9*(3).
27. Miller, D. (2018). Blockchain and the Internet of Things in the industrial sector. *IT Professional, 20*(3), 15–18.
28. Aste, T., Tasca, P., & Di Matteo, T. (2017). Blockchain technologies: The foreseeable impact on society and industry.
29. Henry, R., Herzberg, A., & Kate, A. (2018). Blockchain access privacy: Challenges and directions. *IEEE Security & Privacy, 16*(4), 38–45.
30. Novo, O. (2018). Blockchain meets IoT: An architecture for scalable access management in IoT. *IEEE Internet of Things Journal, 5*(2), 1184–1195.

31. Lv, P., Wang, L., Zhu, H., Deng, W., & Gu, L. (2019). An IoT-oriented privacy-preserving publish/subscribe model over blockchains. *IEEE Access, 7*, 41309–41314.
32. Javaid, U., Aman, M. N., & Sikdar, B. (2018). Blockpro: Blockchain based data provenance and integrity for secure IoT environments. In *Proceedings of the 1st Workshop on Blockchain-enabled Networked Sensor Systems* (pp. 13–18).
33. Valtanen, K., Backman, J., & Yrjölä, S. (2019). Blockchain-powered value creation in the 5g and smart grid use cases. *IEEE Access, 7*, 25690–25707.
34. Javaid, U., Siang, A. K., Aman, M. N., & Sikdar, B. (2018). Mitigating loT device based DDoS attacks using blockchain. In *Proceedings of the 1st Workshop on Cryptocurrencies and Blockchains for Distributed Systems* (pp. 71–76).
35. Sharma, P. K., Chen, M.-Y., & Park, J. H. (2017). A software defined fog node based distributed blockchain cloud architecture for IoT. *IEEE Access, 6*, 115–124.
36. Yu, Y., Li, Y., Tian, J., & Liu, J. (2018). Blockchain-based solutions to security and privacy issues in the Internet of Things. *IEEE Wireless Communications, 25*(6), 12–18.
37. Javaid, U., Aman, M. N., & Sikdar, B. (2019). Drivman: Driving trust management and data sharing in vanets with blockchain and smart contracts. In *2019 IEEE 89th Vehicular Technology Conference (VTC2019-Spring)* (pp. 1–5). IEEE.
38. Dinh, T. T. A., Liu, R., Zhang, M., Chen, G., Ooi, B. C., & Wang, J. (2018). Untangling blockchain: A data processing view of blockchain systems. *IEEE Transactions on Knowledge and Data Engineering, 30*(7), 1366–1385.
39. He, D., Chan, S., & Guizani, M. (2018). Security in the Internet of Things supported by mobile edge computing. *IEEE Communications Magazine, 56*(8), 56–61.
40. Alphand, O., Amoretti, M., Claeys, T., Dall'Asta, S., Duda, A., Ferrari, G., Rousseau, F., Tourancheau, B., Veltri, L., & Zanichelli, F. (2018). IoTchain: A blockchain security architecture for the internet of things. In *2018 IEEE Wireless Communications and Networking Conference (WCNC)* (pp. 1–6). IEEE.
41. Ni, J., Zhang, K., Lin, X., & Shen, X. (2017). Securing fog computing for Internet of Things applications: Challenges and solutions. *IEEE Communications Surveys & Tutorials, 20*(1), 601–628.
42. Sehgal, V. K., Patrick, A., Soni, A., & Rajput, L. (2015). Smart human security framework using internet of things, cloud and FoG computing. In *Intelligent Distributed Computing* (pp. 251–263). Springer.
43. Sarkar, S., & Misra, S. (2016). Theoretical modelling of fog computing: A green computing paradigm to support IoT applications. *Iet Networks, 5*(2), 23–29.
44. Varghese, B., Wang, N., Nikolopoulos, D. S., & Buyya, R. (2020). *Feasibility of FoG computing* (pp. 127–146). Handbook of Integration of Cloud Computing.
45. Yi, S., Li, C., & Li, Q. (2015). A survey of fog computing: Concepts, applications and issues. In *Proceedings of the 2015 Workshop on Mobile Big Data* (pp. 37–42).
46. Alrawais, A., Alhothaily, A., Hu, C., & Cheng, X. (2017). Fog computing for the Internet of Things: Security and privacy issues. *IEEE Internet Computing, 21*(2), 34–42.
47. Pavani, K., & Damodaram, A. (2013). Intrusion detection using MLP for manets.
48. Kulkarni, R. V., & Venayagamoorthy, G. K. (2009). Neural network based secure media access control protocol for wireless sensor networks. In *2009 International Joint Conference on Neural Networks* (pp. 1680–1687). IEEE.
49. Xiao, L., Yan, Q., Lou, W., Chen, G., & Hou, Y. T. (2013). Proximity-based security techniques for mobile users in wireless networks. *IEEE Transactions on Information Forensics and Security, 8*(12), 2089–2100.
50. Xiao, L., Xie, C., Chen, T., Dai, H., & Poor, H. V. (2016). A mobile offloading game against smart attacks. *IEEE Access, 4*, 2281–2291.
51. Xiao, L., Li, Y., Han, G., Liu, G., & Zhuang, W. (2016). Phy-layer spoofing detection with reinforcement learning in wireless networks. *IEEE Transactions on Vehicular Technology, 65*(12), 10037–10047.
52. Ozay, M., Esnaola, I., Vural, F. T. Y., Kulkarni, S. R., & Poor, H. V. (2015). Machine learning methods for attack detection in the smart grid. *IEEE Transactions on Neural Networks and Learning Systems, 27*(8), 1773–1786.

53. Shi, C., Liu, J., Liu, H., & Chen, Y. (2017). Smart user authentication through actuation of daily activities leveraging WiFi-enabled IoT. In *Proceedings of the 18th ACM International Symposium on Mobile Ad Hoc Networking and Computing* (pp. 1–10).
54. Xiao, L., Wan, X., Lu, X., Zhang, Y., & Wu, D. (2018). Iot security techniques based on machine learning: How do IoT devices use ai to enhance security? *IEEE Signal Processing Magazine, 35*(5), 41–49.
55. Xiao, L., Wan, X., & Han, Z. (2017). Phy-layer authentication with multiple landmarks with reduced overhead. *IEEE Transactions on Wireless Communications, 17*(3), 1676–1687.
56. Yan, Z., Zhang, P., & Vasilakos, A. V. (2014). A survey on trust management for internet of things. *Journal of network and computer applications, 42*, 120–134.
57. Li, C., & Wang, G. (2012, June). A light-weight commodity integrity detection algorithm based on Chinese remainder theorem. In *2012 IEEE 11th international conference on trust, security and privacy in computing and communications* (pp. 1018–1023). IEEE.
58. Alias, N. A., & Radzi, N. H. M. (2016). Fingerprint classification using support vector machine. In *2016 Fifth ICT International Student Project Conference (ICT-ISPC)* (pp. 105–108). IEEE.
59. Oulhiq, R., Ibntahir, S., Sebgui, M., & Guennoun, Z. (2015). A fingerprint recognition framework using artificial neural network. In *2015 10th International Conference on Intelligent Systems: Theories and Applications (SITA)* (pp. 1–6). IEEE.
60. Premsankar, G., Di Francesco, M., & Taleb, T. (2018). Edge computing for the Internet of Things: A case study. *IEEE Internet of Things Journal, 5*(2), 1275–1284.
61. Guo, H., Liu, J., Zhang, J., Sun, W., & Kato, N. (2018). Mobile-edge computation offloading for ultra-dense IoT networks. *IEEE Internet of Things Journal, 5*(6), 4977–4988.

Security and Privacy Issues in Cloud and IoT Technology and Their Countermeasures

K. P. Bindu Madavi, Panditi Neelaveni, Pujari Rakesh, and Singamsetty Asish

Abstract One of the most popular technologies in use today is cloud computing. It offers several data storage and access services and uses the internet rather than your computer's hard drive for programs that execute. People can use SaaS, PaaS, IaaS, and other benefits of cloud computing even though many cloud computing providers, like Google, Microsoft, IBM, and Amazon, are moving toward adopting cloud technology, which is causing a noticeable increase in the utilization of various cloud services. There are still numerous difficulties that users must overcome. IoT services with guaranteed security and sequestration are necessary for the real-time relinquishment and perpetration of similar systems for various operations, including smart metropolises, smart grids, smart homes, and smart surroundings. IoT devices produce, process, and communicate vast quantities of data sensitive to privacy and safety. In this way, security has become a major concern that expects many enterprises to use its services. This chapter examines major obstacles and the steps that must be taken to overcome them.

Keywords Cloud computing · Security · IoT · Privacy

K. P. Bindu Madavi (✉) · P. Neelaveni · P. Rakesh · S. Asish
GITAM University, Bengaluru 561203, India
e-mail: bkambam@gitam.edu

P. Neelaveni
e-mail: npanditi@gitam.in

P. Rakesh
e-mail: rpujari@gitam.in

S. Asish
e-mail: asingams@gitam.in

A. Prasad et al. (eds.), *Communication Technologies and Security Challenges in IoT*, Internet of Things, https://doi.org/10.1007/978-981-97-0052-3_6

Introduction

Cloud computing has completely changed how organizations function by offering a scalable, economical, and flexible computing environment. Although cloud computing has numerous benefits, security is still a major concern for businesses considering the switch. Organizations must be vigilant about security risks since sensitive data is often housed on servers owned and operated by third parties. Cloud computing and Internet of Things (IoT) technology are common place in today's digital environment. They provide many advantages, including simple data access and seamless system and device connectivity. But there are also potential privacy and security risks associated with convenience. Sensitive data transmitted by IoT devices and stored in the cloud is susceptible to hacking attempts and security breaches. Steganography makes data security more complex, which involves concealing information within other information. Thus, it is crucial to put into place efficient defenses against these dangers. The security and privacy risks connected with cloud computing and IoT technologies will be examined in this article, focusing on data saved in the cloud and steganography. We will also go through several countermeasures that can be used to reduce these risks.

Security in IOT Technology

The technological revolution, the Internet of Things (IoT), has gained steam in recent years. It refers to a network of physical objects, such as machines, vehicles, buildings, and other things, including sensors, software, connections, and other characteristics that enable them to collect and exchange data. Different sectors use IoT technology, changing how people live, work, and engage with the environment. Here are a few instances of how IoT technology is now being deployed.

- Smart homes: IoT technology automates various tasks, such as turning on lights, adjusting the thermostat, and locking doors. Smart home devices can be controlled remotely through a mobile app or voice commands.
- Healthcare: IoT technology is utilized to manage medication adherence, keep track of patients' vital signs, and notify professionals in an emergency. Wearable devices such as smartwatches and fitness trackers are some examples of IoT technology in healthcare.
- Manufacturing: IoT technology is used in manufacturing to improve productivity, quality, and safety. Connected machines can be monitored remotely, and data can be analyzed to identify trends and patterns that can help optimize operations.
- Energy: IoT technology is used to optimize energy consumption and reduce waste. Smart grid technology can monitor energy usage and adjust supply to meet demand, reducing the need for fossil fuels and many more.

The enormous amount of data that these gadgets produce, however, also poses serious security issues. IoT devices gather and send a lot of data, including sensitive company information, financial data, and personal information. This information is frequently kept on the cloud, susceptible to hackers and online assaults. As a result, protecting IoT devices and their data is essential to preserving people's and organization's security and privacy. The consequences of a security breach can be severe. For example, a hacker could access personal data, such as home addresses or credit card numbers, or take control of connected devices, such as smart thermostats or security cameras. These breaches can also have significant financial implications for businesses and damage their reputation and customer trust.

Security Importance in Cloud

- Data Breaches: Data breaches are a significant concern for businesses and individuals. When sensitive data is stored on cloud servers, it is exposed to potential security threats such as data breaches. Cloud service providers often have robust security measures to protect their customer's data. However, these measures may not be sufficient for sophisticated attacks. Therefore, businesses must take additional measures to safeguard their data.
- Compliance Requirements: Businesses operating in various industries must comply with industry-specific regulations and laws. For illustration, healthcare associations must misbehave with the Health Insurance Portability and Responsibility Act(HIPAA), while financial organizations must comply with the Sarbanes-Oxley Act (SOX). These regulations require organizations to implement specific security measures to safeguard sensitive data. Businesses must ensure that cloud service providers comply with these regulations when using cloud computing.
- Service Disruption: Another significant security concern is service disruption. In the event of a security breach, cloud service providers may shut down services to mitigate the breach's impact. This can result in service disruption for businesses using the cloud. Service disruption can severely impact businesses, leading to financial losses, reputation damage, and even legal issues.
- Data Loss: Data loss is a significant concern for cloud businesses. The loss of sensitive data can lead to financial losses, reputation damage, and even legal issues. Cloud service providers often have backup and recovery mechanisms to protect their customer's data.

Related Work

Data security stored in the cloud has become crucial as cloud services gain popularity. Now, businesses and people seek ways to safeguard their cloud-based data. The methods used to protect data in the cloud will be covered in this article.

- Encryption: Data is transformed into a code through encryption, making it impossible for anyone without the key to decoding it to read. In the cloud, we use encryption to protect the data at rest and in transit. Symmetric and asymmetric encryption are two types in it.
- Access control: Access control is the process of controlling who can access data stored in the cloud. We can achieve it through the use of authentication and authorization mechanisms. Only authorized users can access the data, thanks to access control systems.
- Security monitoring: Security monitoring is the process of monitoring the cloud environment for any security threats. Security information and event management (SIEM) systems may be accomplished using intrusion detection and prevention systems (IDPS).
- Multifactor authentication: Multifactor authentication (MFA) is an authentication process that requires users to provide two or more verification factors to access data. This can include a password, a fingerprint, or a token.

Symmetric and asymmetric algorithms and hash functions are combined for sophisticated cryptographic processes in real-world systems. This is regarded as a cold-blooded plan. Because each family of algorithms has unique strengths and weaknesses, both families are used. Although symmetric ciphers are far quicker than asymmetric ciphers—at least a thousand times faster—they need all participants to consistently participate in the key [1].

Steganography

One of the finest techniques for hiding data, such as messages, among less offensive data or information, such as pictures, audio files, or video files, is steganography. This technique is used for various reasons, such as maintaining the confidentiality of sensitive information, preventing data tampering, and circumventing censorship. With the increasing popularity of cloud-based services, steganography has become a required tool for safeguarding data privacy in the cloud. The eighth bit of a picture, the least significant bit, is modified to include a secret message. With a 24-bit image, one may store three bits in each pixel by changing one bit for each red, green, and blue color component, each represented by a byte. Hence, an image with a resolution of 800 by 600 pixels may include 1,440,000 bits, or 180,000 bytes, of encoded data [2]. One of the key advantages of steganography in the cloud is that it provides an additional layer of security that complements existing security measures. Steganography can assist in limiting unwanted access to sensitive data by encrypting data and concealing it within photographs or other media assets. Moreover, it can aid in preventing data theft or loss. Users can easily store and transfer sensitive information without raising suspicion by hiding data within images or other media files.

Least Significant Bit (Steganography)

The name" steganography" comes from the Greek word" stegos," which means" to conceal." Combining them can result in covered writing. Furthermore, supported by a good steganography system should be encryption. The immunity of concealed information to operations such as loss compression and steganography protects embedded information from scaling, interference, alteration, or removal. Text, image, and audio/ video steganography are the three different forms [3]. Steganography is widely used for hiding data since it is quite secure. A simple and safe way to transfer information over the internet [4].

LSB and MSB are two types of image steganography:

- LSB (Least Significant Bit): LSB embeds data onto covers (Least Significant Bit). The least significant bit of each pixel is changed to indicate that the message will be hidden [3].
- MSB (Most Significant Bit): The MSB (Most Significant Bit) is considered in this method. The image's pixels' MSB bits hide data in each pixel. A picture contains a small secret message [3].

The simplest and most direct method of embedding the message bits in the cover image's least significant bits is known as an LSB-based technique. Steganography hides a hidden message using the least significant bits of digital data on the cover medium, or LSB Steganography. Two techniques for LSB steganography are LSB replacement and LSB matching [5].

There are different stages performed in the LSB technique [2]:

Stage 1: Converting the image to the matrix.
Stage 2: Embedding image.
Stage 3: Converting a matrix to image.
Stage 4: Extraction.

The simplest method for encoding data in a digital audio file is Least Significant Bit (LSB) coding. By substituting a binary message for each sample point's least significant bit in the data to be encoded, LSB coding offers a high degree of flexibility [6]. This technique substitutes a Ms bit for a byte's LSB. Information in photos may be effectively concealed with this technique. To human sight, both the carrier image and the stego image will be identical. The LSB (Least Significant Bit) technique is often employed to conceal information inside the pictures. An image file is nothing more than a file instructing a computer to show different colors and lighting levels on other picture areas. A 24 Bit BMP (Bitmap) picture is the ideal image file for information concealment. Concealing information inside an image with high quality and resolution is simpler. However, 24 Bit graphics work best for information concealment due to their size. Some individuals could choose 8 Bit BMPs or another picture format, like GIF. The rationale is that publishing huge photos online can raise suspicions [7]. This least significant steganography technique is widely accepted since it is straightforward to comprehend, simple to apply, and produces stego-images that conceal data while appearing to have excellent visual fidelity. The time complexity varies with the file sizes we utilize for data encryption. The recommended technique simultaneously places eight bits of secret data in the LSB of the carrier frame's RGB (Red, Green, and Blue) pixel values in the proper 3, 3, and 2 sequences. The R and G pixels receive the message's remaining six (06) bits, while the B pixel only receives two (02) bits [8]. Steganography is used for various purposes, including copyright, intellectual property rights, and military and industrial uses (IPR). It is possible to send and receive messages safely using lossless steganography techniques [8].

Steganography Techniques

DCT

Signal processing uses the DCT (Discrete Cosine Transform) technique to transform signals from the time domain to the frequency domain. DCT is frequently employed in image and video compression techniques like JPEG and MPEG to lessen the data needed to describe an image or video. The DCT operates on a block of data, typically 8×8 pixels in the case of images. The block is divided into smaller sub-blocks, each transformed independently. Each sub-blocks DCT coefficients represent the image's frequency components within that sub-block. The coefficients are sorted in order of significance, and the most significant coefficients are retained while the less significant ones are discarded. The retained coefficients are used to reconstruct the compressed image. The process of applying DCT to an image involves the following steps:

- Partitioning: Little, non-overlapping pieces of pixels, usually 8×8 or 16×16 pixels, are used to divide up the image.
- Transformation: Each pixel block is transformed from the spatial domain to the frequency domain using DCT.
- Quantization: These frequency coefficients produced by the DCT are then quantized by dividing each coefficient by a predetermined value or quantization parameter. This process results in a loss of information, as some of the high-frequency coefficients are discarded.
- Encoding: The quantized frequency coefficients are then encoded using a lossless or lossy compression algorithm to produce a compressed image.

Advantages:

- High compression ratio: DCT is a fantastic choice for storing and transferring many photos since it can achieve high compression ratios while keeping decent image quality.
- Low computational complexity: Due to its effectiveness, the DCT method is appropriate for real-time applications like video streaming.

- Robustness: The compressed image is more resistant to noise and transmission errors than the original image due to the quantization of the frequency coefficients.

Disadvantages:

- Loss of information: The quantization process in DCT leads to a loss of data, which can result in reduced image quality.
- Blocking artifacts: Due to partitioning the image into blocks, DCT can introduce blocking artifacts, which can be visible in the compressed image.
- Difficulty handling high-frequency content: DCT is less effective at compressing high-frequency range, such as sharp edges, resulting in lower image quality in areas with high-frequency content.

DWT

Signal processing uses the DWT (Discrete Wavelet Transform) approach to convert signals from the time domain to the frequency domain. DWT is frequently used in steganography, concealing messages in images and other multimedia files, and image and video compression techniques. The DWT operates on a signal by breaking it into wavelets, each representing a different frequency component. The DWT can analyze a signal at different resolutions, allowing it to detect small details. The DWT is employed in steganography to hide a message inside an image's wavelet coefficients. To encrypt the message, the DWT's least significant coefficients are changed. The alterations to the image are less visible to the naked eye since the low-frequency signal components are related to the least essential coefficients.

Steps:

- Partitioning: The image is divided into discrete, small-pixel blocks, generally 2n × 2n in size, where n is an integer.
- Transformation: To change a block of pixels from the spatial to the frequency domain, DWT is applied to each block of pixels. Decomposing the picture into a collection of wavelet coefficients with various frequency scales and orientations is required.
- Quantization: The wavelet coefficients produced by the DWT are quantized by dividing each coefficient by a predetermined value or quantization parameter. This process results in a loss of information, as some of the coefficients are discarded.
- Encoding: The quantized wavelet coefficients are then encoded using a lossless or lossy compression algorithm to produce a compressed image.

Advantages:

- High compression ratio: DWT is useful for storing and transferring vast photos because it can achieve significant compression ratios while preserving decent image quality.
- Good frequency localization: DWT can provide good frequency localization, meaning it can represent frequency components with high precision and capture localized features in the image.
- Scalability: DWT supports scalable compression, where the compressed image can be decoded at different resolutions or quality levels.

Disadvantages:

- Complexity: DWT can be computationally intensive, making it less suitable for real-time applications such as video streaming.
- The complexity of implementation: Implementing DWT can be challenging and requires high technical expertise.
- Difficulty handling high-frequency content: DWT is less effective at compressing high-frequency content, such as sharp edges, resulting in lower image quality in areas with high-frequency content.

Hashing

A hash algorithm, usually called a hash function, is a mathematical operation that takes as its input (or "message") and outputs a fixed-length string of characters frequently referred to as a hash value or hash code. A hash algorithm's main objective is to convert arbitrary length input data into a fixed-size, often significantly shorter output. Here are some specifics regarding hash algorithms:

- Deterministic: A hash algorithm will always generate the same output from the same input. No matter the input size, a hash algorithm's output always has a predetermined length.
- Pre-image Resistance: Determining the original input from the hash value should be computationally impossible.
- Collision Resistance: Finding two different inputs with the same hash result should be computationally impossible.
- Avalanche Effect: A slight modification to the input ought to produce a noticeably altered hash value.

Advantages:

- Data Integrity: Hash algorithms provide a reliable and efficient way to ensure data integrity. By comparing the hash value of received data with the expected hash value, one can quickly determine if the data has been altered or corrupted.

- Efficient Data Retrieval: Hash algorithms are commonly used in data structures like hash tables and maps to generate unique keys or indexes for efficient data storage and retrieval.
- Password Security: Hash algorithms play a crucial role in password security. Storing hashed passwords instead of plain text helps protect user credentials in case of a data breach. Hashed passwords make it computationally tricky for attackers to reverse-engineer the original passwords.
- Digital Signatures: Hash algorithms are used in digital signature schemes to create a unique hash value of a document or message. This hash value is then encrypted with the sender's private key, producing a digital signature that can be used for authentication and ensuring the integrity of the signed content.

Disadvantages:

- Collision Vulnerability: While hash algorithms strive to be collision resistant collisions can still occur. In some cases, finding two distinct inputs that result in the same hash value is computationally feasible. This poses a security risk in certain scenarios.
- Pre-image Attacks: Hash algorithms are susceptible to pre-image attacks, wherein an attacker tries to find an input that matches a given hash value. While modern hash algorithms aim to be pre-image resistant, advances in computing power and attacks may weaken their security over time.
- Inherent One-Way Function: Hash algorithms are designed to be one-way functions, meaning it is computationally difficult to determine the original input from the hash value. While this is desirable for security purposes, it can also be a disadvantage when the original data needs to be recovered.
- Vulnerability to Quantum Computing: Quantum computers, if developed at a large enough scale, have the potential to break many of the commonly used hash algorithms, including the widely adopted SHA-2 and SHA-3 families. This raises concerns about the long-term security of hash algorithms in the face of quantum computing advancements.

RSA Algorithm

Key generation, which creates the public and private keys, and RSA function evaluation, which examines Encryption and Decryption, are the two phases of the RSA algorithm, commonly known as asymmetric Encryption [9]. Consider PGP, a free program by Phil Zimmerman that uses RSA as its key transportation technique to offer encryption and authentication for email and file storage applications across several platforms, as one example of where RSA is now being utilized [10]. Attacks can be made against the RSA implementation even if it is safe.

- Forward search attack: By encrypting all messages before to the one that corresponds to C, the attacker may quickly decode C if the message space is known.

- Common modulus attack: In some cases, if everyone is given the same modulus, n, but a distinct (e, d) combination, it is possible to decode the message without d.
- Low encryption exponents: The output of me is always strictly less than the modulus n when encrypting with low encryption exponents (for instance, e = 3) and tiny values of the m (i.e., m n1 / e). By finding the eth root of the cipher text over the integers, it is simple to decode cipher communications in this situation.
- A property of the RSA algorithm is that the encrypted result of the two cipher texts is the same as the encrypted result of the corresponding plain texts. In light of this, m1em2e = (m1m2) e (mod n). This multiplicative property makes the prospect of a chosen ciphertext attack possible.

One of its key benefits is that the RSA method for cloud encryption allows for safe communication over an unstable network like the internet. When data is transported via a network, as it often does with cloud computing, it can be intercepted and read by hackers if it is not securely encrypted. Two keys are used by the RSA algorithm: a private key and a public key. Data may be encrypted and decrypted using public and private keys. The public key is freely distributable, so anybody may use it to encrypt data sent to the cloud. However, only the private key owner can decode the data, guaranteeing that only authorized individuals may view the data. Finally, the RSA algorithm is widely used and well-established, making it a reliable choice for cloud encryption. It has been extensively studied and tested, and most encryption libraries and tools support it. It can be easily integrated into cloud computing platforms and applications.

AES Algorithm

The Federal Information Processing Standards (FIPS 197), which contain the AES algorithm's definition, have established the AES algorithm as the industry standard for electronic protection. Joan Daeman and Vincent Rijimen used the Rijndael algorithm to create the AES algorithm, which conceals and reveals the message [11]. The AES algorithm, sometimes called the Rijndael algorithm, is a symmetrical block cipher that uses 128, 192, or 256-bit keys to transform a block of 128 bits of message into 128 bits of cipher text. This is a critical factor in why it is significantly stronger than the 56-bit key DES. The AES algorithm uses the SP, a substitution permutation network, to generate the cipher text over many rounds. The rounds will vary depending on the key length used [12]. The programs used in distributed systems and secure communication [11]. The following four phases make up each encryption cycle: Before replacing bytes, shift rows, mix columns, and add a round key. Each round of Decryption includes the following four steps: All of the next are employed: reverse replace bytes, reverse shift rows, reverse mix columns, and reverse substitute bytes [13].

Literature Review

Wang et al. [14] conducted a study on the issue in 2021. Data security concerns have drawn attention in the large data cloud computing context. For an organization to maintain data security and develop a stable big data cloud computing environment, it is vital to investigate and design appropriate security protection solutions. To employ the Information Protection and security groups services of Big Data Cloud Computing, the author provided a solution to identify the features and types of data security concerns in various phases and periods.

Sun [15] In 2019 study on maintaining security and privacy in cloud computing found that looking at the integration trend of access control, encryption, trust, and other technologies is crucial to strengthen privacy protection. Before you wrap up, give an overall assessment of the problems, challenges, and potential future directions. It was also discovered that to promote the healthy and orderly growth of cloud computing services, and organizations must now take the appropriate safeguards to ensure consumer privacy and declare their privacy policies, such as P3P privacy statements or privacy seal programs.

Zhang et al. [16] carried out research on the subject in 2011. He said we must apply several technologies to secure the data in the cloud. The three components of these technologies are permission, storage protection, and transfer protection. Moreover, he provided a framework for guaranteeing data security in cloud storage systems. Furthermore, it was stated that users should respect authorities to prevent unauthorized access. The storage provider must complete user authorization and access control.

Joshi et al. [17] carried out research into the problem in 2021. By talking about SaaS, PaaS, IaaS, and Data Security, he has conducted a study on the effects of security in cloud computing, including all of its difficulties. He also discussed the numerous security concerns related to cloud computing and proposed solutions, including encryption, masking, virtual private network, and firewall in systems that support the cloud.

Lee et al. [18] carried out an examination of the problem in 2022. Data security, managed through encryption techniques, is one of the main issues with cloud computing, according to him. As a remedy, he suggested the Advanced Encryption Standard, one possible data encryption method (AES). They used Heroku as a cloud platform for their investigation, and to encrypt data on Heroku, they used AES.

AlKhamese et al. [19] carried out research on the subject in 2019. He conducted extensive research on the topic of data security in the cloud and put forth a solution after learning about Cloud Computing, Steganography, Cryptography, LSB, DCT, and information hiding. He concluded that these two security methods could be used in the cloud to protect the transmission of user data.

Ali et al. [20] a study was done in 2021 regarding the problem. On SDN, 5G, MEC, privacy, security, and multi-access edge computing, he has done research. To prevent targeted assaults, the author concluded that MEC providers should include numerous layers of security safeguards. Also included in his proposal was a method for securing data in the cloud using MEC security architecture, security, and privacy strategies, as well as an analysis of current and foreseeable problems, their effects, and solutions.

Nguyen and Le [21] conducted a study on the issue in 2021. Data may become a target of online attackers as cloud storage grows in popularity. Hence, end users need high confidentiality, integrity, and authentication to safeguard their data in the cloud. He offered suggestions for data security in cloud computing, including authentication and encryption methods.

Thabit et al. [22] analyzed the issue in 2021. While it is being sent and kept, the security of sensitive data is one of the major issues the digital world is currently confronting. This is seen as one of the main difficulties with cloud computing. This is a result of distributed system technology's rapid progress. He proposed that we develop a cutting-edge data security solution for cloud computing based on genomics techniques and logical-mathematical operations to ensure data security in the cloud.

Proposed Work

The proposed work includes the following processes, as shown in Fig. 1. The IOT devices like fitness bands, home and TV appliances, cars, and other gadgets are the first data sources. The input consists of a text file and an image file. The text file is encrypted using a Hybrid encryption algorithm, a combination of RSA (Rivest-Shamir-Adleman), UEcrypt, and AES algorithms. Steganography technique's Least Significant Bit algorithm is applied to encrypted files. Stego image, in which we will re-encrypt the encrypted file by calculating the pixels in the stego-image. To decrease the complexity of the encoding process, the LSB method has transformed the encrypted file into an image. The generated file has been encrypted, downloaded, and stored in the cloud. A receiver can download the encrypted stego-image and decrypt the image using the LSB technique. It gave us the output as an original text file. Since the text file is still encrypted. Using a hybrid encryption approach, we were able to decrypt the encrypted text file. Finally, we got the original text file as an encrypted output. The mentioned LSB algorithm 1 and 2 techniques were used at the advanced hybrid encryption and decryption of files i.e. once we encrypted the file using RSA + AES + UEcrypt we again encrypt the file using LSB so that the file is more secure and difficult to access without authorization.

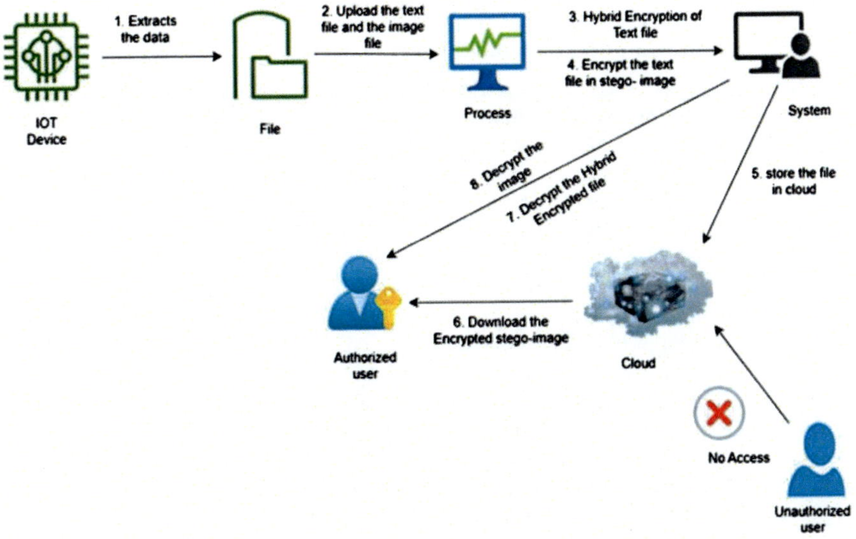

Fig. 1 Proposed architecture

Algorithm 1 LSB Encryption Technique

```
Input: secret message, cover image
secret_message_binary = convert_to_binary(secret_message)
secret_message_size = size_in_bits(secret_message _binary)
for i do each pixel in cover_image:
    red_binary = convert_to_binary(pixel.red)
    green_binary = convert_to_binary(pixel.green)
    blue_binary = convert_to_binary(pixel.blue)
for i do from 0 to secret_message_size - 1:
    red_binary[i] = secret_message_binary[i]
    green_binary[i] = secret_message_binary[i+secret_message_size]
    blue_binary[i] = secret_message_binary[i+2*secret_message_size]
save_stego_image()
```

Algorithm 2 LSB Decryption Technique

```
String secretKey = AEScryption.generateKeyAES();
String encData1 = aes.encryptAES(data);
EncryptDataRSA = new EncryptDataRSA();
String encKey1 = Base64.getEncoder().encodeToString(encKey2);
System.arraycopy(encKey2, 0, result, 0, len1);
System.arraycopy(encData2, 0, result, len1, len2);
String resultStr = Base64.getEncoder().encodeToString(result);
return resultStr;
byte[] result = Base64.getDecoder().decode(resultStr);
System.arraycopy(result, 0, encKey2, 0, len1);
System.arraycopy(result, len1, encData2, 0, len2);
DecryptDataRSA decryptDataRSA = new DecryptDataRSA();
String secretKey = decryptDataRSA.decryptData(encKey2);
AEScryption aes = new AEScryption(secretKey);
String data = aes.decryptAES(encData1);
return data;
```

Results and Discussion

This research paper suggests a method for the hybrid encryption notion. This technique employs several encryption techniques, including RSA (Rivest-Shamir-Adleman), UEcrypt, and AES (Advanced Encryption Standard). The text file will first be encrypted using hybrid Encryption (RSA + AES + UEcrypt). Eventually, we'll use stegano-image to encode the encrypted text file. We compare the algorithm's performance using various stego-image sizes while considering factors such as time complexity, throughput, and data rate for all sizes. The approach we developed is much more effective than others since it combines hybrid Encryption and LSB, a total of 4 separate methods, making it hard for an attacker to steal the data. Other algorithms might only use hybrid encryption or only use LSB. As the file was encrypted twice and with several encryption algorithms, they could not obtain the information even if the attacker managed to decipher one algorithm through trial and error.

Encryption Time:

In this Fig. 2, we have shown the encryption time taken for different file sizes. As you can see, if file size increases, the time required to encrypt the file also increases, resulting in a positive correlation between file size and encryption time. This relationship is likely not linear but exponential or logarithmic on the encryption algorithm.

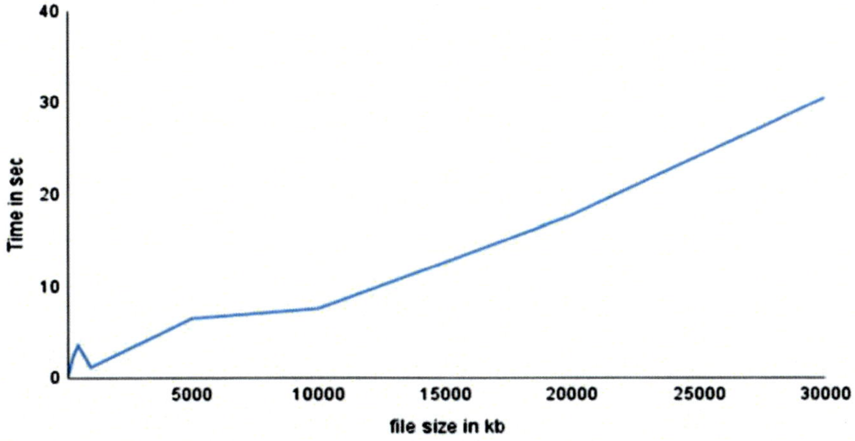

Fig. 2 Encryption time for different file sizes

Encoding Throughput:

$$Throughput = \frac{size\ of\ input\ file}{Time\ taken\ to\ encode} \tag{1}$$

The x-axis represents the different file sizes, while the y-axis represents the throughput of encryption, which is the amount of data that can be transmitted or processed per unit of time, as shown in Fig. 3. As file size increases, the throughput may decrease due to system processing capabilities limitations or the available network bandwidth. Depending on the specific system used, this relationship is likely not linear but exponential or logarithmic.

Encoding Data Rate:

$$Datarate = \frac{Throughput}{Payload\ size} \tag{2}$$

The Fig. 4 show how much data has been encrypted per second based on the payload size. The x-axis represents the different file sizes, while the y-axis represents the data rate of encryption, which is the rate at which data is encrypted per second. Equations 1 and 2 using to calculate the throughput and data rate. As file size increases, the data rate of encryption may decrease due to limitations in the encryption algorithm, the available processing power, or the available network bandwidth. The relationship may vary based on the system configurations.

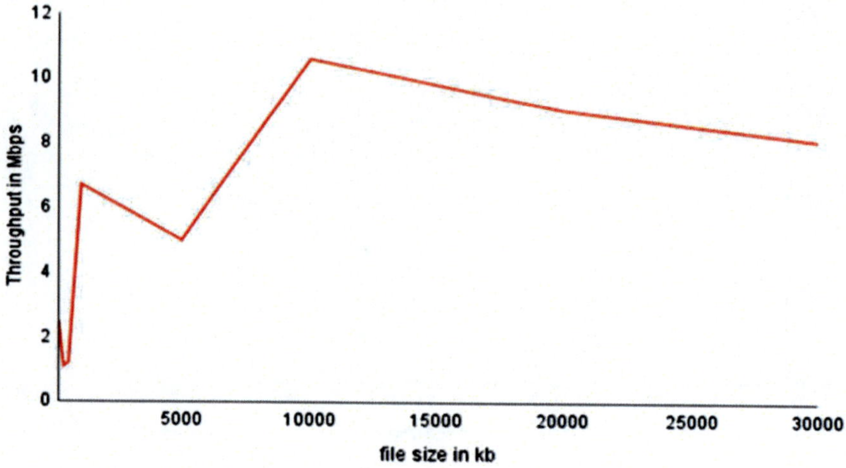

Fig. 3 Throughput for encryption

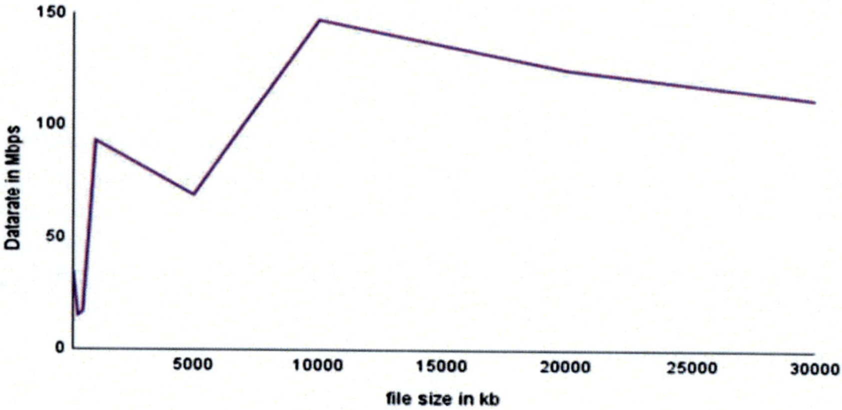

Fig. 4 Data rate for encryption

Decryption Time:

In this Fig. 5, we have shown the decryption time taken for different file sizes. As you can see, if file size increases, the time required to decrypt the file also increases, resulting in a positive correlation between file size and decryption time.

Decoding Throughput:

The x-axis represents the different file sizes, while the y-axis represents the throughput of decryption, which is the amount of data that can be decrypted per unit of time, as shown in Fig. 6. As file size increases, the throughput of decryption may decrease due to limitations in the decryption algorithm, the available processing

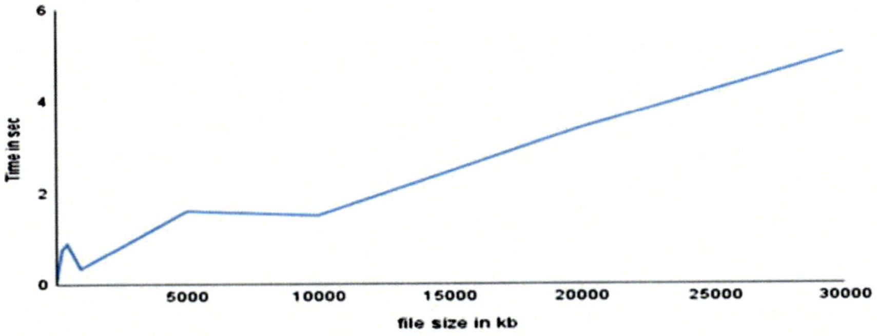

Fig. 5 Decryption time for different file sizes

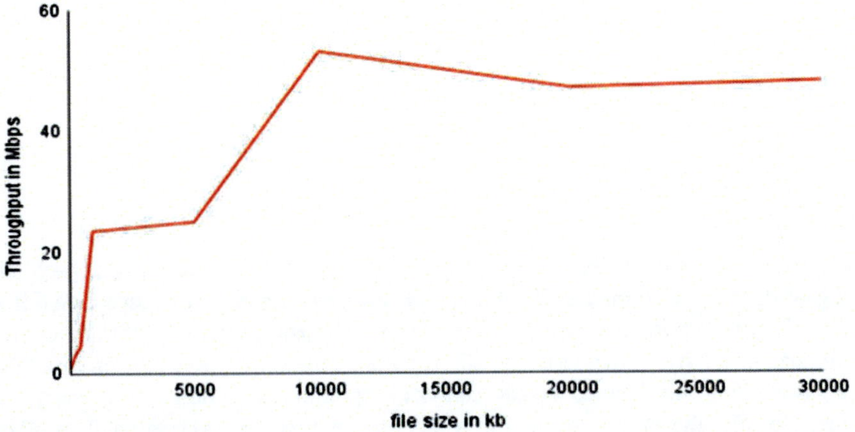

Fig. 6 Throughput for decryption

power, or the available network bandwidth. Depending on the specific system used, this relationship is likely not linear but exponential or logarithmic. By optimizing the decryption algorithm or the hardware and software used for decryption, it may be possible to improve the throughput and increase the amount of data that can be decrypted per unit of time.

Decoding Data Rate:

The Fig. 7 show how much data is decrypted per second based on the payload size. The x-axis represents the different file sizes, while the y-axis represents the data rate of encryption, the rate at which data is decrypted per second. As file size increases, the data rate of decryption may decrease or remain constant due to limitations in the encryption algorithm, the available processing power, or the available network bandwidth. The relationship may vary based on the system configurations.

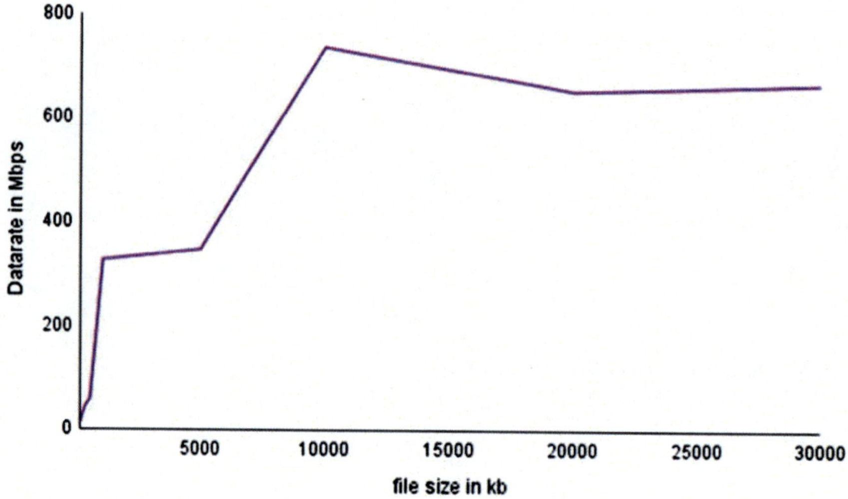

Fig. 7 Data rate for decryption

Comparison Graphs:

See (Fig. 8).

Various algorithms have been extensively employed to address the problem regarding their performance. The Fig. 8 illustrates a comparison of encryption times for different algorithms across file sizes. The x-axis denotes the algorithm name and file size in kilobytes, while the y-axis represents the encryption time in seconds. By examining this Figure, we can analyze each algorithm's duration to encrypt a file. Notably, the proposed algorithm takes more time for encryption than other

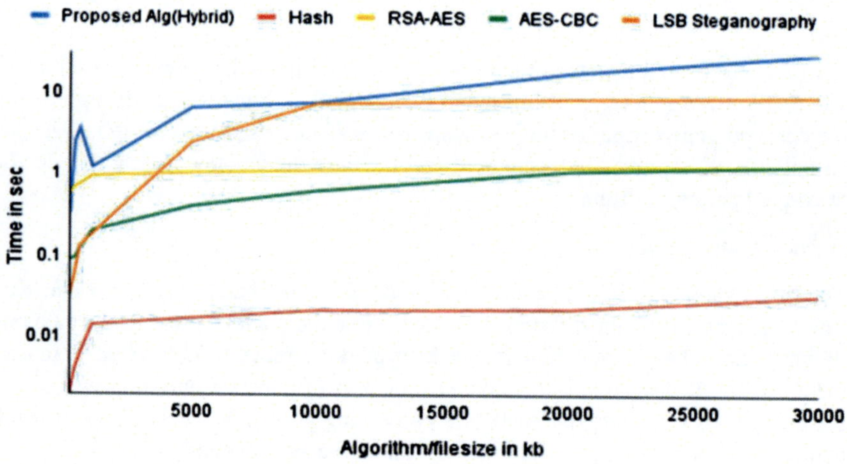

Fig. 8 Comparison of encryption time

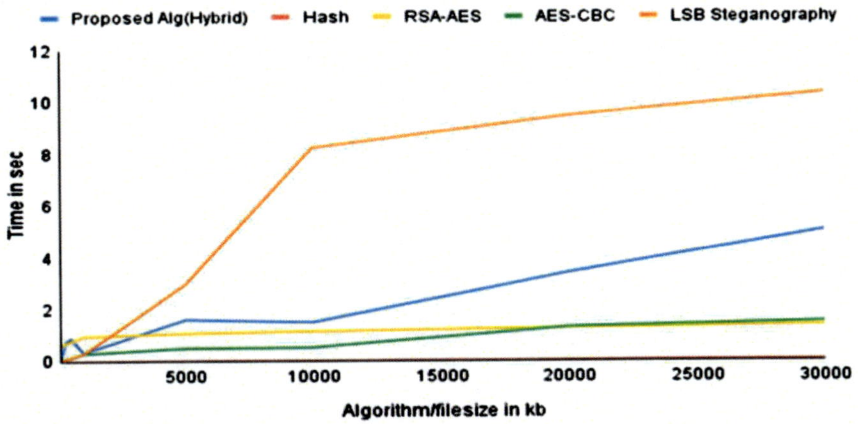

Fig. 9 Comparison of decryption time

algorithms. However, this extended duration may indicate better security than the different algorithms.

Figure 9 compares decryption times for different algorithms across varying file sizes. This Figure enables us to analyze the duration required for an individual to decrypt the data. Notably, the Hash Algorithm takes more time for decryption than other algorithms.

In Figs. 10 and 11, we have compared the throughput values for different algorithms. The x-axis represents the algorithm names and file sizes in kilobytes, while the y-axis indicates the throughput. It is important to note that the throughput may vary based on system configurations. By examining this Figure, we can analyze the throughput, which refers to the rate at which a system or process can complete a task or deliver a product within a specified time frame.

When evaluating algorithmic solutions, the performance and accuracy of various algorithms hold significant importance. In Figs. 12 and 13, a comparison of data rates for different algorithms across different file sizes has been depicted. The x-axis denotes the algorithm names and file size in kilobytes, while the y-axis represents the data rate. By examining Figs. 12 and 13, we can analyze the data rate, which signifies the speed at which digital data is transmitted or processed over a communication channel or data storage medium.

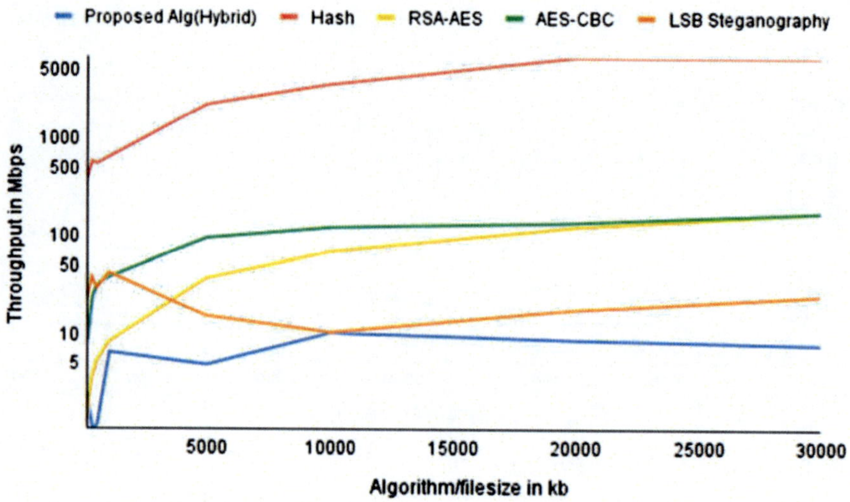

Fig. 10 Comparison of encryption throughput

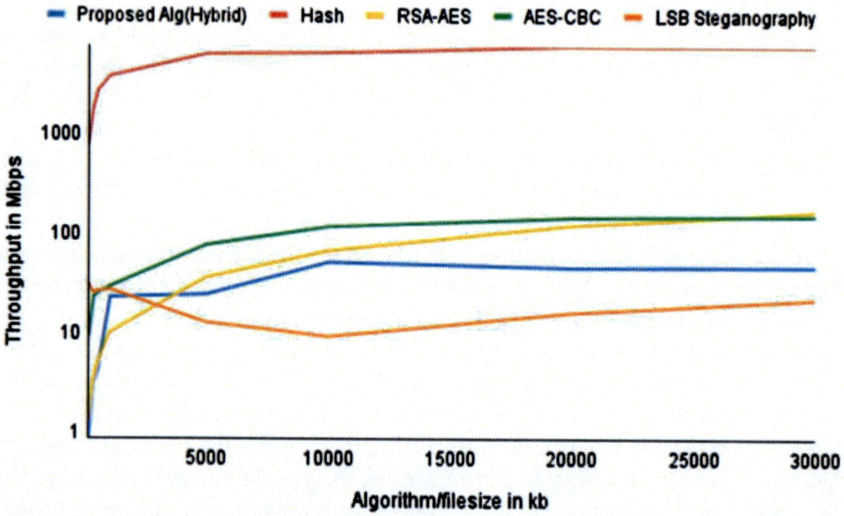

Fig. 11 Comparison of decryption throughput

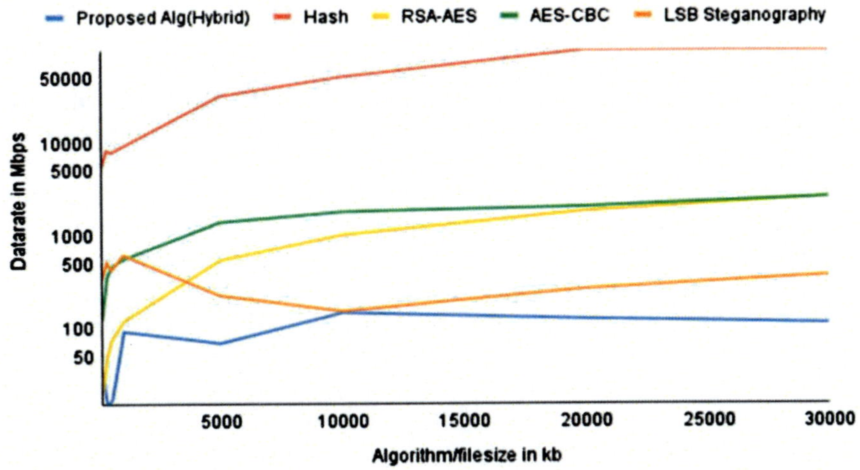

Fig. 12 Comparison of encryption data rate

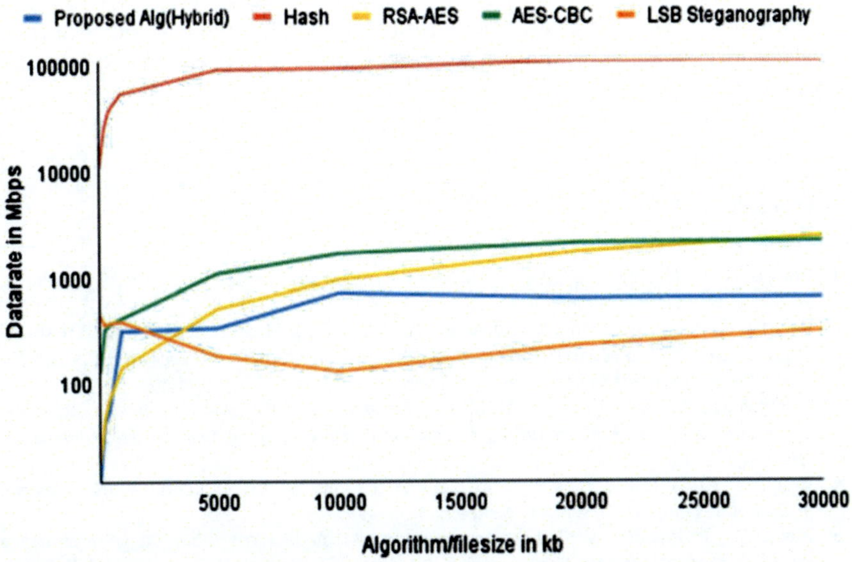

Fig. 13 Comparison of decryption data rate

Conclusion and Future Work

A method is suggested for the hybrid encryption notion in this research study. This method uses several encryption techniques, including UEcrypt, RSA (Rivest-Shamir-Adleman), AES (Advanced Encryption Standard), and LSB approaches. The text file will first be encrypted using hybrid Encryption (RSA + AES + UEcrypt). Eventually, we'll use Stegano-image to encode the encrypted text file. We compare the algorithm's performance using various stego-image sizes while considering factors such as time complexity, throughput, and data rate for all sizes. The approach we developed is much more effective than others since it combines hybrid Encryption and LSB, a total of 4 separate methods, making it hard for an attacker to steal the data. Other algorithms might only use hybrid encryption or only use LSB. As a file was encrypted twice and with several encryption algorithms, even if the attacker managed to decipher one algorithm through trial and error, there'll not be able to obtain the information. The comparative analysis of existing techniques concludes that the proposed algorithm provides enhanced security. This conclusion is based on the observation that the proposed algorithm exhibits a longer encryption time than other techniques. The extended encryption time indicates the implementation of additional security measures within the proposed algorithm, thus ensuring heightened data protection.

References

1. Francis, N. (2018). *An analysis of hybrid cryptographic approaches for information security, 13*, 124–127.
2. Rachael, O., Misra, S., Ahuja, R., Adewumi, A., Ayeni, F., & Mmaskeliunas, R. (2020). Image steganography and steganalysis based on least significant bit (lsb). In *Proceedings of ICETIT 2019: Emerging trends in information technology*, (pp. 1100–1111). Springer.
3. Thangadurai, K., & Sudha Devi, G. (2014). An analysis of lsb based image steganography techniques. In *2014 International conference on computer communication and informatics*, (pp. 1–4). IEEE.
4. Tyagi, V. (2012). Image steganography using least significant bit with cryptography. *Journal of Global Research in Computer Science, 3*(3), 53–55.
5. Rawat, D., & Bhandari, V. (2013). A steganography technique for hiding image in an image using lsb method for 24-bit color image. *International Journal of Computer Applications, 64*(20), 15–19.
6. Nehru, G., & Dhar, P. (2012). A detailed look of audio steganography techniques using lsb and genetic algorithm approach. *International Journal of Computer Science Issues (IJCSI), 9*(1), 402.
7. Gupta, S., Gujral, G., & Aggarwal, N. (2012). Enhanced least significant bit algorithm for image steganography. *IJCEM International Journal of Computational Engineering & Management, 15*(4), 40–42.
8. Aggarwal, A., Sangal, A., & Varshney, A. Image steganography using LSB algorithm.
9. Nisha, S., & Farik, M. (2017). Rsa public key cryptography algorithm. A review. *International Journal of Scientific and Technological Research*, 6, 187–191.
10. Preetha, M., & Nithya, M. (2013). A study and performance analysis of rsa algorithm. *International Journal of Computer Science and Mobile Computing, 2*(6), 126–139.

11. Sinha, A. K., & Jayaraj, N. (2015). Performance analysis of AES cryptographic algorithm. NCRTS.
12. Yehya, D., & Joudi, M. (2020). Aes encryption: Study & evaluation. no. November 2020.
13. Kak, A., & Mayes, K. (2014). Aes: The advanced encryption standard. In *Proceedings of IEEE international conference on computer and network security*, (pp. 140–156).
14. Wang, F., Wang, H., & Xue, L. (2021). Research on data security in big data cloud computing environment. In *2021 IEEE 5th advanced information technology, electronic and automation control conference (IAEAC)*, (vol. 5, pp. 1446–1450). IEEE.
15. Sun, P. J. (2019). Privacy protection and data security in cloud computing: A survey, challenges, and solutions. *IEEE Access*, 7, 147420–147452.
16. Zhang, X., Du, H., Chen, J., Lin, Y., & Zeng, L. (2011). Ensure data security in cloud storage. In *2011 International conference on network computing and information security*, (vol. 1, pp. 284–287). IEEE.
17. Joshi, M., Budhani, S., Tewari, N., & Prakash, S. (2021). Analytical review of data security in cloud computing. In *2021 2nd International conference on intelligent engineering and management (ICIEM)*, (pp. 362–366). IEEE.
18. Lee, B.-H., Dewi, E. K., & Wajdi, M. F. (2018). Data security in cloud computing using aes under heroku cloud. In *2018 27th wireless and optical communication conference (WOCC)*, (pp. 1–5). IEEE.
19. AlKhamese, A. Y., Shabana, W. R., & Hanafy, I. M. (2019). Data security in cloud computing using steganography: A review. In *2019 International conference on innovative trends in computer engineering (ITCE)*, (pp. 549–558). IEEE.
20. Ali, B., Gregory, M.A., & Li, S. (2021) Multi-access edge computing architecture, data security and privacy: A review. *IEEE Access*, 9, 18706–18721.
21. Nguyen, T. N., & Le, T. T. T. (2021). A comprehensive review of authentication and encryption algorithms for data security in cloud computing. RICE, 57–63.
22. Thabit, F., Alhomdy, S., & Jagtap, S. (2021). A new data security algorithm for the cloud computing based on genetics techniques and logical-mathematical functions. *International Journal of Intelligent Networks*, 2, 18–33.

Security Threats in IOT and Their Prevention

Ajay Prasad, Prachi Kapoor, and Thipendra P. Singh

Abstract IoT is making our day-to-day life easy by connecting us to the web which in fact enhance the standard of living. IOT plays a crucial part in our daily life ranging from small devices to massive devices. However, IoT is also quite susceptible to several threats. Due to the increased likelihood and ease with which malicious individuals can target IoT devices in order to satisfy their evil ambitions, IoT safety and confidentiality protection are of the prime importance. User's security gets compromised once an IoT layer is breached. Hackers might simply gain access through a compromised node. Moreover, viruses, hackers and malicious software may compromise information and data integrity, which will expose the entire IoT environment to information anxiety. Like the OSI layers, IoT layers are the best approach for understanding. This chapter describes different kinds of security issues pertaining each layer in IoT and discusses the means and methods to deal with these security threats.

Keywords IoT · Threats · IoT layers · Vulnerabilities · IoT security

Introduction

Due to the lack of a single agreement, numerous IoT architectures have recently been proposed. However, a generic and layered approach of looking at IoT is preferred. Figure 1 illustrates the four levels that make up the fundamental IoT architecture.

A. Prasad (✉)
School of Computer Science UPES, Dehradun, India
e-mail: aprasad@ddn.upes.ac.in

P. Kapoor
UPES, Dehradun, India

T. P. Singh
School of Computer Science Engineering and Technology, Bennett University, Greater Noida, India
e-mail: thipendra@gmail.com

A. Prasad et al. (eds.), *Communication Technologies and Security Challenges in IoT*, Internet of Things, https://doi.org/10.1007/978-981-97-0052-3_7

Fig. 1 Four Layers in IoT

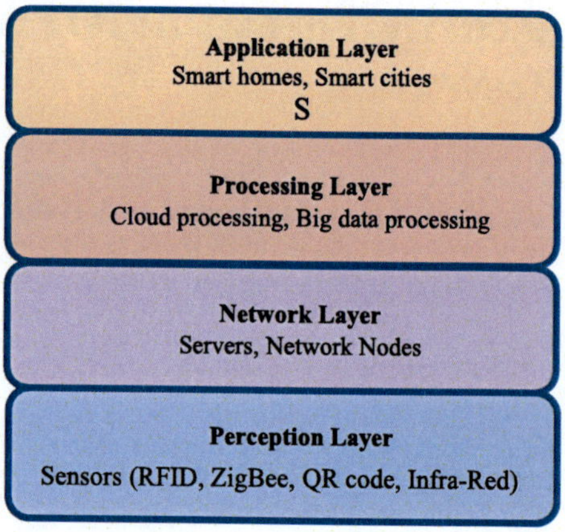

- Perception Layer (PL),
- Network Layer (NL),
- Processor Layer (PL), and
- Application Layer (AL).

Perception Layer (PL): Variety of sensors are included in this layer, including infrared, RFID, QR code, and ZigBee, to collect data on the environment, including humidity, temperature, pH level, pressure, and force [1].

Network Layer (NL): This layer consists of network communication software and physical parts that can be in charge of automatically passing information obtained from the perception layer's sensors to other layers [2].

Processing Layer (PL): This layer, often known as the middleware layer, processes, stores, and analyses a significant amount of data. This layer may automatically compute and process information and also offers various services to the lower layers. Several technologies, including big data processing and cloud computing, use the processing layer [1, 2].

Application Layer (AL): According to the user's needs, this layer offers the services. Smart homes, smart transportation, healthcare, utilities, smart hospitals etc. are together examples of typical IoT applications [1, 2].

Challenges in Internet of Things (IoT)

Internet of Things (IoT) gadgets are currently connecting with one another over internet to give users the data they need. Yet, tackling IoT deployments is not that simple because, in addition to security, there are several problems. In the sessions to follow, we will quickly outline some of the major obstacles [3].

Hardware: The Internet of Things (IoT) hardware is a growing field with a wide range of tough viewpoints. An IoT infrastructure is built and put together by a variety of sensors like light, temperature, or moisture, smart wearables devices for the arm, feet or head and conventional gadgets like smartphones and plamtops. Each of these devices imposes a certain set of requirements that must be met. Due to this diversity and the requirement to incorporate it under the aegis of a safe use, additional hardware-level security issues were brought up [4].

Cost: Standardisation suppresses incompatibilities, device non-integration lowers network reliability, data privacy requires advanced security strategies, energy exhaustion affects the operation, and hardware diversification all directly or indirectly affect and deter- mine cost. As a result, organisations with infrastructures for IOT can face number of issues, including cost estimation, in order to guarantee positive outcomes when making important decisions.

Integration: Lack of efficient standards has a significant impact on the integration of devices in communication networks, and IoT is no different. Internet of Things (IoT) communication seamless integration is harder to accomplish because of the many various languages of programming and a huge variety of distinct parts, utilised in the creation of IoT equipment. "Classic" communication connectivity can be challenging owing to a broad spectrum of accessible technologies making it hard to communicate effortlessly among tasking-vendor devices. With these kinds of conflicts, a network's dependability is drastically reduced, which makes communication unpredictable. Due to these problems, the industry has proposed many solutions like standardising protocols; however, these approaches left aside several mismatched hardware components.

Standardization: As the number of connected gadgets grows quickly, adversaries gain more access points into an infrastructure. Simply put, the area of attack expands when additional Internet of Things devices are connected to a network, increasing the number of devices that are susceptible to malware and compromising the safety of the network as a whole.

Another disadvantage is the inability to independently gather and transfer data and information to another device or network, despite the fact that the content may be crucial but is necessarily vulnerable. For instance, before utilising some IoT devices, users must agree to their terms of service. These agreements could divulge user information, making them vulnerable to attack.

Energy: Internet of Things (IoT) gadgets must effectively fend off an effort to undermine their own power economy. Mini or small ones typically rely on the capacity of a battery and well-charging capabilities with the required periodic servicing for their operation and effectiveness. Software is in charge of monitoring and controlling energy needs as well as continuously optimising energy consumption. However, hardware does not allow software to see energy use, which affects how software fails to correctly perform some tests. Due to power exhaustion, the gadget could then stop functioning. A interesting idea is to make software and hardware creation more energy transparent.

Security and Privacy: As the number of connected gadgets increases rapidly, enemies acquire a greater number of ways to get within a system. In plain English, the attack surface grows with each additional Internet of Things gadget linked to a network since a threat has now got a lot more gadgets that are vulnerable to malware, compromising the security of the entire network.

Furthermore, because the content may be critical but is unavoidably susceptible, the capacity to independently acquire and transfer data and information to different gadget or internet is also a problem. As an illustration, some IoT devices demand that customers accept their terms of service before using them. These kinds of agreements may reveal user information, leaving them open to attack.

Addressing the Vulnerabilities

Addressing the vulnerabilities in the Internet of Things (IoT) is crucial for ensuring the security and privacy of the connected devices and the data they handle. IoT devices are becoming more prevalent in various domains, such as smart homes, healthcare, industrial automation, and transportation. However, these devices often lack adequate security measures, making them vulnerable to exploitation by malicious actors [5]. Addressing these vulnerabilities requires a multi-faceted approach that involves various stakeholders, including manufacturers, users, policymakers, and security experts. Here are some key aspects to consider:

- Secure Design and Development
- Regular Software Updates
- Authentication and Authorization
- Encryption
- Secure APIs
- Secure Boot and Firmware Integrity
- Network Security
- Privacy by Design
- Penetration Testing and Vulnerability Assessments
- User Education and Awareness
- Regulations and Standards

Manufacturers should prioritize security from the very beginning of the device development process. This includes conducting thorough security assessments, following secure coding practices, and adhering to industry best practices and standards. Security should not be an afterthought but an integral part of the device's design. IoT devices must receive regular software updates to patch vulnerabilities and address security flaws. Manufacturers should establish mechanisms to deliver updates automatically and ensure the process is seamless for end-users. Strong authentication mechanisms, such as two-factor authentication, should be implemented to control access to IoT devices. Additionally, proper authorization mechanisms should limit access to specific functionalities and data based on user roles and privileges. Data transmitted between IoT devices and backend systems should be encrypted to prevent eavesdropping and data interception. Encryption protocols like TLS (Transport Layer Security) should be used to secure data in transit. Application Programming Interfaces (APIs) used to communicate with IoT devices should be designed with security in mind. API endpoints should be protected from unauthorized access and potential attacks like injection or cross-site scripting. Implementing secure boot mechanisms ensures that only trusted and authenticated firmware can be loaded onto the device. This helps prevent unauthorized modifications and tampering. IoT devices should be isolated within secure network segments and firewalled appropriately to limit unauthorized access. Network traffic should be monitored to detect any suspicious activities. Data privacy is crucial in IoT systems, especially when dealing with sensitive information. Manufacturers should follow the principle of "Privacy by Design" and minimize the collection of personally identifiable information (PII) to the essential requirements. Regularly conducting penetration testing and vulnerability assessments can help identify weaknesses in IoT devices and their ecosystem. This enables manufacturers to proactively fix issues before they are exploited by attackers. Users should be educated about IoT security risks and best practices for securing their devices. Simple steps like changing default passwords, keeping firmware up to date, and using secure Wi-Fi networks can make a significant difference. Governments and industry bodies should collaborate to establish regulations and standards for IoT security. Compliance with these standards should be enforced, and non-compliant devices should not be allowed in the market.

Addressing vulnerabilities in IoT requires a collaborative effort from all stakeholders involved in the IoT ecosystem. By prioritizing security in the design, development, and usage of IoT devices, we can create a safer and more reliable IoT landscape. The IoT layers are mostly independent and physically separated. Hence the security aspects also does need to be looked in layered fashion. Hereby let's look into each layer of IoT and the prevalent threats and the measures that can minimize them.

Security Threats in Perception Layer (PL)

Perception Layer (PL) (Fig. 2) is prone to hardware attack. PL contains a variety of sensors, including, ZigBee, RFID and WSN (Wireless sensor networks), that may be physically attacked if they remain in one location for a long period of time [4]. Many Internets of Things (IoT) gadgets, like smart wearables and video games, are gathering data about us. Some hackers may share or access this data for nefarious purposes [6]. The following is a description of common perception layer attacks: -

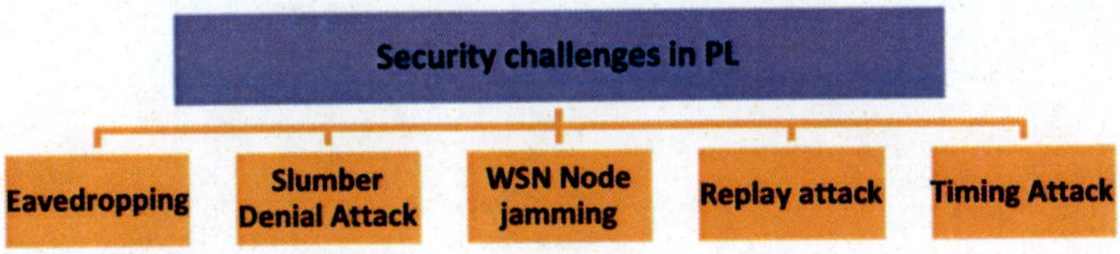

Fig. 2 Security challenges in PL

Eavesdropping

Eavesdropping attacks, additionally referred to simply sniffing attacks or spying attacks, are intrusions during which somebody tries to intercept the data gadgets transmit. To obtain access to the transmitted and received data, it uses an unsecured transfer method [7].

Slumber Denial Attack

In an IoT (Internet of things) communication network, nodes in remote locations are typically powered by disposable batteries; hence, these terminals are configured to sleep when not in use in order to preserve life of battery. By providing false input to the node, the attacker keeps it awake and prevents it from sleeping. This causes the node to use more power, which leads to a node power failure.

WSN (Wireless Sensor Network) Node Jamming

By crowding the WSN (wireless sensor network) signals or by adding some noise signals in the Web, an attacker might cause denial of service for the Internet of Things [8].

Replay Attack

In this assult when a third party eavesdrops on a transmission between a sender and a receiver and steals the sender's real information. Such an attack often takes place during the authentication procedure with the intention of destroying the validity certificates [9].

Timing Attack

Timing attacks work better on systems that have limited processing power. Through this kind of attack, a hacker can reveal vulnerabilities in security and obtain data kept in a system's safety by monitoring how quickly the system responds to various requests, the inputs, encryption techniques, and other actions.

Security Challenges in Network Layer (NL)

Following are the security challenges in Network Layer (NL) (Fig. 3).

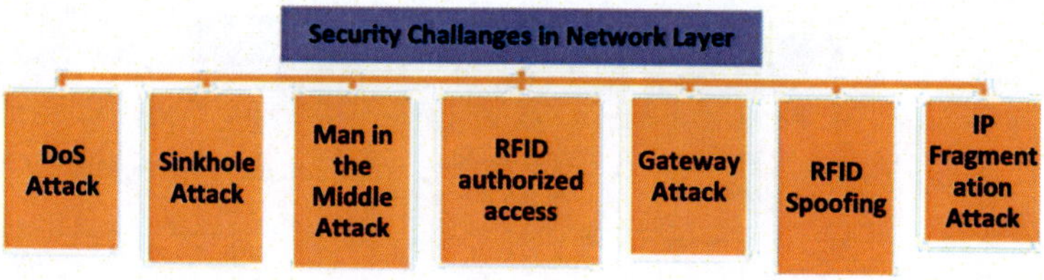

Fig. 3 Security Challanges in network layer

Sinkhole Attack

A sinkhole is created in the middle of the attack by the infected node luring all the traffic from its surrounding area. All of the neighboring node's data is drawn to the intruder or attacked node. The assailant makes an effort to project the image of being the neighbourhood's most appealing relay [10].

Man in the Middle Attack

In man of the Middle attack, the malicious user simply leverages the IoT communication protocol to intervene with the two sensor vertices in order to obtain the sensitive information without having to physically appear on the scene of a network [11].

RFID Authorized Access

Due to the absence of a safe authentication scheme in RFID systems, the tags can be accessed by anybody. This implies that tags are easily influenced.

Gateway Attack

In this attack, the link between the sensors and the internet infrastructure is severed. Routing attacks and Dos attacks in which wrong or nil information is transferred from the internet to nodes or sensors, are also included in this attack.

RFID Spoofing

An RFID spoofing attack targets the RFID signal in order to obtain the data encoded on the RFID tag. The hacker can gain complete access to the IoT system by utilizing this form of attack to transfer his own data by using the original id [9].

IP Fragmentation Attack

In this type of DoS attack, the attacker takes advantage of the Maximum Transmission Unit (MTU) of the internet connection. The internet fails because the size of reconstructed IP packets after transfer is greater than the maximum number of transmission units it can support.

DoS (Denial of Service) Attack

If we talk about attack in DoS, other gadgets and servers are not able to offer the service to the user. Due to a DoS attack, data transfer between sources and devices is stopped.

Security Threats in Processing Layer

Information is delivered over the cloud during this phase, making the cloud attack one of the layer's most exposed attacks. Some potential assaults on this layer include (Fig. 4):

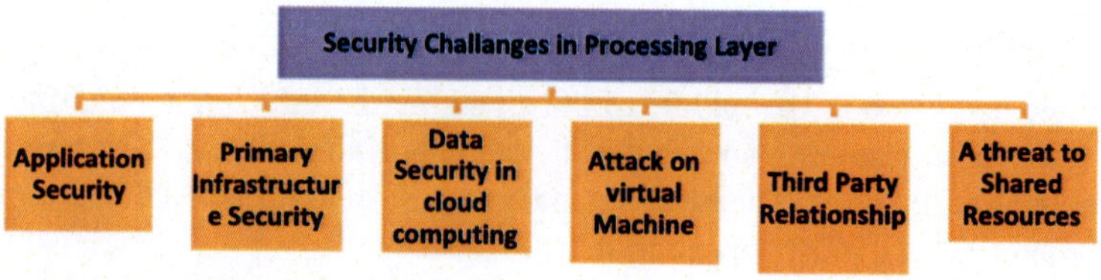

Fig. 4 Security challanges in processing layer

Primary Infrastructure Security

Manufacturers of the application always make efforts to make an application secure but its safety remains in jeopardy because of underlying layers of IoT which can be the wholesome responsibility of service giver.

Data Security in Cloud Computing

A significant security problem is the service provider's data backup. While plain text can be processed and saved in the cloud, it is the SaaS providers' duty to maintain data security so that unauthorized users cannot access sensitive information.

An Attack on Virtual Machine

Virtual Machine (VM) safety is of the prime significance, and any safety breach can result in the collapse of the whole IoT environment.

Third Party Relationships

Mashups, a third-party web service component offered by Platform as a Service (PaaS), raise the security risk because they integrate data from many sources [12].

A Threat to Shared Resources

There is always a security risk in the resources which are being shared because the malicious user can see all the shared items which are between the virtual machines and gathers the necessary information for its illicit purposes.

Application Security

Since web services are typically used to deploy cloud Software as a Service (SAAS) applications, an hijacker can easily use the internet to get control to an IoT network and steal data to further his bad intentions. The Open Web Application Security Project had found a number of security concerns with SAAS [13].

Security Threats in Application Layer

Due to security problems, the service can easily be shuttered as well as compromised. A malicious attack may cause a virus in the software code that activates the application

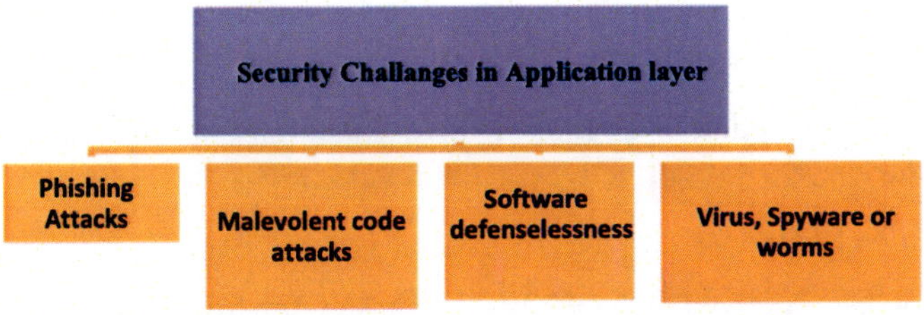

Fig. 5 Security challenges in application layer

to malfunction. Applications can fail to deliver authenticated services that they were intended to deliver or offer the service incorrectly. Following are some broad risks to this layer (Fig. 5).

Phishing Attacks

Attackers have access to restricted information by faking user confirmation identities in emails and on websites.

Software Defenselessness

Software vulnerabilities may increase due to programmers' use of non-standard coding. This technique is used by malicious users to satisfy their unethical urges.

Worms, Virus and Spyware

An malicious user can infect the system with malicious software, that can lead to data corruption, service denial, even theft of information.

Malevolent Code Attacks

This kind of attack might be a hostile "worm" that can target Internet-connected equipment like home routers and security cameras. This kind of cyberattack may disable a car's Wi-Fi and take over the steering wheel, leading to a serious accident [14].

Preventive Measures Taken to Protect the IOT Layers

The following safety precautions can be followed to protect the fundamental four levels of IoT in order to either completely remove or drastically decrease the possibility of malicious attacks by hackers.

Protective Measures for Perception Layer

To defend the perception layer from malicious attacks, the following precautions should be taken: -

Authentication of Devices

To keep hostile gadgets out of the IoT ecosystem and stop forged data from spreading throughout the network, objects must be authenticated prior joining the network.

Safe Booting

Cryptographic hash algorithms may be employed to verify the authenticity and integrity of the software on various IoT connected devices. WH and NH cryptographic algorithms are the best options to this challenge since many hash algorithms cannot be deployed on end network devices since these gadgets have relatively low computational capacity [15].

Integrity of Data

Every gadget used in the Internet of Things (IoT) system ought to include a fault monitoring mechanism, like a digital checksum, a parity bit, etc., to reduce the danger of data tampering. The encryption hash function ought to be employed to create a connected device that remains more safe.

Anonymity

By inserting a node into the IoT network, a hacker can conceal sensitive information like identification, the location, and more. The most effective approach for this issue is the K-anonymity strategy because it performs best on low-processing machines [16].

Secure Physical Designing of End Devices

The cyberattack on the perception layer could be stopped by building end hardware in a safe physical manner. The radio wave circuit, microchip choice, and other parts of gadgets have to be of top standard. For instance, a wireless antenna with a suitable design could enable long-distance transmission.

Preventive Measures for Network Layer

The following precautions should be taken into account to defend the network layer from the hackers attack.

Inside and Outside Data

The nodes of the IoT network may be secured against unauthorized access to preserve data confidentiality. For proof of identity, cryptography can be used between points.

Highly sensitive data is immediately converted into impenetrable cypher code during this process.

Secure Routing

As most routing protocols are not reliable, secure routing is essential for the safe use of sensors. To enable safe routing, information must be routed across many pathways that expose a network to more errors [17].

Integrity of Data

Utilizing an encrypted hash function on the data can assure data integrity by preventing tampering on the side that receives it. Furthermore, the challenge of moderation can be overcome by using error correction mechanisms.

Confidentiality of Data

The IoT network's nodes may be protected from unauthorized entry to maintain data confidentiality. You can use encryption from point to point for identification. During this procedure, sensitive information is quickly transformed into uncrackable cypher code.

Preventive Measures for Processing Layer

The subsequent defense strategies ought to be workable to protect the processor layer from any malicious attacks.

Web Firewall Application

Web firewall software should be used to defend the IoT environment since it can identify potential attackers.

Encryption to Secure Classified Information

Since the primary goal of encryption is to protect sensitive data, which is typically stored or transported to the cloud in encrypted form to prevent security lapses. Today, a variety of encryption techniques are being employed to safeguard the IoT environment and prevent side channel attacks.

Data Fragmentation Redundancy Scattering

With data fragmentation redundancy scattering, the risk of data escape is reduced because each piece of the data does not contain any crucial information. Classified data is split up into several fragments and kept on different servers.

Hyper Safe Lockdown

Hyper safe lockdown and protection against customization of the write-protected memory pages. Additionally, pointing indexes are restricted from converting data into pointer indexes.

Preventive Measures for Application Layer

To shield the application layer from any harmful attacks, the following security precautions should be taken.

User Validation

Reliability and encryption measures are important for a system's safety and confidentiality as any kind of information theft or unauthorized access to the IoT environment might result in a security breach [18, 19].

Risk Assessment Techniques

The application layer shall be secured by risk assessment since it can identify risks to the IoT system. To accomplish this, upgrade the system devices' firmware to strengthen the security mechanisms.

Use of Firewalls

Because a weak password makes it possible to crack both the authentication password and the encryption mechanism, firewalls that keep an eye on both incoming and outgoing network traffic should be utilized.

Use of Antispyware, Anti-Virus or Anti-Adware

All of this software is essential to maintaining the IoT environment's protection, consistency, secrecy, and dependability.

Special Policies and Permissions

For controlling and accessing the IoT framework, specific policies and permissions may be followed. The rules that make up policies are similar to the vocabulary used in XACML, they are composed of a condition and an effect (such as allow or deny). Access control lists provide for the authorization or restriction of outgoing /incoming traffic as well as access requests to the system.

Security Threats in Traditional Network vs Internet of Things (IOT)

Following is the list of differences between Internet of things (IOT) and traditional network safety issues is provided under this section.

Resources

The typical conventional network comprises of a desktop or laptop, servers, and smartphone with abundant resources, but the IoT (Internet of Things) system is made up of FRID and WSN nodes with insufficient resources. Therefore, users can combine lightweight and complicated algorithms to increase safety in a traditional network while using less processing effort. Nevertheless, in the IoT, only lightweight algorithms can be used to preserve the balance among security and processing capacity.

Data Formats

There is not an operating system in the Web of Things, only a straightforward embedded programmed for the chip, although different devices connected to the internet can share data thanks to abstractions of operating systems like Windows or UNIX. The IoT environment's many nodes result in a variety of chip hardware and varied data formats [20].

Communication Ways

The Internet of Things uses wireless connections to connect its nodes, which compromises safety. On the World Wide Web, the majority of interaction is done through quicker and more secure wireless or wired connections. In contrast to IoT nodes, which have limited assets, wireless links are constructed on top of sophisticated secure protocols in mobile internet.

Risk Factors

Internet of Things (IoT) platforms have an increased danger factor than classical networks because a large amount of IoT applications are employed in day to day routine and provide more danger to security if they are lost control over. On the contrary hand, in a conventional network, users cannot be attacked by a malevolent party to obtain their secret information for his illegal purposes if they do not share their own classified information.

Conclusion

IoT has gained appeal to every lifestyle and the interest of researchers over the past few years, comparable to other revolutionary technologies. It additionally presents a number of security and privacy hazards, though. IoT has gained appeal to every lifestyle and the interest of researchers over the past few years, comparable to other revolutionary technologies. It additionally presents a number of security and privacy hazards, though. To stop and protect the IoT system from the security terrorization, preventive steps to safeguard these levels are also recommended. IoT systems face numerous security issues because of resource constraints, increased risk factors, and varied data formats, according to a comparison of safety issues between a standard network and IoT (Internet of things) environment. The aspects presented in the chapter has been summarized in a tabulated form in the Tables 1 and 2.

Table 1 Security Threats in various IoT layers

Security threats in perception layer	Security threats in network layer	Security threats in processing layer	Security threats in application layer
Eve dropping	DoS Attack	Application security	Phishing Attacks
Slumber Denial Attack	Sinkhole Attack	Primary Infrastructure Security	Malevolent code attacks
WSN Node jamming	Man in the Middle Attack	Data security in cloud computing	Software defenselessness
Replay attack	RFID Authorized Access	Attack on virtual machine	Virus, Spyware, or worms
Timing Attack	Gateway Attack	Third Party Relationship	
	RFID spoofing	A threat to shared resources	
	IP Fragmentation Attack		

Table 2 Preventive measures in various IoT layers

Preventive measures for perception layer	Preventive measures for network layer	Preventive measures for processing layer	Preventive measures for application layer
Authentication of devices	Inside and outside data	Web firewall application	User validation
Safe Booting	Secure Routing	Encryption to secure classified information	Risk Assessment techniques
Integrity of data	Integrity of data	Data fragmentation redundancy scattering	Use of firewalls
Anonymity	Confidentiality of data	Hyper safe lockdown	Use of Antispyware, Anti -Virus or anti adware
Secure physical designing of end devices			Special policies and permissions

References

1. Rao, T. A., & Haq, E. U. (2018). Security challenges facing IoT layers and its protective measures. *International Journal of Computer Applications, 975*, 8887.
2. Ren, Z., Liu, X., Ye, R., & Zhang, T. (2017). Security and privacy on internet of things. *2017 7th IEEE International Conference on Electronics Information and Emergency Communication (ICEIEC), Macau, China*, pp. 140–144, https://doi.org/10.1109/ICEIEC.2017.8076530.
3. Sharbaf, M. S. (2022). IoT driving new business model, and IoT security, privacy, and awareness challenges. *2022 IEEE 8th World Forum on Internet of Things (WF-IoT)*, Yokohama, Japan, pp. 1-4,https://doi.org/10.1109/WF-IoT54382.2022.10152044
4. Gerodimos, A., & Maglaras, L. (2023). Mohamed Amine Ferrag. *Nick Ayres, Ioanna Kantzavelou, IoT: Communication protocols and security threats, Internet of Things and Cyber-Physical Systems, 3*, 1–13.
5. Bandi, C., Salehi, S., Hassan, R., Homayoun, S. M. P D, H., & Rafatirad, S. (2021). Ontology-driven framework for trend analysis of vulnerabilities and impacts in IoT hardware. *2021 IEEE 15th International Conference on Semantic Computing (ICSC), Laguna Hills, CA, USA*, pp. 211–214, https://doi.org/10.1109/ICSC50631.2021.00045.
6. Ali Khattak, H., Ali Shah, M., Khan, S., Ali, I., & Imran, M. (2019). Perception layer security in internet of things. *Future Generation Computer Systems, 100*, 144–164, ISSN: 0167-739X.
7. Burhan, M., Rehman, R. A., Khan, B., & Kim, B. S. (2018). IoT elements, layered architectures and security issues: A comprehensive survey. *Sensors, 18*(9), 2796.
8. Farooq, M., Waseem, M., Khairi, A., & Mazhar, S. (2015). A critical analysis on the security concerns of internet of things (iot). *International Journal Computer Application, 111*(7).
9. Ahmed, M. M., Shah, M. A., & Wahid, A. (2017). IoT security: A layered approach for attacks & defenses. In *IEEE International Conference on Communication Technologies (ComTech)*, pp. 104–110.
10. Tumrongwittayapak, C., & Varakulsiripunth, R. (2009). Detecting Sinkhole attacks in wireless sensor networks. In *2009 ICCAS-SICE*, IEEE, pp. 1966–1971.
11. Mohite, S., Kulkarni, G., & Sutar, R. et al. (2013). Rfid security issues. *International Journal Engineering Research Technology, 2*(9).
12. Hashizume, K., Rosado, D. G., Fernandez-Medina, E., & Fernandez, E. B. (2013). An analysis of security issues for cloud computing. *Journal of Internet Services and Applications*, pp. 1–13.

13. Razzaq, A., Latif, K., Ahmad, H. F., Hur, A., Anwar, Z., & Bloodsworth, P. C. (2014). Semantic security against web application attacks. *Information Sciences, 254*, 19–38.
14. Kumar, S. A., Vealey, T., & Srivastava, H. (2016). Security in internet of things: challenges, solutions and future directions. *49th Hawaii International Conference on System Sciences (HICSS)*, pp. 5772–5781.
15. Avoine, G., Bingol, M. A., Carpent, X., & Yalcin, S. B. O. (2013). Privacy-friendly authentication in RFID systems: on sublinear protocols based on symmetric-key cryptography. *IEEE Transactions on Mobile Computing, 12*(10), 2037–2049.
16. Vasilomanolakis, E., Daubert, J., Luthra, M., Gazis, V., Wiesmaier, A., & Kikiras, P. (2015). On the security and privacy of internet of things architectures and systems. *IEEE International Workshop on Secure Internet of Things (SIoT)*, pp. 49–57.
17. Agrawal, P., Ghosh, R. K., & Das, S. K. (2008). Cooperative black and gray hole attacks in mobile ad hoc networks. In *Proceedings of the 2nd International Conference on Ubiquitous Information Management and Communication*, ACM, pp. 310–314.
18. Newsome, J., Shi, E., Song, D., & Perrig, A. (2004). The sybil attack in sensor networks: analysis & defenses. In *Proceedings of the 3rd International Symposium on Information Processing in Sensor Networks*, ACM, pp. 259–268.
19. Singh, M., Rajan, M. A., Shivraj, V. L., & Balamuralidhar, P. (2015). Secure MQTT for internet of things (iot), in. *Fifth International Conference on Communication Systems and Network Technologies, 2015*, 746–751.
20. Jing, Q., Vasilakos, A. V., Wan, J., Lu, J., & Qiu, D. N. (2014). Security of the internet of things: Perspectives and challenges. *Wireless Networks, 20*(8), 2481–2501.

IPV6: Strengths and Limitations

Kingsley Igulu, Friday Onuodu, and Thipendra P. Singh

Abstract The landscape of the internet is ever-changing, and it is crucial to remain informed about the most recent developments. Among these advancements is the integration of Internet Protocol version 6 (IPv6). IPv6 represents the latest iteration of the internet protocol, enabling seamless communication between devices across the internet. Its goal is to improve upon its predecessor's (IPV4) weaknesses, by offering a more expansive address space, enhanced security measures, and improved overall performance. As the internet continues to grow, IPv6 becomes increasingly important, and many sites are now supporting it. In order to support the widespread adoption of 5G communications and the Internet of Things (IoT) in various industries, it is crucial to have a flexible IP technology that can accommodate a large number of addresses and ensure unrestricted connectivity. To meet these demands, the networking industry has initiated a global effort to transition to Internet Protocol version 6 (IPv6). Major governments such as those of the United States of America and China will announce new IPv6 laws and standards. Over 2.5 billion Internet users are currently taking advantage of IPv6 without even being aware of it. While the United States has over 183 million IPv6 users, India has over 519 million, giving it a penetration rate of 76%. Brazil reached 117 million with penetration rate of 42%. In comparison, Germany has more than 32 million people while Japan has 43 million. There are certain countries where the IPv6 penetration rate has above 60%. The enterprise sector is responsible for the remaining 40% of IPv6 adoption, which is necessary for the ultimate switch to an IPv6-only internet. This will make it possible to deprecate the IPv4 internet, which was recently recommended by the United States government, and will result in a reduction in the need to maintain two separate internets. Since the end of 2017, the government of China has made significant efforts

K. Igulu (✉)
Department of Computer Science, Kenule Beeson Saro-Wiwa Polytechnic, Bori, Nigeria
e-mail: igulukt@gmail.com; igulu.kingsley@kenpoly.edu.ng

F. Onuodu
Department of Computer Science, University of Port Harcourt, Port Harcourt, Nigeria

T. P. Singh
School of Computer Science Engineering and Technology, Bennett University, Greater Noida, India
e-mail: thipendra@gmail.com

A. Prasad et al. (eds.), *Communication Technologies and Security Challenges in IoT*, Internet of Things, https://doi.org/10.1007/978-981-97-0052-3_8

to accelerate the deployment of IPv6 across the country, and owing to the participation of Chinese operators, significant headway has been made in this endeavor. According to the latest data, the number of IPv6 users in China has surpassed 760 million. Chinese clients can now access IPv6 connection services, leading to a significant increase in the overall number of IPv6 users. With these tremendous benefits or strength of IPV6, IPV6 is also not immune to limitations. This chapter focuses on the strengths and weaknesses of IPV6. It does give good background of IPv6 and its transitioning intricacies.

Keywords Internet protocol · IPv6 · IoT · Transmission protocol · IP addresses · Security

Introduction

IPv6 is the name for the newest version of the Internet Protocol. It is designed to be an improvement over its predecessor, IPv4, which has been in operation for more than three decades now. The fundamental goal of IPv6 is to solve the problem of address depletion that IPv4 is currently experiencing as a direct result of the tremendous expansion of the internet. This problem has arisen as a direct result of the fact that IPv4 cannot keep up with the exponential growth of the internet. IPv6 features a 128-bit address space, as opposed to the 32-bit address space of IPv4. Because of this, IPv6 can support a significantly larger number of distinct IP addresses, which paves the way for the possibility of providing a one-of-a-kind address to each individual device that connects to the internet. Additionally, IPv6 has improved security features, including IPsec, which is built into the protocol. It also has improved support for mobile devices and auto-configuration, making it easier to set up new devices on the network.

The introduction of IPv6 has brought about many benefits, some of which include improved network performance, better security, and simplified network management. IPv6's larger address space means that it can support more devices on the network, reducing the need for complex network address translation (NAT) configurations. This, in turn, reduces latency and improves network performance. IPv6 also has better support for multicast traffic, which is essential for multimedia applications such as video conferencing and online gaming.

Despite its many strengths, IPv6 also has some weaknesses. One of the main challenges is the lack of support from some legacy devices and applications that are still built around IPv4. This means that organizations that want to adopt IPv6 need to invest in new hardware and software, which can be costly. Additionally, IPv6's auto-configuration feature can be a double-edged sword as it can make it easier for unauthorized devices to connect to the network, compromising security.

History of IPv6

While TCP was initially used to manage packet delivery and routing, TCP/IP eventually split the protocol into distinct operating layers. This separation gave way to the formal creation of IPv4 in 1983, which has since become a foundational element of modern networking. However, as the internet has evolved and expanded, the limitations of IPv4 have become increasingly apparent. This is where IPv6 comes in, offering expanded address space and improved security features. Though it has been slow to gain widespread adoption, many experts believe that IPv6 will be critical for the continued scaling and evolution of the internet. And with TCP still serving as a key component of networking protocols, it's clear that the evolution of these technologies will continue to shape the digital landscape for years to come [1].

The process by which IP addresses are assigned is one of the characteristics of IPv4 that is of the utmost significance. For computers, mobile devices (like smartphones and tablets), home electronics (like security cameras), and other types of machines to communicate with one another across a network, every device must have its own distinct IP address. IP addresses, because of their purpose, are among the fundamental components that make up the Internet [2]. The organization that is in charge of maintaining these numbers is called the Internet Assigned Numbers Authority (IANA). The IANA is saddled with the responsibility of vitally crucial task of administering the IP address pool, and it is this responsibility that makes controlling these addresses so important [3]. The IANA is also in charge of distributing massive blocks of IP numbers to various regional internet registries (RIRs). The RIR is then responsible for registering, distributing, and managing IP addresses within their respective areas in accordance with the policies established by their respective regional communities. These policies are different for each region. These regulations are created by the RIRs themselves. As can be seen in the table below, there are five RIRs, each serving a distinct region somewhere on the planet.

IPv4 addresses, on the other hand, are defined using 32 bits, which means that there can only be around 4.3 billion (2^{32}) distinct IPv4 addresses in use at any given time across the globe. Similarly, the Internet Engineering Task Force (IETF) and the IANA have reserved considerable portions of IPv4 addresses for use in unique applications. For instance, these IP addresses are used for IP multicasting and the development of private networks.

The following factors have contributed to an extremely low number of available IP addresses [2].

The Internet population: It was theoretically possible to allocate a large number of IP addresses to everyone in 1990, when there were approximately 5.3 billion people on Earth but only 0.049% of the population had access to the internet. Nonetheless, both indices have grown considerably since 1990, especially in terms of internet usage; by 2021, the percentage of internet users had risen above 67%, and the world's population had risen above 7.9 billion. Assuming IP addresses be supplied to everyone with Internet connection after 2019, the total number of IPv4 addresses that are theoretically conceivable will not be enough to meet the demand for IP addresses. Adding to the problem is the fact that people may need more than one IP address in order to use all of their Internet-connected gadgets. Computers, and home appliances. The fact that some people need multiple IP addresses only amplifies this problem. The current state of device ownership in developed markets is illustrated by the fact that the typical American household has five connected devices. Given the proliferation of connected devices over the past decade, it is currently extremely difficult to assign a unique IP address to each user of the Internet. This is because it is obvious that there are not enough public IPv4 addresses to satisfy the needs of people with multiple devices.

Always-on connections: Dial-up Internet access via telephone modem was the most popular option for users in the 1990s. Since IP addresses were also shared throughout the modem pools, this is an unfortunate side effect. However, with the advent of always-on internet devices like modems and routers, the availability of free IPv4 addresses has been steadily decreasing in tandem with the technology's meteoric rise in popularity. As a result, always-on connection has become the standard rather than the exception.

Internet of Things gadgets add new requirements: In recent years, there has been a large expansion in the number of internet-connected devices, which is directly attributable to the Internet of Things' quick expansion. In the year 2024, it is anticipated that the number of Internet of Things devices would have grown by 12.2%, reaching a total of 60.7 billion active endpoints [4].

Limited address assigning: Classful addressing of IPv4 was introduced in 1981, and it limited network sizes to 256, 16,777,216, or 16,777,216 hosts, respectively. Since Class C addresses wouldn't be enough for a corporation that required 500 addresses for its devices, Class B was issued to the company to get at least one IP address per device. The business would have to buy IP addresses individually for each device if it wanted to guarantee that each one had its own unique number. It will be shown that the classful addressing system resulted in the assignment of many IP addresses in an inefficient fashion before classless inter-domain routing was introduced in 1993.

As these causes did not become apparent all at once, but rather gradually, the anticipated reduction of IPv4 addresses began to attract attention as early as 1991, and was highlighted in [5]. This is because these variables did not suddenly become apparent. The agencies in charge of administering IP addresses have, over the course of the years, adopted a variety of preventative measures against the continued reduction of unallocated IP addresses. These preventative measures, which are discussed in more detail below, are as follows:

Classless Inter-Domain Routing (CIDR): IETF developed CIDR in 1993 in order to offer an architecture and a approach for the distribution of IP addresses. This was done in acknowledgment of the fact that the space for IP addresses is a finite and shared resource. CIDR replaced the older classful numbering system, which categorized networks according to their size (A, B, or C), and as a result, slowed down the process of IPv4 depletion by reducing the amount of IP addresses that were assigned but not used. This was accomplished by reducing the number of unused IP addresses.

The use of Network Address Translation and private network addressing When the IETF realized that the capacity available for globally unique addresses was likely going to be exhausted, they came up with the idea of assigning IP addresses for use on private Internets [6]. This strategy allowed a wide variety of distinct and self-governing organizations to use the same limited IP address space that was reserved by IANA without the need to collaborate with an Internet registry. Because of this, the private IP address space used for each organization's intranet communications would be completely distinct from the space used by any other organization. In addition, it would be utilized in a much more efficient manner because it would let several enterprises to use these addresses at the same time.

A method known as Network Address Translation (NAT), which is used in this multi-IP space situation, makes it possible for businesses to connect to the global Internet while still reaping the benefits of distinct public IP addresses between different networks. This was made possible by employing NAT. NAT enables the translation of several private addresses into a single public address that can then represent the entire range of addresses [7]. A router is a device used to establish a connection between two or more networks, transform locally assigned IP addresses into globally unique IP addresses, and then forward data packets to their final destinations. The implementation of NAT enables the use of a solitary address for the promotion of the entire network. Both the initiatives to save IP address resources and the attempts to strengthen user security benefited substantially from this development.

Reclaiming and redistributing address spaces that have been unused: There is a large pool of idle IPv4 addresses that were initially allocated to businesses during the network planning stage but are currently being held in reserve. Those IPv4 addresses were distributed to businesses before. The IANA has successfully recovered and redistributed some of these IPv4 addresses to other RIRs [8]. In 2014, however, the IANA disclosed that the worldwide supply of IPv4 addresses had reached a critical level and that the pool of redistributed IPv4 addresses was diminishing. Internet users were understandably alarmed by this development [9]. Local Internet Registries (LIRs) are now able to swap IPv4 address ranges with other LIRs based on inter-LIR surplus or demand (also known as Internet service providers). This is a partial answer to the issue. This ensures that active contact information is disseminated efficiently.

In spite of all of these efforts to slow down the depletion of IPv4 addresses, it was apparent that a different strategy was required. IPv6 was developed as the successor to IPv4 and the next iteration of the Internet Protocol to give a more permanent answer to the problem of IPv4 address exhaustion. This situation had been anticipated for some time. Because IPv6 uses 128-bit addresses, the consequence is a much bigger number of available addresses ($2^{128} = 3.4 \times 10^{28}$ distinct IP addresses). This is the primary benefit that IPv6 offers over IPv4, which uses only 32-bit addresses.

Until far, the mitigating strategies adopted to satisfy the rising demand for IP addresses have been relatively effective. Actually, there exist no definitive date for the extinction of the useful life of IPv4, as it is anticipated that current devices with IPv4 addresses will continue to function long after the IPv4 address pool has been totally depleted. Despite these alternative attempts, acquiring new IPv4 addresses, managing networks, and properly keeping track of the unused IPv4 addresses is becoming increasingly complicated, expensive, and complex. This is because there are a limited number of IPv4 addresses [9]. IPv6 is the only answer that is now available, and it is the only option that can ensure the continued growth of the internet over the long term while still maintaining worldwide connectivity.

Ipv6 Usage

IPV6 Supporting Devices.

According to [10], IPv6 is supported by every single Operating System (OS) that hosts use. The majority of CPEs also have IPv6 functionality, in particular:

- Dual-Stack, which combines stateless and stateful translation and is one of the most popular IPv6 transition mechanisms for Mobile Broadband, and 464XLAT are both supported by mobile devices (e.g., the UEs).
- Fixed CPEs used to support Dual-Stack, and IPv6 Rapid Deployment (6RD), and Dual stack Lite (DS-Lite) as necessitated by the 3GPP. Dual-Stack and 464XLAT are supported by mobile devices (e.g. [11]). However, as of May 2019, the information presented. The 3GPP mandates that fixed CPEs support Dual-Stack, Dual-Stack Lite (DS-Lite), and IPv6 Rapid Deployment (6RD). Dual-Stack and 464XLAT are supported by mobile devices (for example, iPhones) [11]. Nonetheless, as of May 2019, the data reported in [12] has been updated [11]. Therefore, additional transition techniques, such as 464XLAT, must be supported on new fixed CPEs as in [12] which has been revised by [11]. Therefore, other helpful transition strategies like 464XLAT need to be supported on new fixed CPEs as well.
- In addition to other IPv6 transition strategies, Enterprise CPEs enable Dual-Stack.

There are many sites out there that support IPv6, and it is important to know which ones do. Some of the most popular sites that support IPv6 include Google, Facebook, YouTube, and Twitter. These sites have implemented IPv6 to provide better performance, reliability, and security to their users. Other popular sites that support IPv6 include Amazon, Microsoft, Netflix, and LinkedIn. These sites have recognized the importance of IPv6 and have made the necessary changes to support it. Figure 2 is the statistics of sites that support IPv6 as of March, 2023 according to [13].

From Fig. 1, IPv6 is utilized by 20.9% of all sites currently online. 29.7% of the top one million websites are currently utilizing IPV6. IPV6 is used by 40.3% of the top 100,000 websites according to the rankings. IPV6 is utilized by 4,688 out of the top 10,000 websites according to Alexa. As on the 5th of March in 2023, more over fifty percent of the top one thousand sites in terms of rank use IPV6.

Over time, IPV6 has seen steadily rising acceptance rates. According to recent statistics, the global IPV6 adoption rate stands at around 30% [4]. The adoption rate

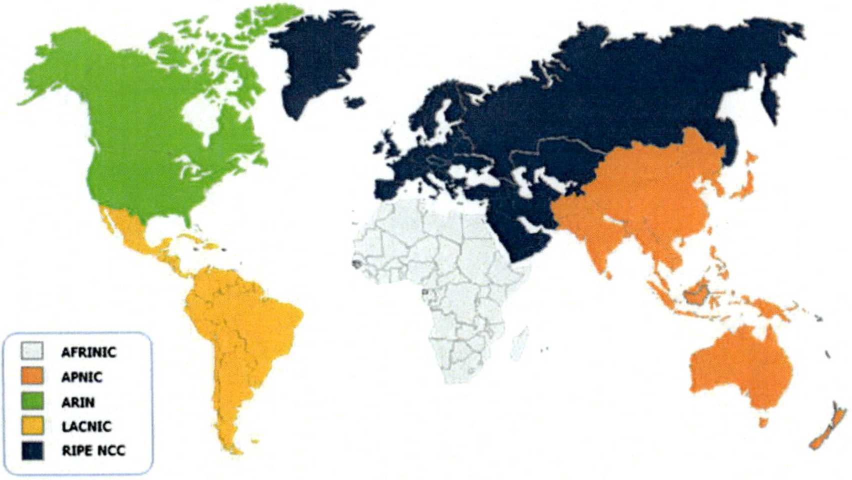

Fig. 1 A map depicting the five RIRs' respective service zones globally [2]

Fig. 2 Usage statistics of ipv6 for websites, March 2023 [13]

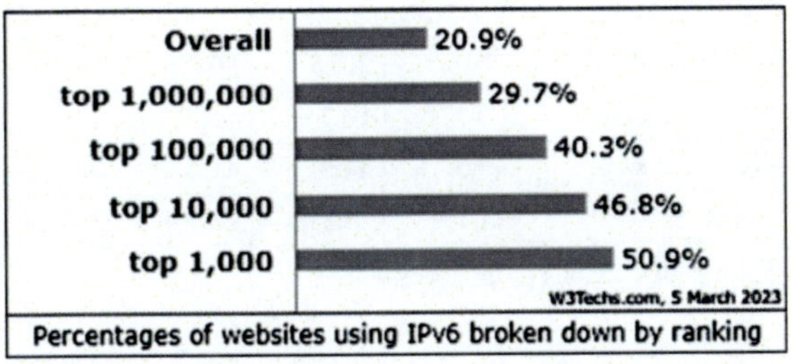

of IPv6 by google as of March 2nd, 2023 is given in Fig. 3. (Note that this figure is for users that access google over IPv6). This may seem low, but it is a significant improvement from just a few years ago. Many organizations and businesses have already switched to IPV6, and more are expected to follow suit in the coming years.

Google measures IPV6 adoption rates and places it at 38.67% as of March, 2023.

The data presented in Table 1 is derived from [15] according to [10]. It demonstrates that the percentage of Autonomous Systems (ASes) that support IPv6 will increase from 21.1% in 2015 to 27.5% in 2020. This equates to a compound annual growth rate (CAGR) of 15.19% for IPv6-enabled networks. Since the overall number of networks (including IPv6 and IPv4) is rising at a compound annual growth rate of 9.23%, this indicates that the number of networks that support IPv6 is increasing significantly more quickly than the networks that support IPv4 [10]. As of March, 2023 the value stands at 2,490 million which gives a CAGR of 55.13% from 2015.

Similarly, according to the data shown in Table 2 [4], the percentage of IPv6 users (also known as the IPv6/total ratio) will expand from 3.22% in 2015 to 24.33% in 2020 at a compound annual growth rate of 67.9%.

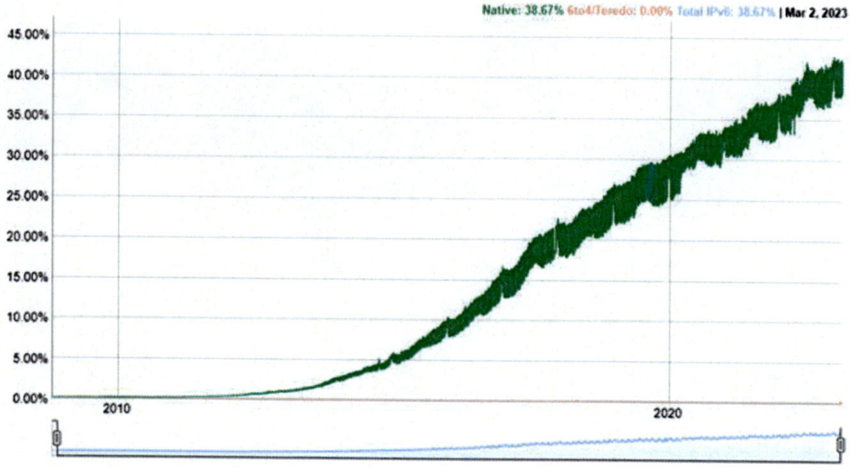

Fig. 3 IPv6 adoption statistics (Google [14])

Table 1 Percentage of ASes supporting IPv6 according to [15]

Advertised ASN	2015	2016	2017	2018	2019	2020	CAGR
IPv6-Capable	9.2	10.7	12.7	14.5	16.4	18.6	15.19%
Total AS's	43.5	44.5	44.4	60.3	63.8	67.7	9.23%
Ratio %	21.1%	24.1%	28.5%	24.1%	25.8%	27.5%	

Source [10] Values in thousands

Table 2 Nos of IPv6 users worldwide (Value in Million) according to [4] 10

Worldwide Users	2015	2016	2017	2018	2019	2020	CAGR
IPv6 users	74.24	179.42	290.68	513.68	574.02	989.25	67.90%
Total user	2303.09	3246.15	3339.37	3410.28	3470.37	4065.21	12.00%
% IPv6 user	3.22%	5.53%	8.70%	15.06%	16.54%	24.33%	

Assigning IPV6 Prefixes and Addresses to CPEs

The ability to apply SLAAC, an abbreviation for stateless address autoconfiguration, at the customer premise equipment (CPE) is one of the most distinguishing characteristics between IPv6 and IPv4 prefix and address assignment [16]. A DHCP server is required for hosts to receive an IP address when using IPv4. SLAAC is used in IPv6 for hosts to construct the "interface identification," which comprises the last 64 bits of their IPv6 address. In the beginning, this was accomplished by adding a prefix to the beginning of the 48-bit MAC address. Without the need for a router or DHCP server, hosts that are connected to the same link can communicate with one another using this method. If a router is connected to the same link as the hosts, the hosts will obtain a prefix from the router, attach it to the interface identification, and generate a globally routable IPv6 address. The SLAAC was developed to enable simple plug-and-play operations [10].

Therefore, the availability of SLAAC is the primary difference between purchasing IPv6 addresses for Mobile broadband (MBB), Fixed broadband (FBB), and enterprise CPEs.

The idea of the stateful DHCP, in contrast to the stateless DHCP, is a distinction that should also be brought to the attention of the audience. Stateful DHCP servers are the most common type of DHCP servers because they store and keep track of data for each host such as IP addresses. However, stateless DHCP servers do not keep track of any data about individual host;. instead, they use information that is shared by all hosts, such as the address of a DNS server. Typically, stateful DHCP servers are used. Stateful DHCP requires the server to maintain information about each host, such as the IP address of each host. In contrast, while utilizing Stateless DHCP, all hosts share the same information, eliminating the need for specific tracking of each host. This is why it is known as a stateless person.

IPV6 Prefix and Address Assignment for MBB UEs

Figure 4 depicts in detail the steps used by a user equipment (UE) to acquire an IPv6 address and establish an IPv6 link with a mobile gateway. It is essential to highlight that IPv6-inapplicable methods are disregarded.

- Step 0: At this stage, the only difference between IPv4 and IPv6 is the PDU session type, which can be either IPv6 or IPv4v6 (Dual-Stack). Other than that, the situation remains same.

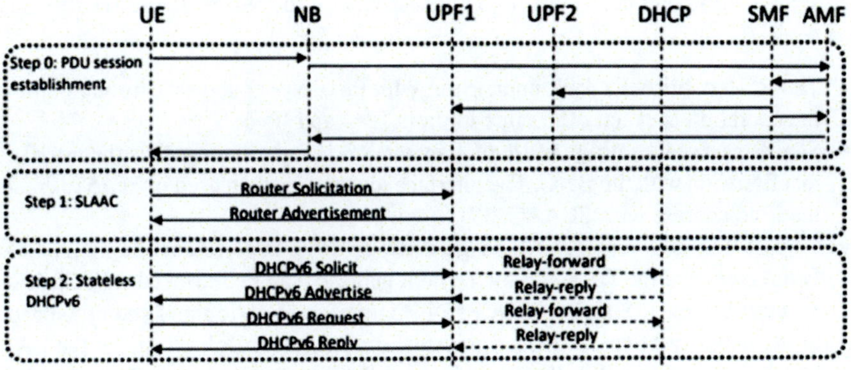

Fig. 4 IPv6 address acquisition (MBB CPE) [10]

- Step 1: Here, the UE will start the SLAAC process. User Equipment (UE) should utilize the Interface ID that is advertised by the User Plane Function (UPF), which is the gateway for the Packet Data Network (PDN) in 4G networks. This would prevent address replication and eliminate the requirement for Duplicate Address Detection (DAD). In instance, the usage of the International Mobile Subscriber Identity (IMSI) as an Interface ID is not permitted. Several important pieces of information, including MTU, the default router, and DNS, would be conveyed to the UE by the UPF in the form of a Router Advertisement for SLAAC.
- Step 2: if the UE needs more settings, it will begin the stateless DHCPv6 protocol. Stateless DHCPv6 has gone from being a "not needed" recommendation to a "essential" one. Because a UE can act as a CPE (hot spot) and have LAN behind it, prefix delegation (PD) is necessary. When doing Prefix Delegation using DHCPv6, the Prefix Exclude Option must be used. The aim is to have both the WAN and the LAN use the same aggregated prefix (LAN). The widespread use of 464XLAT, which requires PREFIX64, which is normally given by DHCP, is yet another argument in favor of giving DHCP more priority.

If the UE gets the same parameters from multiple sources, such as SLAAC and DHCP, the first source that sent them to the UE would be considered authoritative for those parameters.

Only if the WAN is active should the parameters (MTU, DNS, and RA) that were received on the WAN be advertised on the LAN by the UE while it is functioning as a hot spot.

IPV6 Prefix and Address Assignment for FBB RGs

PPPoE (Point-to-point over Ethernet) or IPoE (IP over Ethernet) could be used to establish a connection between an FBB RG (Remote Gateway) and a BNG (Broadband Network Gateway). Following the processes in Fig. 5, RGs will be able to get an IPv6 address and configure IPv6 connectivity on the WAN side to the operator network.

- The RG establishes a PPP connection with the BNG in stages 1 through 3, and the operation is exactly the same for both IPv4 and IPv6.
- Step 4 involves the usage of IPv6CP by the RG in order to negotiate the interface identification with the BNG. The interface identifier is then used in step 5 to create the IPv6 address with SLAAC [17].
- Step 5: The RG use SLAAC to generate an IPv6 address for the WAN link. Parameters like the DNS and the Default gateway can be retrieved with SLAAC as well; however, Stateless DHCPv6 is the preferred option and should be used in the majority of cases. Also significant are the Prefix Delegation for the LAN situated behind the RG, the IP address of the Session Initiation Protocol (SIP)

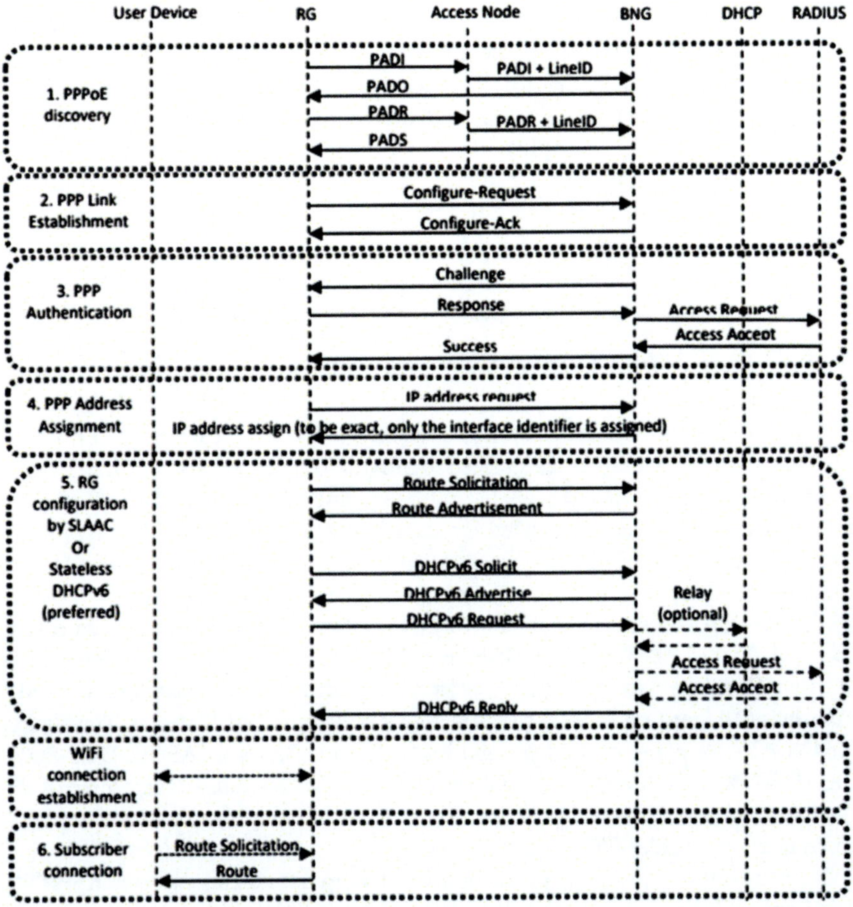

Fig. 5 Prefix and address assignment for FBB RGs [10]

server, and the IP address of the CPE WAN Management Protocol (TR-069) server.

- Step 6: Once the setup of the WAN link is complete, the RG will be able to declare itself as a router to the hosts on the LAN. It is advised to use the streamlined process of SLAAC for assigning user/host addresses, since it requires the setting of only a few parameters (MTU, DNS, and Default Router).

IPV6 Prefix and Address Assignment for Enterprise CPEs

SMBs who use DSL or FTTX to connect to service providers have the same experience with IPv6 address and prefix assignment as FBB businesses.

Manual configuration is used for IPv6 addresses on enterprise CPEs. An entity that utilizes a single Internet service provider (ISP) is permitted to utilize the IPv6 address portion allotted by that ISP. This event is referred to as Provider Aggregable (PA). Larger businesses will not be able to utilize PA space because they frequently utilize many suppliers. An application for a provider-independent prefix allocation (PI prefix allocation) must be made directly to the relevant RIPE NCC by the applicant (RIR). There is a recurring operational expense associated with this kind of allocation [10].

IPv6 Transitioning Solutions

IPv6 transition solutions can be broken down into two distinct categories: AIPv4-to-IPv6 transition technologies are utilized to provide IPv6 connectivity, while BIPv4-to-IPv6 transition technologies enable communication between IPv4 and IPv6.

Various technologies are utilized for transitioning from IPv4 to IPv6 in order to ensure IPv6 connectivity. These technologies include Configured tunnels (6in4), Dual-Stack, Generic Routing Encapsulation (GRE), Native IPv6 behind NAT44 CPEs (6a44), IPv6 Rapid Deployment (6rd), Intra-Site Automatic Tunnel Addressing Protocol (ISATAP), Tunneling IPv6 over UDP through (TSP), and Connection of IPv6 Domains via IPv4 Clouds (6to4). Stateful NAT64, Stateless IP/ICMP Translation Algorithm (SIIT), Combination of Stateful and Stateless Translation (464XLAT), Mapping of Address and Port - Encapsulation (MAP-E), Dual-Stack Lite (DS-Lite), and Mapping of Address and Port - Translation are just some of the IPv4 to IPv6 transition technologies. These tools are employed to facilitate IPv4 communication (MAP-T).

It can be difficult to zero in on the optimal strategy for making the switch to IPv6. It is implausible to assume that all of these technologies will be extensively used. The selection of the IPv6 transition solution will be influenced by various factors, with the market and existing policies being the most significant ones.

Apart from the two broad categories of IPv6 transition solutions, there is another classification based on the different stages of the transition process, namely, (1) IPv6 introduction and (2) IPv6-only. It's essential to note that both these stages are linked to the service delivery aspect rather than the network underlay view.

The goal of the IPv6 introduction stage is to make it possible to launch an IPv6 service in a network that was initially designed for IPv4. It is possible to supply IPv6 services on top of or in conjunction with IPv4 service. This step is for gaining experience with IPv6 and is part of a dual-stack pattern that uses IPv4 and IPv6. It is anticipated at this time that the volume of IPv6 traffic would start off relatively low in comparison to the volume of IPv4 traffic; the exact size of this difference will depend on the IPv6 content that is already available. Even at this early stage, it is expected that the overall volume of IPv6 traffic would gradually increase over time.

The transition to the stage when IPv6 is the sole protocol used might be made when a certain threshold is reached in terms of the volume of traffic that is carried by

IPv6. At this point, the IPv6 protocol is the sole medium via which the service that is offered to subscribers is carried out. This suggests that the customer premise equipment (CPE) communicates with the operator gateway, which could be a Broadband Network Gateway (BNG) or a User Plane Function/Packet Gateway (PGW), over an IPv6 connection. The customer premises equipment (CPE) only possesses an IPv6 address on the wide area network (WAN) side (UPF). IPv4 and IPv6 are still both viable options for hosts and content servers to use. Both protocols are still supported. For instance, NAT64 can open the door for IPv6 hosts to communicate with IPv4 servers, which was not previously allowed. Both IPv4 and IPv6 are viable options for serving as the foundation of the backbone network. IPv6 is the only protocol that is utilized, and IPv4 services are delivered through IPv6. At least for the access element of the architecture, which handles service delivery, IPv6 is the only protocol that is used.

It is vital to remember that deciding whether or not to migrate from IPv6 introduction into IPv6-only can be difficult. This choice is influenced by a wide range of factors, such as economic considerations, legislation, and government regulation, and it is important to keep this in mind.

It is essential to point out, however, that the IPv6-only stage may prove to be more effective from the very beginning in particular scenarios, such as the MBB, due to the fact that the IPv6 introduction phase utilizing Dual-Stack may require large amount of resources. This is something that must be taken into consideration (for example CGNAT costs).

IPv6 Introduction Stage

To allow the establishment of an IPv6 service over an IPv4 infrastructure, either tunneling IPv6 traffic over IPv4 (6RD or Teredo) or enabling dual-stacking at the customer premise equipment (CPE) might be used (CPE).

The following is the recommendation that can be found in Chittimaneni et al. (2014): "dual-stack when you can, tunnel when you must." The Dual-Stack architecture is not only easier to debug and provide support for, but it also provides increased levels of reliability. According to the information that was provided by operators, the Dual-Stack architecture is currently the IPv6 solution that has been implemented the most commonly for MBB, FBB, and corporations. The information that was provided by the operators gives credence to this possibility.

The Dual-Stack IPv6 transition is an excellent option for the duration of the IPv6 introduction phase. nevertheless, in the long run, it does have a few downsides that must be considered. As a result, It is highly suggested that the transition to the IPv6-only stage occur once the number of IPv6 addresses reaches a predetermined threshold.

IPv6-Only Service Delivery

Here, various IPv6-only transition options that are conceivable, as well as the process of choosing one of those solutions to meet the requirements are discussed.

Although it does not offer an explicit recommendation, the article [18] evaluates and contrasts the technical advantages of popular transition options for IPv6-only service delivery, such as 464XLAT, Lightweight 4over6 (lw4o6), MAP-E, DS-lite, and MAP-T.

In addition to Dual-Stack, [19] reveals that 464XLAT is the most extensively implemented IPv6 transition solution for MBB, whereas DS-Lite is the most widely deployed IPv6 transition option for FBB. Both options are limited to IPv6 only [10] recommends that:

1. Use 464XLAT for MBB, FBB, and enterprises if their CPEs can handle it.
2. In the event that neither the FBB nor the enterprise CPEs support 464XLAT, the fallback option is DS-Lite.

How NAT64 and DNS64 Work

With MBB in place, UE-based IPv6 hosts (like applications) can establish native IPv6 connections to remote IPv6 resources. On the other hand, NAT64 and DNS64 are necessities in order to access IPv4 websites. In order to successfully complete the translation, NAT64 [20] is required. If DNS queries are necessary, as shown in Fig. 6, then DNS64 [20] is almost certainly going to be required as well. Take note that Fig. 6 illustrates how a host that only supports IPv6 can access a website that only supports IPv4.

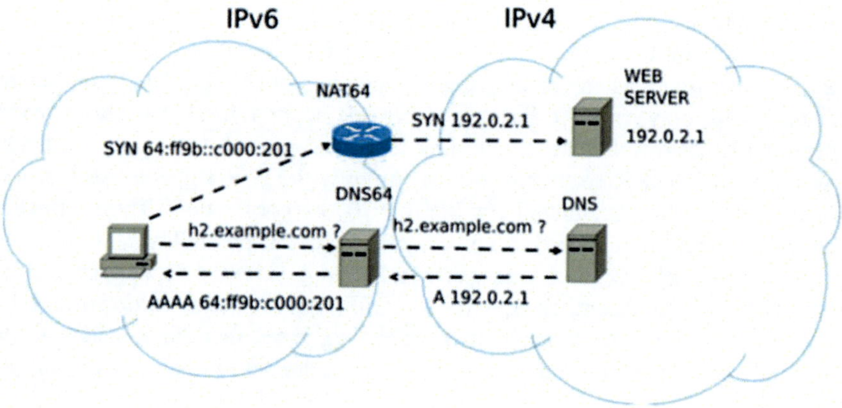

Fig. 6 How NAT64 and DNS 64 work

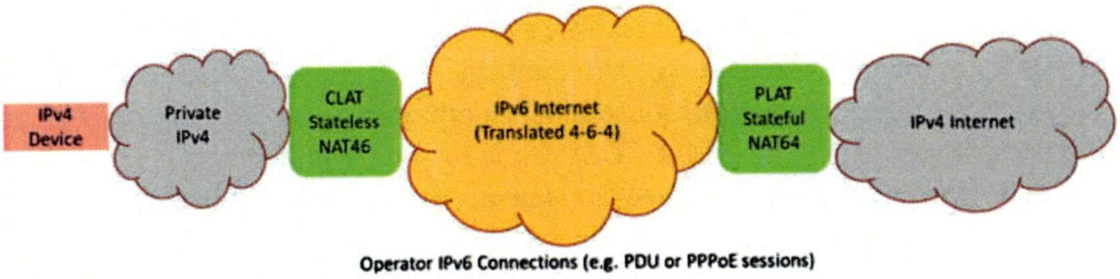

Fig. 7 Overview of the 464XLAT

In addition, NAT64 plus DNS64 isn't adaptable enough to work in all scenarios. Some tethering devices may only support IPv4, for instance, even though the host UE is IPv6-capable. This is where the 464XLAT [21] protocol comes in handy; it allows IPv4 hosts to function behind a CPE that only supports IPv6. As can be seen in Fig. 7, 464XLAT is in fact made up of two NATs, one on the client side (CLAT) at the CPE and one on the provider side (PLAT). IPv4 traffic from IPv4 hosts can be converted to IPv6 traffic using CLAT at the CPE, but the PLAT is still the same as in the previous case. PLAT is the same as the one mentioned in the first case. Therefore, the application of 464XLAT transforms this second case into the first scenario in effect. Any kind of website will be able to access both IPv4 and IPv6 hosts that are located behind an IPv6-only customer premise equipment (CPE), as NAT64 is the sole NAT that the provider uses.

It should be noted that the client component of 464XLAT is supported by the majority of mobile UE OSes (As with mobile operating systems, the provider portion of 464XLAT is irrelevant.). In addition, [18] asserts that mobile operating systems do not, in general, offer support for additional IPv6-only transition alternatives. Because of this, 464XLAT can be regarded as effectively being the only solution for MBB that relies solely on IPv6.

The IPv6-only solution is recommended for FBB and businesses if the customer premise equipment (CPE) supports 464XLAT, CLAT. The same answer will be available for use by MBB, FBB, and businesses, and NAT64 will be the only NAT. As a result, this could reduce OPEX (Operating Expenses) and make network management easier.

Nevertheless, it is essential to point out that Dual-Stack, DS-lite, and 6RD are the mandatory IPv6 transition options according to [22]. This fact should not be overlooked. As a result, it appears that not all retail fixed CPEs are required to provide 464XLAT capability. [11] amended the need for the compatibility of additional IPv6-only transition solutions, such as 464XLAT [22]. As a result of this, 464XLAT should be considered as the primary alternative by network operators in the event that there is a future requirement to implement an IPv6-only solution for FBB and corporations.

It is possible that the DS-Lite IPv6 transition solution will be explored if the operator's CPEs do not support 464XLAT. Both DS-Lite [21] and 464XLAT (Durand et al., 2011) were released in 2013, DS-Lite, on the other hand, was the first IPv6-only transition option made public. It's worth noting that DS-Lite is already in widespread use for IPv6-only fixed deployments. To emphasize this point, many current fixed IPv6-only deployments make use of DS-Lite[21].

Dual-Stack is the IPv6 transition option recommended by [10] for early-stage IPv6 deployment, while 464XLAT / DS-Lite is the preferred method for providing IPv6-only service. Both of these answers depend on DS-Lite and 464XLAT.

IPv6 Deployment and Transitioning Strategies

Here is a quick rundown of the most important considerations for making the switch to IPv6:

- Clearly divide the transition to IPv6 into two phases: the first phase is IPv6 introduction, and the second phase is IPv6 solely. These two stages serve distinct functions and call for distinct approaches to their respective problems. The implementation of IPv6 is for the purpose of gaining expertise with IPv6 and accommodating future services, such as the Internet of Things and Vehicle to X (V2X). As was said before in a publication, Dual-Stack is typically the option that is most suited to the given circumstances. IPv6 traffic may have a slow start relative to IPv4 traffic during this stage, but it is expected to expand more quickly than IPv4 traffic does. The switch to IPv6-only mode can happen after IPv6 traffic reaches a certain critical volume. At this point, IPv6 is used exclusively to deliver service to subscribers. This reduces capital expenditures and operating expenses in the network. As of right now, the 464XLAT seems like the best solution, but the DS-Lite isn't out of the question either for FBB. It's important to remember that, just like in MBB, the IPv6-only phase might be the beginning of a customer's IPv6 service in order to hasten the switch to IPv6 and save down on associated expenses.
- Align the perspectives of individuals and organizations: Many people and departments inside businesses and corporations will be impacted by the shift from IPv4 to IPv6 services. People tend to be hesitant to endorse changes with which they are unfamiliar. Therefore, it is of the utmost importance to communicate often to align the perspectives of both individuals and the business. To ensure that relevant individuals are prepared for the transition to IPv6, it is essential to provide training to those individuals.
- Conduct an audit of the infrastructure's IPv6 capabilityIPv6 affects every link in the communication service chain, including user applications, the information system, terminals, websites, and customer premise equipment (CPE). It's important to determine which devices are IPv6-ready and which need to be upgraded. Any upgrades that are required must also be carefully planned and carried out.
- To cut costs, upgrading to enable IPv6 should be done in conjunction with other types of network improvement. If, for example, a section of the network is getting close to its extinct and needs to be upgraded or replaced, the equipment that takes its place must be able to handle IPv6 capabilities.

- DNS can be used as a switch to turn IPv6 services on or off for end users. This is made possible by the fact that hosts can check for IPv6 records in DNS queries to decide whether or not to utilize IPv6.

When moving from IPV4 to IPV6, there are a few recommended procedures that businesses and other organizations ought to adhere to as a matter of good practice. Taking the time to carefully plan the transition is one of the most critical things that must be done. This entails conducting an audit of the current infrastructure and identifying the components of the system, including hardware and software, that require an upgrade. Additionally, it is essential to make certain that the organization possesses sufficient resources, such as employees and a budget, to successfully manage the shift.

In addition, companies and organizations ought to make certain that they provide sufficient education and assistance to their employees. This can be useful in helping to guarantee that network managers are conversant with IPV6 and are able to successfully manage the transition. It is essential to make certain that the organization has a fallback strategy in place if any problems arise during the transition.

In conclusion, businesses and other organizations need to be certain they are maximizing the potential of IPV6's characteristics. This involves utilizing the improved security features as well as making use of the better routing features to guarantee that mobile devices are able to connect to the network.

Challenges of Transitioning from IPV4 to IPV6

Despite the advantages of IPV6, there are some challenges associated with transitioning from IPV4 to IPV6. IPV6's incompatibility with IPV4 represents one of the main obstacles. This means that organisations will need to upgrade all of their hardware and software to be compatible with the new protocol. This can be costly and time-consuming, and some organizations may not be able to afford it [10].

In addition, transitioning to IPV6 can be technically challenging. This is because IPV6 requires more complex routing rules than IPV4, which can make it more difficult for network administrators to set up and manage networks. It can also lead to slower performance, as the more complex rules can make it difficult for devices to communicate with each other.

Finally, IPV6 is not as widely supported as IPV4. This means that some devices and networks may not be able to take advantage of the benefits of IPV6. This can be especially problematic for organisations that need to be able to communicate with the wider internet.

IPv6 Network Operations

The most important aspects of managing IPv6 networks serve three primary goals, each of which is comparable to a key performance indicator (KPI) for Internet service providers and operators. These goals are as follows: TTM, SLA, & budget.

1. Service Level Agreement (SLA)
   (a) IPv6 Defect Handling: when networks experience challenges, engineers will need to employ novel techniques and resources.
   (b) Management of IPv6 Security IPv6 will provide new security vulnerabilities, which will leave networks and services open to attack. Such dangers require investigation and action to mitigate them.
2. Time to market (TTM)
   (a) Configuration of IPv6 requires training for service delivery teams on IPv6 and the development of new tools for service provisioning.
   (b) IPv6 accounting requires new software (To store, for example, updated user identity information and IPv6 MIBs) as well as new technical skills;
   (c) Monitoring IPv6 performance requires new tools as well as training for people's existing skills.
3. budget compliance (CAPEX & OPEX)
   (a) IPv6 may require new hardware, as well as additional state information in existing hardware. This rise in CAPEX needs to be taken into account. IPv6 adds complexity to networks during the transition period, however it may be "piggybacked" (done in tandem) on equipment renewal so CAPEX need not be a major worry. Therefore, it will cause an increase in operating expenses.

Strengths of IPV6

IPV6 is a newer version of the internet protocol, and it offers many advantages over its predecessor, IPV4. One of the main advantages of IPV6 is its increased scalability. IPV6 allows for more than four billion more IP addresses than IPV4, providing a much larger number of available addresses. This means that more devices can be connected to the internet, and it also makes it easier to allocate resources to different users.

IPV6 also offers improved security features compared to IPV4. For example, IPV6 includes a stronger authentication process and encryption to protect data. This can help prevent unauthorised access to a network, as well as making it easier to detect and respond to cyber-attacks.

In addition, IPV6 has improved support for mobile devices. This is due to its improved routing features, which make it easier for mobile devices to access networks. This can be especially beneficial for businesses that need to stay connected while on the go.

Finally, IPV6 provides better support for multimedia applications. This is because it has larger packet sizes, which makes it easier to transmit audio, video, and other large files over the internet. This can be especially useful for businesses and organisations that use streaming media or other applications that require large amounts of data to be transferred.

Limitations of IPV6

Despite its advantages, IPV6 also has some limitations. One of the main issues with IPV6 is that it is not backward compatible with IPV4 [23–26]. This means that devices and systems that are running IPV4 will not be able to communicate with devices and systems that are running IPV6. This can make it difficult for organisations to transition from IPV4 to IPV6, as they will need to upgrade all of their hardware and software to be compatible with the new protocol.

In addition, IPV6 requires more complex routing rules than IPV4. This can make it more difficult for network administrators to set up and manage networks that are running IPV6. It can also lead to slower performance, as the more complex rules can make it difficult for devices to communicate with each other.

Finally, IPV6 is not as widely supported as IPV4. This means that some devices and networks may not be able to take advantage of the benefits of IPV6. This can be especially problematic for organisations that need to be able to communicate with the wider internet [27–32].

Benefits of IPV6

Despite its limitations, IPV6 still has many benefits that make it worth considering. One of the main benefits of IPV6 is that it provides more efficient internet routing. This is because IPV6 has a more efficient header structure, which means that packets of data can be transmitted more quickly. This can result in faster internet speeds and less latency, which can be beneficial for businesses and organisations that need to transfer large amounts of data.

In addition, IPV6 provides improved security features. The stronger authentication process and encryption that come with IPV6 can help to protect networks from unauthorized access and cyber-attacks. This can be especially important for organisations that need to protect sensitive data or information [33, 34].

The strengths and benefits of IPv6 are due to bandwidth optimization (header simplification, greater size data packets carrying ability, inclusion multicast), prefix aggregation, changes in packet fragmentation, removal of checksum field, inclusion of extension headers, simplified network structure, robust quality-of-service definition, support for emerging and innovative technologies (cloud computing, IoT, blockchain, AI), reduction in energy consumption by networks [35–37].

Finally, IPV6 offers better support for mobile devices. This is because the improved routing features make it easier for mobile devices to access networks. This can be especially beneficial for businesses that need to stay connected while on the go.

How to Use IPV6 More Effectively

Once an organisation has made the transition to IPV6, there are some best practices they can follow to ensure they are using it effectively. One of the most important things to do is to ensure that the network is properly configured. This includes setting up the correct routing rules, as well as ensuring that the network is secure [38–40].

In addition, organisations should ensure that they are taking advantage of the features of IPV6. This includes making use of the improved security features, as well as taking advantage of the improved routing features to ensure that mobile devices can access the network.

Finally, organisations should ensure that they are regularly monitoring their network to ensure that it is functioning properly. This includes monitoring the network for any performance issues, as well as looking for any signs of unauthorised access or cyber-attacks [41, 42].

Advantages of Using IPV6

Overall, there are many advantages to using IPV6, including increased scalability, improved security, better support for mobile devices, and better support for multimedia applications. In addition, IPV6 can provide more efficient internet routing, which can result in faster internet speeds and less latency. This can be especially beneficial for businesses and organisations that need to transfer large amounts of data.

Disadvantages of Using IPV6

Despite its advantages, there are some drawbacks to using IPV6. One of the main issues is that it is not backward compatible with IPV4, which means that organisations will need to upgrade all of their hardware and software to be compatible with the new protocol. In addition, IPV6 is not as widely supported as IPV4, which means that some devices and networks may not be able to take advantage of the benefits of IPV6.

ETSI IPv6 Enhanced Innovation

As the Internet has progressed through its history and new technologies have come into existence, the requirements for networks have shifted to accommodate these developments. Starting in the 1980s with IPv4, the development of IP-based networks accelerated in the 2000s with the advent of MPLS technology, and IPv6 sped things up after 2010 to handle the explosion of connected devices. All of these advancements were made to meet the demands of a growing number of connected devices. In addition to greater address space or fundamental connection, the current development of new services and applications, which has been made possible by 5G networks and cloud computing, has also brought to the emergence of new requirements for networks. Specifically, these include the following:

- Faster, higher-quality connections (e.g.,, higher bandwidth, lower delay, etc.)
- Increased satisfaction from using the site
- We can reduce the time it takes to get services up and running and pinpoint problems much more precisely.

In the era of 5G and cloud computing, various enhancements to IPv6 are planned to suit the increasing and evolving demands of new services and applications (such autonomous vehicles, virtual reality, smart healthcare, etc.). IPv6 Enhanced Innovation (IPE), which is standardized by the European Telecommunications Standards Institute (ETSI), combines IPv6, IPv6-based Protocol Innovation (i.e., next-generation technologies based on IPv6), and artificial intelligence.

IPv6 and Protocol Innovation

Enhanced networks can be built with the help of IPv6 and other protocols that leverage next-generation innovations. These improved networks are the engine that propels the evolution of network service and the development of groundbreaking advances.

The technology known as "Segment Routing over IPv6" (SRv6) is a good example of a product that came forth because of the IPv6 and protocol innovations. SRv6 is a network programming framework that has been standardized by the IETF using various RFCs. This indicates that it makes it possible to define the processing of packets in a network as a separate application. As its name suggests, SRv6 is a protocol that combines the benefits offered by the Segment Routing paradigm and IPv6. Its purpose is to address the networking needs of developing services and applications, surpassing the limitations of traditional packet routing. This is made possible by SRv6, which offers network operators and applications with network programming capabilities via a sequence of IPv6 packet header instructions. SRv6 achieves this objective. SRv6 confers a great number of advantages on a network, including the following [15]:

- Simplified approaches to traffic engineering and management of networks

- Structure of the network that is both more robust and efficient.
- increased usage of available resources
- Scalability that is built in, made possible by the consolidation of IP (i.e., route aggregation enabled by a native IP characteristic) enabling finer-grained traffic diversification solutions to handle the different needs offered by changing technologies and applications.

It is getting greater consumer interest and adoption in large-scale commercial networks thanks to the benefits and opportunities presented by SRv6. Some examples of these networks include LINE Corporation, Softbank, China Telecom Iliad, China Bank, and. Additionally, the open-source community has 11 open-source platforms and apps that support SRv6 [15].

The development of SRv6 has made possible the creation of an intelligent IP network based on IPv6 and other cutting-edge technologies. SRv6 is also designed to foster even more innovation. It is anticipated that increasing the deployment of IPv6 with SRv6 solutions would lead to improvements in important metrics such as reach, scale, simplicity, functionality, and integration, which will ultimately result in an even wider adoption.

In addition to SRv6, there are numerous additional technologies that are now being developed that are made possible by the greater innovation provided by IPv6. Examples include:

- Maintenance, administration, and management (OAM) through telemetry: There are a few drawbacks to using the standard OAM tools, such as their inability to adapt to changing conditions, their lack of automation, and the high cost of processing the vast volumes of statistics generated by certain protocols. In response to these difficulties, the telemetry-based OAM was created. A solution to these problems has been provided in the form of telemetry, which details how data can be obtained via automated communication protocols and then transferred to one or more nodes for analysis. Telemetry elucidates how data can be gathered through mechanized channels of communication. Rich and intricate statistics can be generated when AI is applied.
- Extension headers that are flexible: With this method, we want to fix various problems with the IPv6 packet's "Extension Headers" field, such as limited router adaptability, poor performance (especially for long chains of headers), and vulnerability (particularly routing instructions). It will eventually alter the current standard administrative policy of always disabling Extension Headers under these conditions, and it will add new capabilities to IPv6 headers. The standard procedure in such a situation is to ignore any Extension Headers.
- Service-based Slicing: Different types of 5G services necessitate distinct network capabilities, which are outlined in service level agreements (SLAs).One of the most essential requirements could be to provide assurance that the service that is being provided will not be disrupted in any way by any other traffic that is taking place on the network. (This is what we mean when we talk about 'isolation.') End-to-end network slicing is being developed as part of 5G with the intention of addressing difficulties like these in order to guarantee performance.

- Advanced Multicast: Developed by IETF and referred to as Bit Index Explicit Replication (BIER). This approach seeks to resolve various multicasting constraints. These limitations include difficulties in maintenance and troubleshooting, in addition to the trade-off between scalable flow state and optimal delivery. The term "BIER over IPv6" is used to refer to the technology while discussing IPv6 network solutions. It is now in the draft stage, and it will take between one and two years before it gains widespread adoption on the market. With service-oriented networking, quality metrics like delay, jitter, packet loss, cost, etc. may be scaled independently for each service, allowing for more precise control over the network's performance. This can be done without adding any extra labor or complexity to the control plane of all network nodes while still ensuring user privacy and enabling encryption.

Computing-based networking seeks to maximize resource utilization by distributing user requests (such as for content delivery network (CDN) and public domain name server (DNS) services) to the servers (geographically) closest to the user, utilizing load balancing based on the availability of computing resources in the sites. The objective is to more efficiently distribute user requests (for CDN and public DNS services) across available servers (geographically).

IPv6 and Artificial Intelligence (AI)

IPv6 and AI (machine learning and natural language processing) together open up new possibilities for networks, similar to the way that innovative protocols help them. These new opportunities are in addition to the benefits provided by innovation protocols. IPv6, because of its scalability and flexibility, presents artificial intelligence with increased and improved chances for data collecting from networks. The Internet Protocol version 6 (IPv6) can also be improved with the use of AI by adding more advanced capabilities like flexible networking, network analysis and visualization, and distinguished assurance. These features generate significant benefits for networks in the form of greater operations and maintenance (O&M) efficiency as well as decreased costs associated with related O&M activities. As a result of these advantages, Artificial Intelligence, when combined with IPv6, can generate potential for intelligent application in a variety of domains, such as 5G, private line service, or home broadband, amongst a great deal of other domains. For instance, a bank has implemented IPv6 with artificial intelligence in its network, which enables the bank to detect and pinpoint issues and recover services in a matter of minutes. Because of this, its overall efficiency in operations and maintenance has increased by several times.

In conclusion, the Internet Protocol Evolution (IPE) presents many opportunities for improvement for network operators by fostering potentially fruitful synergies between IPv6, novel technology, and artificial intelligence. IPE is quickly becoming the dominant technology for IP networks in the era of cloud computing and 5G thanks to these advantages, in particular its ability to speed up the innovation of businesses, improve service provision, and enhance the overall quality of the customer experience. As a result of this, it is seen as the next generation of the intellectual property industry. Therefore, the quick implementation of IPv6 is a requirement, as IPv6 will invariably provide the essential foundation for IPv6-based technologies that are presently in development as well as those that are expected to be developed in the near future.

Conclusion

IPV6 is a powerful and versatile protocol that offers many advantages over its predecessor, IPV4. It has improved scalability, security features, and support for mobile devices. However, there are some challenges associated with transitioning from IPV4 to IPV6, including the need to upgrade hardware and software, as well as the more complex routing rules. By following best practices for transitioning and using IPV6, organizations can take advantage of its many benefits.

The benefit of IPv6 is that individual devices have their own address. This simplifies data transmission and frees up router resources so that they can move data as opposed to checking packet integrity. In addition, the fact that IPv6 encrypts traffic and provides VPN-like protection for internet traffic makes it more secure.

IPv6 is a great and important update to an outdated Internet Protocol. But it's not a perfect solution, and it doesn't fix every challenge with the IoT. As of now it's a call of time and a quick migration from the previous version is highly required.

References

1. DARPA-Defense Advanced Research Projects Agency. (1981). Internet Protocol. Defense Advanced Research Projects Agency, Information Processing Techniques Office. Virginia: Information Sciences Institute-University of Southern California. Retrieved February 20, 2023, from https://datatracker.ietf.org/doc/html/rfc791
2. Axion Partners Group. (2021). IPv6: Benefits and best practices of public policies. Axion Partners Group.
3. Internet Asigned Numbers Authority. (2023). Number Resources. Retrieved from iana: http://www.iana.org/numbers
4. APNIC. (2023). IPv6 Statistics. Retrieved Jnauary 28, 2023, from APNIC: https://stats.labs.apnic.net/ipv6/?s=IPv6+Capable&d=Auto&w=120&t=Auto.
5. Clark, D., Chapin, L., Cerf, V., Braden, R., & Hobby, R. (1991). Towards the Future Internet Architecture- RFC 1287. IEFT.
6. Moskowitz, R., Karrenberg, D., Rekhter, Y., Lear, E., & de Groot, G. J. (1996). *Address Allocation for Private Internets* (Issue 1918). RFC Editor. https://doi.org/10.17487/RFC1918
7. Juniper Networks. (2020). Understanding IPv4 and IPv6 Protocol Family. Retrieved from Interfaces User Guide for Security Devices: https://www.juniper.net/documentation/us/en/software/junos/interfaces-security-devices/topics/topic-map/security-interface-ipv4-ipv6-protocol.html

8. Internet Society. (2020). FAQ. Retrieved from Internet society: https://www.internetsociety.org/deploy360/ipv6/faq/#1
9. ICANN. (2014). ICANN Announcements. Retrieved from icann: https://www.icann.org/en/announcements/details/remaining-ipv4-addresses-to-be-redistributed-to-regional-internet-registries--address-redistribution-signals-that-ipv4-is-nearing-total-exhaustion-20-5-2014-en
10. ETSI. (2020). IPv6 Best Practices, Benefits, Transition Challenges and the Way Forward. Antipolis: ETSI. Retrieved January 28, 2023, from https://www.etsi.org/images/files/ETSIWhitePapers/etsi_WP35_IPv6_Best_Practices_Benefits_Transition_Challenges_and_the_Way_Forward.pdf
11. Martinez, J. P., Liu, H. M.-H., & Kawashima, M. (2019). Requirements for IPv6 customer edge routers to support IPv4-as-a-Service (Issue 8585). RFC Editor. https://doi.org/10.17487/RFC8585
12. Singh, H., Beebee, W., Donley, C., & Stark, B. (2013). Basic Requirements for IPv6 Customer Edge Routers (Issue 7084). RFC Editor. https://doi.org/10.17487/RFC7084
13. W3Techs. (2023). Usage statistics of IPv6 for websites. Retrieved from w3techs: https://w3techs.com/technologies/details/ce-ipv6
14. Google Inc. (2023, March 2). Google IPv6. Retrieved from Google: https://www.google.com/intl/en/ipv6/statistics.html#tab=ipv6-adoption
15. Huston, G. (2020). IPv6 report index. Retrieved from https://bgp.potaroo.net/index-v6.html
16. Narten, T., Jinmei, T., & Thomson, S. (2007). Patent No. RFC 4862.
17. Varada, S. V. (2007). IP Version 6 over PPP (Issue 5072). RFC Editor. https://doi.org/10.17487/RFC5072
18. Lencse, G., Martinez, J. P., Howard, L., Patterson, R., & Farrer, I. (2021). Pros and cons of IPv6 transition technologies for IPv4aaS (Issue draft-lmhp-v6ops-transition-comparison-06). Internet Engineering Task Force. https://datatracker.ietf.org/doc/draft-lmhp-v6ops-transition-comparison/06/
19. Heatley, N. (2017). IPV6-ONLY DEVICES EE Mobile Devices. Retrieved February 1, 2023, from https://indico.uknof.org.uk/event/38/contributions/489/attachments/612/736/Nick_Heatley_EE_IPv6_UKNOF_20170119.pdf
20. Matthews, P., Sullivan, A., van Beijnum, I., & Bagnulo, M. (2011). DNS64: DNS Extensions for Network Address Translation from IPv6 Clients to IPv4 Servers (Issue 6147). RFC Editor. https://doi.org/10.17487/RFC6147
21. Mawatari, M., Kawashima, M., & Byrne, C. (2013). 464XLAT: Combination of Stateful and Stateless Translation (Issue 6877). RFC Editor. https://doi.org/10.17487/RFC6877
22. Gont, F. (2014). A Method for Generating Semantically Opaque Interface Identifiers with IPv6 Stateless Address Autoconfiguration (SLAAC) (Issue 7217). RFC Editor. https://doi.org/10.17487/RFC7217
23. Hendriks, L., De Oliveira Schmidt, R., Van Rijswijk-Deij, R., & Pras, A. (2017). On the potential of IPv6 open resolvers for DDoS attacks. Lecture Notes in Computer Science (Including Subseries Lecture Notes in Artificial Intelligence and Lecture Notes in Bioinformatics), 10176 LNCS. https://doi.org/10.1007/978-3-319-54328-4_2
24. Hong, H., Choi, H., Kim, D., Kim, H., Hong, B., Noh, J., & Kim, Y. (2017). When cellular networks Met IPv6: Security problems of Middleboxes in IPv6 cellular networks. *Proceedings - 2nd IEEE European Symposium on Security and Privacy*, EuroS and P 2017. https://doi.org/10.1109/EuroSP.2017.34
25. Loshin P. (2004). *IPv6 -Theory, protocol, and practice.* ELSEIVER Morgan Kaufmann Publishers.
26. Murdock, A., Li, F., Bramsen, P., Durumeric, Z., & Paxson, V. (2017). Target generation for internet-wide IPv6 scanning. *Proceedings of the ACM SIGCOMM Internet Measurement Conference, IMC*, Part F131937.https://doi.org/10.1145/3131365.3131405
27. Cavalcante, J., Trajano, A. F. R., Leite, L., Mariano, M. F., Celestino Junior, J., Patel, A., & De Souza, J. N. (2019). Securing IPv6 wireless networks against malicious router advertisements. *IEEE International Symposium on Personal, Indoor and Mobile Radio Communications*, PIMRC, 2019-September. https://doi.org/10.1109/PIMRC.2019.8904296

28. Cv, R. K., & Goyal, H. (2019). IPv4 to IPv6 migration and performance analysis using GNS3 and Wireshark. *Proceedings - International Conference on Vision Towards Emerging Trends in Communication and Networking*, ViTECoN 2019.https://doi.org/10.1109/ViTECoN.2019.8899746
29. Li, L., Ren, G., Liu, Y., & Wu, J. (2018). Secure DHCPv6 mechanism for DHCPv6 security and privacy protection. *Tsinghua Science and Technology, 23*(1). https://doi.org/10.26599/TST.2018.9010020
30. Livadariu, I., Elmokashfi, A., & Dhamdhere, A. (2017). On IPv4 transfer markets: Analyzing reported transfers and inferring transfers in the wild. *Computer Communications, 111*.https://doi.org/10.1016/j.comcom.2017.07.012
31. Mavani, M., & Asawa, K. (2017). Modeling and analyses of IP spoofing attack in 6LoWPAN network. *Computers and Security*, 70.https://doi.org/10.1016/j.cose.2017.05.004
32. Ravi Kumar, C. V., Venkatesh, K., Sagar, M. V., & Bagadi, K. P. (2016). Performance analysis of IPv4 to IPv6 transition methods. *Indian Journal of Science and Technology, 9*(20). https://doi.org/10.17485/ijst/2016/v9i20/90005
33. Han, Z., Li, Y., & Li, J. (2018). A novel routing algorithm for IoT cloud based on hash offset tree. *Future Generation Computer Systems, 86*.https://doi.org/10.1016/j.future.2018.02.047
34. Su, C., Liu, R., & Li, X. (2019). A study on the distribution of active IPv6 addresses used by websites. *Proceedings of 2019 IEEE 8th Joint International Information Technology and Artificial Intelligence Conference*, ITAIC 2019. https://doi.org/10.1109/ITAIC.2019.8785473
35. Babar, A. K., Zardari, Z. A., Qureshi, S., Han, S., & Hussaini, N. N. (2019). Assessment of IPv4 and IPv6 networks with different modified tunneling techniques using OPNET. *International Journal of Advanced Computer Science and Applications, 10*(9). https://doi.org/10.14569/ijacsa.2019.0100963
36. Lee, K., Kim, S., (Paul) Jeong, J., Lee, S., Kim, H., & Park, J. S. (2019). A framework for DNS naming services for Internet-of-Things devices. *Future Generation Computer Systems, 92*.https://doi.org/10.1016/j.future.2018.01.023
37. Saad, R. M. A., Anbar, M., & Manickam, S. (2018). Rule-based detection technique for ICMPv6 anomalous behaviour. *Neural Computing and Applications, 30*(12). https://doi.org/10.1007/s00521-017-2967-y
38. Pickard, J., Angolia, M., & Chou, T. S. (2018). IPV6 diffusion on the Internet reaches a critical point. *Journal of Technology, Management, and Applied Engineering, 34*(1).
39. Pickard, J., Angolia, M., & Drummond, D. (2019). IPv6 diffusion Milestones: Assessing the quantity and quality of adoption. *Journal of International Technology and Information Management, 28*(1). https://doi.org/10.58729/1941-6679.1375
40. Qin, Z. (2018). Seamless converging system for IPv4/IPv6 transition. *2017 9th International Conference on Advanced Infocomm Technology*, ICAIT 2017. https://doi.org/10.1109/ICAIT.2017.8388898
41. Lu, T. Te, Wu, C. Y., Lin, W. Y., Chen, H. P., & Hsueh, K. P. (2017). Comparison of IPv4-over-IPv6 (4over6) and dual stack technologies in dynamic configuration for IPv4/IPv6 address. *Smart Innovation, Systems and Technologies*, 63.https://doi.org/10.1007/978-3-319-50209-0_32
42. Padole, M., Kanani, P., Raut, L., Jhaveri, D., & Nagda, M. (2017). An insight into IPAddressing. *Oriental Journal of Computer Science and Technology, 10*(1). https://doi.org/10.13005/ojcst/10.01.05

A Survey: Internet of Things (IoTs) Technologies, Embedded Systems and Sensors

Burak Tasci

Abstract The technology known as the Internet of Things, or IoTs, refers to all interconnected computing devices, digital machines, mechanical things, and identifiers that are capable of transferring data across a network without the assistance of humans. Today, IoTs range from small appliances to smart cities. Organizations across various industries are increasingly utilizing IoT technology to enhance their operational efficiency, deliver superior customer service, increase business value, gain deeper customer insights, and enhance their decision-making processes. For this reason, embedded system platforms are used for IoTs solutions developed for different workspaces and applications. In this study, various embedded systems such as Raspberry Pi, ZYNQ, Jetson, Altera Cyclone II, Beagle Bone, Odroid, STM32, PSoC, were investigated, along with IoT communication protocols including IPv6, 6LowPAN, ZigBee, Bluetooth, Wi-Fi, LoRaWAN, Z-Wave, NFC, and Sigfox. Furthermore, a comparative analysis of the sensors utilized in these systems and their respective application areas has been provided.

Keywords Internet of things · Sensors · Embedded systems · Communication protocols · Applıcatıon areas of Iot

Introduction

The impact of the Internet on our lives and the meaning of the term Internet for humanity has gained more meaning especially in recent years. Such that, the researchers have started to work on how the Internet will penetrate our lives and become indispensable, and even to determine the most appropriate use of it. Due to the advancements in the internet and its infrastructure, as well as the evolving technological landscape and the way people engage with it, the meaning of the term "Internet" has significantly transformed compared to previous years. Presently, the Internet is no longer limited to a standalone concept; it has been redefined through

B. Tasci (✉)
Fırat University Vocational School of Technical Sciences, Elazığ, Turkey
e-mail: btasci@firat.edu.tr

A. Prasad et al. (eds.), *Communication Technologies and Security Challenges in IoT*, Internet of Things, https://doi.org/10.1007/978-981-97-0052-3_9

the emergence of the Internet of Things (IoT) that connects various objects and associates them with the digital realm. This concept, which was used for the first time in a presentation in 1999, has reached a wider vision as technology evolves [1]. IoT applications are spreading across all areas of life, such as: industrial manufacturing [2], transportation [3], energy [4], retail [5], smart cities [6], healthcare [7], supply chain [8], agriculture and livestock [9], smart buildings [10]. The main purpose of the IoT concept is to enable objects to communicate with other animate or inanimate objects, regardless of time or place constraints. It enables the collection of data in real time. They deal with a broad chain of thoughts that are complex and intertwined from different perspectives. They provide electronic miniaturization to offer a wide range of new capabilities with networked products, systems and sensors. By saving time and resources, they provide new approaches with the object of improvement in this field. They aim to reduce the transaction cost between the real world and the virtual world. The distribution of 1414 Internet of Things projects introduced to the market in 2020 by sectors is demonstrated in Fig. 1. Accordingly, industrial production, transportation, energy, retail and urban applications constitute the majority of new products [11].

Contributions

In this study, application areas of IoT studies, communication protocols used in application areas, sensors and embedded systems are examined thoroughly. Contributions of this article are as follows:

- First, various application areas are examined.
- Second, communication protocols used in IoT systems are examined.
- Third, the types of sensors used in IoT systems are examined.
- Fourth, embedded systems where IoT studies are carried out are examined.

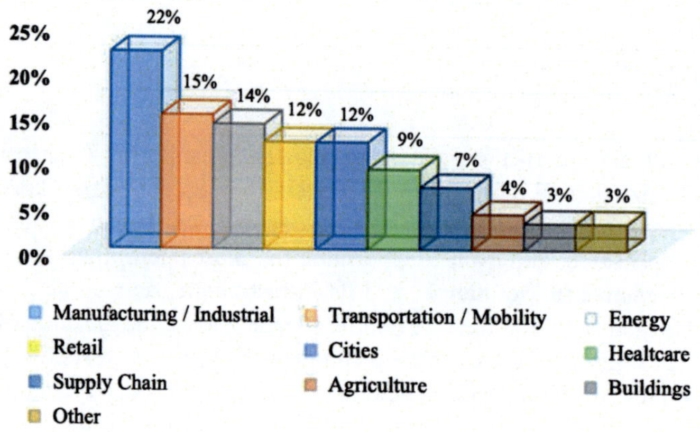

Fig. 1 Top 10 IoTs application areas 2020

Finally, it was discussed which communication protocols would be appropriate to use in IoT application areas.

The survey was organized as follows. Chapter 1 gives an outline of IoT. In Chap. 2 IoT application areas, in Chap. 3 IoT technologies and communication Protocols used, in Chap. 4 Embedded systems where IoT studies are carried out, in Chap. 5 Sensors used in IoT applications and in Chap. 6 the results are given.

Application Areas of IoT

IoT solutions aim at the interaction of people and objects. To this end, IoT contributes to studies in many different areas and facilitates human life. Figure 2 shows the application areas where IoT systems provide solutions.

Transportation/Mobility

The transportation industry is constantly evolving to be safer, faster and more comfortable. Autonomous vehicles equipped with Internet of Things technologies stand out in the transportation sector. IoT devices offer benefits such as weather analysis to help drivers explain potential disruptions. Fuel consumption monitoring helps fleet managers track routine expenses like gas and repairs. These features help companies save money and provide faster delivery to customers. Typical applications in the transportation industry include solutions such as battery monitoring [12], tire pressure monitoring [13], driver monitoring [14] and fleet management [15].

Fig. 2 Main IoT application areas

Manufacturing/Industrial

IoT systems have many benefits for industrial systems as well as in many areas around us. IoT systems used in industry offer positive impact in terms of quality and performance as well as low cost and high efficiency.

By enabling machines to communicate with each other, IoT reduces human effort and shortens the intervention time by predicting potential errors. Real-time monitoring and control of operations are two of the Internet of Things' primary functions in the manufacturing sector [16], Assigning smart machines, smart sensors, smart controllers with special communication and Internet technologies [17], Maximizing security, reliability and dependability through high precision automation and control [18].

Energy

By virtue of sensors and smart devices, energy consumption in facilities can be monitored much more thoroughly. The real-time data obtained can be examined with intensive data processing capabilities and advanced applications. From the perspective of today's technology, unnecessary and costly energy consumption can be detected instantly. With the help of smart meters and sensors, energy consumption can be monitored and controlled in real time. Thanks to the energy management, cost of the energy consumed is monitored instantly and no surprise energy bills are encountered. In either large or small businesses, uncontrollable or unpredictable costs negatively affect business decision-making processes. The real-time determination of the energy consumed, and the calculation of its cost show the instantaneous deviations that may occur in the overhead tables as a warning to the decision makers and allow to take quick measures. Projects in the energy field are focused on smart energy distribution [19], grid optimization [20], remote asset monitoring and management [21], maintenance forecasting and creating greater transparency for customers. Energy consumption and other benefits of energy efficiency:

- Thanks to the advanced automation systems provided by smart technologies, the energy consumption level decreases and the production increases
- More preference for renewable energy sources
- Ensuring uninterrupted energy transfer
- Providing fast access to problems with uninterrupted monitoring of power lines
- Reducing environmental pollution with the increase of renewable energy sources
- Reduction of expenses such as transportation by managing the systems with remote control systems
- Providing economically available, usable and accessible energy.

Retail

The retail industry leverages the vast amount of data collected through these interactions to increase operational efficiency. Using IoT sensors to track the consumer's path in the store improves the merchandise placement strategy.

In the retail industry, IoT application solutions encompass a range of functionalities such as in-store digital signage [22], customer tracking and feedback, product tracking, inventory management, smart vending machines, smart shelves [23], and cashierless stores [24].

Smart Cities

Smart cities can be defined as a city with internet-based solutions to increase the quality of life and work efficiency of people. Intelligent transportation systems are applications that are generated to find solutions to problems such as smart road lighting, smart traffic lights, smart city management, parking lot [25–27]. They enable efficient use of resources. They are successful applications in terms of environmental protection and information systems. Management of solid waste collection process can be cited as an example of smart cities. In a Smart City, waste collection is a crucial issue for the environment and its quality should be seriously considered. Inadequate waste collection systems lead to environmental pollution. This leads to proliferation of insects and rodents and spread of various diseases.

Healtcare

IoT applications in the field of health have recently become widespread. In this field, the demand for certain health applications such as remote health monitoring, digital diagnosis, remote monitoring and robotic assistance are increasing [28–30]. Moreover, in some areas such as sports and personal care, products for the end user that can be described as health products are emerging. In general, these products aim to improve the quality of life.

Agriculture

The greatest contribution of the industrial revolution to agriculture is the transition to production using machine power instead of human power. By making use use of IoT technology in agriculture, the concept of smart agriculture has emerged. Smart agriculture increases the sustainability and improvability of agricultural areas, while reducing the cost. IoT applications can help farmers make decisions to reduce the use of fertilizers and pesticides to provide higher yields, higher quality and more efficient production [31, 32]. It can be used to start automatic irrigation at certain times with smart irrigation systems in agriculture. With the humidity measurement sensor, measures can be taken against flood. With the temperature sensor, measures can be taken against drying out of the plants, and this kind of applications exist.

Supply Chain

A supply chain refers to a system of producers and distributors that is accountable for acquiring raw materials, converting them into intermediate goods and finished products, and distributing those end products to consumers [33]. Supply chain management involves planning and controlling both the flow of information and logistical activities within the company as well as other companies involved in the supply chain. Competition is increasing day by day. It will continue to increase in the future. This will occur not between companies, but between supply chains in which companies are involved. Effective supply chain management is not possible without information sharing and coordination between supply chain members. Information sharing is a key element in supply chain management. Supply chain members must have the necessary information in a timely manner to manage their activities. With the development of information technologies, information sharing between companies has become much easier and it has been seen that companies that share information effectively have achieved significant success in supply chain integration.

Wearable Technologies

The concept of wearable technology is the integration of technology into everyday clothing. Wearable technology can be used to measure and record functions such as number of steps, sleep patterns, eating patterns, heart problems, heart rate, and body temperature. Along with the great interest of people in wearable technology, it is also well-invested economically. For this reason, it can be integrated into any garment, from health to sports. The most common applications: Smart watches, Smart sports products, Smart implants, Smart jewelry can be given as examples.

Technologies and Protocols Using IOT

In Internet of Things applications, different technologies such as IPv6 [34], 6LowPAN [35], ZigBee [36], Bluetooth [37], Wi-Fi [38], LoRaWAN [39], Z-Wave [40], NFC [41] and Sigfox [42] are used for the communication of devices with each other. In Fig. 3, IoT protocols are given.

RFID

RFID (Radio Frequency Identification) is an automated identification technology used to uniquely identify and track objects. An RFID system consists of two main components that facilitate the transfer of information using radio frequency signals: RFID tags and RFID readers.

RFID Tags: RFID tags are small devices comprising a microchip and an antenna. Each tag contains a unique identification number and other stored data on the

Fig. 3 IoT protocols

microchip. These tags can be placed on or integrated into objects. For instance, they can be found on product packaging, animal ears, or even within book pages.

RFID Readers: RFID readers are devices that use radio frequency signals to read or write information from RFID tags. These readers employ an antenna to communicate with the tags, capturing data from them or writing new information onto the tags. Typically, these readers are connected to a network and can process the data in a centralized system.

When combined with IoT systems, RFID enables objects to connect to the internet and communicate with each other. RFID tags allow real-time tracking and monitoring of objects, while IoT platforms facilitate the collection, storage, and analysis of this data. This enables applications such as data sharing between objects, automation of processes, inventory management, tracking, and monitoring.

The integration of RFID with IoT systems offers numerous benefits across various industries and sectors. For example, in the retail industry, RFID tags automate inventory management, making stock tracking more efficient and preventing theft. In logistics and transportation, RFID optimizes the tracking of goods and streamlines the logistics chain. In the healthcare sector, RFID improves patient tracking, inventory management of medications, and enhances patient care processes.

In conclusion, RFID technology, when integrated with IoT systems, forms the foundational infrastructure for data sharing between objects, management of automated processes, and the realization of a smarter world.

IPv6

IP (Internet Protocol) is the common rule that devices on the network use to communicate with each other. IPv6 (Internet Protocol version 6) is a network protocol used for internet communication. It addresses the limitations of IPv4 and offers several advantages. IPv6 expands the address space by increasing the size of IP addresses from 32 to 128 bits. This expansion enables a much larger number of devices to connect to the internet. With approximately 3.4×10^{38} available addresses, IPv6 supports hierarchical addressing levels and simplifies address autoconfiguration.

IPv6 also incorporates enhanced security features. By integrating security protocols like IPsec, it ensures data encryption and integrity, providing a more secure communication environment.

Overall, IPv6 is an evolved network protocol designed to meet the growing demands of the internet, such as the expansion of devices and addressing hierarchies. It offers a larger address space, improved addressing and autoconfiguration mechanisms, and enhanced security features, ultimately delivering a more secure and scalable internet experience.

6LoWPAN

6LoWPAN (Low Power Wireless Personal Area Network Using IPv6) is a developing standard within the Internet realm. It offers easy adaptability to various applications. This open-source protocol, developed by the IETF (Internet Engineering Task Force), is designed to facilitate data transmission over Low-Rate Wireless Personal Area Networks (LR-WPANs) to devices that support IPv6.

Key Features of 6LoWPAN include:

- Compact code size of 22 K (Cage code) and 12 K (Code size without weave)
- RAM requirement of 4 K
- Supports a network size of up to 264 devices
- Compatible with RF radio support of 802.15.4 and beyond
- Provides transport layer options of UDP/TCP
- Internet connectivity is achieved through bridges/routers or similar means
- Utilizes different header stacks to enable its functionality.

IEEE 802.15.4

IEEE 802.15.4 is a standard protocol created for LR-WPAN (Low Speed Wireless Personal Area Network). LR-WPAN is a type of network used in data communication between devices in IoT applications.

Wi-Fi

It is a widely used wireless communication technology for two-way communication in the IEEE 802.11 standard. It has a frequency band between 2.4 GHz and 5 GHz. Current consumption is 350 mA. Due to the high frequency band, many devices can be connected to the same wireless network at the same time. It has a communication network of approximately 35 m for indoor and 100 m for outdoor. It includes the 802.11 standard and three different class frameworks. These are control, management, and data frameworks.

ZigBee

ZigBee is a wireless networking technology that follows industry standards and promotes high interoperability. Its main objective is to facilitate communication between Internet of Things (IoT) devices. ZigBee is a cost-effective wireless networking solution that emerged to address the need for supporting low data speeds, low power consumption, security, and reliability. It is based on the IEEE 802.15.4 standard. ZigBee's network architecture is specifically designed to be self-optimizing.

Bluetooth

Bluetooth technology was first developed by Ericsson Mobile in 1994. Later, the Bluetooth Special Interest Group, established in 1998, continued the development of Bluetooth technology. Bluetooth, which has been standardized under IEEE 802.15.1, is a wireless communication technology utilized for short-range areas and low power consumption. It operates within the 2.45 GHz radio frequency band. It can transfer data between 2.402 GHz and 2.480 GHz. The range is between 1–10 m. Current consumption is 65–170 mA. It is used for both synchronous and asynchronous traffic. It is on the frequency-hopping radio platform. In summary, Bluetooth technology has been developed to enable short-range, mobile and/or fixed devices to communicate with each other without a cable connection.

LoRaWAN

LoRaWAN, short for "Long Range Broad Area Network," is an open-source technology designed for wide area networks that prioritize low power consumption. In the context of Internet of Things (IoT) applications, the longevity of a node's battery is crucial for network capacity, service quality, and security. LoRaWAN networks adopt a star topology configuration, where gateways facilitate communication between end devices and the central network server located at the backend. The technology allows for data transmission rates ranging from 0.3 kbps to 50 kbps, depending on the specific environment. In urban areas, communication with LoRaWAN can span distances of 2 to 5 km, while outside city limits, it can reach up to 15 km.

Z-Wave

It is a protocol that is in the link layer of the protocols, uses low-energy radio waves, and communicates with a wireless network in smart switches, smart home systems, and security systems. It allows home control systems from different manufacturers to be interconnected. It is essential that data is sent in small packets within a 30-m radius with a short delay. Although the data transmission rate is low, the range is higher than technologies such as Bluetooth and BLE.

NFC

Near Field Communication (NFC) is a technology that enables easy data transmission between mobile phones, tablets, laptops, and other electronic devices within a short range. It is an advancement built upon RFID technology. NFC communication operates within a distance of up to 10 cm.

Sigfox

Sigfox is a communication protocol that facilitates extensive signaling between objects. It adopts a single-hop star topology mesh as its network structure. It utilizes two primary modulation techniques: differential binary phase shift keying and Gaussian frequency shifting keying. In Europe, Sigfox operates within the 868 MHz ISM radio band, while in America, it operates within the 902 MHz radio frequency band. The data is encapsulated in packets and transmitted randomly across three different frequencies with diverse source coding. One advantage of Sigfox is its asynchronous nature, which eliminates the need for node synchronization and conserves energy. The design of Sigfox is specifically tailored to fulfill the requirements of IoT networks.

MQTT

MQTT (Message Queueing Telemetry Transport) is a lightweight telemetry messaging protocol that operates over TCP/IP and follows a broadcast-subscribe model. It was initially created in 1999 by Dr. Andy Stanford-Clark and Arlen Nipper to facilitate communication between devices. In 2013, IBM introduced MQTT v3.1 to the OASIS consortium, leading to its standardization. Its current version is V5.0 [43]. This protocol provides the establishment of communication between devices and occupies an important place in the field of Internet of things. MQTT is a protocol that provides uninterrupted, continuous and bidirectional connectivity for devices used in IoT applications. The protocol strives to ensure transmission reliability and accuracy while minimizing bandwidth and device resource requirements.

MQTT (Message Queueing Telemetry Transport) protocol supports various messaging models and incorporates reliability mechanisms that facilitate efficient communication in IoT environments.

Messaging Models:

- Publish-Subscribe Model: MQTT adopts a publish-subscribe messaging model, where devices, known as publishers, send messages (publish) to specific topics. Other devices, called subscribers, can subscribe to these topics and receive the relevant messages. This model enables data sharing and allows multiple devices to receive the same data concurrently.

- Request-Response Model: MQTT also supports a request-response communication model. In this model, devices, referred to as clients, send requests to a specific topic, and devices known as servers respond to these requests. This model facilitates direct communication between devices in IoT applications.

  Reliability Mechanisms:

- Quality of Service (QoS): MQTT offers different levels of Quality of Service (QoS), which determine the reliability of message delivery. MQTT provides three QoS levels: QoS 0 (At most once delivery), QoS 1 (At least once delivery with acknowledgement), and QoS 2 (Exactly once delivery).
- Retain Feature: MQTT's retain feature allows published messages to be stored by the server. This ensures that subscribers can receive the most recent updates when they connect. This mechanism ensures data integrity and prevents disruptions in communication.

These messaging models and reliability mechanisms contribute to MQTT's ability to facilitate efficient communication in IoT environments. By utilizing low bandwidth and optimizing device resources, MQTT enables fast and reliable data exchange among IoT devices.

CoAP

In 2010, the IETF CoRE working group started to develop a protocol that enables restricted devices to communicate over the Internet and interoperability, and CoAP *(The Constrained Application Protocol)* has emerged. The main goal of CoAP is to provide an HTTP-based general communication protocol service for devices with limited special requirements used in IoT applications, considering machine-to-machine (M2M) applications [44].

CoAP (The Constrained Application Protocol) offers a range of features specifically designed for machine-to-machine (M2M) communication. These include support for multicast and asynchronous communication. CoAP is a web protocol that effectively fulfills the requirements of M2M applications operating in restricted environments. It allows for both unicast and multicast support, optionally utilizing UDP as the underlying transport protocol. One of its key features is asynchronous message exchange, facilitating efficient and timely communication. Additionally, CoAP boasts a small packet structure, optimizing resource usage and transmission efficiency.

CoAP incorporates specific messaging models and reliability mechanisms that facilitate efficient communication in IoT environments.

Messaging Models:

- Request-Response Model: CoAP primarily follows a request-response messaging model. IoT devices, referred to as clients, initiate requests to specific resources on other devices, known as servers. The servers respond with the requested data or perform the requested actions. This model enables devices to interact and exchange information in a client–server architecture.

- Observing Model: CoAP introduces an observing model, which allows devices to observe resources on other devices. A client can express interest in a resource and receive notifications whenever that resource changes. This model enables real-time updates and facilitates efficient monitoring and event-driven communication in IoT applications.

Reliability Mechanisms:

- Reliable Transfers: CoAP offers reliable transfers through retransmission mechanisms. When a client sends a request to a server, it expects an acknowledgment. If the acknowledgment is not received within a specified time, the client retransmits the request. This ensures reliable delivery of messages in unreliable networks.
- Message Integrity: CoAP supports message integrity through the use of Message Authentication Codes (MAC) and Cryptographic Message Syntax (CMS). These mechanisms provide data integrity and security by verifying the authenticity and integrity of the messages exchanged between devices.

Facilitating Efficient Communication in IoT Environments: CoAP's design principles and lightweight nature make it well-suited for efficient communication in IoT environments. Some key factors contributing to its efficiency are:

- Low Overhead: CoAP is designed to have a low protocol overhead, making it suitable for resource-constrained IoT devices with limited processing power and memory.
- UDP-based Protocol: CoAP operates over the UDP (User Datagram Protocol), which is a lightweight transport protocol. UDP's connectionless nature and minimal overhead make it efficient for transmitting small-sized IoT messages.
- Caching and Proxying: CoAP supports caching and proxying mechanisms, allowing intermediaries to store and serve frequently accessed resources. This reduces network traffic and improves response times, enhancing the overall efficiency of IoT communication.

In conclusion, CoAP's messaging models, reliability mechanisms, and focus on efficiency make it a viable protocol for facilitating reliable, efficient, and secure communication in IoT applications. It addresses the challenges of constrained environments in the IoT ecosystem.

Embedded Systems Used with IOT

Microprocessors, which contain extremely complex electronic circuits in their internal structure, entered the life of human beings at the beginning of the 1970s. The processing capacity of microprocessors is increasing day by day with the advancement of technology [45, 46]. Contrary to their increasing capacities, the physical dimensions of microprocessors are getting smaller and smaller. The first designed microprocessors had thousands of transistors in their structure and processor speeds were in the kHz range. Microprocessors produced today contain billions of transistors and can operate at GHz-level speeds with their multi-core designs. Microprocessor systems are electronic systems that are based on a central processing unit, that can execute certain algorithms with the help of control circuits, registers, timers and interrupt controller units, and that also perform continuous data communication with memory units and input/output peripherals [47]. When compared to a personal computer (PC), embedded systems have reduced capabilities in either their hardware or their software. An embedded system is a type of computer system that is purpose-built to carry out a particular task and typically has stricter standards regarding its quality and reliabilityIt is possible for very serious issues to arise if, for instance, the electrical system that controls the engine of the vehicle fails to function properly while the driver is operating the vehicle, or if an essential medical gadget malfunctions while the patient is undergoing surgery. On the other hand, there are also embedded systems, such as televisions, gaming devices, and mobile phones, in which the breakdown of the system does not typically result in a situation that is life-threatening [48]. With the advancement of electronics technology, embedded systems have become cheap enough to fit into most electronic devices. The features expected from an embedded system can be listed as follows: Small size, light weight, low power consumption, ability to work under high temperature and pressure, resistance to vibration and chemical exposure, ability to work in harsh field conditions such as electrical noise and electromagnetic interference [49, 50].

Depending on the developments in technology, various embedded system platforms have been created by using FPGA, ARM-based processors or multi-core processors with different architectural structures. There are development platforms on the market with different names, each with its own advantages. Raspberry Pi, ZYNQ, Jetson, Altera Cyclone II, Beagle Bone, Odroid, STM32, PSoC are some of them.

FPGA

Field-programmable gate arrays (FPGAs) are electronic devices consisting of a grid of programmable logic blocks (CLB) interconnected through programmable interconnects. What sets FPGAs apart from application-specific integrated circuits (ASICs) is their ability to be reprogrammed after manufacturing, allowing them to adapt to the specific requirements of the desired application or functionality. This flexibility enables FPGAs to handle a diverse range of design tasks. One-Time Programmable (OTP) Field-Programmable Gate Arrays (FPGAs) do exist; however, the most common varieties are SRAM (Static Random-Access Memory)-based FPGAs, which have the ability to be reprogrammed as the design develops. Due to the fact that FPGAs have been produced with better and newer technology recently, their application areas have changed compared to ASICs and have been used in many different areas [51, 52].

Raspberry Pi

It is a credit card sized computer and was manufactured in the UK with the intention of providing an inexpensive computing device for education. Its origins date back to work done in the Cambridge University computer lab in 2006. The first commercial release of Raspberry Pi was made in February 2012 [53]. Raspberry Pi runs on Linux operating system. It is a low-cost, 32-bit ARM credit card-sized computer that plugs into a computer monitor or TV and uses a standard mouse and keyboard.

Zynq

The system comprises an ARM Cortex A-9 based processor and Xilinx-7 series FPGA, offering hardware programmability [54]. This unique feature allows the CPU to integrate DSP (Digital Signal Processing), ASSP (Application-specific standard parts), and mixed signal functionality. The combination of these capabilities enables hardware acceleration [49, 50]. Development boards featuring the Zynq-7000 chip, which combines FPGA and microprocessor, are available. Examples of such development boards include Zedboard, GENESYS ZU-3EG ZYNQ, Cora Z7-10, and PYNQ-Z1.

STM32

STM32 is a microprocessor that is powered by a CPU based on the Arm Cortex-M architecture and is manufactured by STMicroelectronics. In its most basic form, it is made up of a microprocessor, random access memory (RAM), flash memory for storing the program to be run, and other interfaces, the exact number of which depends on the model. Electronic structures that can be programmed are referred to as interfaces. Each interface within the system is assigned specific memory addresses for programming purposes. The presence of multiple interfaces allows for the implementation of a wide range of applications with minimal electronic components. This not only enables versatile functionality but also helps reduce production costs. The STM32 embedded systems incorporate various components such as ultra-low power comparators, low-power timers (LPTIM), digital-to-analog converters (DAC), analog-to-digital converters (ADC), and 16-bit timers, among others. [55]. The STM32 family has many members. The latest members are listed as STM32F2, STM32H7, STM32F4, STM32F7.

PSoC

It is a family of microprocessor integrated circuits developed by Cypress Semiconductor. These chips contain mixed-signal arrays of CPU cores and configurable integrated analog and digital peripherals.

The PSoC family is distinguished from conventional microcontrollers by the presence of a programmable central processing unit (CPU) core as well as analog and digital blocks. The only component of a conventional chip is the central processing unit (CPU), but an ASIC, FPGA, or CPLD may contain customizable blocks.

PSoC is a one-of-a-kind system since it integrates the architectures of both. PSoCs integrate the capabilities of multiple chips into a single device and include both programmable analog and digital circuitry. These are managed by a microcontroller that is integrated onto the chip, which results in an increased capacity for design revisions as well as a decreased component count [56, 57].

Systems designed with PSoC take up less space on the PCB, consume less power, provide high efficiency and reduce system cost. Peripherals allow interaction with the physical world.

IoT Sensors

In IoT technology, there are many types of sensors with different functions. During the studies, it is determined which sensors will be included depending on the purpose of the study (Fig. 4).

In IoT applications, Sensors such as Heart rate sensor, ECG (Electrocardiography) sensor, oxygen sensor, temperature sensor, acceleration sensor, Water level Sensor, Air quality control system, Audio HD, Video sensors, Video LD, Image, Acoustic sensors, Vibration, Velocity, Humidity, temperature are frequently used. Comparison of sensors, embedded systems and communication protocols used in IoT applications is given in Table 1.

Heart rate sensors: measure the heart rate.

Electrocardiography sensors: allow monitoring of heart rate.

Ultrasonic sensors are non-invasive instruments that can detect solid and liquid objects without physical contact. Detecting object proximity and fluid recognition and concentration are just two of the many functions that these sensors are put to use for. They may also be used to monitor the water level in a tank. Applications related to the identification and concentration of liquids include observing differences in the speed of sound's propagation through various liquids. The sensor calculates the ToF value, often known as the time of flight, of a specific distance. This value can be analyzed by comparing it with values in a lookup table used to identify the liquid.

Light sensors: It is a simple sensor that changes the voltage of Photo resistor or Photovoltaic batteries in parallel with the amount of light detected.

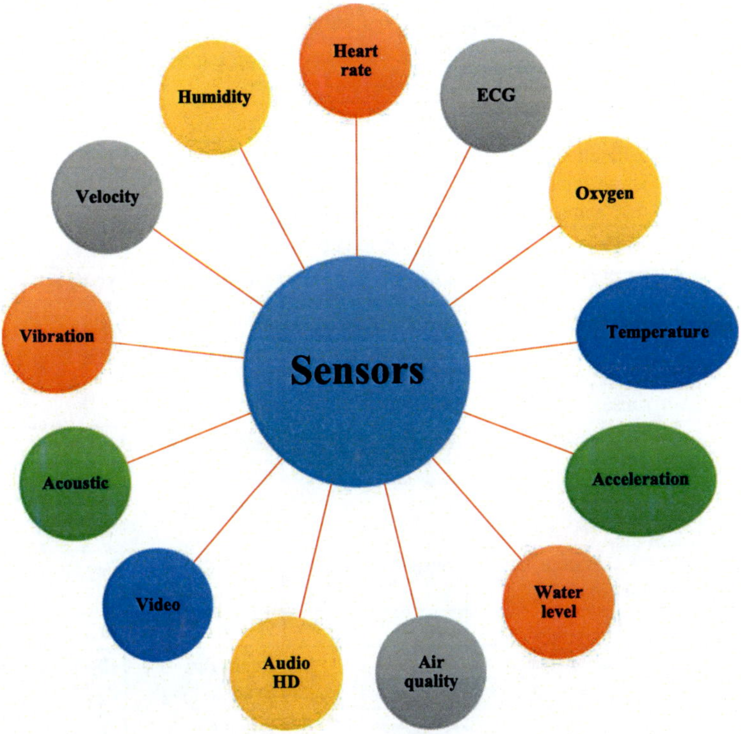

Fig. 4 Sensors used in IoT studies

Color sensors: The reflection of different colors at different intensities is decisive. For example, the color orange reflects red light in a greater amount than the color green. It can be used to detect the presence of an object with a certain color or to detect the type of objects or surfaces.

Touch sensor: They are designed to detect objects by direct contact or at a small distance.

Infrared sensors: measure IR light transmitted through the IR LED to locate objects in an environment. It is widely used in the Internet of Things to create smart products that use sensors such as infrared sensors, proximity sensors, motor control and level sensors.

Sonar sensors: primarily used in navigation applications for object detection.

Laser sensors: used to track and detect a long-range target with laser light.

Pressure Sensors: Pressure sensors detect pressure and convert it into electrical signals.

Water Quality Sensors: These are the sensors used especially in the water distribution networks of cities. With sensors that can instantly send data to the monitoring center via the Internet, water quality and drinking water after disasters are controlled.

Table 1 Comparison of Sensors used in IOT applications

References	Sensors	Board	Application area	Communication system-protocol
[58]	EEG	Raspberry Pi 2	Healtcare	MQTT, XMPP
[59]	Infrared, Ultrasonic, Gass, 3-six compass	Arduino	Waste Management	802.11 wireless LAN, LoRa
[60]	ECG	STM32F103RC	Healtcare	HTTP, MQTT
[61]	DHT-11, YL-69	Arduino	Smart Garden	ESP8266 wifi
[62]	Heart Rate	Raspberry Pi 3	Healtcare	IP
[63]	LDR sensor, IR sensor, temperature sensor	Arduino Uno	Smart Cities	ESP8266 Wifi
[64]	Ultrasonic	WeMos D1	Smart Cities	Wi-fi
[65]	DHT11 sensor, MQ9 gas sensor, FC37 rain sensor	Arduino	Weather İntelligence	ESP8266 wifi
[66]	Temperature, Humidity, Light,	SmartRF05BB	Smart Cities	Zig Bee
[67]	ECG	ZynQ	Healtcare	TCP/IP
[68]	Ultrasonic sensor	Arduino Atmega328	Smart Bin System	ESP8266 wifi
[69]	IR	Raspberry Pi 2	Smart Cities	Wi-Fi Card/GSM
[70]	DHT11	FPGA	Smart Technologies	Internal Chip Bus
[71]	Wireless Sensor Network	PSoC	Smart Technologies	6LoWPAN
[72]	Temperature and humidity	MCU	Agriculture	RS485
[73]	DHT11, PIR, MQ-135	Arduino Mega	Retail	HTTP
[74]	MQ-4, MQ-7, MQ-135, Ultrasonic	NodeMCU	Retail	ESP32 WIFI
[75]	Current	PSoC	Energy	Bluetooth
[76]	LDR, IR	Arduino Uno R3	Smart Cities	ESP8266 WIFI
[77]	Blood Sensors	MCU	Healthcare	Internet
[78]	Moisture, Temperature	Arduino Uno, Raspberry Pi	Agriculture	ESP8266 WIFI
[79]	Wireless sensors	–	Healthcare	RFID
[80]	Wireless sensors	–	Security	Hybrid Key Management Scheme
[81]	Wireless sensors	–	Security	–

Chemical Sensors: They are most commonly used in industrial areas for ambient measurement and inspection. It provides early detection and control of possible leaks in facilities using thousands of chemical types.

Gas Sensors: provide early diagnosis and control of possible gas leaks in facilities. Types such as carbon dioxide, hydrogen, ozone, natural gas, and flue gas meters are the most commonly used.

Smoke Sensors: Optical types detect the shading of rays by smoke, while ionization types work according to the principle of reacting with chemicals in the air.

Infrared Sensors: These are sensors that monitor the changes in the reflections of infrared rays. It can detect motion as well as detect heat absorbed by objects.

Catheter Ablation Sensors: Silicone MEMS Based Pressure Sensors: used in intrauterine pressure sensors to measure contraction pressure and frequency during labor.

Body Cavity Sensors: They measure body temperatures. It has oral and rectal probes.

Conclusion

IoT is gaining popularity as it provides convenience, increases the quality of life, saves time, reduces workload and increases performance in all aspects of human life. In this study, an overview of IoT workspaces has been provided and its applications in many areas have been presented. In IoT applications, there is a wide range of technologies and protocols available for communication between devices. These include IPv6, 6LowPAN, ZigBee, Bluetooth, Wi-Fi, LoRaWAN, Z-Wave, NFC, and Sigfox. The selection of a specific technology depends on factors such as the number of connected devices, data transmission rate, power requirements, and communication distance. Choosing the appropriate technology is crucial and depends on the specific application requirements. For instance, for smart farming applications covering large areas of land, wide area networks like Sigfox and LoRaWAN are preferred. In smart home systems, protocols such as Z-Wave or ZigBee can be chosen, which are specifically designed for this area. The sensors and embedded systems used in the current studies have been examined and their characteristics are presented in Table I.

Declaration of Competing Interest: We declare that we have no conflict of interests.

References

1. Ashton, K. (2009). That 'internet of things' thing. *RFID Journal, 22*, 97–114.
2. Sam, T. H., Wong, W. Y., Gwadabe, Z. L., Balakrishnan, R., Poopalaselvam, R., Adam, A., & Tee, K. S. (2021). The adoption of IoT technology in the Malaysian manufacturing industry. In *AIP Conference Proceedings*, p. 30002. AIP Publishing LLC.

3. Abbas, K., Tawalbeh, L. A., Rafiq, A., Muthanna, A., Elgendy, I. A., El-Latif, A., & Ahmed, A. (2021). Convergence of blockchain and IoT for secure transportation systems in smart cities. *Security Communication Networks.*
4. Sanislav, T., Mois, G. D., Zeadally, S., & Folea, S. C. (2021). Energy harvesting techniques for internet of things (IoT). *IEEE Access., 9*, 39530–39549.
5. Kaur, J., Santhoshkumar, N., Nomani, M. Z. M., Sharma, D. K., Maroor, J. P., & Dhiman, V. (2021). Impact of internets of things (IOT) in retail sector. *Material Today Proceedings.*
6. Abd El-Latif, A. A., Abd-El-Atty, B., Mehmood, I., Muhammad, K., Venegas-Andraca, S. E., & Peng, J. (2021). Quantum-inspired blockchain-based cybersecurity: Securing smart edge utilities in IoT-based smart cities. *Information Processing and Management, 58*, 102549.
7. Marques, G., Bhoi, A. K., Victor Hugo, C., & Hareesha, K. S. (2021). *IoT in healthcare and ambient assisted living.* Springer.
8. Bamakan, S. M. H., Faregh, N., & ZareRavasan, A. (2021). Di-ANFIS: An integrated blockchain–IoT–big data-enabled framework for evaluating service supply chain performance. *Journal Computer Design Engineering, 8*, 676–690.
9. Babu Loganathan, G. (2021). Smart agriculture system with E-carbage using iot. *International Journal Model Agricuture, 10*, 928–931.
10. Majeed, U., Khan, L. U., Yaqoob, I., Kazmi, S. M. A., Salah, K., & Hong, C. S. (2021). Blockchain for IoT-based smart cities: Recent advances, requirements, and future challenges. *Journal Network Computer Application*, 103007.
11. P., S. (2020). Top 10 IoT applications in 2020. IoT Analytics: Market Insights for the Internet of Things, https://iot-analytics.com/top-10-iot-applications-in-2020/
12. Bharathi, S. H. (2021). IoT-based battery monitoring system for electric vehicle. *Turkish Journal Computer Mathmatics Education, 12*, 3524–3528.
13. Ingemarsdotter, E., Diener, D., Andersson, S., Jonasson, C., Mellquist, A.-C., Nyström, T., Jamsin, E., & Balkenende, R. (2021). Quantifying the net environmental impact of using IoT to support circular strategies—the case of heavy-duty truck tires in Sweden. Circular Economy and Sustainability, 1–38.
14. Kong, Q., Lu, R., Yin, F., & Cui, S. (2021). Blockchain-based privacy-preserving driver monitoring for maas in the vehicular IoT. *IEEE Transactions on Vehicular Technology, 70*, 3788–3799.
15. WORK, I. I. P., Securing a Vehicle Fleet Management Through Blockchain and Internet of Things.
16. Bhaskaran, P. E., Maheswari, C., Thangavel, S., Ponnibala, M., Kalavathidevi, T., & Sivakumar, N. S. (2021). IoT Based monitoring and control of fluid transportation using machine learning. *Computers & Electrical Engineering, 89*, 106899.
17. Rathee, G., Balasaraswathi, M., Chandran, K. P., Gupta, S. D., & Boopathi, C. S. (2021). A secure IoT sensors communication in industry 4.0 using blockchain technology. *Journal Ambient Intelligent Humaniz. Computer, 12*, 533–545.
18. Coman, C. M., D'amico, G., Coman, A. V., & Florescu, A. (2021). Techniques to improve reliability in an IoT architecture framework for intelligent products. *IEEE Access, 9*, 56940–56954.
19. Avancini, D. B., Rodrigues, J. J. P. C., Rabêlo, R. A. L., Das, A. K., Kozlov, S., & Solic, P. (2021). A new IoT-based smart energy meter for smart grids. *International Journal of Energy Research, 45*, 189–202.
20. Hameed, Z., Raza, N., Sial, M.R.K., Ghafoor, Z., & Ahmad, F. (2021). Smart grid optimization by implementing the algorithm of artificial neural network. In *2021 National Computing Colleges Conference (NCCC)*, pp. 1–7. IEEE.
21. Parrino, S., Peruzzi, G., & Pozzebon, A. (2021). LoPATraN: Low power asset tracking by means of narrow band IoT (NB-IoT) technology. *Sensors, 21*, 3772.
22. Lasrado, S., Ghodinde, A., David, S., & Nivelkar, M.: *Surveillance using digital signage.*
23. Li, R., Song, T., Capurso, N., Yu, J., Couture, J., & Cheng, X. (2017). IoT applications on secure smart shopping system. *IEEE Internet of Things Journal, 4*, 1945–1954.

24. Schögel, M., & Lienhard, S. D. (2020). Cashierless stores–the new way to the customer? *Marketing Review St. Gallen.*
25. Hammoudeh, M., Epiphaniou, G., Belguith, S., Unal, D., Adebisi, B., Baker, T., Kayes, A. S. M., & Watters, P. (2020). A service-oriented approach for sensing in the Internet of Things: intelligent transportation systems and privacy use cases. *IEEE Sensor Journal.*
26. Tripathy, A. K., Mishra, A. K., & Das, T. K. (2017). Smart lighting: Intelligent and weather adaptive lighting in street lights using IOT. In *2017 International Conference on Intelligent Computing, Instrumentation and Control Technologies (ICICICT)*, pp. 1236–1239. IEEE.
27. Frank, A., Al Aamri, Y. S. K., & Zayegh, A. (2019). IoT based smart traffic density control using image processing. In *2019 4th MEC International Conference on Big Data and Smart City (ICBDSC)*, pp. 1–4. IEEE.
28. Pathinarupothi, R. K., Durga, P., & Rangan, E. S. (2018). Iot-based smart edge for global health: Remote monitoring with severity detection and alerts transmission. *IEEE Internet of Things Journal, 6*, 2449–2462.
29. Mishra, S. S., & Rasool, A. (2019). IoT health care monitoring and tracking: A survey. In *2019 3rd International Conference on Trends in Electronics and Informatics* (ICOEI), pp. 1052–1057. IEEE.
30. Verma, V., Mishra, A., Bisht, V., & Bala, J. (2020). Internet of diagnostic things: emerging horizon towards precision and digital health care. *Journal Health Science Research*, 51–61.
31. Hnatiuc, M., Geman, O., Avram, A. G., Gupta, D., & Shankar, K. (2021). Human signature identification using iot technology and gait recognition. *Electron, 10*. https://doi.org/10.3390/electronics10070852
32. Suma, V. (2021). Internet of Things (IoT) based smart agriculture in India: An overview. *Journal ISMAC, 3*, 1–15.
33. Lee, H. L., & Billington, C. (1992). Managing supply chain inventory: Pitfalls and opportunities. *Sloan Management Review, 33*, 65–73.
34. Savolainen, T., Soininen, J., & Silverajan, B. (2013). IPv6 addressing strategies for IoT. *IEEE Sensors Journal, 13*, 3511–3519. https://doi.org/10.1109/JSEN.2013.2259691
35. Gündoğan, C., Kietzmann, P., Schmidt, T. C., & Wählisch, M. (2018). ICN-LoWPAN: header compression for the constrained IoT. In *Proceedings of the 5th ACM Conference on Information-Centric Networking*, pp. 184–185.
36. Ronen, E., Shamir, A., Weingarten, A.-O., & O'Flynn, C. (2017). IoT goes nuclear: Creating a ZigBee chain reaction. In *2017 IEEE Symposium on Security and Privacy (SP)*, pp. 195–212. IEEE.
37. Chang, K.-H. (2014). Bluetooth: A viable solution for IoT?[Industry Perspectives]. *IEEE Wireless Communications, 21*, 6–7.
38. Tozlu, S., Senel, M., Mao, W., & Keshavarzian, A. (2012). Wi-Fi enabled sensors for internet of things: A practical approach. *IEEE Communications Magazine, 50*, 134–143.
39. Rizzi, M., Ferrari, P., Flammini, A., & Sisinni, E. (2017). Evaluation of the IoT LoRaWAN solution for distributed measurement applications. *IEEE Transactions on Instrumentation and Measurement, 66*, 3340–3349.
40. Babun, L., Aksu, H., Ryan, L., Akkaya, K., Bentley, E. S., & Uluagac, A. S. (2020). Z-iot: Passive device-class fingerprinting of zigbee and z-wave iot devices. In *ICC 2020–2020 IEEE International Conference on Communications (ICC)*, pp. 1–7. IEEE.
41. Vagdevi, P., Nagaraj, D., & Prasad, G. V. (2017). Home: IOT based home automation using NFC. In *2017 International Conference on I-SMAC (IoT in Social, Mobile, Analytics and Cloud) (I-SMAC)*, pp. 861–865. IEEE.
42. Mekki, K., Bajic, E., Chaxel, F., & Meyer, F. (2018). Overview of cellular LPWAN technologies for IoT deployment: Sigfox, LoRaWAN, and NB-IoT. In *2018 IEEE International Conference on Pervasive Computing and Communications Workshops (Percom Workshops)*, pp. 197–202. IEEE.
43. Banks, A., Briggs, E., Borgendale, K., & Gupta, R.: (2019). MQTT Version 5.0. OASIS Stand. 7.

44. Bormann, C., Castellani, A. P., & Shelby, Z. (2012). Coap: An application protocol for billions of tiny internet nodes. *IEEE Internet Computing, 16*, 62–67.
45. Toulson, R., & Wilmshurst, T. (2007). *Fast and effective embedded systems design.*
46. Heath, S. (2002). *Embedded systems design.* Elsevier.
47. Ganguly, A. K. (2014). *Embedded systems: Design, programming and applications.* Alpha Science International, Ltd.
48. Peckol, J. K. (2019). *Embedded systems: A contemporary design tool.* Wiley.
49. Chu, E.T.-H., & Fang, C.-W. (2015). CALEE: A computer-assisted learning system for embedded OS laboratory exercises. *Computers & Education, 84*, 36–48.
50. Valvano, J. W. (2012). Embedded systems: Real-time interfacing to ARM® Cortex (TM)-M Microcontrollers. Jonathan W. Valvano.
51. Kothari, D. P., & Vasudevan, S. K. (2014). *Embedded systems.* New Academic Science Ltd.
52. Leong, P. H. W., & Tsoi, K. H. (2005). Field Programmable Gate Array technology for robotics applications. In *2005 IEEE International Conference on Robotics and Biomimetics-ROBIO*, pp. 295–298. IEEE.
53. Teach, Learn, and Make with Raspberry Pi, https://www.raspberrypi.org/
54. Crockett, L. H., Elliot, R. A., Enderwitz, M. A., & Stewart, R. W. (2014). The Zynq Book: Embedded Processing with the Arm Cortex-A9 on the Xilinx Zynq-7000 All Programmable Soc. Strathclyde Academic Media.
55. STM32F0—Arm Cortex-M0—USB Microcontrollers—STMicroelectronics, https://www.st.com/en/microcontrollers-microprocessors/stm32f0-series.html
56. TAŞCI, B., & Yavuz, E. (2019) Wireless elevator call system design with PSoC. In *2019 International Conference on Applied Automation and Industrial Diagnostics (ICAAID)*, pp. 1–5. IEEE.
57. Tasci, B., & Erol, Y. (2020). PSoC based embedded Rf link design. In *2020 13th International Conference on Communications (COMM)*, pp. 381–385. IEEE.
58. Konstantinidis, E., Conci, N., Bamparopoulos, G., Sidiropoulos, E., De Natale, F., & Bamidis, P. (2015). Introducing Neuroberry, a platform for pervasive EEG signaling in the IoT domain. In *Proceedings of the 5th EAI International Conference on Wireless Mobile Communication and Healthcare*, pp. 166–169.
59. Chen, W.-E., Wang, Y.-H., Huang, P.-C., Huang, Y.-Y., & Tsai, M.-Y. (2018). A smart IoT system for waste management. In *2018 1st International Cognitive Cities Conference (IC3)*, pp. 202–203. IEEE.
60. Yang, Z., Zhou, Q., Lei, L., Zheng, K., & Xiang, W. (2016). An IoT-cloud based wearable ECG monitoring system for smart healthcare. *Journal of Medical Systems, 40*, 1–11.
61. Kumar, Y., & Rufus, E. (2018). Smart Kitchen garden using "BIoThrough" at a low cost. In *2018 Fourteenth International Conference on Information Processing (ICINPRO)*, pp. 1–3. IEEE.
62. Kazi, S. S., Bajantri, G., & Thite, T. (2018). Remote heart rate monitoring system using IoT. Tech. *Sensor Heartb. Using IoT. 5.*
63. Singh, H., Pallagani, V., Khandelwal, V., & Venkanna, U. (2018). IoT based smart home automation system using sensor node. In *2018 4th International Conference on Recent Advances in Information Technology (RAIT)*, pp. 1–5. IEEE.
64. Memon, S. K., Shaikh, F. K., Mahoto, N. A., & Memon, A. A. (2019). IoT based smart garbage monitoring & collection system using WeMos & Ultrasonic sensors. In *2019 2nd International Conference on Computing, Mathematics and Engineering Technologies (iCoMET)*, pp. 1–6. IEEE.
65. Parmar, J., Nagda, T., Palav, P., & Lopes, H. (2018). IOT based weather intelligence. In *2018 International Conference on Smart City and Emerging Technology (ICSCET)*, pp. 1–4. IEEE.
66. Sung, W.-T., Chen, J.-H., & Tsai, M.-H. (2016). Applications of wireless sensor network for monitoring system based on IOT. In *2016 IEEE International Conference on Systems, Man, and Cybernetics (SMC)*. pp. 613–617. IEEE.
67. Vijendra, V., Kulkarni, M., & Murgan, R. (2019). ECG pattern recognition and beat classification using internet of things and hardware acceleration on ZynQ (SOC) platform with high performance computational PCIe protocol. In *BIODEVICES*, pp. 261–268.

68. Srinivas, M., Benedict, S., & Sunny, B. C. (2019). IoT cloud based smart bin for connected smart cities-A product design approach. In *2019 10th International Conference on Computing, Communication and Networking Technologies (ICCCNT)*, pp. 1–5. IEEE.
69. Singh, A., Aggarwal, P., & Arora, R. (2016). IoT based waste collection system using infrared sensors. In *2016 5th International Conference on Reliability, Infocom Technologies and Optimization (Trends and Future Directions) (ICRITO)*, pp. 505–509. IEEE.
70. Huang, S.-Z., & Chen, R.-Q. (2018). FPGA-based IoT sensor HUB. In *2018 International Conference on Sensor Networks and Signal Processing (SNSP)*, pp. 139–144. IEEE.
71. Altayeb, M., & Abdella, S. (2018). The internet-of-things and integration with wireless sensor network comprehensive survey and system implementation. In *2018 International Conference on Computer, Control, Electrical, and Electronics Engineering (ICCCEEE)*, pp. 1–6. IEEE.
72. Zhao, J., Zhang, J., Feng, Y., & Guo, J. (2010). The study and application of the IOT technology in agriculture. In *2010 3rd International Conference on Computer Science and Information Technology*, pp. 462–465. IEEE.
73. Mahbub, M., Hossain, M. M., & Gazi, M. S. A. (2020). IoT-cognizant cloud-assisted energy efficient embedded system for indoor intelligent lighting, air quality monitoring, and ventilation. *Internet of Things, 11*, 100266.
74. Kasat, N. N., Gawande, P. D., & Gawande, A. D. (2019). Smart city solutions on drainage, unused well and garbage alerting system for human safety. In *2019 9th International Conference on Emerging Trends in Engineering and Technology-Signal and Information Processing (ICETET-SIP-19)*, pp. 1–6. IEEE.
75. Singha, S., Singh, A. S., Prasad, S., & Alam, A. (2019). A study on power optimization techniques in PSoC. In *2019 International Conference on Communication and Signal Processing (ICCSP)*, pp. 225–229. IEEE.
76. Nagamani, S., Reddy, J. M., Lavanya, G., & Chandana, A. L. H. (2019). Smart street light management system using internet of things. In *2019 International Conference on Intelligent Computing and Control Systems (ICCS)*, pp. 103–107. IEEE.
77. Anuradha, M., Jayasankar, T., Prakash, N. B., Sikkandar, M. Y., Hemalakshmi, G. R., Bharatiraja, C., & Britto, A. S. F. (2021). IoT enabled cancer prediction system to enhance the authentication and security using cloud computing. *Microprocessors and Microsystems, 80*, 103301.
78. Podder, A. K., Al Bukhari, A., Islam, S., Mia, S., Mohammed, M. A., Kumar, N. M., Cengiz, K., & Abdulkareem, K. H. (2021). IoT based smart agrotech system for verification of Urban farming parameters. *Microprocessors and Microsystems, 82*, 104025.
79. Bhushan, S., Bohara, B., Kumar, P., & Sharma, V. (2016). A new approach towards IoT by using health care-IoT and food distribution IoT. In *2016 2nd International Conference on Advances in Computing, Communication, & Automation (ICACCA) (Fall)*, pp. 1–7. IEEE.
80. Sharmila, Kumar, P., Bhushan, S., Kumar, M., & Alazab, M. (2023). Secure key management and mutual authentication protocol for wireless sensor network by linking edge devices using hybrid approach. *Wireless Personal Communications*, 1–23.
81. Singh, A. K., Alshehri, M., Bhushan, S., Kumar, M., Alfarraj, O., & Pardarshani, K. R. (2021). Secure and energy efficient data transmission model for WSN.

Non-IP Based Communication in IoT: Handling Security Challenges

Sanjeev Kumar, Thipendra P. Singh, and Saurabh Kumar

Abstract This chapter provides an in-depth exploration of non-IP-based IoT communication and the associated security challenges. It commences by elucidating non-IP communication systems and juxtaposing them with IP-based communication. A detailed examination of the merits and drawbacks of non-IP communication ensues. Moreover, the chapter delves into the realm of security issues inherent in non-IP-based IoT communication. In this context, we delve into the various threats and attacks that could jeopardize the confidentiality and integrity of non-IP-based communication technologies, exposing their inherent vulnerabilities and risks. To illustrate these concerns, we scrutinize real-world case studies that depict instances of non-IP-based IoT communication, dissecting the security issues and proposing potential solutions for each scenario. Furthermore, this work incorporates mathematical models designed to assess non-IP communication protocols and IoT device security issues. These models find practical application in a case study, where we employ them to analyze and predict the behavior of a non-IP communication protocol. We also emphasize the significance of adhering to security standards and best practices when establishing non-IP-based IoT connectivity. These principles furnish a comprehensive framework for constructing secure communication protocols, underscoring the importance of proactive security measures. Lastly, as we peer into the future of non-IP-based communication, we advocate for ongoing vigilance and comprehensive study to ensure its seamless and secure integration into IoT ecosystems.

S. Kumar (✉)
University of Petroleum and Energy Studies, Dehradun, India
e-mail: sanjeev.kumar@ddn.upes.ac.in

T. P. Singh
School of Computer Science Engineering and Technology, Bennett University, Greater Noida, India
e-mail: thipendra@gmail.com

S. Kumar
Department of CSE, Sharda School of Engineering and Technology, Sharda University, Greater Noida, India
e-mail: saurabh.kumar1@sharda.ac.in

A. Prasad et al. (eds.), *Communication Technologies and Security Challenges in IoT*, Internet of Things, https://doi.org/10.1007/978-981-97-0052-3_10

Keywords Non-IP-based IoT communication · Security challenges · IoT vulnerabilities · Threats and attacks · Confidentiality and integrity · Security standards · Proactive security measures · IoT ecosystem integration

Introduction

The Internet of Things (IoT) is reshaping our lives, work, and interactions with the world. This transformation has revolutionized how devices communicate among themselves and with the internet. Nonetheless, a substantial portion of IoT devices relies on Internet Protocol (IP) for communication, introducing notable security concerns. Non-IP communication provides an alternative, yet it carries its own distinct challenges. This chapter aims to delve into the realm of non-IP-based communication within IoT and address the security issues it presents. It will furnish a comprehensive grasp of diverse non-IP communication technologies and their security ramifications. Additionally, the chapter will proffer practical remedies for mitigating security threats in non-IP-based IoT communication. This discussion caters to IoT professionals, researchers, educators, and students seeking to comprehend the security intricacies of non-IP-based IoT communication. It is also relevant for individuals responsible for fortifying the security of IoT systems and networks [1, 2].

Non-IP Based Communication in IoT

The advent of the Internet of Things (IoT) has brought about a revolutionary shift in how gadgets communicate with the internet. Traditional connection technologies like TCP/IP and HTTP, with their high overhead and power consumption, often prove impractical for certain IoT devices. Consequently, alternative communication technologies have emerged, offering increased efficiency and reduced weight. Among these alternatives, non-IP based communication technologies have gained prominence.

Non-IP based communication technologies employ protocols optimized for low-power, low-data-rate interactions. These methods excel in shorter-range scenarios and are particularly well-suited for low-powered Internet of Things devices. Notable examples of such technologies include ZigBee, Bluetooth Low Energy (BLE), and LoRaWAN, all of which circumvent the need for reliance on the Internet Protocol (IP) [3, 4].

Internet Protocol (IP) serves as the cornerstone of contemporary networking, facilitating data packet routing among internet-connected devices. However, the overhead and power consumption associated with IP make it less than ideal for some IoT devices. In contrast, non-IP based communication technologies prioritize lightweight, efficient operations, rendering them a favorable choice for IoT devices.

A fundamental distinction between IP-based and non-IP based communication lies in the addressing of data. IP-based communication employs a hierarchical addressing scheme encompassing network, subnet, and host addresses. In contrast, non-IP based communication employs unique short addresses specific to each device.

Furthermore, the transmission of data differs substantially between the two. IP-based communication adheres to a packet-switched approach, breaking down data into packets for transmission over a network. Non-IP based communication, conversely, adopts a message-based approach, sending data as complete messages [1–4].

Advantages and Disadvantages of Non-IP Based Communication

Non-IP communication technologies present a multitude of unique benefits when contrasted with IP-based communication, rendering them exceptionally well-suited for diverse Internet of Things (IoT) applications. One of their primary advantages lies in their minimal power consumption. These non-IP communication methods are precisely engineered to conserve power, a critical feature for IoT devices that rely on batteries for prolonged operation. This quality ensures that these devices can operate efficiently over extended durations without requiring frequent battery replacements or recharging [5–8].

Beyond their power efficiency, non-IP communication technologies are finely tuned for low data rates. This characteristic makes them ideal for IoT devices transmitting small data volumes sporadically. Their adeptness at managing compact data packets establishes them as a cost-effective choice for applications such as environmental sensors, smart home devices, and wearable technology, where the handling of substantial data streams is unnecessary.

Cost-effectiveness stands as another prominent advantage of non-IP communication technologies. Their implementation and maintenance costs are often lower compared to their IP-based counterparts. This affordability positions them as a pragmatic option for deploying economical IoT devices in various applications, encompassing agriculture, logistics, and smart city infrastructure.

Short-distance communication is a defining attribute of non-IP communication technologies. They are engineered to function over relatively limited distances, rendering them suitable for IoT devices situated in close proximity to one another. This short-range capability aligns with scenarios where devices necessitate communication within localized environments, such as home automation systems or industrial sensor networks.

Nevertheless, non-IP communication technologies come with their own set of constraints. These limitations encompass restricted range, confining their utility to localized settings. Their optimization for low data rates can constrain bandwidth, rendering them unsuitable for applications that demand high-speed data transmission, such as high-definition video streaming. Furthermore, these technologies often employ proprietary protocols, which can impede interoperability with other devices and systems, curtailing the adaptability of IoT ecosystems.

Arguably, the most pressing concern associated with non-IP communication technologies is their limited security features. They may not provide the same level of security as IP-based communication methods, rendering them more susceptible to attacks and threats. Consequently, they may not be the best choice for applications necessitating stringent security measures, including critical infrastructure or healthcare systems.

In summary, non-IP communication technologies furnish a lightweight, power-efficient, and cost-effective alternative to IP-based communication within the IoT landscape. Their merits, encompassing low power consumption, efficient handling of low data rates, and affordability, render them a viable selection for numerous IoT applications. However, careful consideration and mitigation of the associated trade-offs, such as limited range, bandwidth, interoperability, and security, are essential to ensure the secure and effective deployment of IoT devices and infrastructure. Striking a balance between these advantages and limitations is paramount when choosing the most fitting communication technology for specific IoT applications.

Security Challenges in Non-IP Based Communication in the IoT

Security takes center stage in the rapidly expanding Internet of Things (IoT) landscape, particularly concerning secure communication among devices. Non-IP based communication technologies introduce a distinctive array of security challenges, demanding meticulous attention and a proactive stance to guarantee the secure and reliable deployment of IoT devices and systems [8–10].

In addition to the challenges previously mentioned, several other pivotal security considerations come into play. Device Authentication assumes a central role in this context, as non-IP communication technologies may lack the robust mechanisms required for authenticating devices, potentially opening the door to unauthorized access and data or device manipulation. Upholding Data Integrity holds paramount significance in preventing unauthorized alterations during data transmission. Privacy Concerns may surface due to the limited security features, emphasizing the necessity for robust data protection measures and adherence to privacy regulations. Effective management of Firmware and Software Updates proves essential to fortify devices against emerging threats, and Post-Deployment Security practices become imperative to sustain device security over time, as new vulnerabilities may surface. User Access Control emerges as a critical aspect in preventing unauthorized configuration changes and data breaches. Moreover, Compliance with Legal and Regulatory requirements stands as a vital obligation, as different industries are subject to specific regulations governing data protection, security, and privacy.

In summary, addressing these multifaceted security challenges is not only crucial but also an ongoing process. The implementation of robust security measures, continuous monitoring, and adherence to industry-specific regulations are indispensable for the IoT to fully harness the benefits of non-IP based communication technologies while effectively mitigating the associated security risks.

Threats to Non-IP Based Communication in the IoT

Non-IP based communication technologies within the IoT face an array of vulnerabilities, including the following [11, 12].

Man-in-the-middle (MitM) attacks

In a MitM attack, an assailant intercepts communication between two devices and assumes the identity of one of them. This permits the attacker to pilfer data, manipulate data, or even obstruct communication entirely. Non-IP based communication is especially susceptible to MitM attacks because these protocols often lack encryption and authentication.

Illustration: An attacker could execute a MitM attack against a ZigBee network by inserting a rogue ZigBee coordinator between two communicating devices. The attacker could then impersonate one of the devices and intercept or modify the transmitted data.

Replay attacks

A replay attack entails the interception of a legitimate message, which is subsequently replayed later. This tactic can deceive a device into undertaking actions it should not, such as transmitting sensitive data or unlocking a door. Non-IP based communication is notably vulnerable to replay attacks due to the absence of message integrity checks.

Illustration: An attacker could conduct a replay attack against a Bluetooth Low Energy (BLE) lock by intercepting the command used to unlock the lock and then replaying it later, granting unauthorized access.

Denial-of-service (DoS) attacks

In a DoS attack, a malefactor inundates a device or network with traffic to render it unavailable. This can be achieved by sending numerous messages or messages designed to consume substantial resources. DoS attacks pose a significant threat to non-IP based communication, as these protocols often lack congestion control mechanisms.

Illustration: An attacker could launch a DoS attack against a LoRaWAN network by transmitting a multitude of empty messages, depleting available bandwidth and hindering legitimate communication.

Eavesdropping

Eavesdropping entails the covert interception of communication between two devices. This can be accomplished through the use of a directional antenna or software-defined radio (SDR). Non-IP based communication is susceptible to eavesdropping because of the absence of encryption.

Illustration: An attacker could eavesdrop on communication between a Z-Wave thermostat and a Z-Wave hub, pilfering temperature settings and potentially gaining control over the thermostat.

Signal jamming

Signal jamming involves transmitting interference to disrupt communication between two devices, using techniques like transmitting on the same frequency as the legitimate signal or using a high-power transmitter to overwhelm the receiver. Non-IP based communication is particularly susceptible to signal jamming as these protocols often operate on unlicensed frequencies.

Illustration: An attacker could jam the communication between a drone and its controller, leading to a loss of control and potential accidents.

Mitigation Strategies

To mitigate the security risks posed by these threats, several measures can be adopted, including:

- Implementing encryption and authentication: These measures can protect against MitM attacks, replay attacks, eavesdropping, and signal jamming.
- Using congestion control mechanisms: These mechanisms can help mitigate the impact of DoS attacks.
- Utilizing frequency hopping or spread spectrum techniques: These techniques can reduce the impact of signal jamming.
- Implementing intrusion detection systems/intrusion prevention systems (IDS/IPS): These systems can detect and block malicious traffic.
- Keeping devices up to date: Regular software updates often include security patches that mitigate known vulnerabilities.

By taking these steps, organizations can safeguard their non-IP based communication systems from a wide spectrum of security threats.

Mitigating Security Challenges in Non-IP Based Communication

Mitigating security challenges in non-IP (Internet Protocol) based communication systems can be challenging but is essential, especially in environments where traditional IP-based communication is not suitable or secure enough. Non-IP communication systems vary widely in terms of their nature and requirements, so the specific security measures and strategies you need to implement will depend on the unique characteristics of your system. It's essential to conduct a thorough risk assessment and work with security experts to tailor your security approach accordingly. Mitigating security challenges in non-IP (Internet Protocol) based communication systems requires a multifaceted approach. Here are some strategies to address security challenges in non-IP based communication systems [6–12, 18, 20].

Encryption

Encryption techniques fall into two primary categories: symmetric and asymmetric. Symmetric encryption employs the same key for both encryption and decryption, while asymmetric encryption employs a public key for encryption and a private key for decryption. Common symmetric encryption algorithms include AES, RC4, and Blowfish, while common asymmetric encryption algorithms comprise RSA, Diffie-Hellman, and ECC. In non-IP based communication technologies, encryption can be applied using various methods. For instance, ZigBee packets can be encrypted using AES, and ECC can secure BLE messages.

Authentication

Authentication methods are divided into three key categories: knowledge-based, possession-based, and biometric. Knowledge-based authentication involves users entering a secret, such as a password or PIN, while possession-based authentication requires users to have a physical item, like a security token or smart card. Biometric authentication necessitates the provision of biometric identifiers like fingerprints or facial scans. In non-IP based communication, authentication methods encompass:

- Pre-shared keys: Utilized for device authentication, commonly in ZigBee and BLE networks.
- Digital certificates: Employed to authenticate devices and users, typically in LoRaWAN and Wi-Fi networks.
- Biometric authentication: Including fingerprint scanners and facial recognition, used to authenticate users in non-IP based communication systems.

Access Control

Access control mechanisms are categorized as discretionary access control (DAC) and mandatory access control (MAC). DAC empowers users to manage access to their owned resources, while MAC enforces security policies defined by system administrators. Access control methods in non-IP based communication involve:

- Role-based access control (RBAC): A DAC mechanism assigning users to roles with specific permissions, commonly in ZigBee and BLE networks.
- Attribute-based access control (ABAC): A DAC mechanism making access decisions based on user, resource, and environmental attributes, often in LoRaWAN and Wi-Fi networks.
- Mandatory access control (MAC): Enforces security labels on objects and subjects, particularly in military and government applications.

Physical Security

Physical security measures can be preemptive or detective. Preventive measures are designed to deter and thwart attacks, while detective measures aim to detect and investigate breaches. Preventive physical security measures in non-IP based communication systems involve:

- Tamper-evident packaging: Prevents unauthorized tampering with devices without detection.
- Locking mechanisms: Deter unauthorized access to devices.
- Environmental monitoring systems: Detect environmental changes like temperature, humidity, and vibration. Detective physical security measures include:
- Intrusion detection systems (IDS): Detect suspicious activity in networks.
- Video surveillance systems: Monitor the physical environment around devices.
- Access control systems: Track device access and usage.

Monitoring and Logging

Monitoring and logging systems encompass network monitoring and security information and event management (SIEM) systems. Network monitoring systems, like Wireshark, Nmap, and tcpdump, track network traffic and identify anomalies. SIEM systems, such as Splunk, QRadar, and LogRhythm, collect and analyze security events from various sources.

Standardization

Standardization is the process of establishing consistent technical specifications or criteria to ensure that materials, products, processes, and services meet their intended purpose. It plays a crucial role in the IoT due to several reasons, including:

- Interoperability: Facilitating seamless interaction among products and services from different manufacturers.
- Security: Enhancing non-IP based communication technologies' security by offering guidance on implementing security measures like encryption and authentication.
- Innovation: Fostering innovation by providing a common foundation for new products and services.

Key standardization organizations in the non-IP based communication technology realm include the International Organization for Standardization (ISO), International

Electrotechnical Commission (IEC), IEEE Standards Association (IEEE-SA), 3rd Generation Partnership Project (3GPP), LoRa Alliance, and ZigBee Alliance. These organizations develop standards for various non-IP based communication technologies, such as ZigBee, LoRaWAN, LTE-M, and NB-IoT. Standardization is pivotal in ensuring the secure and interoperable deployment of non-IP based communication technologies in the IoT, providing a reliable and secure ecosystem for these technologies.

Interoperability

Standardization plays a pivotal role in ensuring interoperability, a critical requirement in the IoT ecosystem. With a multitude of devices and technologies from different manufacturers and vendors, it's essential to have common standards that enable seamless communication and collaboration. By adhering to established standards, IoT devices and systems can work together cohesively, regardless of their origins, thus promoting the growth and expansion of the IoT landscape.

Security Enhancement

Security is a paramount concern in the IoT, and standardization is a key enabler in this regard. By defining security protocols, encryption methods, and authentication standards, standardization organizations help bolster the security of non-IP based communication technologies. These standards provide clear guidelines on implementing robust security measures, mitigating vulnerabilities, and safeguarding sensitive data. As a result, standardization not only enhances security but also fosters trust among users and stakeholders, encouraging the widespread adoption of IoT technologies.

Innovation Facilitation

Standardization is not merely about adhering to rigid rules; it also fosters innovation. By providing a common foundation and a level playing field, standards create an environment where new products and services can flourish. Entrepreneurs and innovators can build upon existing standards, confident that their creations will seamlessly integrate with the broader IoT ecosystem. This promotes a culture of innovation and competitiveness, driving the development of cutting-edge solutions and ensuring the continuous growth of the IoT.

Examples of Standardization Organizations:

The International Organization for Standardization (ISO) is renowned for developing and promoting international standards. These standards cover a wide range of fields, including technology and IoT-related areas, contributing to global interoperability and consistency.

- The International Electrotechnical Commission (IEC) focuses on standards for electrical and electronic devices and systems, providing essential guidance for the development and implementation of IoT technologies.

- The IEEE Standards Association (IEEE-SA) specializes in standards for electronics and electrical engineering, ensuring the compatibility and reliability of IoT devices.
- The 3rd Generation Partnership Project (3GPP) is responsible for setting standards for mobile communications, including those crucial to IoT devices and connectivity.
- The LoRa Alliance and ZigBee Alliance are industry-specific groups that concentrate on standardizing non-IP based communication technologies, such as LoRaWAN and ZigBee, respectively. These alliances play a vital role in defining the parameters that enable seamless communication and security in their respective technologies.

In conclusion, standardization is a linchpin in the IoT ecosystem, enabling interoperability, enhancing security, and fueling innovation. Standardization organizations are at the forefront of developing and promoting the standards that make the IoT a more reliable, secure, and dynamic landscape. As the IoT continues to evolve, standardization will remain instrumental in shaping its future.

Non-IP Based Communication Technologies in IoT

Non-IP Based Communication Technologies in the IoT [11–20], non-IP based communication technologies encompass a diverse array of communication protocols and standards tailored for IoT devices and systems. In contrast to traditional Internet Protocol (IP) communication, these technologies employ varying networking topologies and communication methods. Some of the prevalent non-IP based communication technologies integral to the IoT landscape include.

Zigbee

Zigbee stands as a low-power wireless communication standard extensively employed in IoT applications, including sensors, home automation systems, and smart lighting. It adheres to the IEEE 802.15.4 standard and primarily operates within the 2.4 GHz frequency band. Zigbee networks adopt a mesh networking topology, allowing devices to communicate directly or relay through intermediary devices. This characteristic renders Zigbee networks highly scalable and dependable. Moreover, Zigbee's energy-efficiency is particularly valuable for battery-powered IoT devices.

Z-Wave

Z-Wave, a proprietary wireless communication technology, finds utility in home automation systems and IoT devices. Operating within the 900 MHz frequency band, Z-Wave relies on a mesh networking topology. Renowned for its reliability and user-friendliness, Z-Wave is also lauded for its energy-efficient nature, making it an ideal choice for IoT devices reliant on batteries.

Bluetooth

Bluetooth, a ubiquitous wireless communication standard, is versatile and is extensively employed across a wide spectrum of devices, including IoT sensors and wearables. Operating on short-range radio waves within the 2.4 GHz frequency band, Bluetooth is favored for its cost-effectiveness and ease of implementation. Its energy efficiency is another compelling feature, rendering it well-suited for battery powered IoT devices.

Near Field Communication (NFC)

Near Field Communication (NFC) is a short-range wireless communication technology harnessed in IoT applications like payment systems and access control. NFC operates within the 13.56 MHz frequency band. Its hallmark is secure communication, necessitating devices to be in close proximity for data exchange. NFC is also characterized by its speed, making it a preferred choice for contactless payment solutions.

Radio Frequency Identification (RFID)

Radio Frequency Identification (RFID) serves as a wireless communication technology widely used in IoT devices, including inventory management systems and access control. RFID operates across various frequency bands, encompassing low-frequency (LF), high-frequency (HF), and ultra-high-frequency (UHF). RFID technology involves affixing RFID tags to objects, enabling their tracking and identification. It is particularly valuable in scenarios requiring user authentication and asset monitoring.

Non-IP based communication technologies hold a pivotal role in the IoT ecosystem, offering a multitude of advantages, including low power consumption, scalability, and reliability. These technologies are pivotal in a broad spectrum of IoT applications, encompassing home automation, asset tracking, and industrial automation. Their versatility and energy-efficient nature make them indispensable components of the IoT landscape, addressing various communication needs and fostering innovation in the realm of connected devices.

Comparison of Non-IP Based Communication Technologies

Non-IP based communication technologies vary in terms of their range, power consumption, data rates, and other technical specifications. Table 1 provides a comparison of some of the key characteristics of Zigbee, Z-Wave, Bluetooth, NFC, and RFID [11–14].

Table 1 Comparison of Non-IP based communication technologies

Technology	Range	Power consumption	Data rate	Topology
Zigbee	10–100 m	Low	20–250 kbps	Mesh
Z-Wave	30–100 m	Low	9.6–100 kbps	Mesh
Bluetooth	10–100 m	Low-Moderate	1–3 Mbps	Point-to-point or Mesh
NFC	<10 cm	Low	424 kbps	Point-to-point
RFID	Up to 10 m	Low	10–100 kbps	Point-to-point or RFID reader

Security Considerations, Challenges and Solutions for Non-IP Based Communication Technologies

Security considerations and challenges for non-IP based communication technologies involve robust authentication, encryption, secure protocols, physical security, network segmentation, patch management, monitoring, redundancy, key management, awareness, compliance, and disaster recovery. Tailoring these measures to the system's unique characteristics is essential.

Security Challenges

Non-IP based communication technologies in the IoT introduce distinct security challenges that necessitate comprehensive solutions to ensure the secure deployment of IoT devices and systems. This section delves into some of the critical security challenges associated with non-IP based communication in the IoT and explores potential remedies [11–16].

Encryption Gap

A prominent security challenge linked to non-IP based communication in the IoT is the absence of encryption. Many non-IP based communication technologies might not support encryption or employ weak encryption methods, rendering them susceptible to security breaches. Malicious actors can intercept communication between IoT devices, compromising the confidentiality of transmitted data.

Feeble Authentication

Another pressing security concern associated with non-IP based communication in the IoT pertains to weak authentication. Numerous non-IP based communication technologies may lack robust authentication mechanisms or rely on feeble authentication methods, potentially exposing them to unauthorized access. Attackers can exploit these vulnerabilities to gain access to IoT devices and systems, either by impersonating legitimate devices or exploiting authentication weaknesses.

Interoperability Hurdles

Non-IP based communication technologies might suffer from interoperability issues, creating security vulnerabilities and augmenting the risk of attacks. Incompatibilities between technologies can lead to misconfigurations and vulnerabilities that attackers can leverage. Furthermore, non-interoperable devices and systems can give rise to security blind spots, providing fertile ground for potential attacks.

Proprietary Protocol Complexity

Non-IP based communication technologies may employ proprietary protocols, which can obfuscate security assessments. Proprietary protocols often lack rigorous security testing, and their vulnerabilities may remain obscure. Attackers can exploit these protocol-specific weaknesses to compromise IoT devices and systems, potentially evading conventional security measures.

Physical Vulnerabilities

Non-IP based communication technologies may be susceptible to physical attacks, such as signal jamming and eavesdropping. Attackers can disrupt wireless signals between IoT devices by jamming, rendering them incapable of communication. Additionally, eavesdropping on IoT device communication can yield sensitive information that malicious actors can exploit to orchestrate attacks.

In summary, non-IP based communication technologies in the IoT present an array of unique security challenges. Addressing these challenges requires robust solutions that encompass encryption implementation, authentication enhancement, resolving interoperability issues, evaluating proprietary protocols, and fortifying defenses against physical attacks. By proactively tackling these security challenges, the IoT can achieve a more robust and secure deployment of non-IP based communication technologies.

Security Solutions

Encryption: To address the lack of encryption in non-IP based communication technologies, IoT device manufacturers can implement strong encryption algorithms to protect the data being transmitted between IoT devices. Encryption can prevent attackers from intercepting and reading the data being transmitted, which can protect the confidentiality of the data.

Authentication

To address the weak authentication in non-IP based communication technologies, IoT device manufacturers can implement strong authentication methods, such as two-factor authentication or biometric authentication. Strong authentication can prevent attackers from gaining unauthorized access to IoT devices and systems, which can protect the integrity and availability of the data.

Interoperability Standards

To address interoperability issues in non-IP based communication technologies, IoT device manufacturers can adopt interoperability standards such as the Thread protocol or the AllJoyn framework. Interoperability standards can ensure that IoT devices and systems can communicate with each other seamlessly, which can reduce security vulnerabilities and increase the overall security of IoT systems.

Open Protocols

To address the issue of proprietary protocols in non-IP based communication technologies, IoT device manufacturers can adopt open protocols such as Zigbee or Bluetooth. Open protocols are subjected to rigorous security testing, and their security vulnerabilities are well known, which can make them more secure than proprietary protocols.

Physical Security

To address the issue of physical attacks in non-IP based communication technologies, IoT device manufacturers can implement physical security measures such as tamper-resistant hardware and secure enclosures. Tamper-resistant hardware can prevent attackers from physically accessing the IoT device and tampering with it, while secure enclosures can protect the IoT device from physical attacks such as signal jamming or eavesdropping.

Non-IP based communication technologies in the IoT present unique security challenges that must be addressed to ensure the secure deployment of IoT devices and systems. In this section, we have discussed some of the key security challenges associated with non-IP based communication in the IoT and explored some of the solutions to these challenges. By implementing strong encryption, authentication, interoperability standards, open protocols, and physical security measures, IoT device manufacturers can improve the overall security of non-IP based communication technologies in the IoT.

Case Studies on Non-IP Based Communication Security in IoT

In this section, we will present case studies on the security of non-IP based communication in the IoT. These case studies will illustrate the security challenges associated with non-IP based communication technologies in the IoT and highlight some of the solutions that can be used to address these challenges.

Case Study 1: WirelessHART [17]

Wireless HART is a non-IP based communication technology used in industrial control systems (ICS). It is designed to provide wireless communication between sensors, actuators, and controllers in industrial environments. WirelessHART uses a proprietary wireless protocol that operates in the 2.4 GHz frequency band [17].

The security of WirelessHART has been the subject of several studies, which have identified some security vulnerabilities. One of the main security vulnerabilities of WirelessHART is the lack of encryption. WirelessHART does not provide end-to-end encryption, which means that data transmitted between devices can be intercepted and read by attackers.

To address this security vulnerability, some manufacturers have implemented encryption on top of WirelessHART. For example, Honeywell has developed an encryption layer that can be used with WirelessHART to provide end-to-end encryption.

Case Study 2: Z-Wave [19]

Z-Wave is a non-IP based communication technology used in smart homes and buildings. It is designed to provide wireless communication between devices such as lights, thermostats, and security systems. Z-Wave uses a proprietary wireless protocol that operates in the 900 MHz frequency band.

The security of Z-Wave has also been the subject of several studies, which have identified some security vulnerabilities. One of the main security vulnerabilities of Z-Wave is weak authentication. Z-Wave uses a shared network key to authenticate devices, which can be easily compromised by attackers.

To address this security vulnerability, Z-Wave has introduced a new security protocol called Security 2 (S2). S2 provides stronger authentication mechanisms, such as asymmetric encryption, which can prevent attackers from impersonating devices and gaining unauthorised access to the network.

Case Study 3: EnOcean [21]

EnOcean is a non-IP based communication technology used in energy harvesting devices. It is designed to provide wireless communication between devices that use energy harvesting technologies such as solar, thermal, and mechanical energy. EnOcean uses a proprietary wireless protocol that operates in the sub-GHz frequency band.

The security of EnOcean has also been the subject of several studies, which have identified some security vulnerabilities. One of the main security vulnerabilities of EnOcean is its lack of encryption. EnOcean does not provide end-to-end encryption, which means that data transmitted between devices can be intercepted and read by attackers.

To address this security vulnerability, EnOcean has introduced a new security protocol called EnOcean Security (EOS). EOS provides end-to-end encryption and message authentication, which can prevent attackers from intercepting and reading data transmitted between devices.

Non-IP based communication technologies in the IoT present unique security challenges that must be addressed to ensure the secure deployment of IoT devices and systems. The case studies presented in this chapter illustrate some of the security vulnerabilities associated with non-IP based communication technologies in the IoT and highlight some of the solutions that can be used to address these vulnerabilities. By implementing strong encryption, authentication, and other security measures, IoT device manufacturers can improve the overall security of non-IP based communication technologies in the IoT.

Mathematical Models for Non-IP Based Communication and Security Challenges in IoT Devices

This section explores mathematical models for non-IP based communication protocols and security challenges in IoT devices. It discusses how mathematical models can be used to analyze and evaluate the security of non-IP based communication protocols in IoT devices, and to identify potential vulnerabilities and threats.

Mathematical models can be used to analyse and evaluate the performance of non-IP based communication protocols on IoT devices. These models can help identify potential performance issues such as delays, packet loss, and bandwidth limitations.

For example, network modelling techniques such as queueing theory can be used to model the performance of non-IP based communication protocols in IoT devices. Queueing theory can be used to analyse the queuing behaviour of IoT devices and estimate the delay and throughput of the system.

Mathematical models can also be used to analyse and evaluate the security challenges of IoT devices. These models can help identify potential security vulnerabilities and threats and evaluate the effectiveness of security measures and protocols.

For example, formal methods such as model checking can be used to analyse the security of non-IP based communication protocols in IoT devices. Model checking can be used to verify that the communication protocol satisfies certain security properties, such as confidentiality, integrity, and availability.

Case Study: Mathematical Modelling of a Non-IP Based Communication Protocol in IoT Devices

This section presents a case study on the mathematical modelling of a non-IP based communication protocol for IoT devices. It describes how mathematical models were used to evaluate the performance and security of the protocol and identify potential vulnerabilities and threats.

Mathematical models can be a powerful tool for analyzing and evaluating non-IP based communication protocols and security challenges in IoT devices. By using mathematical models, IoT device manufacturers can identify potential vulnerabilities and threats, evaluate the effectiveness of security measures and protocols, and design more secure and efficient non-IP based communication protocols for IoT devices.

Some examples of mathematical formulations and equations related to the discussion on mathematical models for non-IP based communication and security challenges in the IoT.

Queueing Theory Model for Non-IP Based Communication Protocols

In a queuing system, the arrival rate of packets is denoted by, and the service rate (i.e., the rate at which packets are processed and sent) is denoted by μ. The average number of packets in the system (including those waiting in the queue and those being processed) is given by:

$$L = \lambda / \mu - \lambda$$

The average waiting time of a packet in the queue is given by:

$$Wq = L / \mu$$

The average time a packet spends in the system (including waiting time and processing time) is given by:

$$W = 1 / \mu - \lambda$$

Model Checking Verification for Security Properties

To verify that a non-IP based communication protocol satisfies a security property such as confidentiality, integrity, or availability, we can use temporal logic formulas. For example, to verify that a protocol satisfies confidentiality, we can use the following formula:

$$-P(PQ)$$

where P represents the presence of sensitive information and Q represents the possibility of an attacker being able to access that information. The formula states that it is not possible for P to be true and Q to be false (i.e., an attacker cannot access sensitive information).

Network Flow Model for Data Routing

In a network flow model, the flow of data through a network is represented by a graph, where nodes represent devices and edges represent communication links. The flow of data from one device to another is represented by a flow variable, and the capacity of each communication link is represented by a capacity variable.

To find the optimal data routing through the network, we can use the following optimisation problem:

maximise $\sum f_{ij}$ is subject to: $\sum f_{ij} - \sum f_{ji} = s_i - d_i$ for all i $0 <= f_{ij} <= c_{ij}$ for all i,j

where f_{ij} is the flow variable from node i to node j, s_i is the source data rate at node i, di is the destination data rate at node i, and c_{ij} is the capacity variable for the communication link between nodes i and j. The first constraint ensures that the flow balance is maintained at each node, and the second constraint ensures that the flow variable does not exceed the capacity of the communication link.

Popular Algorithms for Non-IP Based Protocol Security

In this section, we delve into prominent security algorithms tailored for non-IP based communication protocols, showcasing their suitability for bolstering IoT device security. These algorithms serve as the linchpin for implementing vital security features, including encryption, authentication, and integrity verification, within the realm of non-IP based communication protocols.

Symmetric Key Encryption Algorithms

Among the arsenal of security measures, symmetric key encryption algorithms stand out as a prevalent choice for fortifying non-IP based communication protocols on IoT devices. These algorithms operate on the premise of a shared key, which plays a dual role in both encrypting and decrypting data. This approach guarantees data confidentiality and acts as a formidable bulwark against unauthorized data access.

Noteworthy examples of symmetric key encryption algorithms encompass:

- Advanced Encryption Standard (AES): As one of the cornerstones of symmetric key encryption, AES boasts widespread usage. It operates with a 128-bit block size and key sizes of 128, 192, or 256 bits, exemplifying its versatility and robustness.
- Data Encryption Standard (DES): Although relatively older, DES is a stalwart encryption algorithm with a 64-bit block size and a 56-bit key size. Its legacy continues to influence security practices.
- Triple Data Encryption Algorithm (TDEA): Often referred to as 3DES, this algorithm leverages three rounds of DES encryption with either two or three distinct keys. This layered approach enhances security, particularly in legacy systems.
- Blowfish: Recognized for its swiftness, Blowfish deploys a variable-length key ranging from 32 to 448 bits, making it an agile choice for encryption.

The discourse further delves into a comprehensive evaluation of these algorithms, scrutinizing their strengths and vulnerabilities. Additionally, it explores best practices for the judicious implementation of symmetric key encryption within non-IP based communication protocols. Paramount considerations, such as selecting an apt key size commensurate with the requisite security level and adopting secure key distribution methodologies to stave off unauthorized key access, are central to the discussion.

Asymmetric Key Encryption Algorithms

Asymmetric key encryption algorithms, also recognized as public key encryption algorithms, operate on the principle of employing a pair of keys—a public key and a private key—to facilitate data encryption and decryption. These algorithms introduce an elevated tier of security, given that the private key is closely guarded while only the public key is openly shared.

Prominent examples of asymmetric key encryption algorithms encompass:

- RSA (Rivest-Shamir-Adleman): Among the stalwarts of asymmetric encryption, RSA hinges on a key pair featuring a public key and a private key. The key sizes span from 1024 to 4096 bits, underpinning its prevalence in the field.

- Elliptic Curve Cryptography (ECC): A contemporary encryption algorithm, ECC harnesses elliptic curves to generate key pairs. Notably, ECC achieves a similar level of security as RSA but with the advantage of smaller key sizes, a boon for resource-constrained environments.
- Diffie-Hellman Key Exchange: This methodology serves as a means for securely exchanging keys across potentially insecure channels, a salient component in non-IP based communication protocols.

The ensuing discussion embarks on a comprehensive exploration of the strengths and vulnerabilities inherent to these algorithms. It also delves into best practices for implementing asymmetric key encryption within non-IP based communication protocols. Crucial factors, such as the judicious selection of key sizes and algorithms that strike a balance between the requisite security level and the constraints posed by performance and memory considerations, come under scrutiny.

Hashing Algorithms

Hashing algorithms serve the purpose of validating the integrity of data transmitted within non-IP based communication protocols. These algorithms undertake the task of generating a fixed-size hash value from the data, thereby enabling the verification of data against any unauthorized alterations or corruption during its transmission.

Prominent hashing algorithms encompass:

- Secure Hash Algorithm (SHA): SHA comprises a family of hashing algorithms, each offering distinct security levels. Among them, SHA-256 and SHA-512 emerge as the most prevalent choices in practice.
- Message Digest (MD): This older hashing algorithm continues to see widespread use, with MD5 being its most commonly adopted variant.

The ensuing discussion ventures into a comprehensive examination of these algorithms, scrutinizing their merits and demerits. It also delves into best practices for the implementation of hashing within non-IP based communication protocols. Critical considerations encompass the selection of a hashing algorithm capable of delivering the requisite security level, alongside ensuring that the chosen hash function boasts collision resistance. Collision resistance, a pivotal property, thwarts the identification of two inputs yielding the same hash value, further fortifying the data's integrity.

Digital Signatures

Digital signatures play a pivotal role in authenticating the sender of data within non-IP based communication protocols. They are founded on asymmetric key encryption algorithms, furnishing a means for the recipient to corroborate both the data's legitimacy and the sender's identity.

Prominent Digital Signature Algorithms Encompass

- RSA
- Elliptic Curve Digital Signature Algorithm (ECDSA)
- Digital Signature Algorithm (DSA)

RSA and ECDSA emerge as widely employed digital signature algorithms, with ECDSA particularly tailored for deployment in resource constrained IoT devices due to its modest key sizes. DSA, while older, continues to maintain its relevance within the realm of digital signatures.

A thorough exploration of these algorithms is required to comprehend their strengths and shortcomings, alongside the adoption of best practices for integrating digital signatures within non-IP based communication protocols. Crucial considerations encompass safeguarding the private key used for data signing, ensuring that the recipient can authenticate the digital signature's legitimacy using the corresponding public key.

These algorithms offer a versatile suite of security features for IoT devices operating within non-IP based communication protocols, encompassing encryption, authentication, and integrity verification. The prudent selection and meticulous implementation of these algorithms are essential for preserving the security of data exchanged through these protocols. By gaining an understanding of their strengths and limitations and adhering to best practices during implementation, IoT devices can be significantly fortified against potential vulnerabilities and security breaches.

Security Challenges and Solutions for Low-Power Wide Area Networks (LPWANs)

In this section, we will introduce Low-Power Wide Area Networks (LPWANs) and delve into their advantages and disadvantages within the realm of IoT applications. Furthermore, we will provide an overview of various LPWAN technologies, such as Sigfox, LoRaWAN, and NB-IoT.

LPWANs present distinctive security challenges due to their long-range, low-power communication capabilities. In this section, we will examine the security challenges associated with LPWANs, including the potential for eavesdropping, spoofing, and replay attacks.

Encryption stands as a crucial security facet for LPWANs. Within this section, we will explore the array of encryption algorithms applicable to LPWANs, including the likes of the Advanced Encryption Standard (AES) and ChaCha20. Additionally, we will delve into best practices concerning key management and secure key exchange.

Authentication holds another pivotal role in securing LPWANs. In this section, we will elucidate the various authentication mechanisms adaptable to LPWANs, including pre-shared keys (PSKs), digital certificates, and public key infrastructure (PKI). The following authentication mechanisms find application in LPWANs:

- Pre-shared key (PSK): PSK serves as a straightforward yet effective authentication mechanism where two devices share a secret key. To authenticate itself to another device, a device transmits a message encrypted with the shared key. The recipient can decipher the message and verify the sender's identity if it possesses the same shared key.

- Digital certificates: Digital certificates furnish a more secure authentication mechanism than PSKs. They leverage public-key cryptography to create a unique digital signature for each device. When a device seeks authentication with another, it sends a message signed with its digital signature. The recipient can authenticate the sender's identity by verifying the digital signature using the sender's public key.
- Mutual TLS (mTLS): mTLS, a variant of TLS, employs digital certificates to authenticate both the client and the server, thereby yielding a high level of security. However, it is relatively more intricate to implement compared to PSK or digital certificates alone.
- Identity-based cryptography (IBC): IBC is a contemporary authentication mechanism that employs identity-based encryption for device authentication. In IBC, devices need not store or exchange any secret keys; instead, they employ their public keys to authenticate themselves to one another. While IBC is exceptionally secure, its implementation is more complex compared to PSK or digital certificates.
- OAuth2: OAuth2, an open standard for authorization, can be adapted for implementing authentication in LPWANs by allowing devices to delegate their authentication to a trusted third-party. This can simplify the authentication implementation and bolster security.

The choice of the most suitable authentication mechanism within an LPWAN hinges on the specific requirements of the application. For instance, if security takes precedence, mTLS or IBC could be deemed fitting choices. Conversely, if simplicity is of paramount importance, PSK or digital certificates may be more apt.

Securing Firmware Updates in LPWANs

Ensuring the security of firmware updates is paramount in the realm of IoT devices operating within LPWANs. In this section, we will delve into the best practices for securing firmware updates in LPWANs, encompassing elements such as digital signatures, secure boot, and more. The following best practices are instrumental for safeguarding firmware updates in LPWANs:

- Employ a Secure Authentication Mechanism: LPWANs offer an array of authentication mechanisms, as previously discussed. When it comes to firmware updates, opting for a robust authentication mechanism is crucial to guarantee that only authorized devices can receive firmware updates.
- Leverage a Secure Firmware Update Protocol: Several secure firmware update protocols tailored for LPWANs are available. These protocols utilize various techniques to ensure the secure and reliable delivery of firmware updates.
- Implement Digital Signatures: The use of digital signatures plays a pivotal role in verifying the authenticity of firmware updates. This measure acts as a deterrent against malicious actors delivering counterfeit firmware updates.
- Incorporate Encryption: Encryption is an essential layer of protection to shield firmware updates from interception or unauthorized modification by potential attackers.

- Adopt a Staging Process: Employing a staging process is an effective strategy for reducing the risk associated with deploying faulty firmware updates. In a staging process, the firmware update is initially rolled out to a limited subset of devices. Once the update is thoroughly tested and validated, it can be subsequently deployed to all devices.

LPWAN Security Mechanism and Best Practices

In this section, we will delve into the security standards and best practices essential for LPWANs. We will explore established security standards, such as the NIST Cybersecurity Framework, and provide comprehensive guidance on the implementation of security best practices in LPWANs. The following summarizes the most prevalent LPWAN security standards and best practices:

LPWAN Security Mechanism

- LoRaWAN: LoRaWAN incorporates a wide array of security mechanisms, including pre-shared keys (PSKs), digital certificates, and mutual TLS (mTLS). The selection of specific security mechanisms depends on the network topology and the unique security requirements of the application.
- Zigbee: Zigbee supports a diverse set of security mechanisms, encompassing PSKs, digital certificates, and mTLS. The choice of security mechanisms hinges on the Zigbee network's topology and the specific security needs of the application.
- NB-IoT: NB-IoT relies on mTLS for both authentication and encryption.
- LTE-M: LTE-M adopts the same security mechanisms as LTE, featuring IPsec and TLS.

LPWAN Security Best Practices

- Secure Authentication Mechanism: Employ a robust and secure authentication mechanism to ensure that solely authorized devices gain access to the LPWAN network and communicate with one another.
- Secure Communication Protocol: Select a communication protocol that supports encryption and authentication to protect data from potential interception or manipulation by malicious actors.
- Firmware Updates: Prioritize the prompt application of firmware updates, as they frequently include security patches that address known vulnerabilities within LPWAN devices.
- Continuous Network Monitoring: Continually monitor LPWAN networks for any indications of suspicious or unauthorized activity, enabling the timely detection of potential security threats.
- Implementation of Security Policies and Procedures: Implement comprehensive security policies and procedures to govern the utilization of LPWAN networks and effectively mitigate security risks, establishing a robust security framework.

The amalgamation of these security standards and best practices collectively serves to fortify the security of LPWANs and the IoT devices that rely upon them, mitigating the associated risks of potential security breaches.

Over-The-Air (OTA) Updates for Non-IP Based Communication Protocols

Over-the-Air (OTA) updates are a convenient and efficient way to deliver firmware and software updates to embedded devices. OTA updates can be delivered over a variety of wireless communication protocols, including Wi-Fi, cellular, and Bluetooth. However, many embedded devices use non-IP based communication protocols, such as ZigBee, LoRaWAN, and BLE[.

OTA updates for non-IP based communication protocols can be more challenging to implement than OTA updates for IP-based protocols. This is because non-IP based protocols often have limited bandwidth, packet size, and reliability. Additionally, many non-IP based protocols do not support standard OTA update mechanisms, such as TFTP and HTTP.

Despite the challenges, there are a number of ways to implement OTA updates for non-IP based communication protocols. One approach is to use a proprietary OTA update protocol. This approach gives the device manufacturer the most flexibility in terms of how the OTA update process is implemented. However, it also requires the device manufacturer to develop and maintain their own OTA update infrastructure.

Another approach to OTA updates for non-IP based communication protocols is to use an open source OTA update framework. There are a number of open source OTA update frameworks available, such as OTAFirmware and OTA-Agent. These frameworks provide a common set of APIs and functionality for implementing OTA updates. This can simplify the development and maintenance of OTA updates for non-IP based communication protocols.

Regardless of the approach used, OTA updates for non-IP based communication protocols typically involve the following steps:

- The device manufacturer hosts the OTA update firmware or software on a remote server.
- The device periodically checks the remote server for new OTA updates.
- If a new OTA update is available, the device downloads the update from the remote server.
- The device verifies the integrity of the OTA update.
- The device installs the OTA update.

The OTA update installation process may involve restarting the device, updating the device's bootloader, or updating the device's firmware or software.

Here are some examples of how OTA updates are being implemented for non-IP based communication protocols:

ZigBee: The ZigBee Alliance has developed a ZigBee Over-the-Air (ZOTA) specification. ZOTA provides a standard way to deliver firmware and software updates to ZigBee devices. ZOTA updates can be delivered over a variety of ZigBee network topologies, including mesh and star networks.

LoRaWAN: The LoRa Alliance is developing a LoRaWAN Over-the-Air (LOTA) specification. LOTA will provide a standard way to deliver firmware and software updates to LoRaWAN devices. LOTA updates will be delivered over the LoRaWAN network, which uses a star-of-stars topology.

BLE: There are a number of proprietary OTA update protocols for BLE devices. Additionally, some BLE chip vendors provide their own OTA update solutions.

OTA updates for non-IP based communication protocols are becoming increasingly important as more and more embedded devices are deployed in remote and difficult-to-access locations. OTA updates can help to ensure that these devices are kept up to date with the latest security patches and bug fixes.

Challenges and Limitations

OTA updates for non-IP based communication protocols can be more challenging to implement than OTA updates for IP-based protocols. This is because non-IP based protocols often have limited bandwidth, packet size, and reliability. Additionally, many non-IP based protocols do not support standard OTA update mechanisms, such as TFTP and HTTP.

Some of the challenges and limitations of OTA updates for non-IP based communication protocols include:

Limited bandwidth: Non-IP based communication protocols often have limited bandwidth. This can make it difficult to download large OTA updates.

Small packet size: Non-IP based communication protocols often have small packet sizes. This can make it difficult to implement OTA update protocols that require large packets, such as TFTP.

Unreliable communication: Non-IP based communication protocols are often not as reliable as IP-based protocols. This can make it difficult to ensure that OTA updates are delivered successfully.

Lack of standard OTA update mechanisms: Many non-IP based protocols do not support standard OTA update mechanisms, such as TFTP and HTTP. This can make it necessary to develop proprietary OTA update protocols.

OTA updates for non-IP based communication protocols are becoming increasingly important as more and more embedded devices are deployed in remote and difficult-to-access locations. OTA updates can help to ensure that these devices are kept up to date with the latest security patches and bug fixes.

Despite the challenges and limitations, there are a number of ways to implement OTA updates for non-IP based communication protocols.

Implementing Secure Boot and Firmware Updates in Non-IP Based Communication Protocols

In this section, we introduce the concepts of secure boot and firmware updates for IoT devices, discussing the importance of secure boot and firmware updates for maintaining the security of IoT devices and preventing unauthorized access [17–22].

Secure boot is a mechanism that ensures that only authenticated firmware can be loaded onto a device. In this section, we discuss how secure boot can be implemented in IoT devices that use non-IP based communication protocols, exploring the use of digital signatures and secure boot loaders to ensure that only authenticated firmware is loaded onto a device.

Firmware updates are critical for maintaining the security and functionality of IoT devices. In this section, we discuss how firmware updates can be implemented on IoT devices that use non-IP based communication protocols, exploring the use of digital signatures, secure boot, and other security features to ensure that firmware updates are authentic and secure.

Best Practices for Secure Boot and Firmware Updates

Secure boot and firmware updates are essential for protecting embedded devices from cyberattacks. Secure boot ensures that only trusted firmware is loaded during the boot process, while firmware updates allow device manufacturers to deliver security patches and bug fixes to devices in the field. There are some best practices for secure boot and firmware updates as followings.

Secure Boot

- **Use a trusted platform module (TPM):** A TPM is a hardware security module that can be used to store and manage cryptographic keys. TPMs can be used to implement secure boot by verifying the digital signature of the firmware before it is loaded.
- **Use a secure boot loader:** A secure boot loader is a piece of firmware that is responsible for verifying the digital signature of the firmware and loading it into memory. Secure boot loaders should be designed to be resistant to tampering and attack.
- **Keep the secure boot keys secure:** The secure boot keys should be stored securely and only accessible to authorized personnel. If the secure boot keys are compromised, an attacker could use them to load malicious firmware onto the device.

Firmware Updates

- **Use a secure firmware update mechanism:** Firmware updates should be delivered using a secure mechanism that prevents unauthorized updates. Secure firmware update mechanisms typically use encryption and digital signatures to verify the authenticity of the firmware update.

- **Sign all firmware updates:** All firmware updates should be signed with a cryptographic signature. This allows the device to verify the authenticity of the firmware update before it is installed.
- **Require user authorization for firmware updates:** Firmware updates should only be installed with the user's authorization. This prevents unauthorized firmware updates from being installed on the device.
- **Test all firmware updates thoroughly:** Before deploying a firmware update, it should be thoroughly tested to ensure that it is stable and secure.

In addition to the above best practices, it is also important to keep the firmware of the device up to date. Firmware updates often include security patches that can help to protect the device from known vulnerabilities.

Additional tips for secure firmware updates:

- **Use a staging process:** When rolling out a new firmware update, it is a good idea to use a staging process. This involves deploying the update to a small number of devices first to test it and make sure there are no problems. If the update is successful, it can then be rolled out to the remaining devices.
- **Use a backup and restore process:** Before installing a firmware update, it is a good idea to back up the device's configuration. This will allow you to restore the configuration if the firmware update fails.
- **Monitor the device after the update:** After installing a firmware update, it is a good idea to monitor the device to make sure that it is working properly. If you notice any problems, you can contact the device manufacturer for support.

By following these best practices, you can help to protect your embedded devices from cyberattacks.

Security Standards and Best Practices for Non-IP Based Communication in IoT

The Internet of Things (IoT) is rapidly growing, with billions of devices connected to the internet. Non-IP based communication protocols are becoming increasingly popular for IoT devices because they offer a number of advantages over IP-based protocols, such as low power consumption, long range, and low cost. However, non-IP based protocols also pose a number of security challenges [22–25].

This paragraph discusses the security standards and best practices for non-IP based communication in IoT. It will cover the following topics:

- Security standards for non-IP based communication protocols
- Best practices for implementing secure non-IP based communication in IoT
- Case studies of secure non-IP based communication in IoT

Security Standards for Non-IP Based Communication Protocols.

There are a number of security standards for non-IP based communication protocols. Some of the most common standards include:

- ZigBee Over-the-Air (ZOTA): ZOTA is a security standard for ZigBee networks. It provides a secure way to deliver firmware and software updates to ZigBee devices, as well as to encrypt communication between ZigBee devices.
- LoRaWAN Over-the-Air (LOTA): LOTA is a security standard for LoRaWAN networks. It provides a secure way to deliver firmware and software updates to LoRaWAN devices, as well as to encrypt communication between LoRaWAN devices.
- Bluetooth Low Energy (BLE) Security: BLE Security provides a number of features to protect BLE communication from unauthorized access and eavesdropping. These features include encryption, authentication, and authorization.

In addition to these specific standards, there are also a number of general security principles that should be followed when implementing non-IP based communication in IoT. These principles include:

- Use strong encryption: All data transmitted over non-IP based networks should be encrypted. This will help to protect data from unauthorized access, even if the data is intercepted.
- Implement authentication: Authentication mechanisms should be used to ensure that only authorized devices can communicate with each other. This can be done using a variety of methods, such as X.509 certificates or device IDs.
- Implement authorization: Authorization mechanisms should be used to control which devices can access which resources. This can be done using a variety of methods, such as role-based access control (RBAC) or access control lists (ACLs).

Best Practices for Implementing Secure Non-IP Based Communication in IoT.

In addition to following the security standards and principles discussed above, there are a number of best practices that can be followed to implement secure non-IP based communication in IoT. These best practices include:

- Use a layered security approach: Security should be implemented at all layers of the communication stack, including the physical layer, the data link layer, and the network layer. This will help to protect against a wider range of attack vectors.
- Keep the firmware up to date: Firmware updates often include security patches that can help to protect against known vulnerabilities. It is important to keep the firmware of all IoT devices up to date.
- Monitor the network for suspicious activity: Network monitoring tools can be used to detect suspicious activity on the IoT network. This can help to identify and respond to attacks quickly.

Case Studies of Secure Non-IP Based Communication in IoT.

Here are a few examples of how secure non-IP based communication is being used in IoT:

- Smart home devices: ZigBee ZOTA is being used to deliver firmware and software updates to smart home devices, such as thermostats, light bulbs, and door locks. This helps to ensure that these devices are always protected from the latest security vulnerabilities.
- Industrial control systems: LoRaWAN LOTA is being used to deliver firmware and software updates to industrial control systems (ICS) sensors and actuators. This helps to ensure that these systems are always operating securely and that the data collected by the sensors is protected from unauthorized access.
- Wearable devices: BLE Security is being used to protect communication between wearable devices, such as smartwatches and fitness trackers, and smartphones. This helps to ensure that the data collected by the wearable devices is protected from unauthorized access.

By following the security standards, best practices, and case studies discussed in this chapter, you can help to protect your IoT devices from cyberattacks and ensure that they are always operating securely.

Conclusion

- Non-IP-based communication technologies play a vital role in the Internet of Things (IoT), offering several advantages over IP-based communication, such as low power consumption, scalability, and reliability. However, these technologies also pose unique security challenges due to their inherent characteristics and the diversity of IoT applications.
- This chapter has provided a comprehensive overview of non-IP-based IoT communication and its related security issues. It has highlighted the key security challenges and solutions, presented mathematical models for non-IP communication protocols and IoT device security issues, and emphasized the importance of proactive security measures.

The Following Are the Key Takeaways from the Chapter

- Non-IP-based communication technologies are essential for IoT, but they pose unique security challenges.
- A comprehensive understanding of the security challenges of non-IP-based IoT communication is crucial for designing and deploying secure IoT systems.
- Mathematical models can be used to analyze and predict the security of non-IP communication protocols.
- Proactive security measures are essential for protecting non-IP-based IoT systems from cyberattacks.
- As the IoT ecosystem continues to evolve, it is imperative to remain vigilant and continuously study the security challenges and solutions of non-IP-based communication technologies. By doing so, we can ensure the safe and successful

inclusion of these technologies into IoT ecosystems, enabling a more connected and intelligent future.

Future Directions

The following are some promising directions for future research on the security of non-IP-based IoT communication:

- Development of new security standards and protocols: Existing security standards and protocols may not be sufficient to address the unique security challenges of non-IP-based IoT communication. New standards and protocols need to be developed that are tailored to the specific needs of non-IP-based IoT systems.
- Design and implementation of secure non-IP-based IoT devices and networks: New secure non-IP-based IoT devices and networks need to be designed and implemented that leverage the latest security technologies and best practices.
- Development of tools and techniques for security assessment and monitoring: New tools and techniques need to be developed to assess the security of non-IP-based IoT systems and monitor them for suspicious activities.

By pursuing these research directions, we can contribute to the development of a more secure and reliable IoT ecosystem.

References

1. Edwards, J., & Bramante, R. (2015). *Networking self-teaching guide*. Wiley.
2. Hartpence, B. (2011). *Packet guide to core network protocols*. "O'Reilly Media, Inc."
3. Rayes, A., & Salam, S. (2018). *Internet of things from hype to reality*. Springer.
4. Gao, H., Yin, Y., Yang, X., & Miao, H. (2019). *Testbeds and research infrastructures for the development of networks and communities*. Springer.
5. IoT for Non-IP Devices | Amazon Web Services. (2021). Amazon Web Services. https://aws.amazon.com/blogs/iot/iot-for-non-ip-devices-2/
6. Non-IP device Communication Protocols. (n.d.). Internet of Things Stack Exchange. https://iot.stackexchange.com/questions/2253/non-ip-device-communication-protocols
7. Agarwal, T. (2021). Communication Protocols: Basics and Types with Functionality. ElProCus—Electronic Projects for Engineering Students. https://www.elprocus.com/communication-protocols/
8. Non-IP protocols—Sunflower-CISSP.com. (n.d.). https://www.sunflower-cissp.com/glossary/cissp/3827/non-ip-protocols
9. Grant, J. (2020). Non-IP Networking: Why the Industry Needs a Framework—In Compliance Magazine. In Compliance Magazine. https://incompliancemag.com/article/non-ip-networking-why-the-industry-needs-a-framework/
10. Shue, C. A., & Gupta, M. (2010). An Internet without the Internet protocol. *Computer Networks, 54*(18), 3232–3245. https://doi.org/10.1016/j.comnet.2010.06.009
11. Thompson, L. M. (2008). Industrial Data Communications. ISA.
12. Badenhop, C., & Ramsey, B. (2016). Carols of the Z-Wave security layer; or, robbing keys from Peter to unlock Paul Int J PoC, 12, pp. 6–12.
13. Palattella, M. R., et al. (2016). Internet of things in the 5G Era: enablers, architecture, and business models. *IEEE Journal on Selected Areas in Communications, 34*(3), 510–527. https://doi.org/10.1109/JSAC.2016.2525418

14. Orfanos, V. A., Kaminaris, S. D., Papageorgas, P., Piromalis, D., & Kandris, D. (2023). A comprehensive review of IoT networking technologies for smart home automation applications. *Journal of Sensor and Actuator Networks, 12*, 30. https://doi.org/10.3390/jsan12020030
15. Badenhop, C. W., Graham, S. R., Ramsey, B. W., Mullins, B. E., & Mailloux, L. O. (2017). The Z-Wave routing protocol and its security implications. *Computers & Security, 68*, 112–129. https://doi.org/10.1016/j.cose.2017.04.004
16. Sicari, S., Rizzardi, A., Grieco, L., & Coen-Porisini, A. (2015). Security, privacy and trust in Internet of Things: The road ahead. *Computer Networks, 76*, 146–164. https://doi.org/10.1016/j.comnet.2014.11.008
17. Chen, D., Nixon, M., & Mok, A. (2010). WirelessHARTTM. Springer Science & Business Media.
18. Zohourian, A., Dadkhah, S., Neto, E. C. P., Mahdikhani, H., Danso, P. K., Molyneaux, H., & Ghorbani, A. A. (2023). IoT Zigbee device security: A comprehensive review. *Internet of Things, 22*, 100791. https://doi.org/10.1016/j.iot.2023.100791
19. Braghin, C., Lilli, M., & Riccobene, E. (2023). A model-based approach for vulnerability analysis of IoT security protocols: The Z-Wave case study. *Computers & Security, 127*, 103037. https://doi.org/10.1016/j.cose.2022.103037
20. LoRa Alliance®. (2023). LoRa Alliance®. https://lora-alliance.org/
21. Woodward, E. (2022). EnOcean ultra-low power wireless networks: What they are and how they work. Pressac Communications. https://www.pressac.com/insights/enocean-ultra-low-power-wireless-networks-what-they-are-and-how-they-work/
22. Internet Engineering Task Force (IETF). (2012). Datagram Transport Layer Security (DTLS) Version 1.2. RFC 6655. https://tools.ietf.org/html/rfc6655
23. Internet Engineering Task Force (IETF). (2014). The Constrained Application Protocol (CoAP). RFC 7252. https://tools.ietf.org/html/rfc7252
24. Organization for the Advancement of Structured Information Standards (OASIS). (2019). MQTT Version 5.0. https://docs.oasis-open.org/mqtt/mqtt/v5.0/mqtt-v5.0.html
25. National Institute of Standards and Technology (NIST). (2017). Guidelines for Securing Internet of Things (IoT) Devices. NIST Special Publication 800–181. https://nvlpubs.nist.gov/nistpubs/SpecialPublications/NIST.SP.800-181.pdf

Security Challenges to IOT and Cloud-Based Systems in the Era of Quantum Attacks

V. R. S. Mani

Abstract All cryptographic algorithms and security measures are prone to different kinds of novel Cyberattacks in the era of quantum mechanics. This is a great threat to the survival of complex technological systems. Cryptographic codes of networked vehicles and complex systems will be decrypted using quantum computers. The evolution of quantum computers and algorithms is a boon for attackers to breach the security of modern internet and IoT-based connected Vehicles and complex systems. Computation using Qubits, which have a large amount of processing capability that can swiftly decrypt the current encryption algorithms is the direction that technology is gradually headed in. technology is gradually advancing in Therefore, protecting our systems from security issues is urgently needed. IoT will also be one of the disciplines, which need to be secured to prevent any malicious activity. IoT device security requirements "can be very complex". More complex and sophisticated Security algorithms need to be developed. Hence, the security provider must, for instance, take into account security measures for supply chain security, protected patch management, defense against cyberattacks, authentication and validation process, and secure transaction. This chapter discusses the threats to internetworked devices and their mitigation in the era of quantum computers.

Keywords Quantum computing · Cyberattack · IoT

Introduction

The World Economic Forum has cautioned that modern technologies like AI, IoT, and quantum computing have the potential to change human existence. IoT and AI are examples of contemporary technologies that can assist us in making wise choices and implementing them at our discretion from a society's susceptibility to cyber-attacks. The use of IoT-based applications and solutions is growing exponentially. According

V. R. S. Mani (✉)
Department of Electronics and Communication Engineering, Francis Xavier Engineering College, Tirunelveli, India
e-mail: vrsece@rediffmail.com

A. Prasad et al. (eds.), *Communication Technologies and Security Challenges in IoT*, Internet of Things, https://doi.org/10.1007/978-981-97-0052-3_11

to IoT Analytics, there will be 28 billion connected IoT-based sensors globally by 2025, or about 3.4 for every person on the planet. (UN world population estimate for 20,258.18 billion). IoT applications will begin producing enormous amounts of data that, if supported by a robust configuration made up of appropriate connectivity, gateways, and analytics, will be helpful for our daily applications. Numerous initiatives unite companies and governments to foster trust, advance cyberspace security, evaluate the consequences of cyberattacks, and aid victims, according to the study. Employing cyberattacks Due to catastrophic hacks and damaging breaches, billions of private files, account details, and confidential information have been exposed since 2010 The growing number of IoT devices on the market is evidence of the IoT industry's success, however many of these devices have resource issues. Many IoT devices are unable to employ conventional security solutions as a result, hence it is crucial to provide IoT devices with access to lightweight security solutions. Security limitations are defined as restrictions dependent on the networking, software, and hardware of IoT devices. Hardware-based restrictions include algorithmic, battery life, and storage area. Software-based restrictions include those related to integrated software. Mobility, scalability, and sluggish intermittent network connections caused by the use of low-power radios and low data rates are just a few networking-related limitations (Fig. 1).

IoT equipment might be broadly dispersed and open to physical attacks [1, 2]. These may include side-channel attacks, which seek to discover confidential information about encryption passwords by examining how a security algorithm is used. A timing attack, for instance, might attempt to take advantage of the fact that a key-generation algorithm's execution time can vary depending on the key value produced. A key's storage time can reveal information about the proportional population of 0 s and 1 s within it because it may take longer to write a 1 into memory than a 0. Another choice is power analysis. If writing a 1 into memory uses more energy than writing a 0, this could reveal sensitive information. IoT fragmentation makes it more difficult to secure an environment. A chip's ability to locate and connect to the planned IoT network depends on the silicon vendor managing the data that must be embedded in

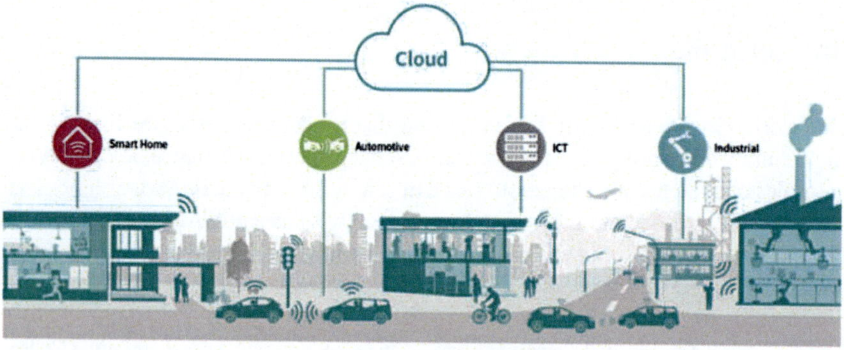

Fig. 1 Cyber-physical system (https://www.erp-information.com/cyber-physical-systems

each chip. The manufacturer of the devices that use these processors must make sure that cryptographic operations are carried out correctly. Manufacturers and integrators of IoT hubs must offer software for controlling, gathering, and organizing data collected by devices. These service companies might also be in charge of overseeing authentication.

IoT ecosystems and the data that flows through them are accessible to many parties, but no one assumes an overall security obligation. It won't be effective to hire outside experts to complete that if the task's parameters aren't specified.

Concerns relating to the proliferation of Quantum computers might invalidate current encryption methods have sparked a lot of research. These PQC techniques could be useful for allowing IoT security due to their valuable properties. It is difficult to manage encryption keys and digital signatures that are long enough to provide adequate security on devices with less amount of space, communications infrastructure, and power in the age of quantum computing. For server and device verification, digital signatures are essential. NIST USA is looking into new digital signature methods. The problem is revealed by a comparison with ECDSA, a prevalent technique used today: ECDSA requires sending a public key of 256 bits and a signature with 587 bits of security to communicate a signature with 128-bit security. An 896-byte public key and a 690-byte signature make up the NIST analysis's last remaining PQC digital signature scheme. In other words, the PQC implementation of a digital signature requires about 15 times the amount of bandwidth, as well as more processing power and memory to hold cryptographic keys, than the ECDSA implementation. There may be additional PQC digital signature protocols that use the same capacity as ECDSA. If not, IoT devices will need to depend on alternative methods of server authentication, like greater reliance on key-encapsulation techniques and pre-shared keys. Additionally, NIST is seeking PQC algorithms that are naturally more resistant to physical attacks than those currently in use. Practitioners of IoT security must be conscious of standardization procedures and determine which PQC strategies will be effective given the limited resources of IoT devices. NIST is investigating the post-quantum computing world's bandwidth needs for digital signatures as well as how resistant the new algorithms will be to side-channel attacks. The IoT needs to be secured in the future era of quantum computing because today's signature schemes are too cumbersome. Currently, 90% of the data collected by devices is lost or discarded for a variety of reasons, such as bandwidth restrictions and privacy and security mandates. Additionally, a wealth of unstructured data can be found from a variety of sources, including news Web sites, contact centers, and social networks. Users can combine all of these different data sources in real time, gain a deeper understanding of what's happening, and generate insightful data by making use of new cognitive IOT capabilities and cloud-based infrastructure.

Today the IoT portfolio is enhanced with many cognitive technologies, including text analytics, video and picture analytics, machine learning, and natural language processing. A different range of customers and business partners will be able to use these cloud skills to improve current systems.

Post-Quantum Cryptography

Two major categories can be used to categorize the hunt for future cryptosystems. Quantum encryption and post-quantum cryptography are the first and second, respectively. Finding mathematical puzzles that a quantum computer can't answer, then creating cryptosystems based on these puzzles is the goal of post-quantum cryptography. These issues can be seen in multivariate, hash-based, and lattice-based encryption. Utilizing quantum mechanical principles to develop novel cryptosystems is known as quantum cryptography. Quantum key distribution is a type of quantum encryption. It is based on a few basic principles of quantum mechanics, which can make cryptosystems very secure against an observer attempting to learn secret messages or keys. Quantum cryptography is a novel area that is being studied. In addition to the zeros and ones present in classical computing, superposition also makes infinitely many states between zero and one conceivable in Quantum computing. These qubits in a quantum program can be placed in an unlimited number of locations between zeros and ones. A qubit in superposition has a chance of being in state one and a chance of being in state zero. When one attempts to read a qubit, its superposition collapses to either a zero or a one. Although we will never know the state or orientation of this bit before it collapsed, the measurement's chance will depend on the qubit's initial state (Fig. 2).

Quantum key distribution (QKD) is a technique for doing secure communication utilizing the concepts of quantum physics. Here, the sender and receiver create a private shared key that can only be used by them to communicate securely. Although it is a regular task in quantum cryptography, quantum key distribution is not the same as quantum cryptography. The ability of the two communicating parties to recognize the eavesdropper trying to steal the key is the most important aspect of QKD and is based on the fundamental idea of quantum mechanics: any measurement of a

Fig. 2 Quantum computer (https://www.pngmart.com/image/tag/quantum-computing)

quantum system affects it. To try to eavesdrop on the key, a third party must measure it in some way, which introduces observable irregularities. A communication system that recognizes listening devices.

Only a key is produced and distributed using QKD; no communication data is sent. This message can then be encrypted (and decrypted) using any encryption algorithm of your choice and transmitted over a common communication route. The one-time pad algorithm is the one most often linked with QKD because, when combined with a secret, random key, it is proven to be secure. It is frequently combined with cryptography that uses symmetric key algorithms, such as the Advanced cryptography Standard algorithm, in real-world scenarios.

Quantum Key Exchange Technique

Quantum communication uses the encoding of information in quantum states, or qubits, as opposed to the usage of bits in conventional communication. Particles are frequently used to symbolize these quantum states. Quantum key distribution makes use of particular properties of these quantum states to provide security. Although there are many ways to distribute quantum keys, they can be divided into two groups based on the traits they share.

i ***Construct and evaluate protocols:***

   Contrary to conventional physics, measurement is a key part of quantum mechanics. When a quantum state is measured, it usually experiences some form of change.

ii ***Protocols that make use of entanglement***

   It is possible for the quantum states of two (or more) different entities to merge together to the point that they can no longer be distinguished as separate objects but rather as a single quantum state. When two things are entangled, for instance when measuring one thing has an impact on the other. Anyone with the ability to intercept either of the entangled items that are being traded by two people and reveal their identify (together with the information they have learnt) has the ability to change the entire system. These two methods can be further divided into the discrete variable, continuous variable, and distributed phase reference coding families of protocols.

Implementation/Experimentation in QKD

A QKD between two of its laboratories was tried by the Indian Defence Research and Development Organization in Hyderabad in December 2020. The setup also showed how to identify when a third party was attempting to intercept the communication. The deployed system's quantum-based security against listening in was verified at a distance of over 12 km (7.5 mi) and with a large reduction in power level in an optical communication channel. To produce photons without depolarization effects, an analog laser source was employed, and the setup's timing precision was quite low—between 10 and 12 s. Single-photon avalanche detectors (SPADs) have been able to attain key rates in the kbps range when recording photon arrivals because of the low qubit error rate.

The Indian Space Research Organization also showed a 300-m-long free-space quantum communication in March 2021. At the Space Applications Centre (SAC), Ahmadabad, a free-space QKD was performed between two of their laboratories which are opposite each other for an online meeting. Time synchronization is done using a NAVIC receiver. Indian researchers succeeded in developing a Free space channel for the transmission of encoded data and images later in January 2022. India intends to create Satellite-Based Quantum Communication (SBQC) after demonstrating QKD between two land-based systems.

In July 2022, researchers published the experimental results of a device-independent quantum key distribution (DIQKD) protocol that employs quantum entanglement (as suggested by Ekert) to assure resilience to quantum hacking assaults. They performed the following steps to create two ions that were about two meters apart and extremely entangled: An 88Sr + qubit can be found in each of Alice's and Bob's ion trap components. When the atoms are first stimulated to an electronic state, they become entangled. Two photons are also created during this process. A Bell-basis approach is then used to transmit, collect, and measure these photons. The ions are subsequently projected into a highly entangled state as a result.

Commercial Quantum Key Distributions

Seven companies currently provide commercial quantum key distribution systems: ID Quantique (Geneva), Magi Q Technologies, Inc. (New York), QNu Labs (Bengaluru, India), (Australia), Q Rate (Russia), SeQure Net (Paris), Quantum Optics Jena (Germany), and KEE Quant (https://www.keequant.com). (Germany). Toshiba, HP, IBM, Mitsubishi, NEC, and NTT are further companies with ongoing study programs. KETS Quantum Security (UK) is another. The first-ever quantum key distribution cash transfer took place in Vienna, Austria, in 2004. The results of the October 21, 2007, national election in Switzerland's canton (state) of Geneva were sent to the nation's capital using quantum encryption technology provided by the Swiss business Id Quantique. In 2013, Battelle Memorial Institute installed a QKD system at its main campus in Columbus, Ohio, as well as its manufacturing site in Dublin, Ohio. Field tests for the Tokyo QKD network have been ongoing for some time.

Quantum Key Distribution Networks

QKD networks are networks that generate quantum keys and distribute quantum keys among numerous nodes or locations using optical quantum cryptography. Using traditional encryption techniques, such as the one-time pad, quantum keys can be used to encrypt and decrypt data. Quantum keys are random, un crackable sequences of bits. Banking, government, the military, and the healthcare industry are just a few examples of the many applications and use cases that can benefit from the high level of security and anonymity that QKD networks can offer. To take advantage of the current infrastructure and technology, QKD networks can also be combined with existing fiber optic networks. In numerous nations and regions, including China, Europe, Japan, and Canada, QKD networks have already been installed and tested.

i. ***Quantum Network of the DARPA***

   A 10-node quantum key distribution network called the DARPA Quantum Network ran continuously in Massachusetts from 2004 to 2007 for four years, 365 days a year. It was developed in collaboration with Harvard University and Boston University by International Business Machines' Research Division. It made it possible to create a standards-based, quantum key distribution-protected Internet computer network.

ii. ***SECOQC***

   The First Quantum Cryptographic network became live at a scientific meeting in Vienna in October 2008. The money for this network came from the EU. Using a 200 km optical cable, this quantum network connected six locations in Vienna and another town 70 km to the west of Vienna.

iii. ***Swiss Quantum***

   Field testing of a very long Quantum Key Distribution (QKD) was successfully done by Swiss Quantum Id Quantique. The main goal of the Swiss Quantum network project, which was launched in Geneva City in April 2010, was to confirm the dependability and resilience of QKD in continuous operation over an extended length of time in a field setting. The experiment was stopped in January 2012, just a few months after it had been operational for almost two years.

iv. ***Chinese Quantum network***

   In May 2009, a multilevel network based on QKD became live. Using a relay contact, the several subnets are linked to a backbone network made up of four nodes, which forms the hierarchical network. The backbone nodes were connected by an optical switching quantum router. Each subnet's nodes were also connected by optical switches, which were in turn connected to the backbone network by reliable relays.

v. ***Tokyo QKD Network***

The Tokyo QKD Network was introduced on the first day of the UQCC2010 conference [3]. NEC, Mitsubishi Electric, NTT, and NICT from Japan, Toshiba Research Europe Ltd. (UK), Id Quantique (Switzerland), and All Vienna from Europe (Austria) are the seven companies that are a part of the network. The image shows researchers from the University of Vienna, the Institute for Quantum Optics and Quantum Information (IQOQI), and the Austrian Institute of Technology (AIT) as "All Vienna".

vi. ***Los Alamos National Laboratory***

Since 2011, Los Alamos National Laboratory has operated a hub-and-spoke network. To route all communications, use the hub. Each network node is equipped with laser-based quantum transmitters thanks to the technology, but not with expensive, substantial photon detectors. The receiver of quantum communications is the hub alone. A one-time pad is transmitted by a network station to the central hub, to make safe transactions through the network link. Using a second one-time pad from the second node, another secure transaction can be done to another node. Only if the central hub is safe is the network as a whole secure. The only tool needed for a single component is a laser: The size of a prototype node is comparable to a case of matches.

vii. ***Eagle-1***

A satellite-based QKD system will be launched in 2024, by the European Space Agency using satellite Eagle-1(6).

Future Outlook

The high cost of hardware and lack of a clear threat to current key exchange methods have an impact on the widespread usage of QKD outside of highly protected zones. However, most nations have access to fiber optic cable networks. To handle issues with standardization in quantum cryptography, the ETU has established a Standardization task force. In the context of focused projects, European Metrology Institutes are creating the measurements necessary to characterize QKD system components. Developed over two decades of study, Toshiba's ground-breaking QKD technology safeguards communication infrastructure from current and emerging cyber threats and enables the sale of UK-made goods that pave the way for the quantum internet.

QKD Challenges

QKD's dependence on hardware makes it less adaptable to upgrades and security fixes. Equipment specifically designed for quantum key dissemination is required. The security of QKD is derived from particular physical layer communications, and it is founded on physical properties. Users must either rent exclusive fiber links or physically maintain free-space transmitters to accomplish this (7). It cannot be readily integrated into already installed network hardware and designed using Firmware or using other techniques like network as a service.

The risks of insider threats and infrastructure expenses are increased by QKD. Trusted relays are frequently required by Key distribution systems, adding to the expense of secured infrastructure and attacks from people present within the organization. This rules out a lot of potential use scenarios. It is very challenging to protect and verify the distribution of quantum keys. A QKD system offers stricter conditional protection that can be offered through equipment and system design in place of the traditional, unconditional security derived from the laws of physics (as modeled and widely recommended). However, the tolerance for mistake in cryptographic security is many orders of magnitude lower than in most physical engineering contexts, making it very difficult to validate. The precise equipment utilized to carry out Quantum Key Distribution may introduce vulnerabilities.

The danger of DDOS attacks is increased by QKD. The theory related to QKD protection schemes and unauthorized user accessing the network service ensures DOS attack is a key risk for QKD.

Techniques to Mitigate QKD Challenges

The following are some of the techniques used to mitigate the challenges posed by the exorbitant cost of hardware and the threat to current key exchange protocols.

i. No trust architecture

   The No trust security model, also known as No trust architecture (ZTA), zero trust network architecture, or zero trust network access (ZTNA), as well as occasionally known as perimeter-less security, defines a procedure for creating and implementing IT systems [4, 5]. According to the zero trust security model, devices shouldn't be trusted by default, even if they are connected to a permissioned network like a corporate LAN and even if they have already been confirmed. The model's guiding philosophy is "Never trust, always verify". In order to implement ZTNA, strong identity verification is developed, device compliance is checked prior to access being allowed, and least privilege access is guaranteed to only explicitly permitted resources.

ii. Hash-based Cryptography

   Several aspects of the hash-based approach are advantageous to the IoT ecosystem. Cryptographic algorithms using hash functions don't depend on any other preconceived notions of network security, such as number-theory-based hardness, but only on cryptographic hash functions. As a result, it makes crypt analysis much more difficult. This lessens the system's overall complexity. To attain the desired performance, the hash-based scheme must be flexible in the hash function it chooses because it is intrinsically dependent on the application-specific environment. The collision resistance, pre-image resistance, and second-pre-image resistance properties of the hash function provide defense against various attacks [6]. The hash-based approach offers a solution to defend IoT applications from attackers that take the credentials from mission-critical or long-running devices with resource constraints. There are lightweight hash function variants that give IoT applications the choice of appropriate device parameters for resource-constrained devices, enhancing network performance. Since hash functions in hash-based schemes only work in one direction, they are secure with both backward and forward secrecy.

iii. Lattice-based Cryptography

Lattice-based algorithms are more secure and very hard to solve, due to the inherent nature of a lattice structure. Lattice-based cryptographic systems can be decrypted only with the knowledge of the correct key. IoT- product development businesses are primarily concentrating on creating lighter, more affordable, more intelligent products [7] while paying little attention to security concerns. Quantum risks to encryption in networks and services based on intelligent IoT devices are potentially plausible. As a result, post-quantum security considerations must be made while developing security solutions for smart Internet of Things devices. The lattice-based cryptosystem is a great choice for the future due to its broad applicability, high level of security, and effectiveness in thwarting attacks. Numerous cryptosystems typically use the random key generation method. Random key generation is an intrinsic characteristic of Lattice-based cryptography. The closest Vector Problem and Shortest Vector Problem are two complex problems that are very hard to solve and algorithms based on these problems are resistant to quantum attacks. Lattice-based techniques are energy efficient, flexible, and easy to implement.

iv. Optical Quantum Cryptography

By employing quantum keys to encrypt and authenticate the transactions and the participants, as well as to create and verify the signatures and hashes, optical quantum cryptography can enhance the speed and security of blockchain and cryptocurrencies. Blockchain and cryptocurrencies can also be shielded from quantum attacks that could jeopardize the cryptographic protocols and algorithms that support them via optical quantum cryptography. By using quantum keys to encrypt and decode data and services, as well as to authenticate and authorize devices and users, optical quantum cryptography can improve the security and privacy of smart cities and IoT. Additionally, optical quantum cryptography can facilitate effective and secure coordination and communication between devices and users as well as between devices and networks. Quantum random number generation (QRNG) is a method for creating unbiased, unpredictable random numbers using optical quantum cryptography. Numerous applications and use cases that involve encryption, security, simulation, gaming, lotteries, and statistics all depend on random numbers. However, it can be difficult to produce random numbers using traditional techniques, such as software algorithms or physical devices, as they may be influenced by predictable elements or outside sources. By leveraging the inherent unpredictability of quantum states, such as the polarisation or phase of photons, to produce truly random and verifiable random numbers, optical quantum cryptography can get over these restrictions. Additionally, optical quantum cryptography can facilitate dependable and secure random number distribution over a wide area network and quick and efficient random number production.

Conclusion and Research Directions

Post-quantum cryptography makes use of several cryptographic primitives, including hashes, digital signatures, symmetric and asymmetric keys, and random number generators, which are frequently used in block chain-based technologies. A wide range of security properties, such as resistance against different fault conditions, protection against cyber assaults, and adaptation to a wide range of enterprises, can be provided through the integration of block chain and post-quantum cryptography components. Some hybrid architectures proposed using the above in this era are found in [8–10]. To find the optimum answer for IOT devices and components that are quick and power-efficient, more technical solutions and analyzes are needed. Few projects and activities for post-quantum cryptography standardization have recently been identified [6, 9]. More research has to be done in areas that are not yet addressed to improve against threats involving more sophisticated systems and techniques to IOT infrastructure.

Additionally, the contributions made in this direction can be strengthened if standardization parameters are identified early on.

Infrastructure needs for combined post-quantum cryptography and IoT present another significant hurdle. Affordable hardware that supports complex high-performance computing, complex networks, and standard procedures with an integrated environment is required for IoT applications. In the network of intelligent digital cities, IoT networks are involved in a wide range of sectors and challenges that go along with them. Long-range communications for industrial IoT networks, resource-constrained devices, circuits, and networks, information protection in the clinical and family welfare industry, and accessibility and traceability problems in industrial operations. Post-quantum cryptography will meet the essential protection criteria while computations using qubit can increase the speed of information processing in all circumstances. Therefore, it would be fascinating to investigate the combined quantum-IoT ecosystem for domain-specific problems in the future. Risk analysis in various fields, such as quantum, cyber, and cloud infrastructure, is very essential to the analysis of the Cloud integrated IOT era. Continuous evaluations based on the risk assessment report during the development and deployment of IoT applications is an interesting research topic. With the development of technologies and the related concerns, this needs to be investigated. The cost of developing IoT applications may rise due to the dearth of quantum resources. As a result, cost estimation is required for IoT applications. It is necessary for those people knowing related areas to design integrated platforms for quantum computing using current technical innovations. Also, it is necessary to discover or teach knowledgeable individuals about handling IoT infrastructure. Critical infrastructure based on the Internet of Things needs more interconnectedness and interdependencies. Consider the networks for healthcare and public transport. The system is more vulnerable due to the rise in users and data volume. Investigating threat models that can understand the requirements of living systems is therefore required. The best candidates for IoT networks were lattice-based and hash-based techniques. These plans are all discovered to be equally effective and strong as conventional cryptosystems. In the upcoming years, compact and Qubit-encrypted small devices will be produced to meet the demands of next-generation internet-connected networks as the interconnection of devices in IoT networks expands globally. These devices will have an optimal bit architecture, optimal qubit-based protection, will run quickly, and will be immune to threat vectors [7, 11].

Cyberattacks on industrial facilities may result in the loss of production efficiency or the theft of information about how products are made. Information security will be important as electronic systems become more and more networked over time. Post-quantum cryptography now lacks standards and understanding. The standards are necessary to allow for communication between various systems, such as an internet-connected device and a large connected system. Also, device makers need to be aware that the security of their products may be seriously threatened by quantum computers. They ought to adopt the necessary security measures as soon as possible and adequately identify any potential future dangers. Efficiency plays a vital role in employing a particular security technique in some application contexts apart from cyberattacks. In addition to processing speed, performance takes into account memory requirements, computation complexity, different data size, etc. For example, more structured mathematical puzzle-based methods often have smaller keys. Different conditions must be met for quantum technology, like quantum computers or quantum sensors, to be commercially viable. For quantum computers to be extensively used, we need a computer that can truly show a benefit for practical tasks (such as chemical analysis, AI, etc.) over currently used cloud techniques. Promoting the market adoption of cryptographic algorithms immune to attacks in the post-quantum era requires raising public awareness.

References

1. Resul, D. A. S., & Gündüz, M. Z. (2019). Analysis of cyber-attacks in IoT-based critical infrastructures. *International Journal of Information Security science, 8*(4), 122–133.
2. Humayed, A., Lin, J., Li, F., & Luo, B. (2017). Cyber-physical systems security-a survey. *IEEE Internet of Things Journal, 4*, 1802–1831.
3. Sasaki, M., et al. (2011). Field test of quantum key distribution in the Tokyo QKD Network. *Optics Express, 19*(11), 10387–10409.

4. Yao, Q., Wang, Q., Zhang, X., & Fei, J. (2021). Dynamic access control and authorization system based on zero-trust architecture. In: Proceedings of the 2020 1st international conference on control, robotics, and intelligent systems. CCRIS '20 (pp. 123–127). New York, NY, USA: Association for Computing Machinery. https://doi.org/10.1145/3437802.3437824. ISBN 978–1–4503–8805–4. S2CID 230507437.
5. Higgins, K. J. (2021). Forrester pushes 'zero trust' model for security. Dark Reading. Informa.
6. Xu, R., Cheng, C., Qin, Y., & Jiang, T. (2018). Lighting the way to a smart world: lattice-based cryptography for the internet of things. arXiv preprint arXiv:1805.04880.
7. Thakor, V. A., Razzaque, M. A., & Khandaker, M. R. A. (2021). Lightweight cryptography algorithms for resource-constrained IoT devices: a review, comparison and research opportunities. *IEEE Access, 9*, 28177–28193. https://doi.org/10.1109/ACCESS.2021.3052867
8. Shahid, F., Khan, A., & Jeon, G. (2020). Post-quantum distributed ledger for the Internet of Things. *Computers & Electrical Engineering, 83*, 106581.
9. Gill, S.S. (2021). Quantum and blockchain-based serverless edge computing: A vision, model, new trends and future directions. *Internet Technology Letters*, e275.
10. Meng, Y., Qu, Z., Muhammad, G., & Tiwari, P. (2023). Secure and efficient data transmission based on quantum dialogue with hyperentangled states in cloud office. *Internet of Things, 24*, 100911 (ISSN 2542–6605).
11. Palmieri, P. (2018). Hash-based signatures for the Internet of things: position paper. In *Proceedings of the 15th ACM international conference on computing frontiers* (pp. 332–335).
12. Kimani, K., Oduol, V., & Langat, K. (2019). Cyber security challenges for IoT-based smart grid networks. *International Journal of Critical Infrastructure Protection, 25*, 36–49.
13. Gabriel, A.J., et al. (2014). Post-quantum cryptography based security framework for cloud computing. Journal of Internet Technology and Secured Transactions
14. Korzh, B., Lim, C. C. W., Houlmann, R., Gisin, N., Li, M. J., Nolan, D., & Zbinden, H. (2015). Provably secure and practical quantum key distribution over 307 km of optical fibre. *Nature Photonics., 9*(3), 163–168.
15. Collier, Z. A., & Sarkis, J. (2021). The zero trust supply chain: Managing supply chain risk in the absence of trust. *International Journal of Production Research*, 59(11), 3430–3445. https://doi.org/10.1080/00207543.2021.1884311. ISSN 0020–7543. S2CID 233965375.

Overview of Internet of Things-Based Smart Logistics Systems

B. Ravi Chandra, Krishan Kumar, Ajay Roy, and I. Sharath Chandra

Abstract The challenges brought on by quickly shifting customer expectations, the opportunities presented by new technology, and the facilitation of new business models may all be effectively handled by smart logistics. This review article provides a thorough summary of recent developments in the Internet of Things (IoT)-based smart logistics, including intelligent freight handling, storage, and delivery. The essential technologies and effects of IoT-based smart logistics research and implementations, as well as their industry and regional distributions, are shown through a bibliometric analysis of articles published between 2008 and 2019. The purpose and value of IoT in smart logistics are highlighted in the content analysis, along with the challenges that come with IoT-based smart logistics. Additionally, it points up areas that need more research. The current challenges include the limited technological capabilities and proliferation of the Internet of Things, problems with standardized data collection and processing, and security and privacy concerns. They also include technical issues with RF identification and wireless sensing networks. There is a big need for research to get over the fundamental technical hurdles of IoT to boost Intelligent, diverse systems in logistics applications and collaboratively build cutting-edge technology for information and communications and systems for management.

Keywords Machine intelligence · Cloud computing · Internet of things (IoT) · Big data · And smart logistics (AI)

B. Ravi Chandra (✉)
G.Pullaiah College of Engineering and Technology, Kurnool, AP, India
e-mail: chandrabrc11@gmail.com

K. Kumar · A. Roy
ECE, Lovely Professional University, Punjab, India
e-mail: krishan.22397@lpu.co.in

A. Roy
e-mail: ajayroy@live.com

I. S. Chandra
ECE, Matrusri Engineering College, Hyderabad, India
e-mail: Sharath.Inguva@gmail.com

A. Prasad et al. (eds.), *Communication Technologies and Security Challenges in IoT*, Internet of Things, https://doi.org/10.1007/978-981-97-0052-3_12

Introduction

The logistics industry is experiencing tremendous change as a result of technical advancements, new market entrants, start-ups, and changing customer expectations. This transition has both danger and opportunity (PwC 2016). Modern technologies such as the Internet of Things (IoT), cloud computing, big data, and blockchain considerably simplify and improve the logistics process. New rivals replace the market leaders with creative business tactics based on cutting-edge technology. For instance, Uber Freight has a detrimental effect on the American freight sector because it links carriers with the best shipments that are accessible and creates, through its platform and app, an on-demand supply chain. Customers want to be able to purchase things with more flexibility, quicker delivery times, and increased safety. The abovementioned advancements encourage more severe competition while increasing production and customer delight. Some shipping companies run the danger of going out of business if they don't upgrade their current logistical systems [1]. Using modern technology to create intelligent logistics systems is an effective method to stay on top of the times.

There are various diverse views on what makes for good logistics, and no consensus has yet been established. Contrarily, smart logistics refers to a variety of logistical operations that are controlled, planned, and regulated more clearly than the alternative to traditional ways. These activities include transportation, storage, and service and support. Different sorts or levels of intelligence may exist based on the purpose or emphasis. Smart logistics can take many different forms, including autonomous logistics [2], intelligent client logistics [3], the Physical Internet (PI) [4], intelligent transportation systems (ITS) [5], smart freight [6], and autonomous logistics [2]. Table 1 provides a summary of them.

Even if there are several forms of smart logistics, each one depends as it relates to the use of technology for communication and information (ICTs) to accomplish its objectives. Here, we offer a new definition of intelligent transportation that places emphasis on the transportation of goods and is based on advanced technology and skilled management. Transporting goods, storage, and the main logistical components that represent the movement of items are shipping and delivery. Figure 1 displays the smart logistics map.

Cargo handling, Logistics for storing and delivery operations may be turned into sharing information, rapid response, and manpower consolidation by combining connectivity, big data, cloud services, deep learning (AI), and modern governance, as shown in Fig. 1. Cooperative progress has resulted in an integrated and collaborative system that is smart logistics. Additionally, Compared to the conventional method, it improves the intelligence of administrative activities. Finally, smart logistics employs cutting-edge ICTs and upkeep to transform the primary logistics cooperative effort into one that is intelligent and integrated and achieves material integration, exchange of information, and quick reaction. From the consumer's point of view, intelligent logistics offers more effective, flexible, accurate, and secure logistical solutions.

Table 1 Intelligent logistics professionals

Classifications	Interpretation	Preference
Independent logistics	Windt and Hülsmann [3] define independent logistics as the autonomous cooperation and management of logistical activities. It makes decentralized judgments by using embedded systems and IoT technologies such as radio frequency identification (RFID)	Logistics object decision-making is decentralized
Transportation intellect systems	By integrating different data and communication methods (ICTs) into all forms of cargo and passenger transportation, ITS is utilized to make road transport better, more efficient, and more sustainable [6]	The management of traffic and transportation
The real world internet	The real-world internet is an open system for the logistics chain that is based on operational, digital, and real interconnectedness. It uses interfaces, protocols, and containers to do this [5]. A new age of dense transportation, distribution networks, and transportation is brought about by multi-layered interconnectivity, which ultimately occurs everywhere and anytime (Montreuil and Crainic 2016)	The networking of goods-encapsulating containers
Intelligent transport	Intelligent transportation merges traffic organization with cutting-edge technology for cargo monitoring and automobile tracking, allowing for better cargo management and implementation [8]	The integration of ICT with mechanisms for moving freight
Intelligent logistics focused on the customer	Consumer smart transportation aims for more client orientation in terms of accessibility, adaptability, and flexibility in managing inventories, transport networks, and order processing [4]	Activities in logistics that focus on customers

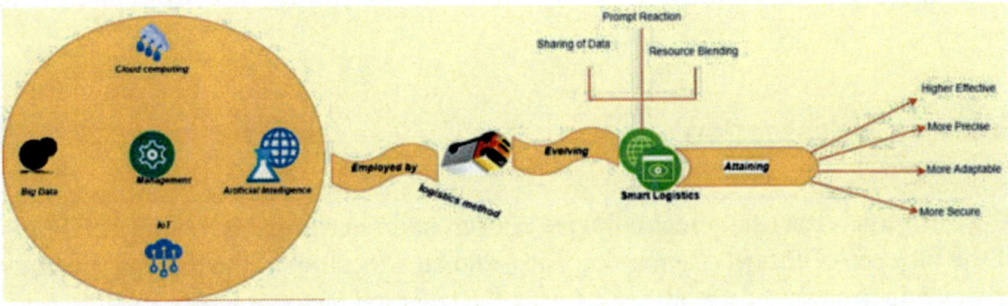

Fig. 1 Shows the smart logistics concept map

The secret to developing smart transport is to use cutting-edge methods as much as possible. Out of all of them, the supply chain has used IoT technology the most [7–9]. IoT is the new name for information and communication technology (ICT) based on the Internet [10–13]. To perform data transmission, exchange, and control with unique IDs among objects, it appears to employ a variety of wearable sensors, including RFID, machine-to-machine (M2M), and wireless sensor networks (WSN), implement embedded systems, and integrate network communication technologies. The Iot, which aims to link the real world to the internet, is primarily driven by item self-identification, sharing of information, and interactive processing, according to [14]. IoT applications enable machine-to-machine (M2M) and human-to-machine communication [15, 16]. The technological components of the Internet of Things have received extensive discussion; interested readers may consult [15, 17].

There aren't many publications that, to the most of our knowledge, thoroughly and effectively show the potential of Internet-of-things intelligent transportation studies and implementations in literary works, even though IoT technology has been extensively used only for digital wellbeing supply chain, particularly in cargo handling, storage of goods, and transportation. Recent literature in this area has either concentrated on a particular IoT technology subset or murky logistical explanations.

The following are the three literary contributions made by this essay. First, we provide a form of intelligent transportation that will be used for the rest of the study. The importance of the technology and its influence on smart logistics are highlighted in the second section, which provides an overview of IoT research and applications in logistics. We conclude by discussing the difficulties and future research requirements for IoT-based smart logistics.

The following portions of this essay are arranged as follows. Section 2 presents the review process and bibliometric analyses. In Sect. 3, a review and analysis of research and applications of IoT in cargo handling, storage, and delivery are provided. In Sect. 4, the difficulties of IoT-based intelligent transportation are covered. The research requirements for creating smart logistics are outlined in Sect. 5. Section 6 provides a conclusion to this article. A dictionary of terms and acronyms is included in the appendix.

Statistical Evaluation

A set of terms describing technologies chosen and their impacts, commercial uses of the Internet-of-things efficient logistics, and an allocation of the nations, areas, or organizations analyzed are all included in the research methodology.

The most important industrial uses are shown in Fig. 2. As we see, fresh items, farming, pharmaceuticals, e-commerce, medications, fisheries, and smoking sectors are among those that employ industrial applications. The majority of research—61 percent—concerns cold chain logistics in the fresh items, agricultural, and fishing sectors. The chemical sector comes in second, accounting for 17% points of research.

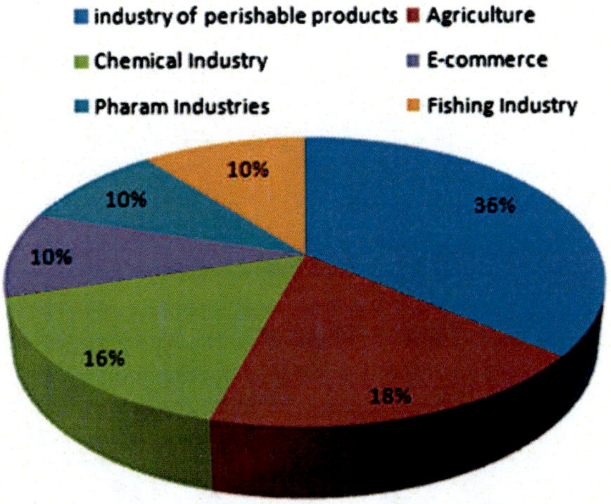

Fig. 2 Some of the most important industrial uses

These sectors emphasize quality and safety in the delivery process, which requires additional oversight and control.

The most popular terms for accepted technology are listed in Fig. 3, together with their results. The primary logistics-related IoT technology features are RFID, WSNs, and ZigBee. Smart logistics powered by IoT's fundamental characteristics and function are RFID's trackable actual logistics data capabilities. To increase the effectiveness of goods management, RFID also offers precise interior placement. WSNs are effective for ensuring vehicle and cargo safety during condition monitoring. WSN-enabled smart containers feature wireless connection and sensor systems that allow them to measure and communicate inside data while being sealed. They accomplish autonomous surveillance, observation, and warning reporting. The advancement of smart logistics is supported by other technologies, such as AI, case-based reasoning (CBR), and context-aware data mining. One of the most amazing Internet-of-things intelligent transportation ongoing research is ensuring the security of logistical operations and the integrity of commodities.

Figure 4 depicts the spread of the primary research countries, areas, and organizations. They are all wealthy or have scientific and technological expertise. Hong Kong, the U.S., Europe, and China are the four largest countries, areas, or organizations. Asia, North America, and Europe are the regions that are most engaged in smart logistics research. China, the largest developing nation, has a sizable trading market, a sizable industrial base, and strong applications in technology and science. all of which substantially encourage the research and implementation of intelligent transport. The reality that all rising countries—aside from China—are well-advanced countries, areas, or organizations indicates that the majority of rising nations don't do research or employ smart logistics.

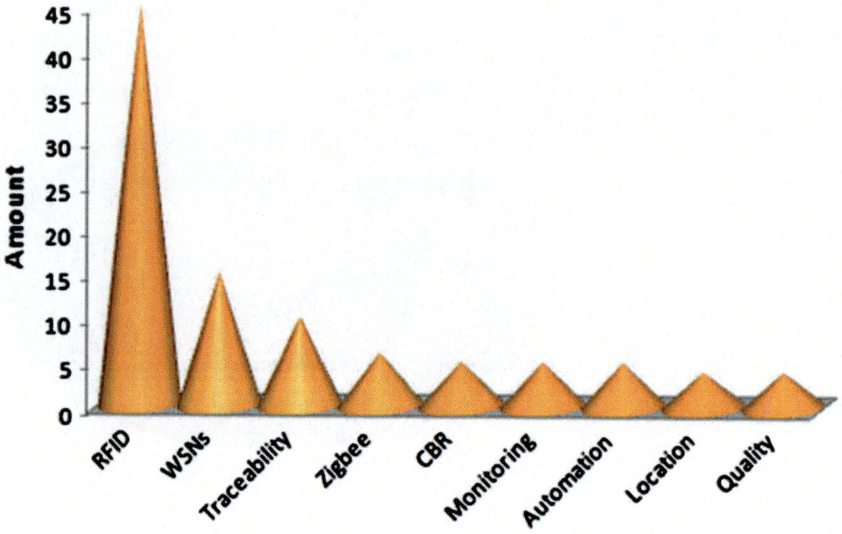

Fig. 3 Effects of technologies adopted and their statistics

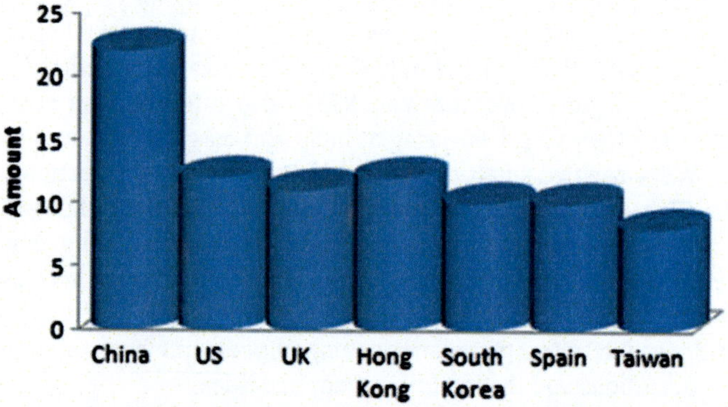

Fig. 4 Significant nations' distribution

Smart Logistics Powered By IoT Intelligent Cargo Transportation

The physical movement of goods and cargo by ship, airplane, rail, or motor vehicle is referred to as freight transportation. It is concerned with loading capacity waste, loading work errors, wasteful procedures, shipping protection, and product safety. To solve such difficulties, smart freight using IoT is a methodology for embedding Information and communication technologies into transport systems and cargo.

IoT technology improves the efficiency, convenience, and visibility of freight transportation. RFID technology is extensively used in transportation to track and identify goods and vehicle information. RFID, for example, has the potential to record and monitor logistical data required for mission loading/unloading, cargo location, and cargo details [18], and data from client orders, such as customer name, volume, and stock-keeping unit (SKU) types throughout transit [19]. Such information is utilized to optimize vehicle configuration and route planning [18], as well as vehicle routing [19]. The electronic product code (EPC) on the RFID tag holds data about the product and can track its movement [18]. On the long journey, RFID tags on containers, cars, and drug containers would be read. To offer navigation, route mapping, and passenger transportation, other auxiliary technologies [20], such as a Global positioning system and a geographical information system (GIS), are frequently utilized in conjunction with RFID. A global positioning system is used to pinpoint and trace a location of a vehicle instantly; GIS gives operators access to information on the geographic distribution of roadways, basic services, and driving conditions.

WSN, in particular, an IoT technology, enhances transportation security and monitoring. WSN-based remote product core temperature measurement is crucial to ensuring nutrition and minimizing losses for fresh produce [21]. This is because the storage environment has a direct impact on the perishability of perishable commodities. Using ambient sensors, which keep an eye on the storage environment of the items while they are being transported, respiration rate and temperature (oxygen and carbon dioxide) data are gathered [18]. With the monitoring methods discussed above, non-invasive detection and data collecting is more efficient and secure. RFID has the potential to be utilized to track load changes, identify drivers, maintain WSN setup, and maintain the correct environmental conditions [22]. As a consequence, it is possible to track, trace, and keep an eye on products as they travel. It should be highlighted that using WSN to monitor railroad conditions improves the rail freight transportation industry's security, reliability, and safety [23–25]. WSNs are capable of observing and analyzing railway networks, including rails, track crossings, tunnels, and machinery mounted on trains. Users are alerted in real-time by this safety tracking system when a freight railcar is malfunctioning.

IoT-based smart freight transportation is summarised and categorized in Table 2 according to the main method of transportation. It should be mentioned that WSN and RFID technologies are widely employed in freight transportation by road, river, and rail. All three modes of transportation are impacted by the Internet of Things, involving the observation, locating, and monitoring of vehicles in real time. According to the primary mode of transportation, Table 2 summarises and groups IoT-based smart freight transportation. It should be mentioned that the transportation of commodities by road, river, and rail makes substantial use of the technologies of RFID and WSN. Among the three modes of transportation affected by the Internet of Things are actual monitoring, tracking, and observation of transportation systems and cargo. Planning vehicle routes and fleet management will be made easier as a result, which has ramifications for the movement of freight by road. It is difficult to examine freight containers in canals. While RFID may give unique identification and electronic container sealing, WSNs may allow non-intrusive measuring and monitoring. The Internet of Things will have the most impact on rail freight transportation when it comes to the monitoring and alerting of various railway assets. There has been little research on the strategic implications of IoT for freight delivery, such as the architecture of transportation networks and mode selection. Optimization of vehicle routes and fleet management is made easier by the implications for road freight transportation in particular. Opening canal freight containers for inspection is challenging. While WSNs can provide non-intrusive monitoring and measurement, RFID can provide unique IDs and electronic container closing. IoT has a considerable influence on rail freight transportation, with a focus on the condition tracking and alerting of various railway assets. However, there is currently a dearth of studies on the strategic effects of IoT on freight delivery, including the design of transportation networks and mode selection.

(i) **Intelligent delivery**

Delivery is the quick transportation of goods from a warehouse or fulfillment center to the designated location. A more effective, secure, and exact courier service is required. However, there are several problems with the present transportation system, such as low operational effectiveness [26], delivery failures [27], cargo theft [28], and degradation of perishable goods [29]. Research is being conducted to address the aforementioned issues using IoT-based smart delivery.

The distribution process has been tracked and observed, as well as exchanged and interacted with using Internet of Things (IoT) technology. To promote information flow and exchange among delivery service providers, integrated devices, RFID, detector, ad hoc wireless networking, and wireless ad hoc networking technologies

Table 2 IoT-based smart cargo mobility in brief

Transportation mechanism	Effect of the internet of things	Internet of things innovation
Logistics via roadway	Transportation resource mapping, identifying, and watching in real-time; enhancing traffic performance; improving vehicle design and path design; transferring and exchanging led to the following; vehicle maintenance and improvement	QR codes, WSNs, RFID, barcodes, EPC, M2M, ZigBee, 4G, and other wireless communication networks
Transportation via waterways	Surveillance, detecting, and watching with an emphasis on quality; data pooling; giving out special identifications; and putting digitally sealed containers	3G,4G, GPRS, RFID and other WSNs,
Railroad transport	Railroad facility identity, supervision, and defect warning promptly, as well as genuine cargo visibility, as well as a reduction in the requirement for human replacement and record-keeping errors	RFID, Wi-Fi,WSNs, ZigBee,, ZigBee, and other WSNs

are used [30]. Sharing or cooperative distribution is made possible by IoT integration of physical and logistical resources, improving delivery effectiveness and maximizing resource utilization [31]. Delivery vehicles are outfitted with a range of intelligent detectors, Radio frequency identification (RFID), and Satellite devices that are linked to databases and the smart devices of other users, ensuring that sensing data is sent and shared. A delivery worker carrying a smartphone outfitted with a GPS module and an operational RFID technology may be monitored at all times, both inside and outdoors, in addition to delivery trucks. While RFID is employed for internal location, GPS is used for exterior geographic coordinates. Deliveries may be monitored using internet GIS, GPS, and digital measuring instruments linked to various detectors and cloud providers [32].

IoT technology aids in reducing shipping failure and ensures delivery dependability. To increase delivery dependability, such as proper ordering and on-time delivery, RFID is widely recognized as a tool for determining and monitoring the status of goods and vehicles. With the deployment of RFID, product losses and incorrect destinations may be drastically decreased. Container robbery may be significantly reduced by combining readily available mobile devices, RFID, and geofencing technologies [33] or RFID hierarchical clustering evidence techniques [29].

Mobile network technology, GPS, RFID-based e-seals, and other technologies could help with car rescheduling after an accident [34]. By using WSN to track the status of the vehicles, goods loss during delivery may be avoided as well. For instance, the ambient light sensor may detect unauthorized truck door openings and offer details about events like unloading and loading [20]. To perform safe driving, sensor data such as speed limit and engine load may be employed, and sleepy steering may be identified using a continuous sleepiness detection sensor [35].

Barcode scanners are used to provide private courier services and courier verification. By gathering, monitoring, and recording exterior environmental variables, Iot solutions, particularly Wireless sensors and Tagging with detectors, can guarantee the security and integrity of commodities, particularly delicate ones, during shipping. To track items throughout the distribution process, WSN may automatically gather ideal environmental information such as temperature and humidity [36]. Customers are welcome to review this material and provide staff criticism and improvements. In the stringent control-required cold storage facilities supply of medicines, chilled vegetables, and fresh seafood, thermostats are often utilized. To track ambient control, anticipate item perishability, and monitor environmental indicators and product status, WSN sensor data may be used [32]. Through actuators, feedback is managed by a predetermined criteria function. To ensure that delivery products are preserved in the right environmental conditions, several researchers have used RFID and Wireless sensor technologies [37].

IoT improves the safety and efficacy of delivery services. The task and data may be tracked, examined, and shared. Efficient supply improves reliability while reducing loss. Table 4 outlines and classifies IoT-based smart delivery based on the products given. IoT technology performs somewhat varied functions in various commodities delivered. Smart cargo handling delivery emphasizes the delivery status and guarantees that things arrive safely and appropriately. Due to the nature of expensive materials, it is important to take into account the chance that goods may spoil while being delivered. As a result, environmental factors affecting perishable goods must be tracked, quantified, and gathered. Environmental measures should be put in place in the interim. Few studies, however, focus on using IoT to optimize supply capabilities and speed up delivery rather than to control the supplier base.

Actual Issues of Rfid Technical Glitches

Even when functioning regularly, RFID is sensitive to electromagnetic fields [16]. Noise, for instance, could have an impact on RFID. Steel containers piled on top of one another during container transit may cause damage that affects location and identification [38]. Tag collision and tag identification loss are problems with RFID [39]. Although RFID enhances stocking and supply chain management, a barrier is provided by massive tag classification statistics [40]. Available capacity errors and deletion mistakes are the main technical problems with the implementation of RFID in storage [37]. When stock items or aluminum wrappings attenuate or impede the RFID tag's spectrum, receiving errors happen. RFID tags and sensors have prices, thermal efficiency, and network implementation, according to [41]. Since, for example, RFID data can only be obtained when Tags are available, which is frequently at transportation hubs like ports or warehouses, RFID usage is restricted, and tracking resolution is compromised.

(a) **WSNs' technical issues**

The significant signal absorption of moisture products, consistency of communication channels, sensor battery life, and bandwidth and density expenses of communications infrastructure are major technical barriers to Wireless sensor condition or quality control in freight transit [42]. The industrial surveillance system may be impacted by a connectivity breakdown brought on by several sensors operating at once [43]. Delivery of alarms is delayed by monitoring network overload, which also results in data loss and postponement [23]. ZigBee, which is a popular technology for data transmission in WSNs, has several limitations, such as high power consumption, inefficiency, significant transmitting data lags, intra- and inter-interfering, and issues with security and privacy [1]. Additionally, the Wireless sensor node design has to be enhanced for cost savings and efficiency [44].

(b) **IoT's limited expansion**

RFID and WSNs are already widely used in the logistics business, while other IoT methods, such as Z-Wave, Bluetooth low energy (BLE), and machine-to-machine (M2M), have yet to gain traction. The majority of IoT-based smart home practices Logistical are confined to the integration of incomplete logistics systems and intelligent assistance. Furthermore, dynamic delays and connection concerns occur in data transmission during public transport using today's wireless communication systems such as General packet radio service [24]. It is neither conducive to emergency response and outcome throughout public transport, like as control of temperature and alert transmission.

(c) **IoT's technical potential is constrained**

IoT is useless in complex logistics applications due to its insufficient processing capability and information skills—. IoT has a limited impact on how shipping optimizes freight weights and vehicle selection. The difficult problems of storage allocation and container/truck loading are obstacles that IoT technology cannot simply solve [45]. On the other hand, dynamic WMS cannot be achieved with IoT. Without the aid of sophisticated software, the Internet of Things (IoT cannot provide truck route optimization for the delivery of fresh produce. IoT devices may gather enormous amounts of information regarding providing material and demand, but it is unable to explain how to optimize route optimization and maximize the use of delivery services. The problems of fuel waste and high costs in last-mile delivery are not sufficiently addressed.

(d) **IoT standardisation difficulties**

Trab et al. [46] claims that the development of the IoT normal design and endpoint naming are both difficult tasks. For instance, successful implementation necessitates the use of a recognized RFID protocol. The improved visibility of logistical operations is one of the biggest advantages of IoT, and it depends on the consistency of file types, technology, and standards. The Internet of Things also has to deal with issues caused by various applications, surroundings, and gadgets [7]. The heterogeneity results from a variety of demands for multi-supplier services and applications, in addition to disparities in device capabilities and features [8]. As a result, there is now a sizable barrier to feedback, communication, and information perception. [47] asserts that it would be challenging to develop several Internet of Things apps and services due to the variety of devices and networks.

(e) **IoT data gathering and processing challenges**

These qualities may be seen in the observable and accessible transportation data: growth in data volume, the coexistence of structured and unstructured data, noisy data that is incomplete, erroneous, or redundant, and dispersed data containing useful information. Particularly, the collected and collected data is intimately connected to the transport company, yet it is so vague the difficulty of it to display for strategic planning, and information categorization and assessment are also highly challenging. Even though a huge quantity of data includes a variety of Informational material, the usable sections are always restricted. Aside from the complicated aspects of big data, dealing with it is a tremendous task. Furthermore, while amassing over time and significant volumes of information at a rapid pace, additional data processing and storage capacity requirements must be met [48].

(f) **Concerns about IoT confidentiality and privacy**

IoT concerns regarding security and privacy are preventing the adoption of Internet-of-things intelligent transportation, particularly in industrialized nations. Because of the increased confidentiality and security dangers posed by IoT, there is a need for regulatory actions such as assault detection and data evaluation.

User privacy and password protection should be improved. IoT may be attacked using all three layers: the application layer, the network layer, and the perception layer. Issues with data privacy may reveal information, allowing for the improper use of surveillance of user preferences and schedules. Users are concerned about privacy and information security breaches in the monitoring of administrative procedures, particularly traders and consumers. Moreover, logistics companies worry that their data may be stolen by competitors. Security and privacy issues with information exchange are especially serious. Confidentiality is the most pressing issue for collaborative delivery due to the nature of accessibility and collaboration in connected cars.

Research Requirements Iot Technology Advancement

IoT technology has advanced in terms of functionality and performance. Numerous anti-collision techniques for Radio frequency identification have been created to avoid tag collisions, and numerous novel methods have been proposed to find lost tags. By enhancing communication protocols and preprocessing sensor data in-network, the sensor's battery life may be increased. IoT is progressing, and its meet specifications are also making considerable progress. These almost unlimited addresses greatly facilitate addressing any quantity of devices needed for IoT. Cellular networks of the fifth generation (5G) will be quicker than ever before (to transport high information), be more responsive, have smaller latency, and consume less energy and money (to be more cost-effective). The goal of the 5Generation design is to support a large variety of IoT applications with heightened requirements for delay, robustness, range, and power. But both real demand and technical innovation know no boundaries. Further research is required to address the aforementioned challenging IoT challenges in Sect. 4.

Advanced Management and Collaborative Development of Many Technologies Computing in the Cloud

Developments like AI, IoT, and big data are advanced with the help of cloud computing. It houses vital infrastructure that is used by several applications. It might be used for crucial problems such as how to collect and store and utilize IoT data. Moreover, it may simplify ideal docking, effectively integrate diverse operational resources, and encourage cooperative development between transport parts. Depending on the logistics data gathered by IoT, a smart supply chain information network based on the cloud may provide actual time and flexible solutions. Additionally, it can cater to the requirements of each logistics component before creating effective logistics plans that facilitate communication and the exchange of information. Additionally, clouds increase device dependability by enabling the offloading of heavy processes, which prolongs battery life or enables the development of a modularized architecture.

The combination of the web with IoT (or CoT) is referred to as cloud IoT. This new IT paradigm takes into account both the IoT's constrained processing and storage capacities as well as the nearly endless possibilities of cloud technologies in these areas. Cloud IoT enables the construction of IoT-based shared systems, enables the logistics system to handle complex situations and changes automatically, and supports the development of IoT-based smart logistics. A cloud-based logistics cold chain system, for example, connects customers and logistics service providers and provides shared data and tools for managing logistics businesses. Flexibility and scalability are features of intelligent CoT-based supply chain integration and management solutions. Agile methodologies are employed for resource sharing and participant participation throughout the whole supply chain life cycle.

(a) **Massive amounts of data**

If the enormous amounts of data generated by IoT were managed properly and right away, info crisis and energy waste might be transformed into data wellness and resource utilization. Big data technology may be used to model and analyze historical data to forecast future occurrences, such as delivery delays. To assist in decision-making, it may also pre-process and mine heterogeneous data. Optimizing truck, train, and airplane flight patterns may use big data to improve logistics while lowering energy use and environmental effect. With the fewest resources and lowest logistics costs, it can also satisfy the most demanding logistical needs. For instance, IoT and/or the Internet might be used to gather big data on delivery requests from logistics delivery organizations, which could then be utilized to determine the best delivery route based on the delivery requirements. With this approach, logistical delivery resources are used more effectively at a lesser cost. Based on specifics like client position, operation scheduling, traffic bottlenecks, and cargo, big data technologies may suggest optimal routes and other circuits. In other words, the optimum mode of transportation, the optimal container capacity, and the most effective delivery route may all be determined by a thorough study of relevant data.

(b) **Artificial intelligence**

AI allows computers to replicate, learn from, and develop human intellect. Robots using AI may mimic human behavior, learn from their errors, and adapt to new inputs. Computers might potentially be trained to do certain activities using deep learning and natural language processing. On a vast quantity of historical data and current knowledge, it can foresee future events and make exact judgments. It can also access data and discern good information from bad. IoT devices may use AI judgments once IoT data collection has helped AI become more potent, accurate, and intelligent. Additionally, AI enhances IoT's capability for sensing. Thanks to machine vision, for example, robots can now collect more real-time data and carry out identification and detection tasks with greater accuracy. Biometric technologies like face and fingerprint recognition help enhance human–machine connection.

Smart logistics with AI help may improve logistics decision-making and provide the most effective operating plan. Issues with logistics route optimization have been addressed using certain complex algorithms. Transport and delivery routes have been optimized using genetic algorithms. The ant colony algorithm was used to guide the movement of autonomous guided vehicles in an automated warehouse. Intelligent algorithms can also be used to handle issues with ships or trucks.

(c) **Advanced administration**

In addition to working together to create and maintain the previously mentioned technologies, smart logistics need enhanced management. The development of smart logistics makes use of advanced management to implement efficient logistics planning, organization, leadership, and control. A management information system is required to facilitate information interchange and collaboration in smart delivery and transportation. A smart warehouse system that uses big data, cloud computing, and IoT is preferable.

The architecture of a digital logistics chain for the joint development of IoT, cloud technology, big data, AI, and improved governance is shown in Fig. 5. In the context of smart logistics, IoT devices like RFID readers and sensors collect a lot of data, which is then processed via edge computing or big data analysis technologies via cloud computing. AI uses cloud computing to deep understand some of these details and make better conclusions. Actuators included in the Internet of Things complete the decision-making process. To organize, plan, and coordinate the utilization of facilities and equipment, smart logistics makes use of modern management. It also collaborates with these technologies to help the logistics system. Finding out how to integrate advanced management with ICTs to carry out smart logistics requires research.

(d) **Smart logistics development**

Intelligent freight movement, storage, and delivery are all parts of intelligent logistics. The usage and blending of IoT, big data, cloud computing, and AI allow the logistics system to exchange information, react fast, and combine resources. Through the provision of a more agile, accurate, efficient, and secure logistics system, it seeks to provide clients with more adaptable, reliable, speedier, and safer services related to logistics.

(e) **IoT Application Extension in Logistics**

IoT-based smart logistics have allowed for the communication, sharing, and visualization of data and information. As a result, consumers and delivery and transportation service providers are more likely to communicate, increasing process transparency and accountability and enhancing client satisfaction. Decentralized decision-making and quick responses to store events are also made possible by this. However, the IoT-based smart logistics system falls short in terms of intelligence. IoT devices

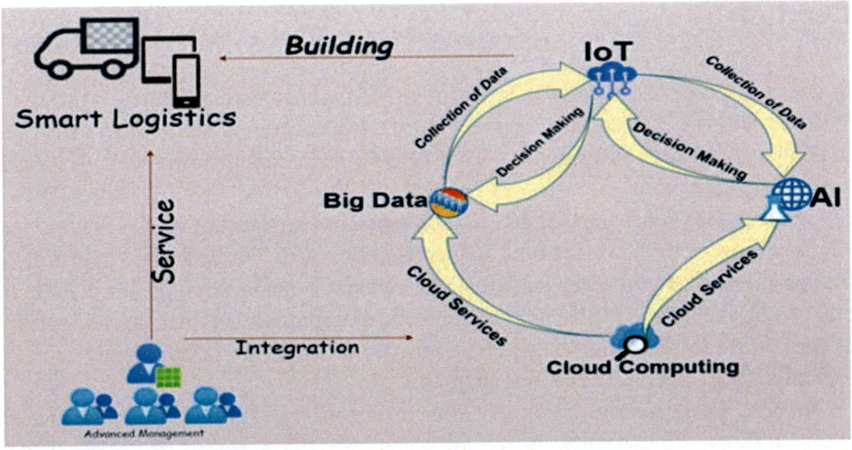

Fig. 5 Designing a smart logistics system to collaborate on the creation of various technologies and sophisticated management

with insufficient processing capability might render erroneous judgments quickly. Two IoT technology components, RFID and WSNs, are the focus of current research and applications. Therefore, we must get a greater knowledge of and implement additional Internet of Things (IoT) logistics technologies, including M2M, BLE, and Z-Waze.

(f) **Logistics ICT development collaboration**

Additional technologies that greatly enhance IoT-based smart logistics intelligence include big data, cloud computing, and artificial intelligence (AI). By combining these complex ICTs, it may be possible to overcome a variety of challenging problems, such as loading challenges, dynamic routing problems, interior placement problems, and MHE process failures, and thus increase the effectiveness, adaptability, and accuracy of logistics operations. Huge volumes of data are collected by IoT devices, which big data, artificial intelligence, and cloud computing analyze and comprehend to assist the logistics system in making the best decisions. IoT, cloud computing, big data, and AI used together might create a collaborative system for smart logistics. Smart logistics must thus research ways to promote teamwork in the development of complex ICTs. The collaborative application, however, cannot be used to construct an integrated smart logistics system.

Conclusions

In-depth coverage of the study and innovative IoT-based logistics software is provided in this article. We concentrated on studying and evaluating IoT studies and their uses in delivery, logistics, and shipping. The review's main conclusions and suggestions are as follows:

Currently, the most popular IoT technologies in logistics are RFID, WSNs, and ZigBee. One of the core features and capabilities of IoT-based smart logistics is the monitoring of real-time logistics data, which is made possible via RFID. In emergency scenarios, WSNs help monitor the environment and ensure the security of cargo and vehicles. Smart logistics should be used in the perishable food business, agriculture, fisheries, and chemical logistics.

For both Western economies and China, the creation and application of IoT-based smart logistics are essential. In the great majority of emerging countries, this is lacking. Even while studies have not shown that it may change how transportation networks are built or which modes are employed, the usage of IoT may increase the effectiveness, visibility, security, and traceability of freight transportation. Smart warehousing powered by IoT would be more effective, transparent, precise, and secure. The warehousing process and storage options would also be made simpler. IoT enables decentralized decision-making in a centralized WMS. On the other hand, choosing a strategy of storage allocation and optimizing internal routing continue to be difficult for Smart warehouses powered by IoT.

Smart delivery involves the interactivity and transmission of delivery information between the customers and the delivery service providers. With IoT, delivery would be more controlled, secure, and transparent. On the other hand, smart delivery cannot be sped up or made more effective only via the usage of IoT.

The present obstacles to Smart logistics based on IoT comprise technical challenges with RFID and WSN, limited IoT growth and technological proficiency, IoT likeness difficulties, data collection of IoT and handling difficulties, and Internet of Things security and privacy concerns.

By combining modern ICTs, it may be possible to handle certain specific Smart logistics issues based on IoT, such as route planning, container filling, and facility allocation.

References

1. Aju, O. G. (2015). A survey of Zigbee wireless sensor network technology: Topology, applications, and challenges. *International Journal of Computer Applications, 130*(9), 47–55.
2. Akgul, A. (2019). Soon, most logistics companies may go out of business due to technology. Retrieved from https://www.morethanshipping.com/soon-most-logistics-companies-may-go-out-of-business-due-to-technology/.
3. Windt, K., & Hülsmann, M. (2007). Changing paradigms in logistics—understanding the shift from conventional control to autonomous cooperation and control. In M. Hülsmann & K. Windt (Eds.), *Understanding autonomous cooperation and control in logistics: The impact*

of autonomy on management, information, communication, and material flow (pp. 1–16). Springer.

4. Mcfarlane, D., Giannikas, V., & Lu, W. (2016). Intelligent logistics: Involving the customer. *Computers in Industry, 81*, 105–115.
5. Montreuil, B., Meller, R. D., & Ballot, E. (2013). Physical internet foundations. In T. Borangiu, A. Thomas, & D. Trentesaux (Eds.), *Service orientation in holonic and multi agent manufacturing and robotics* (pp. 151–166). Springer.
6. European Commission. (2019). Intelligent transport systems. Retrieved from https://ec.europa.eu/transport/themes/its_en.
7. Evans, D. (2011). The Internet of Things: How the next evolution of the internet is changing everything. Retrieved from https://www.cisco.com/c/dam/en_us/about/ac79/docs/innov/IoT_IBSG_0411FINAL.pdf.
8. Sternberg, H., Hagen, A., Paganelli, P., & Lumsden, K. (2010). Intelligent cargo-enabling future's sustainable and accountable transportation system. *World Journal of Science Technology & Sustainable Development, 7*(3), 253–262.
9. Xiao, X., He, Q., Fu, Z., Xu, M., & Zhang, X. (2016). Applying CS and WSN methods for improving efficiency of frozen and chilled aquatic products monitoring system in cold chain logistics. *Food Control, 60*, 656–666.
10. Sharma, V., You, I., Pau, G., Collotta, M., Lim, J. D., & Kim, J. N. (2018). LoRaWAN-based energy-efficient surveillance by drones for intelligent transportation systems. *Energies, 11*(3), 573.
11. Chow, H. K. H., Choy, K. L., Lee, W. B., & Lau, K. C. (2006). Design of an RFID case-based resource management system for warehouse operations. *Expert Systems with Applications, 30*(4), 561–576.
12. Ashton, K. (2009). That 'Internet of Things.' *RFID Journal, 22*(7), 97–114.
13. Weber, R. H. (2010). Internet of Things–new security and privacy challenges. *Computer Law & Security Review, 26*(1), 23–30.
14. Weber, R. H., & Weber, R. (2010). *Internet of Things*. Springer.
15. Feng, X., Yang, L. T., Wang, L., & Vinel, A. (2012). Internet of Things. *International Journal of Communication Systems, 25*(9), 1101–1102.
16. Uckelmann, D., Harrison, M., & Michahelles, F. (2011). An architectural approach towards the future Internet of Things. In Architecting the Internet of Things (pp. 1–24). Springer, Berlin.
17. Miorandi, D., Sicari, S., De Pellegrini, F., & Chlamtac, I. (2012). Internet of Things: Vision, applications and research challenges. *Ad Hoc Networks, 10*(7), 1497–1516.
18. Musa, A., & Dabo, A. A. A. (2016). A review of RFID in supply chain management: 2000–2015. *Global Journal of Flexible Systems Management, 17*(2), 189–228.
19. Naskar, S., Basu, P., & Sen, A. K. (2020). A literature review of the emerging field of IoT Using RFID and its applications in supply chain management. In Securing the Internet of Things: Concepts, methodologies, tools, and applications (pp. 1664–1689). IGI Global, Hershey, PA.
20. Bendaya, M., Hassini, E., & Bahroun, Z. (2019). Internet of Things and supply chain management: A literature review. *International Journal of Production Research, 57*(15–16), 4719–4742.
21. Liu, S., Zhang, Y., Liu, Y., Wang, L., & Wang, X. V. (2019). An 'Internet of Things' enabled dynamic optimization method for smart vehicles and logistics tasks. *Journal of Cleaner Production, 215*, 806–820.
22. Cheung, B. K. S., Choy, K. L., Li, C. L., Shi, W., & Tang, J. (2008). Dynamic routing model and solution methods for fleet management with mobile technologies. *International Journal of Production Economics, 113*(2), 694–705.
23. Ruiz-Garcia, L., Barreiro, P., Robla, J. I., & Lunadei, L. (2010). Testing ZigBee motes for monitoring refrigerated vegetable transportation under real conditions. *Sensors, 10*(5), 4968–4982.
24. Jedermann, R., Pötsch, T., & Lloyd, C. (2014). Communication techniques and challenges for wireless food quality monitoring. *Philosophical Transactions of the Royal Society A: Mathematical, Physical and Engineering Sciences, 372*(2017), 1–18.

25. Santa, J., Zamora-Izquierdo, M. A., Jara, A. J., & Gómez-Skarmeta, A. F. (2012). Telematic platform for integral management of agricultural/perishable goods in terrestrial logistics. *Computers and Electronics in Agriculture, 80*, 31–40.
26. Shrestha, P. L., Hempel, M., Sharif, H., & Chen, H. H. (2013). Modeling latency and reliability of hybrid technology networking. *IEEE Sensors Journal, 13*(10), 3616–3624.
27. Fraga-Lamas, P., Fernández-Caramés, T. M., & Castedo, L. (2017). Towards the internet of smart trains: A review on industrial IoT-connected railways. *Sensors, 17*(6), 1457.
28. Hodge, V. J., O'Keefe, S., Weeks, M., & Moulds, A. (2015). Wireless sensor networks for condition monitoring in the railway industry: A survey. *IEEE Transactions on Intelligent Transportation Systems, 16*(3), 1088–1106.
29. Cagliano, A. C., De Marco, A., & Rafele, C. (2017). E-grocery supply chain management enabled by mobile tools. *Business Process Management Journal, 23*(1), 47–70.
30. Fu, H. P., Chang, T. H., Lin, A., Du, Z. J., & Hsu, K. Y. (2015). Key factors for the adoption of rfid in the logistics industry in Taiwan. *International Journal of Logistics Management, 26*(1), 61–81.
31. Yang, M. H., Luo, J. N., & Lu, S. Y. (2015). A novel multilayered RFID tagged cargo integrity assurance scheme. *Sensors, 15*(10), 27087–27115.
32. Shi, J., Zhang, J., & Qu, X. (2010). Optimizing distribution strategy for perishable foods using RFiD and sensor technologies. *Journal of Business and Industrial Marketing, 25*(8), 596–606.
33. Yu, J., Subramanian, N., Ning, K., & Edwards, D. (2015). Product delivery service provider selection and customer satisfaction in the era of Internet of Things: A Chinese e-retailers' perspective. *International Journal of Production Economics, 159*, 104–116.
34. He, Y., Zhou, F., Qi, M., & Wang, X. (2019). Joint distribution: Service paradigm, key technologies and its application in the context of Chinese express industry. *International Journal of Logistics Research and Applications*, 1–17.
35. Bogataj, D., Bogataj, M., & Hudoklin, D. (2017). Reprint of 'Mitigating risks of perishable products in the cyber-physical systems based on the extended MRP model.' *International Journal of Production Economics, 194*, 113–125.
36. Oliveira, R. R., Cardoso, I. M. G., Barbosa, J. L. V., da Costa, C. A., & Prado, M. P. (2015). An intelligent model for logistics management based on geofencing algorithms and RFID technology. *Expert Systems with Applications, 42*(15–16), 6082–6097.
37. Kido, T., & Nakamura, M. (2016). New SaaS-based operations management system to realize safe driving support and improve transport quality: Logifit TM-NexTR. *Fujitsu Scientific and Technical Journal, 52*(4), 92–97.
38. Tsang, Y. P., Choy, K. L., Wu, C. H., Ho, G. T. S., Lam, H. Y., & Tang, V. (2018). An intelligent model for assuring food quality in managing a multi-temperature food distribution centre. *Food Control, 90*, 81–97.
39. Trebar, M., Lotrič, M., & Fonda, I. (2015). Use of RFID temperature monitoring to test and improve fish packing methods in styrofoam boxes. *Journal of Food Engineering, 159*, 66–75.
40. Kim, K., Kim, H., Kim, S. K., & Jung, J. Y. (2016). I-RM: An intelligent risk management framework for context-aware ubiquitous cold chain logistics. *Expert Systems with Applications, 46*, 463–473.
41. Garrido-Hidalgo, C., Olivares, T., Ramirez, F. J., & Roda-Sanchez, L. (2019). An end-to-end Internet of Things solution for reverse supply chain management in industry 4.0. *Computers in Industry, 112*, 103127.
42. Chen, G., Dong, W., Zhao, Z., & Gu, T. (2019). Accurate corruption estimation in ZigBee under cross-technology interference. *IEEE Transactions on Mobile Computing, 18*(10), 2243–2256.
43. Huang, J., Wen, Z., Kong, L., Ge, L., Wu, M. Y., & Chen, G. (2019). Accelerate the classification statistics in RFID systems. *Theoretical Computer Science, 788*, 39–52.
44. Pereira, P. P., Eliasson, J., Kyusakov, R., Delsing, J., Raayatinezhad, A., & Johansson, M. (2013). Enabling cloud connectivity for mobile internet of things applications. *2013 IEEE seventh international symposium on service-oriented system engineering* (pp. 518–526). Redwood City.

45. Jabbar, S., Khan, M., Silva, B. N., & Han, K. (2018). A REST-based industrial web of things' framework for smart warehousing. *Journal of Supercomputing, 74*(9), 4419–4433.
46. Trab, S., Bajic, E., Zouinkhi, A., Thomas, A., Abdelkrim, M. N., Chekir, H., & Ltaief, R. H. (2017). A communicating object's approach for smart logistics and safety issues in warehouses. *Concurrent Engineering Research and Applications, 25*(1), 53–67.
47. Zacharewicz, G., Deschamps, J. C., & Francois, J. (2011). Distributed simulation platform to design advanced RFID based freight transportation systems. *Computers in Industry, 62*(6), 597–612.
48. Vanderroost, M., Ragaert, P., Verwaeren, J., De Meulenaer, B., De Baets, B., & Devlieghere, F. (2017). The digitization of a food package's life cycle: Existing and emerging computer systems in the logistics and post-logistics phase. *Computers in Industry, 87*, 15–30.

Fault Sensor Detection and Authentication Mechanism for Improving Quality of Services in Smart Homes

A. Rajavel, Praveen Kumar Premkamal, and A. Anandh

Abstract Smart city improves the welfare of the citizens using information and communication technology. The Smart city also helps to increase the efficiency of government operations as well as the quality of government operations. Smart homes are an integral component of a smart city, empowering individual home owners to contribute a sustainable and livable urban environment. A Smart home uses various Internet of Things (IoT) sensors. Sensors are critical components of the system because they collect valuable informations and the accompanying communication networks that convey the data in real-time. The precise values measured by sensors indicate the sort of regulating measure that must be done. Malfunction will likely lead to a fluctuation in the measured value of sensors, which will have an impact on the smart home's overall optimization. It is therefore essential to determine if all of the sensors attached to a smart home system are functioning properly as a result of this. Furthermore, the data measured from the sensor must be confirmed to ensure that it is received from the actual device. Data hacking by another device may cause unintended complications. Thus, in this chapter, we propose a fault sensor detection and authentication mechanism for improving the quality of services in smart homes. This chapter presents a fault identification mechanism of various sensors that help to resolve the fault and get the proper and enough information for various services. Further, our chapter validates the data source which helps to improve the data integrity.

Keywords Smart home · Sensors · Fault-detection · Authentication

A. Rajavel (✉) · P. K. Premkamal · A. Anandh
Kamaraj College of Engineering and Technology, Virudhunagar, India
e-mail: rajaveleie@kamarajengg.edu.in

P. K. Premkamal
e-mail: praveencse@kamarajengg.edu.in

A. Anandh
e-mail: anandhcse@kamarajengg.edu.in

A. Prasad et al. (eds.), *Communication Technologies and Security Challenges in IoT*, Internet of Things, https://doi.org/10.1007/978-981-97-0052-3_13

Introduction

A "smart city" is a technologically enhanced urban environment that uses a variety of electrical gadgets and sensors to collect specific data. The information is then used to improve operations across the city. Effective asset, resource, and service management is achieved through the use of information derived from such data. Data is gathered from people, devices, buildings, and assets to monitor and manage traffic and transportation systems, power plants, utilities, water supply networks, garbage, criminal investigations, information systems, schools, libraries, hospitals, and other community services. This data is subsequently processed and analyzed. The idea of a "smart city" combines information and communication technology (ICT) with numerous physical devices connected to the Internet of Things (IoT) to interact with residents and improve the efficiency of local operations and services. City authorities can directly engage with the community and city infrastructure thanks to smart city technology, which also allows them to monitor what is happening and changing in the city. Information and communication technologies are used to improve the quality, efficiency, and interactivity of city services, as well as to save costs and resource consumption and to enhance public involvement in government.

Smart homes are an important aspect of smart cities, as they allow individual residents to participate in the larger goal of creating a more sustainable and efficient urban environment. Smart home technology can provide numerous benefits to city dwellers, such as energy efficiency, improved security, and convenience. Smart home technology can be used to control lighting, heating, and appliances, as well as to monitor the home environment. Smart home systems can also be used to automate tasks such as watering plants, controlling pet feeders, and even controlling window blinds. Additionally, smart home systems can be used to monitor the security of a home, allowing the user to be alerted in the event of an intruder or other emergency.

A smart home uses a variety of sensors to track and manage various aspects of the home environment. The performance of the smart home device may be impacted when a sensor malfunctions, which can result in inaccurate readings or no readings at all. As a result, it is critical to spot and fix defective sensors as soon as feasible. Various researchers addressed the same problem in the literature [1–8]. The majority of the literature detects the faulty sensors alone which will not give the instant solution for the problem. This is the disadvantage of the present methods. On the other hand, IoT sensor authentication plays an important role in a smart home by providing a secure way to identify and authenticate authorized IoT devices. It aids in preventing unauthorized entry into the house by malicious actors. Authentication enables the network to be secured and guarantees that only authorized devices can send and receive data [9]. Additionally, authentication aids in preventing malicious actors from accessing private information or taking control of the home's systems by ensuring that only authorized devices are linked to the network.

Hence, in this chapter, we propose "Fault Sensor Detection and Authentication Mechanism for Improving Quality of Services in Smart Cities". The major contribution of this chapter is as follows:

1. Our scheme detects faulty sensors similar to many existing algorithms. In addition, our solution suggests the use of a dummy sensor to aid in continued measurement even if one of the sensors fails. So, the functionality is not at all stopped even though the main sensors failed.
2. Our system guarantees that data received from IoT sensors are from authorized IoT devices using device labels and key. It ensures the security of the smart home.

Literature Review

This section describes the various research works conducted to detect faults in the literature. Zhu et al. [1] suggested an augmented descriptor system. In this system, the original states, actuator faults, and sensor faults were combined to form the augmented state vector. The relationship between the preconditions relative to these two kinds of systems was discussed after transforming the augmented descriptor system into a general system. Gaddam et al. [2] dealt with a complete summary of the discovery of sensor faults, irregularities, outliers in the Internet of Things, and challenges. Gill et al. [3] addressed the problem of slow adaptation of home automation by using flexible home automation architecture. They integrated Zigbee through a shared home gateway.

The automation of the light switch, radiator valve, and safety sensor was done using a Zigbee network. Ruba et al. [4] used FPGA to detect faults in current sensors in the measurement of 3 phase inverter. Li et al. [5] proposed Adjacent Information Recovery Filter to detect faults in the sensor with incomplete data. An event-triggered scheduling mechanism was used by Jin et al. [6] to detect the faults in sensors. Here, two independent Markonic jumps were introduced for random switching between normal condition to faulty condition and temporal occurrence of a mistake. Using machine learning in the Internet of Things, Uppal et al. [7] recommended solutions for the error pattern. Cheng et al. [8] proposed Throughput and Time delay distribution to detect faults in home devices. Karmaker et al. [9] used Discrete Cosine Transform and average Discrete Cosine Transform to trust the numerical data from different sensors.

Tang et al. [10] integrated different sensors from smart homes and cities to predict their present and future values. Jennifer et al. [11] proposed a new architecture to handle big data and analysed the data in real-time with minimum delay. Rajalakshmi et al. [12], proposed a method to process and analyze the sensor data where the hidden data were present by using rapid decision-making. Saeed et al. [13] used multiple wireless sensor networks to detect fire in a house. He also used GSM and GPS to avoid false alarms. Zhang et al. [14] proposed an extended Kalman filter algorithm to detect faults in sensors. Wu et al. [15] discussed sensor validation approaches such as data consistency check, data reconciliation, and data fusion, as well as their use in smart homes. Ahmad et al. [16] gave a thorough assessment of the existing state-of-the-art in smart house EMSs, emphasizing the need for more research in this field to build more effective and sustainable energy management systems for smart homes. Singh et al. [17] provided an overview of several defect detection and diagnosis methodologies, including rule-based, model-based, and data-driven approaches. They also reviewed the benefits and drawbacks of each method, as well as instances of its use in smart home systems.

Kumari et al. [18] examined the difficulties and concerns connected with defect detection and diagnosis in smart homes, including data quality, sensor location, and system complexity. They also talked about how important it is to integrate problem detection and diagnostic techniques with other smart home systems like energy management and security. Zhao et al. [19] conducted a literature analysis on defect detection and diagnosis in smart homes, concentrating on three major categories: (1) sensor-based approaches, (2) model-based approaches, and (3) hybrid approaches that integrate sensor and model-based techniques. Sensor-based techniques entail the use of sensors to identify irregularities in the behavior of smart home devices or appliances. They addressed a variety of anomaly detection approaches, including statistical analysis, machine learning, and fuzzy logic. They also mentioned some of the drawbacks of sensor-based techniques, such as the requirement for a significant volume of data and the difficulty in interpreting sensor signals.

Naser et al. [20] presented a thorough examination of the problem detection and diagnostic methodologies utilized in residential smart grid systems. They emphasized the need for ongoing research in this area to overcome the issues and increase the dependability and stability of smart grid systems. Lu et al. [21] emphasized the need for good defect detection and diagnostic approaches for smart homes, which were becoming increasingly popular as interest in home automation and the Internet of Things (IoT) grew. Lazaroa et al. [22] emphasized the problems of Building Management Systems and provided a technique for counterfeiting them. Anam Abid et al. [23] presented advances in Fault Detection and Diagnosis (FDD) algorithms as well as estimates of exploration work in this field. The evaluation included both classic model-based and comparatively recent signal processing-based FDD approaches, with an emphasis on artificial intelligence-based FDD methods. Li et al. [24] introduced typical problem kinds and causes in sensors, as well as distinct types of fault detection methods utilized in industry and agriculture. They reviewed the benefits and drawbacks of various methodologies, as well as the present obstacles and ideas for future research areas.

Mallick et al. [25] addressed the problem of identifying a range of sensor defects that could occur in IoT-equipped smart homes configured to derive activities of daily living (ADL) from the activation of a set of sensors. Sathiyavathi et al. [26] have classified various fault detection approaches performed in wireless sensor network. Singh et al. [27] performed a comparative study on four different FDD systems and suggested a FDD method to reduce the energy consumption of HVAC systems. Yuneela et al. [28] discussed various technologies such as ZigBee, Bluetooth, RF module, cloud, mobile, Raspberry Pi and Arduino (central controller), SMS, and web for handling sensor data to a cloud platform. Antony Vigil [29] used IR for controlling electrical appliances, LDR to turn ON and OFF lights, and LED to indicate the status. Majumder et al. [30] described the technical comparison of IoT Based Home automation systems using smart phone.

Proposed Methodology

The emergency shutdown of a system is not necessarily the best course of action if one or more of its components fail. To provide an ongoing operation to handle the catastrophic failure, simultaneous reconfiguration of the drive control strategy is necessary. The proposed methodology uses a dummy sensor along with the actual sensor. In case of failure of the actual sensor, the measurement can be done with the help of a dummy sensor until the failed actual sensor is replaced with the newer one. A detailed explanation of the proposed scheme is given in this section.

The smart home network must first register all of the sensors used in the system. A device label and key are issued by the Smart Home network. The device label and key aid in making sure that the sensor data is only available within registered networks. This suggested approach makes use of both the original measuring sensor and a dummy sensor. Both sensor values are monitored, and any significant marginal reading variation is utilized to determine which sensor is at fault. Here, the Ultrasonic sensor, IR sensor, and LDR sensor are used to identify the type of door lock (Single lock and Double lock), identify the person near the automatic door, and to measure the intensity of light in the room. Figure 1 depicts the suggested method's block diagram.

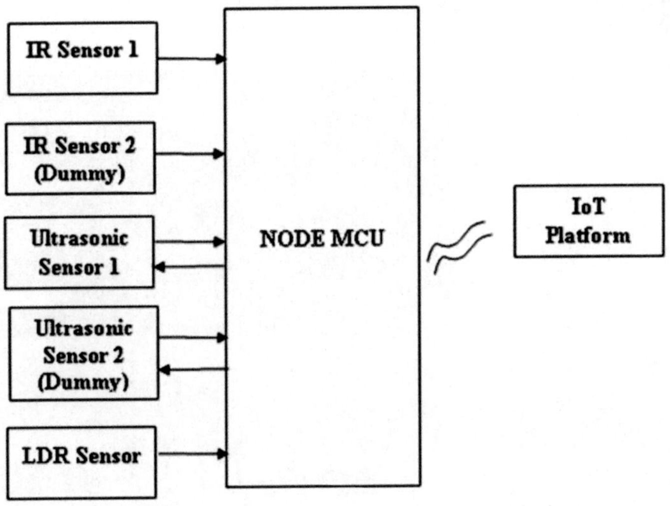

Fig. 1 Block diagram of proposed method

Ultrasonic Sensor

The HC-SR04 ultrasonic sensor uses SONAR to determine an object's distance, just like bats. It offers superior non-contact range detection from 2 to 400 cm, or 1 inch to 13 feet, in a small design that is simple to use and consistently produces accurate results. The process is unaffected by sunlight or dark materials, even though soft materials like cloth might be difficult to detect acoustically. It fully consists of a transmitter and receiver for ultrasonic signals. The ultrasonic sensors are placed in the door lock arrangements. It measures the distance of the lock from the lock position. If there is no lock on the door, then it shows a greater distance. If a single lock is provided in the door, the distance value is reduced. If the double lock is provided in the door, the distance value will become zero. By measuring the value of the distance from the ultrasonic sensor, it gives information on whether the door is unlocked or single-locked or locked. The ultrasonic sensor has 4 pins where the two pins serve as positive and negative power supply, one pin is used as ECHO (input pin), and the other pin is used as TRIGGER (output pin).

The 10-microsecond pulse is given to the trigger pin to transmit 8 pulses of frequency 40 kHz. These 8 pulses travel in the air and return after hitting any object. The distance is calculated based on the wind speed and time taken by the signal. Figure 2 shows the ultrasonic sensor module and Fig. 3 shows the pulses generated in ultrasonic pins.

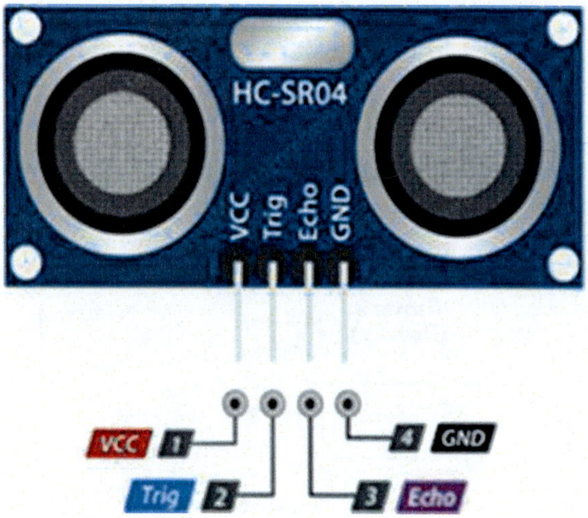

Fig. 2 Ultra sonic sensor

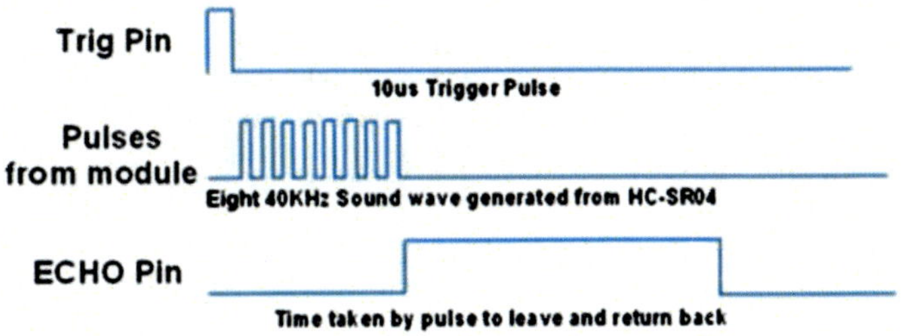

Fig. 3 Pulses in ultrasonic pins

IR Sensor

An electrical device known as an infrared proximity sensor, sometimes referred to as an IR sensor, emits infrared light to detect movement in an item and to sense some aspect of its surroundings. Since it is passive, this sensor can only detect infrared radiation.

The IR sensor module is used to identify the presence of any person near the door. If any person is present near the door, it detects and automatically open the door and if there is no person nearby then it closes the door. An active IR sensor module has both transmitter and receiver in one side. It uses light sources either laser diode for imaging sensor or LED for non-imaging sensor. The light source sends the light signal and reflects back to the receiver when it hits the object. The received intensity of light will be stronger if the object is nearby and weaker if the object is far away. Based on the output value, the presence of person is detected. The working of IR sensor is illustrated in Fig. 4.

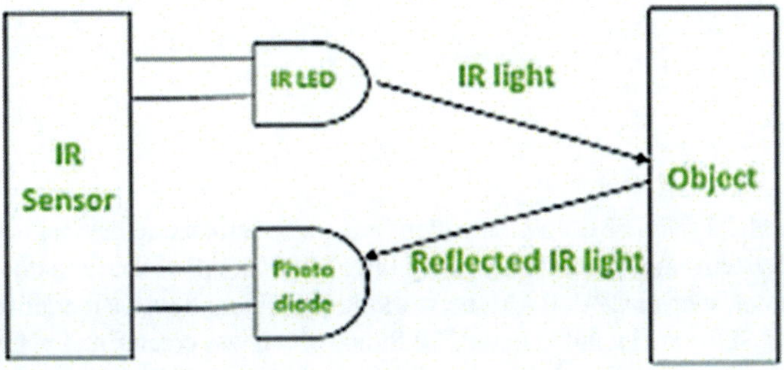

Fig. 4 Working of IR sensor

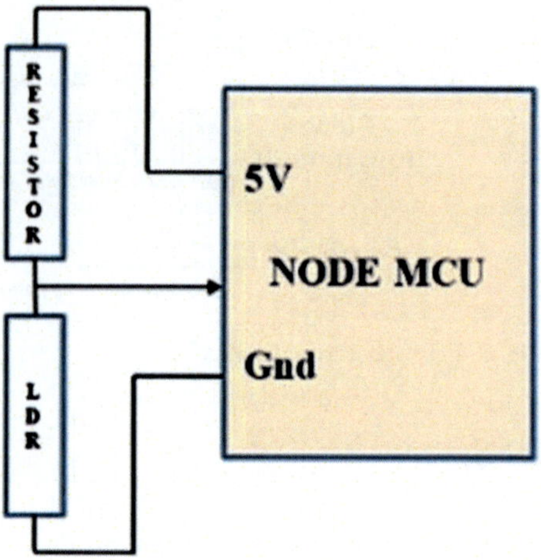

Fig. 5 Working principle of LDR sensor

LDR Sensor

An LDR, or light-controlled resistor, is a type of variable resistor that modifies its resistance in response to the quantity of light it receives. It shows that resistance decreases as light intensity increases.

The light-dependent resistor (LDR) is used to automate the light intensity in the room by measuring the light intensity in the room. The resistance value of LDR gets changed when the amount of light that falls on the sensor is changed. By measuring the output voltage of LDR, the bulb luminance can be altered in a room thereby maintaining a good ambiance. The connection diagram of the LDR sensor is shown in Fig. 5. Here, the field control method is used to change the luminance of the light intensity.

NodeMCU

The NodeMCU (Node Micro-Controller Unit), an open-source device, is built on a low-cost System-on-a-Chip (SoC) called the ESP8266. All of the fundamental parts of a computer, such as CPU, RAM, networking (Wi-Fi), and even a modern operating system and SDK, are included in the ESP8266, which was created and manufactured by Espressif Systems. For all kinds of Internet of Things (IoT) projects, this makes it a terrific solution. It is programmed using Arduino IDE. Here, the sensor value measured using an ultrasonic sensor, IR sensor, and LDR sensor is transmitted to the IoT Remote app cloud platform. Table 1 shows the pin configuration of the Node MCU with different sensors.

In addition, the Node MCU initially checks whether the data received from sensors are already registered with the smart home network using the received device label

Table 1 Pin configuration of nodeMCU

Sensor	Pin number of node MCU
IR sensor 1	D0
Dummy IR sensor	D1
Ultrasonic sensor trigger	D2
Ultrasonic sensor echo	D3
Dummy ultrasonic sensor trigger	D4
Dummy ultrasonic sensor echo	D5
LDR sensor	A0

and key. If yes, then it proceeds further, otherwise, it never proceeds further. In this way, the source of data is ensured. The hackers may try to hack the smart home network by sending data from some other sensors (out of the home network). In this case, their authentication failed because they could not give the exact device key.

IoT Remote App

IoT Remote App is compatible with Arduino to access, monitor and control the sensor value in the dashboard. In the data field, the data can be read or written from or to the sensor. It shows the graphical representation of sensor values. In IoT Remote App, the cloud variables are created in Things. In the dashboard, the display modes and settings of each variable are created. Things cloud variables are assigned to the dashboard variables. The network Wi-Fi name, password, and security key are assigned in the Things. Figure 6 shows the definition of cloud variables in things. Figure 7 shows how are the cloud variables associated with the device variables. Figure 8 shows the Wi-Fi connection configurations. Figure 9 shows the display of variables in the Dashboard of the IoT Remote app.

After configuring the variables in IoT Remote App, the coding is written in the Arduino IDE and is downloaded in NodeMCU hardware. The sensor values are updated in the IoT Remote App. The working of NodeMCU is explained as a flowchart in Fig. 10. Here is the description of the algorithm:

Step 1: Connect the NodeMCU with a pre-defined Wi-Fi network

Step 2: Read the actual and dummy sensor value from the ultrasonic sensor, IR sensor and LDR with the help of NodeMCU

Step 3: Calculate the ultrasonic sensor value into the distance with the help of calibration formula and convert it into cm unit

Step 4: Map all the measured sensor values to the IoT cloud variable.

Step 5: IoT cloud Remote App checks the device ID, and network ID for authentication

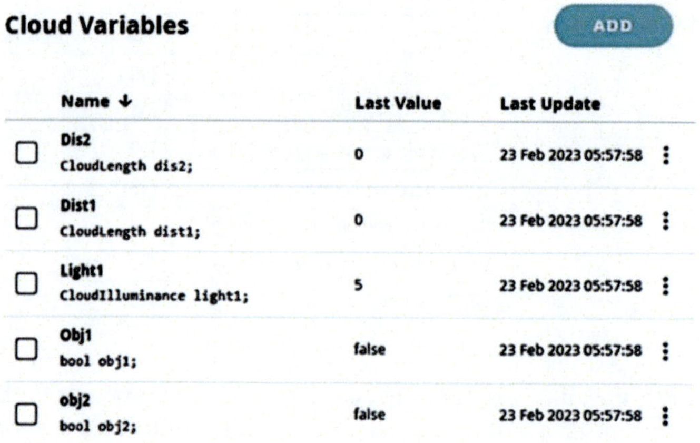

Fig. 6 Cloud variable definition

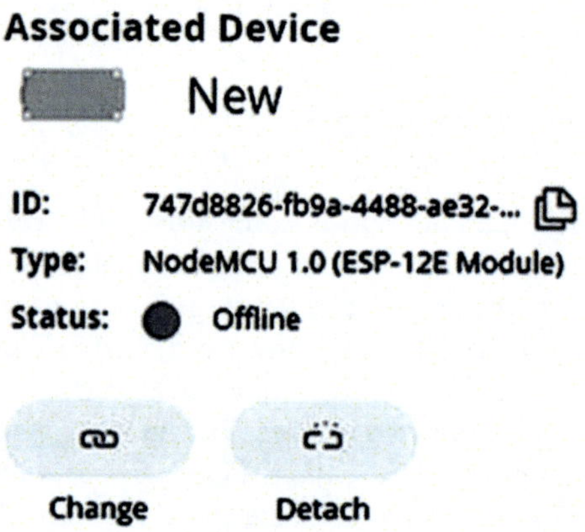

Fig. 7 Associate device configuration

Fig. 8 Network configuration in IoT remote app

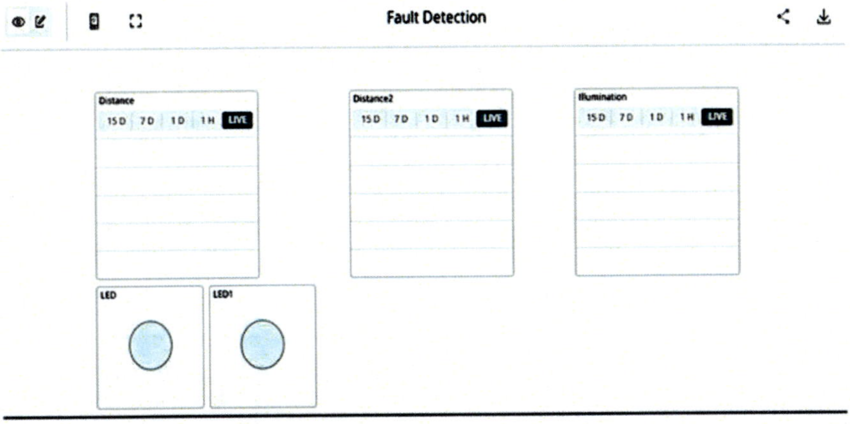

Fig. 9 Dashboard view in IoT remote app

Step 6: If authentication is confirmed, the cloud variables are updated and displayed in the dashboard

Step 7: The actual and dummy sensor values are analyzed for deviations. If the deviation is large, the fault condition is detected

Step 8: Do manual testing of actual and dummy sensors to identify which sensor is faulty and replace the sensor.

Result and Discussion

The connections of different sensors are given to Node MCU as per the block diagram and Arduino code is downloaded to monitor the output of the different sensors. Different case studies are made. In the first case, all the sensors are working properly without any fault. Figure 11 shows the output of all sensors in the IoT dashboard. Distance and Distance2 parameters are used to measure the lock value of the door using the actual ultrasonic sensor and dummy ultrasonic sensor respectively. The illumination parameter measures the intensity of light in the room using an LDR sensor. LED and LED1 are the parameters that indicate the presence of a person near the door from the original IR module and dummy IR module. The glowing of the LED indicates, the presence of a person near the door. Since there is no fault in the sensor, the sensor and dummy sensor show the same values. In the second case, a small change is made in the positioning of the ultrasonic sensor.

As a result, there are some minor values changes occur in original and dummy sensor values. Since the deviations are within the acceptable limit, the system shows no fault in any sensors. Figure 12 shows slight variations in the sensor output. In the third case, a dummy ultrasonic sensor is made faulty. The deviation of actual and

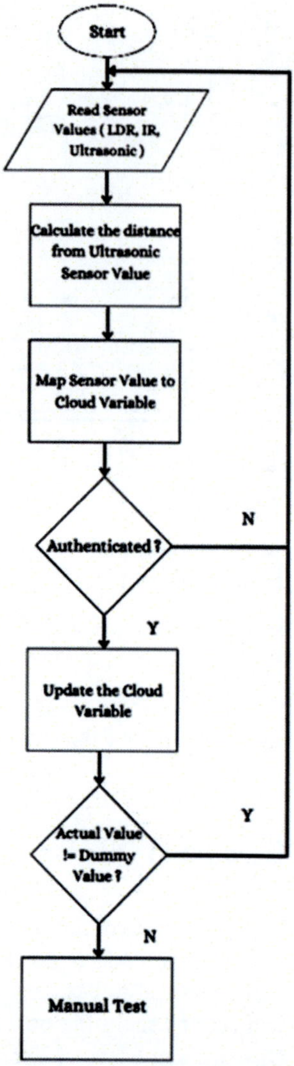

Fig. 10 Flowchart for proposed methodology

dummy sensor values is large and the information of a fault in the ultrasonic sensor is noted. Hence a manual verification is done to identify whether the fault is in the original or dummy sensor. The faulty sensor is replaced with the fresher sensor. The faulty ultrasonic sensor output is depicted in Fig. 13. In the fourth case, the fault in both the ultrasonic sensor and IR sensor is measured and it is shown in Fig. 14.

The novelty of the proposed algorithm is explained as follows: the actual sensor reads the signal and displays the results which is not guaranteed to be a correct value. Sometimes the failure of the sensor also displays any random results which are not correct. In each automation process, accurate sensor measurement plays an

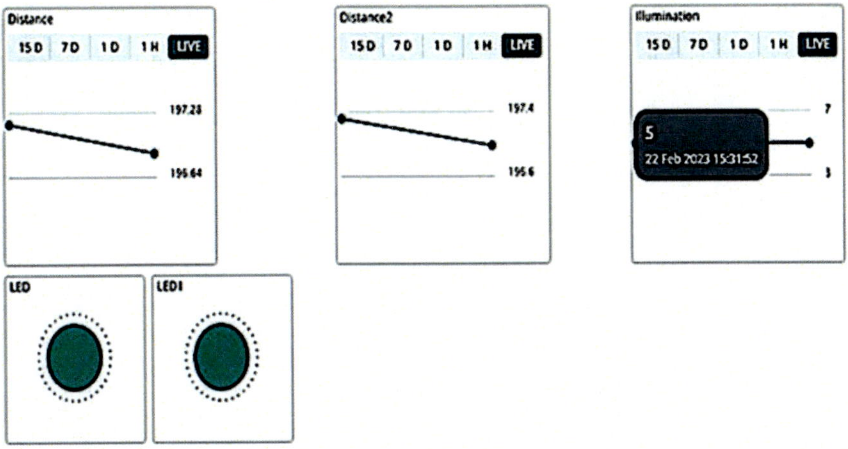

Fig. 11 Output without any fault

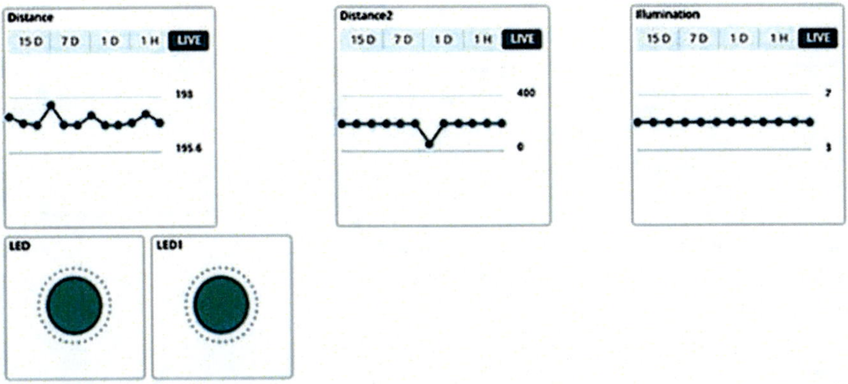

Fig. 12 Variation in ultrasonic sensor but without any fault

important role. The additional use of a dummy sensor with the same characteristics measures the same parameter with two results. If both sensor values are equal or with a small deviation, then it is guaranteed that the actual sensor measures the correct measurement. If there is more deviation between the actual sensor and the dummy sensor, then there is a fault in any one of the sensors. If there is any fault in the actual sensor, it needs to be replaced. During the replacement period, the dummy sensor measurement will be helpful in the automation process. Thus, a dummy sensor helps check the correct value of the actual sensor and it also acts as a replacement during the fault condition of the actual sensor.

The comparison of the actual sensor and dummy sensor is explained as follows: the actual ultrasonic sensor and the dummy ultrasonic sensor are compared in three

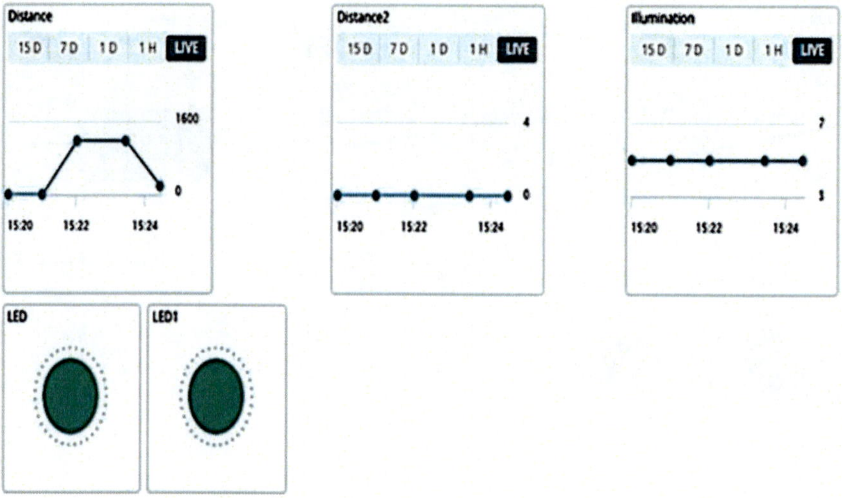

Fig. 13 Fault in dummy ultrasonic sensor

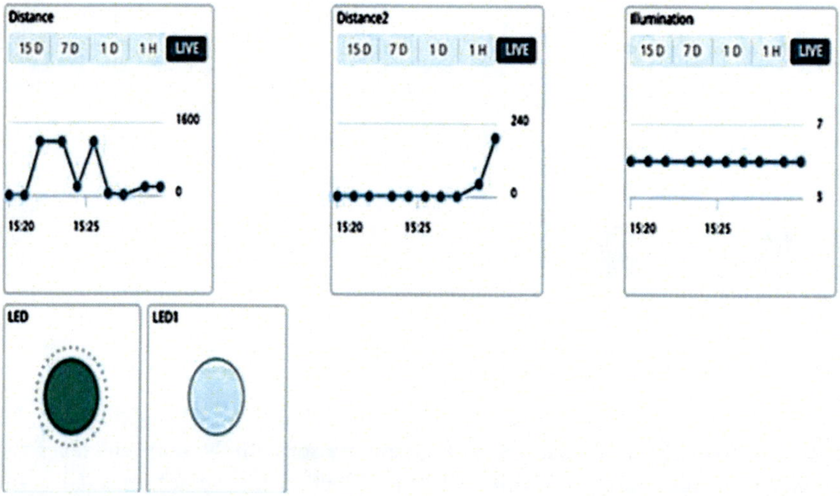

Fig. 14 Fault in ultrasonic sensor and IR sensor

different conditions, without lock, single lock, and double lock. Table 2 shows the result of actual and dummy ultrasonic sensors in three different conditions.

The experiment of actual and dummy ultrasonic sensors is also carried out at different conditions. The graphical representation of the actual and dummy ultrasonic sensor is shown in Fig. 15.

Table 2 Comparison of actual and dummy ultrasonic sensor

Different conditions	Actual ultrasonic sensor (cm)	Dummy ultrasonic sensor (cm)
Without lock	5.5	5.45
Single lock	3.5	3.46
Double lock	0.5	0.45

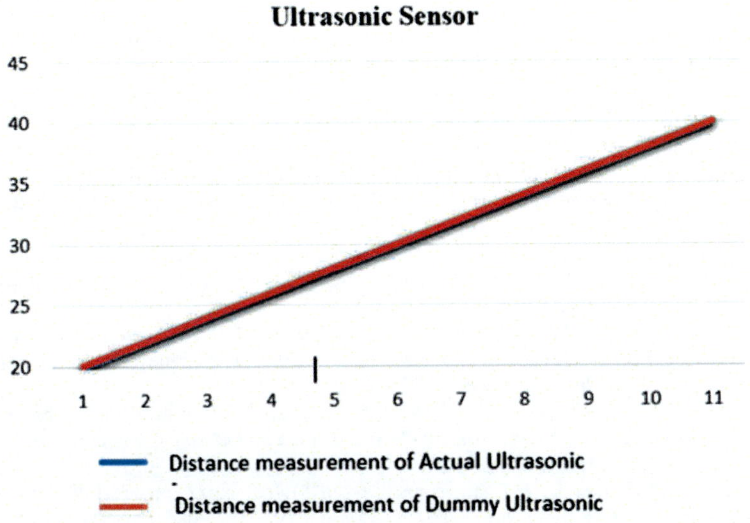

Fig. 15 Comparison of actual and dummy ultrasonic sensors

Conclusion

In this chapter, we proposed the fault sensor detection and authentication mechanism for improving the quality of services in smart homes. A dummy sensor is added along with the original sensor which helps the system for continuous functioning. The authentication of the device is carried out using the device label and key. The smart home network accepts the data from sensors if the device label and key are correct which ensures the integrity of the data. Our implementation results confirm that our scheme identifies the faulty sensors correctly.

References

1. Fanglai Zhu, Y., & Shan, Y. T. (2021). Actuator and sensor fault detection and isolation for uncertain switched nonlinear system based on sliding mode observers. *International Journal of Control, Automation and Systems, 19*, 3075–3086.
2. Gaddam, A., Wilkin, T., Angelova, M., & Gaddam, J. (2020). Detecting sensor faults, anomalies and outliers in the internet of things: A survey on the challenges and solutions. *Electronics, 9*(3), 511. https://doi.org/10.3390/electronics9030511
3. Gill, K., Yang, S.-H., Yao, F., & Xin, L. (2009). A zigbee based home automation system. IEEE Transactions on Consumer. *IEEE Transactions on Consumer Electronics, 55*(2), 422–430.
4. Ruba, M., Nemes, R.O., Ciomei, S.M., & Martis, C. (2020). Simple and robust current sensor fault detection and compensation method for 3-phase inverters. *IEEE Access, 8*. https://doi.org/10.1109/ACCESS.2020.2974769
5. Li, D., Zhou, Y., Guoqiang, H., & Spanos, C. J. (2019). Handling incomplete sensor measurements in fault detection and diagnosis for building HVAC systems. *IEEE Transactions on Automation Science and Engineering, 17*, 833–846. https://doi.org/10.1109/TASE.2019.2948101
6. Jin, Z., Yanyan, H., Li, C., & Sun, C. (2019). Event-triggered fault detection and diagnosis for networked systems with sensor and actuator faults. *IEEE Access, 7*, 95857–95866. https://doi.org/10.1109/ACCESS.2019.2928473
7. Uppal, M., Gupta, D., Juneja, S., Dhiman, G., & Kautish, S. (2021). Cloud- based fault prediction using IoT in office automation for improvisation of health of employees. *Journal of Health Engineering*, 1–13. https://doi.org/10.1155/2021/8106467
8. Cheng, K., Xu, J., Zhang, L., Xu, C. X., & Cui, X. (2022). Fault detection method for Wi-Fi based smart home devices. *Wireless Communication and Mobile Computing, 2022*(1), 12. https://doi.org/10.1155/2022/4328307
9. Karmakar, C. G., Das, R., & Kamruzzaman, J. (2019). IoT sensor numerical data trust model using temporal correlation. *IEEE Internet of Things, 19*, 3075–3086. https://doi.org/10.1109/JIOT.2019.2957201
10. Tang, S., Shelden, D. R., Eastman, C. M., Pishdad-Bozorgi, P., & Gao, X. (2019). A review of building information modeling (BIM) and the internet of things (IoT) devices integration: Present status and future trends. *Automation in Construction, 101*, 127–139. https://doi.org/10.1016/j.autcon.2019.01.020
11. Raj, D. J. S., & Zhang, Y. (2020). A novel information processing in IoT based real time health care monitoring system. *Journal of Electronics and Informatics, 2*, 188–196.
12. Krishnamurthi, R., Kumar, A., Gopinathan, D., Nayyar, A. (2020). An overview of IoT sensor data processing, fusion, and analysis techniques. *Sensors, 20*(21). https://doi.org/10.3390/s20216076.
13. Saeed, F., Paul, A., & Rehman, A., Hong, W. H., & Seo, H. (2018). IoT-based intelligent modeling of smart home environment for fire prevention and safety. *Journal of Sensor and Actuator Networks, 7*(1). https://doi.org/10.3390/jsan7010011.
14. Zhang, X., Foo, G., & Kingjet, M. (2013). Sensor fault detection, isolation and system reconfiguration based on extended Kalman filter for induction motor drives. *IET Electric Power Applications, 7*, 607–617.
15. Wu, Y., Wu, Z., & Gen, X. (2020). A review of fault diagnosis and sensor validation techniques for smart homes. *IEEE Access, 8*, 10311–10326.
16. Badar, A. Q. H., Anvari-Moghaddam, A. (2020). Smart home energy management system. A review. *Advances in Building Energy Research, 16*(4), 118–143. https://doi.org/10.1080/17512549.2020.1806925.
17. Yang, J., & Sun, L. (2022). A comprehensive survey of security issues of smart home system: "'Spear'" and "'Shields",' theory and practice a survey on fault detection and diagnosis techniques in smart homes. *Digital Object Identifier, 10*, 124167–124191.

18. Chen, J., Zhang, L., Li, Y., Shi, Y., Gao, X., & Hu, Y. (2022). A review of computing-based automated fault detection and diagnosis of heating, ventilation and air conditioning systems. *Renewable and Sustainable Energy Reviews, 161*. https://doi.org/10.1016/j.rser.2022.112395.
19. Zhao, Y., Lu, H., & Hu, G. (2019). Fault detection and diagnosis in smart homes: A survey. *Journal of Ambient Intelligence and Humanized Computing, 10*, 673–686.
20. Rivas, A. E. L., & Abraoa, T. (2020). Faults in smart grid systems: Monitoring, detection and classification. *Electric Power Systems Research, 189*. https://doi.org/10.1016/j.epsr.2020.106602.
21. Gao, Z., Cecati, C., & Steven, X. (2015). A survey of fault diagnosis and fault-tolerant techniques part I: Fault diagnosis with model based and signal-based approaches. *IEEE Transactions on Industrial Electronics, 62*(6), 3757–3767.
22. Lazaroa Molnar, S., Shaker, H. R., Mohmed, N. (2016). Fault detection and diagnosis for smart buildings: State of the art, trends and challenges. In *3rd MEC international conference on big data and smart cities*. IEEE Xplore. https://doi.org/10.1109/ICBDSC.2016.7460392
23. Abid, A., Khan, M. T., Iqbal, J. (2020). A review on fault detection and diagnosis techniques: Basics and beyond. *Artificial Intelligence Review, 54*, 3639–3664.
24. Li, D., Wang, Y., Wang, J., & Wang, C. (2020). Recent advances in sensor fault diagnosis: A review. *Sensors and Actuators, A Physical, 309*, 112–123. https://doi.org/10.1016/j.sna.2020.111990
25. Mallick, Y. M., Misra, A., Ganguly, N., Lee, Y. (2020). Unified detection and correction of IoT faults in smart homes. IEEE Explore. https://doi.org/10.1109/WoW-MoM49955.2020.00028.
26. Sathiyavathi, R., Bharathi, B. (2017). A review on fault detection in wireless sensor network. In International conference on communication and signal processing. IEEE Explore. https://doi.org/10.1109/ICCSP.2017.8286633.
27. Singh, V., Mathur, J., Bhatia, A. (2022). A comprehensive review: Fault detection, diagnostics, prognostics, and fault modeling in HVAC systems. *International Journal of Refrigeration, 144*, 283–295.
28. Yuneela, K., Sharma, A. (2022). A review paper on technologies used in home automation system. In 6th International conference on computing methodologies and communications. IEEE Explore. https://doi.org/10.1109/ICCMC53470.2022.9753928.
29. Vigil, A. (2018). A review on RC-home automation using LDR and IR sensors. *International Journal of Pure and Applied Mathematics, 118*.
30. Majumder, S., Ray, S., Datta, S. (2021). Use of Internet of Things in home automation: a review. In *Modeling, simulation and optimization, proceedings of CoMSO* (pp. 55–72).

Security in IOT-Enabled Smart Agriculture Systems

Chandrasekaran Senthil kumar and Rajasekaran Vijay Anand

Abstract The agricultural sector accounts for a substantial portion of the GDP and provides employment, which is among the most important industries that contribute significantly to economic development. Precision agriculture and the Internet of Things (IoT) significantly changed how the agricultural sector operates over the current period. The implementation of such developments depends heavily on data transmission between different farming equipment. Through the investigation of potential attacks and risks, this research seeks to explore the security situations relevant to agriculture. There has been research done on IoT-based simulators for agriculture. The criteria for security were spelled forth based on a proposed architecture for smart agriculture that is irrespective of any core technologies that could be deployed. To obtain this objective, we confer the technology of IoT-based smart ecosystem for agriculture by estimating their framework, and their claim. In addition, we came across the opportunities and trends of IoT claims for precise agriculture and specify the open challenges and issues of IoT in agriculture. We hope that the conclusions of this study will significantly constitute the research strategies and increment of IoT-based resolutions suggesting to enhance quality and productivity. It was also researched how far the industry has progressed in developing IoT-based systems and applications.

Keywords Agricultural sector · Smart farming · Internet of Things (IoT) · Attacks · Security

C. Senthil kumar · R. Vijay Anand (✉)
School of Computer Science Engineering and Information Systems (SCORE), Vellore Institute of Technology, Vellore, India
e-mail: vijayanand.r@vit.ac.in

C. Senthil kumar
e-mail: senthilkumar.c@vit.ac.in

A. Prasad et al. (eds.), *Communication Technologies and Security Challenges in IoT*, Internet of Things, https://doi.org/10.1007/978-981-97-0052-3_14

Introduction

The rise in human beings shows how important farming and agriculture are in enhancing and producing huge food for everyone. The functioning of the farming industry is a vital parameter in the economies of various nations. A country's economy is supported by improved efficiency in the agricultural sector. Farming has relatively lower margins of profit on the commodities as well as goods they obtain. Farmers are thus losing interest in working in agriculture as it takes a significant duration to form food or agricultural items in various industries it does in a variety of other industries, and because employment rewards for farmers are significantly less than ideal. Since unanticipated natural disasters can directly affect agribusiness if landowners are not prepared for unknown circumstances, the financial danger connected with farming or ranching is larger. The outline of technologies like the Internet of Things (IoT) altered the way farming is carried out and improved operational skills in the agricultural industry [1].

Agriculture is one of the prominent industries of the Indian economy. Agricultural production is crucial for the financial well-being of the nation since it supplies food for everybody [2]. Fresh approaches and innovations are continually developed and deployed to fulfill humanity's present global demands. This has resulted in the emergence of the Internet of Things [3, 4]. The Internet of Things (IoT) is defined as the collection of all items contained inside products, detectors, computers, applications, and individuals that connect, distribute information, and engage through the Internet environment to give a full solution between the real and virtual worlds [5]. Smart village implementation is more difficult than introducing smart cities due to weak architecture, limited connection, a high number of diverse physical items, and changing cultural traditions in rural regions [6]. Agriculture may be considered a field of study that involves the exploitation of the soil, the growth of crops, the care of animals, and eventually the selling of the end products. By combining technology with agricultural methods, it is possible to meet the rising demand for food in both quantity and quality criteria [7]. Smart agriculture is the term used to describe modern agricultural methods that use technological advances. A type of agriculture termed organic agriculture also referred to as smart agriculture, uses some modern technology in both the agricultural and animal industries to increase both quantity and quality of output while minimizing the impact on the environment [8]. Thus, it is essential to safeguard and advance agriculture. Due to its favorable was developed with the significance of agriculture in mind. Agriculture and information technology monitor various agricultural outputs [9]. Recent IoT technologies provide a significant opportunity for creative thinking and intelligent systems integration that may enhance many facets of the agriculture industry. The structure and complexity IoT based communication cover a broad value of constraints both in the kind of information and the settings that the IoT devices [10]. The Internet of Things is primarily focused on environmental monitoring, automation of machine and process control, precision agriculture, and traceability. IoT is a term that describes a network of interconnected technologies that may offer unique IDs for people, animals, objects, computers, and other digital and technical technologies [11]. The contemporary smart city scenario is greatly facilitated by the growing creation and use of Internet of Things technology, pushing the idea of intelligent cities to the level of big data [12]. Extremely precise agriculture uses data on moisture content, climate, rainfall, and type of crop to calculate the precise water quantity and administration duration required to increase crop yields while also cutting down on farmers' labor expenditures [13]. Connected devices, intelligent healthcare systems, smart automobiles, smart farms, smart factories, smartphones, and even people are examples of IoT devices. Intelligent home appliances include, for instance, a smart TV, smart doorbell, smart lighting, smart display, smart toaster, and smart refrigerator. Allowing everyone accesses to things now, regardless of place or time, has altered human existence [14]. The IoT is a new paradigm that integrates the physical and digital worlds to address pressing issues in daily living. The IoT predicts the connectivity of many different intelligent entities and devices in our environment, which can gather processes and send data [15]. Figure 1 depicts an extensive and comprehensive array of IoT applications for smart farming. The IoT is a network of devices of everything that are associated with electronic gadgets, actuators, servers, applications, and persons through into the Internet context to connect, exchange data, and engage in a way that provides a complete explanation of the real world and the virtual world. Smart buildings, green infrastructure, intelligent buildings, driverless vehicles, precision agriculture, workplace planning, nursing, and commerce are just a few of the areas where IoT has been used.

With the limited amount of land and water available due to the growing population, agricultural practices that maximize crop output and resource efficiency are encouraged to be used [16]. To address the issues that contemporary agriculture is experiencing, using IoT technology is a workable approach for greater efficiency.

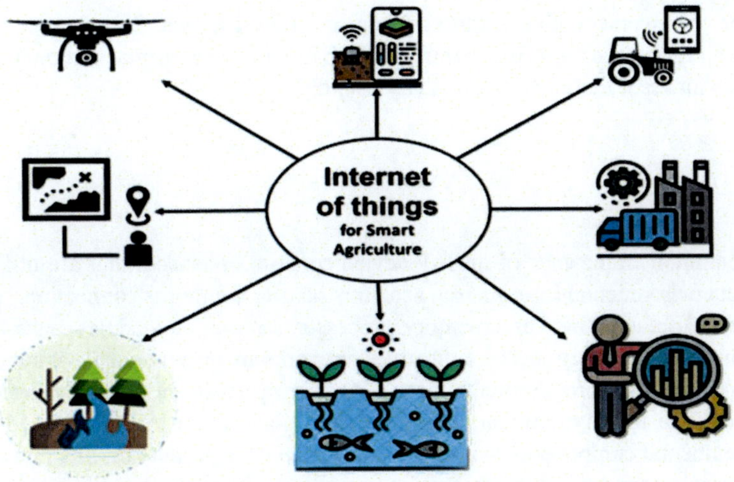

Fig. 1 An overview of the Internet of Things

Benefits of Smart Agriculture

In the subsequent discussion, we will explore a number of advantages that are attained through the use of smart agriculture.

- The implementation of intelligent technology in the agricultural industry has the potential to significantly enhance the quantity of agricultural output. This will contribute to the provision of sustenance for a more extensive demographic.
- The quality of food produced has a significant impact on the health and nutrition of individuals across different socioeconomic groups within the nation. The enhancement of food quality has a positive impact on the health and longevity of the population, hence facilitating a more substantial contribution of individuals to the economic development of the nation.
- The integration of smart technologies into conventional agricultural operations has the potential to enhance operational efficiency. Consequently, this facilitates enhanced utilization of agricultural resources throughout the procedure.
- The cultivation cost can be optimized by the implementation of high quantities, superior quality, and efficient procedures. These factors contribute to a reduction in the overall cost of cultivation, ultimately leading to a rise in the remuneration received from the agricultural product.
- The agriculture sector, as a significant contributor to the economy, is associated with substantial levels of food and resource wastage. Smart technologies have the potential to be employed for the purpose of monitoring and mitigating this wastage.
- Time efficiency is a key advantage of smart agriculture, as it enables the timely provision of necessary pesticides, fertilizers, and other chemicals. This timely provision contributes to the production of high-quality agricultural produce while minimizing losses.
- The enhancement of agricultural operations' efficiency and the reduction of agricultural wastage result in a significant drop in the environmental carbon footprint. This makes it more environmentally friendly.

IoT Device

The common framework of an IoT device consists of sensors that are utilized to collect environmental information, actuators based on wireless connections, and an integrated network that has a memory, processor, association modules, input–output systems, and an energy source. Embedded networks are programmable collaborating modules, such as programmable arrays. Sensor systems are normally designated to operate in the environment, in soil, in nature, and in air to estimate and gather environmental components that significantly affect the production like nutrients of soil, temperature, humidity, etc. Precise agriculture operations provide solutions that are uniquely developed on huge farmlands so that systems support some unique parameters like temperature, humidity, and lifecycle of the service.

Based on the operational requirements, there exist various sensors applied in precise agricultural sections. Sensors can be classified into various phases like sensor location, optical sensor, electrochemical device, and airflow sensors. These devices are utilized to gather information like the temperature of the air, the temperature of the soil, humidity of the air, moisture of the soil, wind speed, direction of wind, solar radiation, and pressure of barometric.

IoT Applications in Agriculture

The agriculture sector has witnessed significant utilization of the Internet of Things (IoT) in many applications. Some notable applications include:

- The proposed smart soil cultivation system encompasses many tasks such as ploughing, weeding, seedbed preparation, and sowing, which are conducted prior to the harvest to prepare the field soil.
- Smart irrigation systems are designed to automate the artificial provision of an optimal quantity of water for the purpose of promoting plant growth in a controlled and regulated way.
- Smart fertilizer systems refer to the automated process of applying fertilizers to agricultural fields, wherein there is precise control over factors such as the quality and quantity of fertilizer used, as well as the timing of the application.
- Smart pest detection and control systems encompass a comprehensive approach to monitoring and identifying the presence of pests, evaluating the extent of crop damage caused by these pests, and implementing effective strategies to manage and mitigate infestations.
- Smart livestock farming encompasses the utilization of advanced technologies to enhance livestock breeding practices and optimize agricultural precision, hence augmenting both the quality and quantity of animal produce.
- The smart harvesting system employs Internet of Things (IoT) methodologies to effectively gather the yield from a given agricultural field.
- The proposed smart farm management system aims to utilize data analytics to enhance the productivity of agricultural production.
- The management of ground water quality is crucial for ensuring optimal quantity and quality of groundwater, which significantly impacts the final agricultural output. Hence, the implementation of Internet of Things (IoT) is employed in this system to employ strategies aimed at preserving adequate levels of groundwater.

Networking Technology

The networking technologies for IoT show that to embed IoT into the precise agriculture section, networking technology must significantly enhance the device evolution of IoT. they take a significant role in the improvement of IoT networks. The present statement can be derived as spectrum, architecture, and protocols.

Protocols: Various wireless networking protocols have been suggested for precise agriculture. Various protocols, and systems in a smart agriculture network can be significant, interchange data, and provide choices to handle and monitor farming scenarios and enhance production and yield efficiency.

Related Works

The research [17], disclosed that the use of drones and a variety of devices connected to the Internet of Things (IoT) can be used in Intelligent Precision Agriculture (IPA) to track an area of agriculture. Drones may also be used to gather information from mobile phones and transmit it to a Ground Station Server (GSS). Moreover, the IoT farm ecosystem is exposed to a variety of possible dangers to gadgets, drones, and the GSS. Therefore, the process produces heavy replay attacks in the system as the main drawback. To overcome this issue, a novel identification and management of keys system for IoT-based IPA, known as AKMS-Agri IoT, was recently put forward in conjunction with a confidential blockchain-based approach. Although, cloud computers generate blocks to validate and add them to the secure blockchain center. Hence, the blocks of information are created by combining encoded operations and their corresponding fingerprints by the GSS.

The era of environmentally conscious agriculture has begun, and its effects on society and nature are anticipated to be profound said by [18]. The Internet of Things (IoT) technology has arisen as the main route for innovative ways of farming in this setting. The software stage procedure employed by IoT nodes, pathways as well as application server hardware is one of the many parts of the underneath network of communications structure and technology that contribute to the uncommon potential for gathering and organizing data provided by IoT. Data Distribution Service (DDS) and WebSocket. All these parameters are used to overcome the protection issues and the major drawback of this paper is it will transfer a very less amount of data.

The Internet of Things has advanced quickly in the past few years and has had a significant impact on the way various sector functions are discussed [19]. As a result, IoT technology has raised production based on the huge value of industries. Smart agriculture significantly profited through IoT. Farming precision, excellent harvest rates, and the effective use of earth's resources may all be sustained for an extended amount of time frame because of clever farming. Monitoring abilities, technology for communication for transmitting information from sensors, and information investigation for gleaning insights from the information obtained from intelligent agriculture.

Producers will be able to exercise wise judgments and benefit from them thanks to all of these courses. Moreover, this also produces disadvantages. To overcome this issue, an intelligent agricultural risk tracking system uses cloud computing to efficiently track sensor deviations, equipment circumstances, and safety risks. Furthermore, the blockchain relies upon clever-contract software that is tested to track and react to safety threats.

In [20], reveals that by utilizing cutting-edge equipment and methods, such as smart farm machinery, highly accurate detectors, maps, and complex computer networks, intelligent farming seeks to maximize the financial rewards from farming while minimizing the related labor costs. The web and the Internet of Things is a significant tool for getting, tracking, manufacturing, and organizing agricultural resource information of agricultural precision and environmental monitoring sectors due to the increasing need for individualized agricultural operations. Here, an architecture based on the Internet of Things that incorporates the newest technology is offered for building a management system for crop machinery fitted with automated navigational systems. In the preliminary production stage, different agricultural equipment cyber-models and the appropriate nodes for sensing were built. Therefore, a significant problem in this situation involves how to link the FMIS to truly intelligent agriculture.

In [21], the author discusses the growing need for water and food from an increasing population that is expanding worldwide, precision farming is currently attracting a lot of interest. As a result, producers will require access to water as well as land suitable for farming. Producers require an answer that alters their methods of operation because these resources are scarce. The answer to producing greater, superior, as well as more lucrative harvests with fewer assets is precise watering. Though, the most important aspect is cost reduction in the effective utilization of groundwater throughout the cultivation of agricultural land. Therefore, to utilize water more effectively, various watering systems are constructed using artificial intelligence. Those systems' poor capacity to absorb information makes them unsuitable for unstable environments. In this setting, this study suggests a deep learning neural network-based smart system for precision farming (DLiSA) that is enabled by the Internet of Things (IoT).

Architecture

The protection criteria were laid out in light of the proposed architecture, which is autonomous of any possible underlying technology for smart farming. A multi-layered based farming architectural style that can be used for any smart farming application, independent of the enabling technology [22]. Also, each layer of the architecture's security needs has been established. Based on an analysis of the costs associated with computing decryption and communicating, the performance of the authentication methods is evaluated. a layered design that is adaptable to any farming environmental deployment. To prevent the content of the sensor information from being revealed to any third-party organization, devices from the network transit layer must secure sensor information using encryption techniques [23]. Interaction between the classification purposes and the module detects the layers and the data access layer, the data transmission level and data gathering layer, and the online storage layer and the data harvesting mechanism will be made simpler. To protect against unit destruction, access point capture assaults, and attribution, the locations of the sensors, immovable equipment, and the genuine whereabouts of the mobile platforms must also stay hidden. All components in the architecture's levels will need this security device. Between the user layer and the module detects, the isometric view and the data transfer layer, the transmission layer, the data collection layer, and the data capture layer and the digital storage layer, this protection objective is necessary [24]. To provide safety regarding device destruction, access point capture assaults, and traceability, the locations of the detectors, and stationary equipment, together with the interesting locations of the moveable devices, must be kept secret. All components inside the architecture's levels will need this facility. The standards for cybersecurity have been spelled forth based on a suggested architecture for smart farming that is autonomous of any underlying technology that may be deployed. The design is meant to be used for ongoing remote field observation. The plan makes use of an encryption process, mixing, and stream cipher algorithms [25] (Fig. 2).

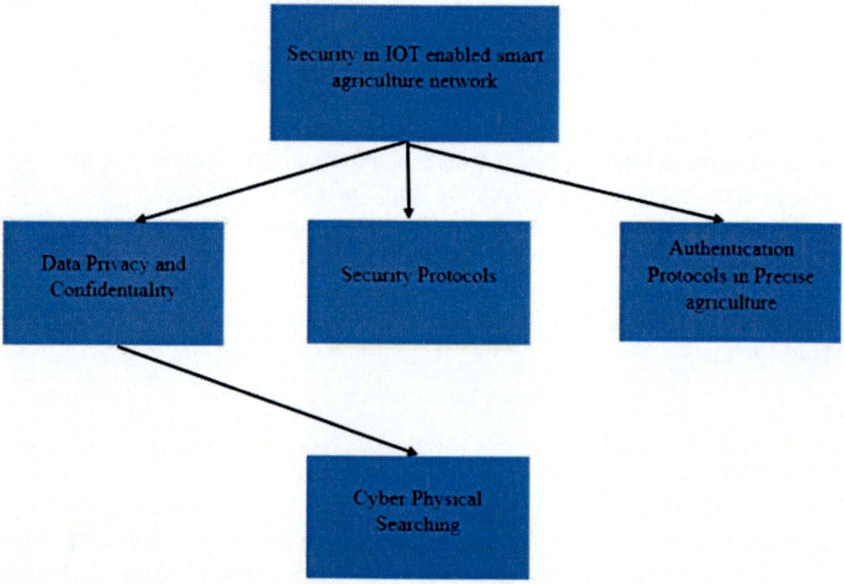

Fig. 2 A flow model based on security in IoT-enabled smart agriculture systems for the processes and procedures involved in ensuring system security

IoT-Based Farming

High crop yields and resource-efficient practices in the agricultural sector may increase production [26]. Smart farming is one of the IoT's possible applications for significant enhancements to processes, according to recent IoT developments. The monitoring of soil conditions, insect management, crop growth, animal herds, water level control, atmospheric tracking, and other Internet of Things applications in intelligent agriculture are only a few examples. To sense the working and outside surroundings and transform the sensed analog signals into a digital format, sensor equipment is deployed on the farm close to the premises and nearby the agriculture area. The most common sensors in smart agriculture are those that measure moisture, temperature, light, pressure, and closeness. In this part, we provide an overall structure for an IoT system for smart agriculture, which is built on three key elements: IoT gadgets, communication technologies, and information processing and storage options. Figure 3 provides a representation of the IoT infrastructure for smart agriculture.

To attach objects to a cloud for interaction and exchange of data, IoT is utilized. IoT has the potential to produce billions of linked or "smart objects." These intelligent items are capable of gathering environmental data and corresponding with other systems online. Applications built on the IoT are being created that enable devices for monitoring and controlling functions in a variety of domains, such as processes, household appliances, medical monitoring software, smart homes, smart cities, smart agriculture, etc. Applications for the Internet of Things are particularly important for

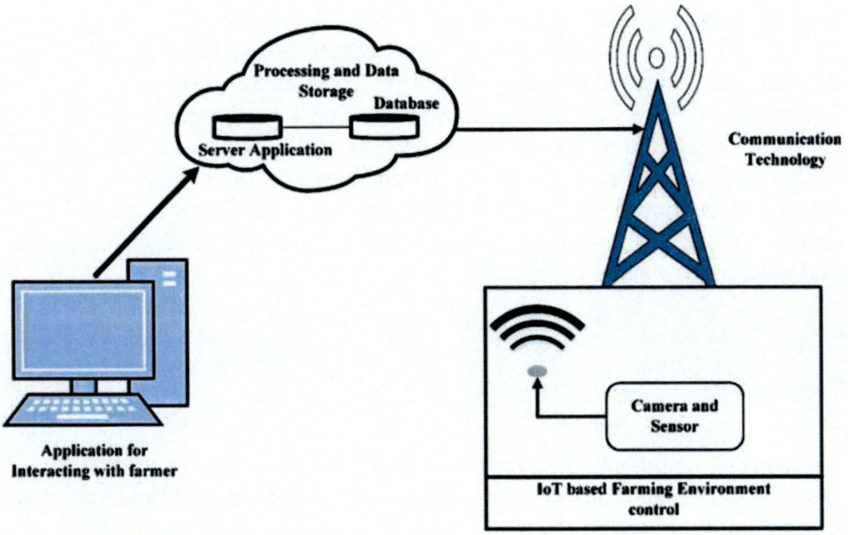

Fig. 3 IoT-based smart agriculture

harvesting agricultural yields, water supply, post-harvesting, processing, transportation, crop storage, leaf wetness sensors, and sales throughout the agriculture industry. Numerous specialized sensors, including soil moisture sensors, humidity detectors, infrared sensors, rain forecast sensors, solar radiation sensors, etc., may be utilized in agriculture applications [27].

Security Protocols

The term "agricultural supply chain" refers to the "farm-to-fork" conceptual process that includes the operational phases of harvesting, preparation, transportation, preservation, and consumption of diverse agricultural goods. The organizational, psychological, technical, and scientific hurdles must be overcome to include internet media that has the potential to alter the food production and distribution chain of custody [28]. Using computational data gathering to foresee any various sensors that could be absent and providing a free circuit to measure humidity levels, the biometric traceability technique, demonstrates enhanced data stream efficiency and precision [29]. The relationship's role in recognizing a description structure of all potential interactions between the physical components. According to this framework, many use cases for access control that is continuously determined depending on the properties of the object are developed for reading, listening, and executing rights [30]. To monitor the process, visibility, and discoverability of the agriculture value chain, A system constructed around intelligent contact information and a similar technique was described. This framework was then realistically applied to accomplish fragmentation and monitoring of farmed product descriptions using QR codes providing a test case of leveraging smart contracts for blockchain-based soybeans provenance to demonstrate increased system visibility [31]. This framework developed an infrastructure for food safety traceability that uses big data, two-dimensional RFID codes, and rice to store information in an IoT network and provide a uniform 3rd certification for packaged foods [32]. The 2D code remains the dominant traceability technique even though more sophisticated technologies are being used, and health and safety duplicity coupled with random anomalies are highly widespread in Indian food traceability systems [33].

Cyber-Physical Searching

A system that gives landowners the ease of searching for the data they need to handle agricultural operations on a communications system. data structure made up of data aggregation nodes with power generation and the capacity to gather sensor data within the network area [34]. In a bid to keep assailants from collecting details from the interest and incoming packets, the name identification, which is normally present in ICN, is camouflaged as rubbish by introducing disinformation that is appropriately adjusted so that outsiders cannot recognize these as waste [35]. The data gathered from agriculture networking may be seen in a variety of ways thanks to big data analysis, which can also aid in improved data storage and handling, which improves search functionality [36].

Data Confidentiality and Privacy

A connection of sensing devices is used by the platform to collect sensor data from agricultural systems while also ensuring its privacy, security, and verification. Examine individuality packing, indication transmission, and broadcasted authentication as tools for managing security and anonymity in agriculture relationships [37].

Access Control

Three modules with ontology for allusion identity management have been suggested for a smart farm environment. A framework for smart farming with three layers is shown in the first module: a practical property level, an automation overlay, and an operations surface [38]. To track the transmission of data amongst them, each unique commodity has online counterparts in the automation module. A representative matrix of all possible interactions between the physical components is recognized thanks to the linkages. This method allows for the establishment of various user types with read, enter, and provides a way to execute periodic security controls depending on the characteristics of the material [39]. Table 1 illustrates the security protocol in IoT architecture.

Table 1 Security protocols in IoT

Protocols	Functions
Message queue telemetry transport (MQTT)	It is a significant protocol for connecting distinct devices with minimal bandwidth
Constrained application protocol (CoAP)	It is utilized for resource-restricted, minimal-power device in lossy networks
Data distribution service (DDS)	This protocol is utilized for quality information communication
Advanced message queue protocol (AMQP)	It is utilized for enabling interoperable and encrypted messaging amongst applications and organizations
Extensible messaging and presence protocol (XMPP)	This creates a persistent association between a server and a client or amongst many servers from one server

Data Management

Cryptocurrencies, digitalization, and data centers should all be part of a data organization structure for IoT-based agricultural systems with authenticity monitoring systems. To offer a key-value pair for the statistics from peripheral various sensors, a data integration approach has been created [40]. If the amount exceeds a certain limit, a cryptocurrency client receives a data array with the most current stamp. Alternatively, the data link layer gets it through the network and updates it to the flow board or routes the packet following the most recent routing specification in the data frame if there is no appropriate flow labeling for the datagram [41]. Ensuring that there are no mistakes in the communication or controlling transmission, monitors the system's integrity. After implementing Permission Sharp and initiating a DDoS assault, the outcome was evaluated. The results indicated that the operation was terminated before the number of blocks increased [42]. The supercomputer calculates the total of all cipher texts included in that interior once the device driver first selects a collection sector. This combined cryptographic hash sum and the identification codes are sent to the individuals knowing are in that accumulation region. If the aggregated unencrypted result is legitimate, a translation component and an authentication code are calculated by the human and submitted to the software. The web server determines the overall decoded amount after collecting decoded parts from multiple users [43]. Table 2 illustrates the issues in security and the solutions to them.

Solution and Challenges

The pairing of cellphones and electrical gadgets Software is being innovated, and cellphones are seen as an IoT enabler. Different hardware and software have been developed to make mobile phones flexible in the agricultural area. A thorough analysis of online services that provide agricultural solutions has been provided [4]. The three fundamental criteria of verification, remote access, and investor privacy make

Table 2 Security issues in IoT and solutions

Threat	Solutions
Replay attack	A framework for replay attack identification for battery-based IoT systems designed for healthcare
IoT malware	A recurrent network-based methodology for IoT malware hunting threat
Man-in-the-middle occurrence	Intrusion detection system for IoT
Android malware linear	Malware identification based on SVM for significant IoT services
Node capture attack	IoT-based scanner for identifying threats

the safety of IoT-based agricultural production. The connection, however, has to be protected from outside threats at the device level, and the data fusion at the network topology needs to be safe. Only authorized users will be able to access information collected by the app layer because certain organizations have been granted permission. Intrusion detection, which includes the security of hardware and information acquisition reliability, is the most prevalent public safety issue at the device layer [45]. In the Internet of Things, data is transmitted to other connections, including the cloud, when it is sent from a bridge to an edge device. There are many security requirements in place for edge devices, including biometric authentication, data cleansing, cryptography, data flow methods, etc. Hacking, spying, middle assault, and deceit are some examples of privacy concerns. As a result, throughout the data-collecting phase, secrecy, identification, and fidelity must be used. Identification, security systems, and investor confidentially are the three fundamental needs for IoT-based smart agricultural protection. The connection, however, has to be protected from outside threats at the higher layers, and the data fusion at the protocol stack needs to be safe. Defense in depth, which includes the security of hardware and information acquisition protection, is the most prevalent security concern at the perceptual layer. Security breaches, which include the sharing of private data and locations, are another potential security risk. The core network is essential for enhancing security in the Internet of Things [46]. Machinery, which is charged with handling data and serving as an interface for communication between the implementation layer and the data, is positioned between them. Private and confidential data storage is necessary for the service layer. IoT-based infrastructure is vulnerable to risks including susceptibility, denial of service, unauthorized access, and many virus injections since secure communication over wireless media is extremely difficult to implement in IoT applications. Those assaults have an impact on data privacy and reliability in this manner. Among the top levels, the IoT visual application framework is thought to be tightly connected to the cloud due to its storage and computational capacity. Hence, control methods should be used to maintain and administrate possession and exposure has been linked for physical people, computers, and organizations [47].

In the context of smart cultivation, interoperable means the capability of two or more different networks, options, and components to work together to share data, advance workflows, incorporate technical breakthroughs, and conform to legal and regulatory requirements. The four kinds of interoperability include technological or system interoperability, semantic/data connection, operational interoperability, and legal interoperability. To establish data/semantic interoperability, all parties must agree on the purpose, nature, and context for the exchange of data and exploitation. Smart farming networks are more adaptable because of interoperability, which also makes it possible for farmers to collaborate to enhance farming practices by exchanging data. Standardization is crucial for interoperability. Making the most of data with processing would prove very time- and resource-intensive without ways to collect data from various digital agricultural solutions' sources. It is essential to use procedures and best practices that transform knowledge and data into standardized formats to support integrating data, communication, and sharing practices, as well as collaborative decisions [48].

Along with interoperability in smart agricultural technologies and equipment, managing reliable information is a critical requirement in a smart farm network. The extent to which the knowledge and its source are dependable, reliable, real, genuine, and symptomatic of the problem is known as data reliability, often referred to as data quality. Reliability may be evaluated at some levels, covering both on- and off-farm levels. It is essential to make sure that the data output is legitimate and precise from the viewpoint of a farm A smart agricultural system should reassure farmers about the legitimacy and dependability of the analytics and data from far away. Since it is hard to confirm the accuracy of every data source, problems with the quality of data are also expected to arise throughout the integration process. Poor judgments made based on inaccurate data cause agricultural businesses to lose thousands of dollars annually [49].

Scalability is a factor that should be considered while developing a platform for intelligent farming. In smart farming, scalability refers to how easily a system may increase its capacity, particularly the number of technological elements like actuators and sensor arrays while enabling fast analysis. Scalability problems in a smart farming design could lead to substandard system performance in this data processing environment. It has been advised to take into account the potential for smart agricultural system uses to expand in size and complexity from the very beginning of system design. As an example, take into account a major chicken industry that operates farms in several regions. To integrate all of the supply chains from feed to hatching facilities and farms with poultry to shipment, the organization requires a platform [50].

Smart farming systems also call for quick and accurate decisions and actions. Research needs systems that enable farmers to perform the correct tasks at the proper times. The ability to analyze massive amounts of data inflows as they are being generated or archived is known as real-time data analytics. For instance, if a farmer wants to sell their products or buy equipment, a prolonged delay in accessing data and making decisions may cause them to miss out on a contract or make a presumption based on outdated data. Rapid collection of information, manufacturing, and export are goals of real-time processing. Instantaneous processing of data also enables intelligent players in farming to make suitable choices at the right moment, minimizing risks and unfavorable outcomes. Real-time processing is additionally made possible by a smart farming platform, which is necessary for bolstering farming techniques and preparing farmers for unforeseen events. Real-time data collection is also required for real-time data processing to be possible. While real-time data enables real-time judgments and actions, results obtained through the analysis of old data may outcome in erroneous decisions [51].

A smart agricultural platform ought to emphasize safety and confidentiality and offer adequate protection methods, just like any other smart platform. Modern technology creates a framework where distantly internet-connected gadgets can be accessed, enabling attackers to organize cyberattacks. The attackers may want to steal data or impair the systems for smart farming. For instance, Yang et al. illustrated how security breaches were caused by adversaries using IoT devices in computer systems. As an example, a bad actor could use an unprotected smart device to gain access to a farm's data and acquire information regarding its goods, monetary decisions, and plans. The intrusive party may potentially cause equipment and product damage. Security is another major problem because smart agricultural systems have access to a variety of data sources, including corporate and personal data. Security measures and data protection could be ensured by smart agricultural systems provision of procedures. From collecting data to the delivery of data-driven services, these procedures should safeguard data from beginning to end and at all phases of data processing [9].

Smart farming solutions are designed and developed in response to regulations for agricultural practices. More recent research, for example, has explored farmers' worries about confidentiality and information sharing. The policies in place to safeguard their proprietary knowledge, commercial profit margins, and privacy are not thought to be enough by farmers. Additionally, there are issues with the technology suppliers' non-compliance with farm data contracts and security best practices. Standards and best practices that specify the rights and obligations of various parties in smart farming systems are necessary to allay these worries and foster trust. Additionally, there is a need to establish norms for data-sharing procedures and implement regulations to establish responsibility and impose punishments in the event of potential conflicts. Agriculture data processing and administration may be impacted by additional laws relating to food safety, managing supply chains, disease tracking, and other purposes. These regulations might be updated to take into account brand-new agri-food requirements and advancements in technology. The technology could incorporate policy needs for tracking, conformity, and sustainability [49].

Attacks in Agriculture

A full description of cybercrime dangers, assaults, and mitigations that may be used in the area extends when used in conjunction with IoT-enabled smart devices. A Wi-Fi Spoofing attack exploits that caused a Microcontroller Board to be forcibly disconnected from the connection and prohibited it from rejoining was conducted experimentally on a smart farm design along with other kinds of network assaults [54]. The author [55] suggested a learning-based and convolutional network-based intrusion detection system against distributed DoS (Denial of Service) assaults. It has been determined that connection and Data transmission and digital farming are opportunities and chances qualities, and it provides the structure to anticipate security flaws in a setting conducive to digital agriculture.

Threat Model

The contents are anticipated to be sent via regular channels, according to the security analysis of the systems examined in this work. Several trustworthy persons want to send communications covertly. An enemy A within this scenario is presumptively capable of the accompanying [56]. The attacker has total control over the channels that are used to communicate data, which gives him the ability to intercept, erase, and modify signals already present in the network. The antagonist can create fresh materials and spread them across the network. An enemy can pose as one of the truthful parties so that the other truthful parties won't be able to tell who it is. A may also resend a letter that has previously been sent across the network to the same recipient. The end-point nodes are regarded as unreliable. Any communication-related equipment may be remotely taken over to conduct power analysis attacks [57] and retrieve any private credentials kept in the storage space of the equipment.

Enhanced Security, Assaults, and Countermeasures in Smart Agriculture

In a wirelessly device distant agronomic observing system, A method of authentication has been described that allows a user to get real-time environment data from sensor nodes. The entry point transmits the information of the sensor to the central station after receiving it from the sensor nodes through a wireless router [58]. Upon registration and verification by the access point, the individual gets permission to view the information. After enrollment, a user submits biometric data that is processed using the curved extractor's physiological purpose of planning. The user then gets a smart card from the base station that contains safe provable variables as well as the unique accessible variable. Together with the card reader, the user signs in using their identification and username [59]. The confidential fingerprint curved characteristic is provided by the biometric reproduction function and is used to validate the private smart card settings. Making the intelligent card's unchanging privacy options responsive to prevent tracking improves the technique.

In addition to preventing user impersonation attacks and ensuring perfect rearward secrecy and user request, it makes sure that a private key that is a component of the cryptographic keys and shared with all participants is confined to being accessible only to the base station and sensor node [60]. IoT sensors and actuator devices that are positioned in and out of a P2P greenhouse smart farm have been suggested as part of an authentication mechanism. Once access is granted, a user must first confirm them to gain entry to the equipment. Identical to this, any two devices should verify each other before beginning any kind of connection. The research several security measures that may be used to protect the information flow set up in an IoT-based smart agriculture system. The architecture is used to develop an authenticator on the agricultural equipment that continually polls login details via RFID (radio frequency identification) and accomplishes constant identity management. recommended a different remote client detection algorithm for a wireless array of sensors for agriculture such that a gateway node enables the field-based sensory nodes to communicate online with remote users, cell phone users, and a server application located well outside the greenhouse environment. The design is meant to be used for ongoing global field observation. The method makes use of symmetrical decoding and encryption, hashing, and stream cipher security. User credentials and user biometrics are the only two elements used in this approach for user identification. The most recent advancements in precision agriculture are covered in this part from the viewpoint of the industry, including the technologies that are now available and how they are used.

IoT-enabled smart agriculture systems have become increasingly popular in recent years. They allow farmers to optimize their crop yields, reduce waste, and conserve resources such as water and energy. However, as with any IoT system, security is a critical concern for smart agricultural systems. Essential security measures can be taken to ensure the safety and reliability of IoT-enabled smart agricultural systems.

Secure Communication

Communication between IoT devices and the cloud-based system should be encrypted and secure to prevent unauthorized access or data breaches. Data should be encrypted end-to-end, and communication protocols should be secure and reliable. Authentication and Access Control: IoT devices should be authenticated before being allowed to connect to a network. Access control should be in place to ensure that only authorized devices can access the network. This can be achieved using secure passwords, biometric authentication, or other advanced authentication methods. Regular Software Updates: Regular updates should be performed to ensure the system is updated with the latest security patches and fixes. This can be achieved by implementing an automatic update system that checks for updates and installs them whenever necessary. Data Protection: Data collected by IoT devices should be protected to prevent unauthorized access, tampering, or theft. This can be achieved by implementing data encryption and access-control measures. Data backups should also be performed regularly to ensure data can be recovered during a data loss event. Physical Security: Physical security measures such as access control, video surveillance, and intrusion detection should be implemented to protect IoT devices from physical threats, such as theft, vandalism, or tampering. Risk Assessment: Regular risk assessments should be conducted to identify potential security risks and vulnerabilities in the system. This can help proactively identify and address security issues before they become a problem. Securing IoT-enabled smart agriculture systems is crucial for ensuring their safety, reliability, and effectiveness. Implementing these measures can help mitigate security risks and protect IoT devices, data, and networks from unauthorized access, tampering, and theft.

Authentication Protocols in Smart Agriculture

A design that is adaptable to any farm environmental deployment. A connected thing IoT- enabled agro landscape consists of a network of sensors in a field with the potential to detect prominent climate characteristics that aid in understanding the present requirements of the products. The best detector is selected according to external aspects including energy usage, receptivity to microwave intrusions, maximum temperature, and dryness range. Besides this, harsh environments with direct exposure to sun, chemicals, and animals combined with severe weather may make industrial sensors at risk of harm. As a result, field devices must be often repaired. As a result, environmental instruments must be repaired. Moving or stationary objects which are made to collect data from sensors in the field make up the data transit layer. Network nodes, portal centers, and core networks are examples of units in the data gathering layer that may organize and prepare data for storage after receiving it during the data transit layer. The massive volume of data that is received from diverse domains is stored steadily by the data storage layer. When necessary, it may include methods like data stores, large data storage, or blockchain. The computational layer comprises methods for analyzing and handling the data to extract important information that may be put into a machine learning-based life choice system. The outcomes of this layer must be transformed into simple instructions that work with the equipment in the ground. Machines at the actuating layer can accurately and timely accept instructions from the data processing layer and carry them out. Agriculture gear needs repair and uses fuel to function precisely. Machines at the actuating layer can accurately and timely accept instructions from the data processing layer and carry them out. Agriculture gear needs repair and uses fuel to function precisely. All parties involved in the cultivation, including those involved in pre-harvest, production and farming, harvest, storage, transport, and distribution of the finished farm products, make the user layer. Agriculture, distributors, merchants, and customers are all involved.

Conclusion

The integration of Internet of Things (IoT) enabled smart agriculture systems initiates a significant transformation in the agricultural domain, presenting a multitude of advantages and prospects for improved effectiveness, output, and ecological viability. The incorporation of Internet of Things (IoT) devices and networking technologies into agricultural practices holds the promise of fundamentally transforming food production, resource management, and crop health monitoring. Nevertheless, the presence of these chances also brings out security issues that necessitate attention in order to guarantee the ongoing prosperity and dependability of these systems. The advantages of smart agriculture are readily apparent, encompassing enhanced resource management, diminished environmental footprint, and heightened crop productivity. The utilization of Internet of Things (IoT) devices empowers farmers with up-to-date data and valuable insights, facilitating the adoption of informed decision-making processes. Consequently, this enables the implementation of more efficient agricultural practices. Moreover, the capability to remotely monitor and regulate agricultural processes offers increased convenience and decreased operational expenses. Internet of Things (IoT) devices have a significant impact on the field of smart agriculture, providing a diverse range of applications that encompass soil monitoring, irrigation control, animal management, and precision farming. These gadgets serve to enhance the acquisition of data from diverse sensors and empower intelligent decision-making, ultimately leading to a notable enhancement in agricultural output. Networking technology plays a pivotal role in facilitating the integration of Internet of Things (IoT) in the domain of smart agriculture. This technology enables efficient and uninterrupted communication among various devices and centralized control systems.

The utilization of advanced networking solutions is of utmost importance in order to guarantee the dependability of data interchange, minimize latency, and enhance resilience under challenging environmental circumstances. Nevertheless, the agriculture sector has many security challenges as it progressively depends on the Internet of Things (IoT). Ensuring the preservation of data security and privacy is of utmost importance, given the transmission and storage of sensitive information on digital platforms. The implementation of secure communication channels, robust authentication methods, and access control mechanisms is of utmost importance in order to protect against potential threats and prevent unauthorized access to the system. In order to effectively tackle these difficulties and enhance security in IoT-enabled smart agriculture systems, it is imperative to use a comprehensive approach. This includes the deployment of resilient security mechanisms, preemptive steps against future cyber-physical threats, and efficient data management procedures. Furthermore, it is imperative to engage in ongoing research and development in order to maintain a competitive edge against emerging threats and weaknesses.

When considering the future, it is expected that the security landscape in IoT-enabled smart agriculture will undergo further development and transformation. As technological advancements progress, it is inevitable that new risks and attack vectors will arise. In order to address these concerns, it is imperative to foster a joint endeavor including various stakeholders, such as farmers, technology suppliers, and governmental entities. Collectively, it is possible to cultivate and execute pioneering security measures in order to guarantee the enduring viability and triumph of intelligent agricultural systems.

References

1. Raghuvanshi, A., Singh, U. K., Sajja, G. S., Pallathadka, H., Asenso, E., Kamal, M., Singh, A., & Phasinam, K. (2022). Intrusion detection using machine learning for risk mitigation in IoT-enabled smart irrigation in smart farming. *Journal of Food Quality, 2022*, 1–8.

2. Quy, V. K., Nam, V. H., Linh, D. M., Ngoc, L. A., & Gwanggil, J. (2022). Wireless communication technologies for IoT in 5G: Vision, applications, and challenges. *Wireless Communications and Mobile Computing, 2022*, 3229294.
3. Sinche, S., Raposo, D., Armando, N., Rodrigues, A., Boavida, F., Pereira, V., & Silva, J. S. (2020). A survey of IoT management protocols and frameworks. *IEEE Communications Surveys Tutorials, 22*, 1168–1190.
4. Elijah, O., Rahman, T. A., Orikumhi, I., Leow, C. Y., & Hindia, M. N. (2018). An overview of internet of things (IoT) and data analytics in agriculture: Benefits and challenges. *IEEE Internet of Things Journal, 5*, 3758–3773.
5. Aazam, M., Harras, K. A., & Zeadally, S. (2020). Health fog for smart healthcare. *IEEE Consumer Electronics Magazine, 8*(2), 96–101.
6. Devi, R. D., Nandhini, S. A., Hemalatha, R., & Radha, S. (2019). IoT enabled efficient detection and classification of plant diseases for agricultural applications. In *2019 International conference on wireless communications signal processing and networking (WiSPNET)* (pp. 447–451). Chennai, India.
7. Monteiro, A., Santos, S., & Gonçalves, P. (2021). Precision agriculture for crop and livestock farming—Brief review. *Animals, 11*, 2345.
8. Shadrin, D., Menshchikov, A., Somov, A., Bornemann, G., Hauslage, J., & Fedorov, M. (2020). Enabling precision agriculture through embedded sensing with artificial intelligence. *IEEE Transactions on Instrumentation and Measurement, 69*, 4103–4113.
9. Glaroudis, D., Iossifides, A., & Chatzimisios, P. (2020). The survey, compare, and research challenges of IoT application protocols for smart farming. *Computer Networks, 168*, 107037.
10. Balasubramaniyan, M., & Navaneethan, C. (2021). Applications of Internet of Things for smart farming–A survey. *Materials Today: Proceedings, 47*, 18–24.
11. Bellini, P., Nesi, P., & Pantaleo, G. (2022). IoT-enabled smart cities: A review of concepts, frameworks, and key technologies. *Applied Sciences, 12*(3), 1607.
12. Kashyap, P. K., Kumar, S., Jaiswal, A., Prasad, M., & Gandomi, A. H. (2021). Towards precision agriculture: IoT-enabled intelligent irrigation systems using deep learning neural network. *IEEE Sensors Journal, 21*(16), 17479–17491.
13. Ahmed, S. H., & Zeebaree, S. (2021). A survey on security and privacy challenges in smart home-based IoT. *International Journal of Contemporary Architecture, 8*(2), 489–510.
14. Farooq, M. S., Sohail, O. O., Abid, A., & Rasheed, S. (2022). A survey on the role of IoT in agriculture for the implementation of smart livestock environment. *IEEE Access, 10*, 9483–9505.
15. Vangala, A., Sutrala, A. K., Das, A. K., & Jo, M. (2021). Smart contract-based blockchain-envisioned authentication scheme for smart farming. *IEEE Internet of Things Journal, 8*, 10792–10806.
16. Hassija, V., Chamola, V., Gupta, V., Jain, S., & Guizani, N. (2021). A survey on supply chain security: Application areas. *Security Threats, and Solution Architectures, IEEE Internet of Things Journal, 8*(8), 6222–6246.
17. Alfian, G., Syafrudin, M., Farooq, U., Ma'arif, M. R., Syaekhoni, M. A., Fitriyani, N. L., Lee, J., & Rhee, J. (2010). Improving the efficiency of RFID-based traceability system for perishable food by utilizing IoT sensors and machine learning model. *Food Control, 110*.
18. Wang, L., Xu, L., Zheng, Z., Liu, S., Li, X., Cao, L., Li, J., & Sun, C. (2021). Smart contract-based agricultural food supply chain traceability. *IEEE Access, 9*, 9296–9930.
19. Salah, K., Nizamuddin, N., Jayaraman, R., & Omar, M. (2019). Blockchain based soybean traceability in agricultural supply chain. *IEEE Access, 7*, 73295–73305.
20. Zheng, M., Zhang, S., Zhang, Y., & Hu, B. (2021). Construct food safety traceability system for people's health under the internet of things and big data. *IEEE Access, 9*, 70571–70583.
21. Dandage, K., Badia-Melis, R., & Ruiz-Garc´ıa, L. (2017). Indian perspective in food traceability: A review. *Food Control, 71*, 217–227.
22. Ding, L., Wu, J., Zhang, X., Li, J., & Ma, J. (2021). Privacy preserved cyber physical searching for information-centric intelligent agriculture. *IEEE Open Journal of the Computer Society, 2*, 106–116.

23. Wazid, M., Bera, B., Mitra, A., Das, A. K., & Ali, R. (2020). Private blockchain-envisioned security framework for AI-enabled IoTBased drone-aided healthcare services. In *2nd ACM MobiCom workshop on drone assisted wireless communications for 5g and beyond (DroneCom'20)*, pp. 37–42.
24. Vidyashree, L., & Suresha, B. M. (2019). Methodology to secure agricultural data in IOT. In *Emerging technologies in data mining and information security* (pp. 129–139). Springer Singapore, Singapore.
25. FIPS PUB 197. (2021). *Advanced encryption standard*. National Institute of Standards and Technology (NIST), U.S. Department of Commerce. Retrieved from June 2021, http://csrc.nist.gov/publications/fips/fips197/fips-197.pdf.
26. Chukkapalli, S. S. L., Piplai, A., Mittal, S., Gupta, M., & Joshi, A. (2020). A smart-farming ontology for attribute based access control. In *IEEE 6th intl conference on big data security on cloud (BigDataSecurity), ieee intl conference on high performance and smart computing, (HPSC) and IEEE Intl conference on intelligent data and security (IDS)* (pp. 29–34).
27. Friha, O., Ferrag, M. A., Shu, L., & Nafa, M. (2020). A robust security framework based on blockchain and SDN for fog computing enabled agricultural internet of things. In *International conference on internet of things and intelligent applications (ITIA)* (pp. 1–5).
28. Yousefi, S., Karimipour, H., & Derakhshan, F. (2021). Data aggregation mechanisms on the internet of things: A systematic literature review. *Internet of Things, 15*, 100427.
29. Song, J., Zhong, Q., Wang, W., Su, C., Tan, Z., & Liu, Y. (2020). FPDP: flexible privacy-preserving data publishing scheme for smart agriculture. *IEEE Sensors Journal* https://doi.org/10.1109/JSEN.2020.3017695.
30. Dasaklis, T. K., Casino, F., Patsakis, C. (2019). Defining granularity levels for supply chain traceability based on IoT and blockchain. In *International conference on omni-layer intelligent systems* (pp. 184–190). Crete, Greece.
31. Zhou, M., Zheng, Y., Guan, Y., Peng, L., & Lu, R. (2021). Efficient and privacy-preserving range- max query in fog-based agricultural IoT. *Peer-to-Peer Networking and Applications, 14*, 2156–2170.
32. Sontowski, S., Gupta, M., Laya Chukkapalli, S. S., Abdelsalam, M., Mittal, S., Joshi, A., & Sandhu, R. (2020). Cyber attacks on smart farming infrastructure. In *IEEE 6th international conference on collaboration and internet computing (CIC)* (pp. 135–143).
33. Ferrag, M. A., Shu, L., Djallel, H., Choo, K.-K. R. (2021). Deep learning based intrusion detection for distributed denial of service attack in agriculture 4.0. *Electronics, 10*(11)
34. Dolev, D., & Yao, A. (1983). On the security of public key protocols. *IEEE Transactions on Information Theory, 29*(2), 198–208.
35. Yang, X., Shu, L., Chen, J., Ferrag, M. A., Wu, J., Nurellari, E., & Huang, K. (2021). A survey on smart agriculture: Development modes, technologies, and security and privacy challenges. *IEEE/CAA Journal of Automatica Sinica, 8*(2), 273–302.
36. Ali, R., Pal, A. K., Kumari, S., Karuppiah, M., & Conti, M. (2018). A secure user authentication and key-agreement scheme using wireless sensor networks for agriculture monitoring. *Future Generation Computer Systems, 84*, 200–215.
37. Chen, M., Lee, T.-F., & Pan, J.-I. (2019). An enhanced lightweight dynamic pseudonym identity based authentication and key agreement scheme using wireless sensor networks for agriculture monitoring. *Sensors, 19*(5).
38. Chae, C.-J., & Cho, H.-J. (2018). Enhanced secure device authentication algorithm in P2P-based smart farm system. *Peer-to-peer networking and applications, 11*(6), 1230–1312.
39. Bothe, A., Bauer, J., & Aschenbruck, N. (2019). RFID-assisted continuous user authentication for IoT-based smart farming. In *IEEE international conference on RFID technology and applications (RFID-TA)* (pp 505–510).
40. Rangwani, D., Sadhukhan, D., Ray, S., Khan, M. K., & Dasgupta, M. (2021). An improved privacy-preserving remote user authentication scheme for agricultural wireless sensor network. *Transactions on Emerging Telecommunications Technologies, 32*(3), e4218. https://doi.org/10.1002/ett.4218

41. Jawad, H. M., Nordin, R., Gharghan, S. K., Jawad, A. M., Ismail, M. (2017). Energy-efficient wireless sensor networks for precision agriculture: A review. *Sensors, 17*(8).
42. Barh, A., & Balakrishnan, M. (2018). Smartphone applications: Role in agri-information dissemination. *Agricultural Review, 39*(1), 82–85.
43. Kamienski, C., Soininen, J.-P., Taumberger, M., Dantas, R., Toscano, A., Cinotti, T. S., et al. (2019). Smart water management platform: IoT-based precision irrigation for agriculture. *Sensors, 19*(2), 276.
45. Agrawal, H., Dhall, R., Iyer, K. S. S., Chetlapalli, V. (2019). An improved energy efficient system for IoT enabled precision agriculture. *Journal of Ambient Intelligence and Humanized Computing, 11*, 1–12.
46. Balaji, S., Nathani, K., & Santhakumar, R. (2019). IoT technology applications and challenges: A contemporary survey. *Wireless Personal Communications, 108*(1), 363–388.
47. Minh, Q. N., Nguyen, V. H., Quy, V. K., Ngoc, L. A., Chehri, A., & Jeon, G. (2022). Edge computing for IoT-enabled smart grid: The future of energy. *Energies, 15*(17), 6140.
48. Al-Turjman, F., Zahmatkesh, H., & Shahroze, R. (2020). Security in grid and IoT-enabled cities. In: *Smart grid in IoT-enabled spaces*, pp. 249–279. CRC Press.
49. Chaganti, R., Varadarajan, V., Gorantla, V. S., Gadekallu, T. R., & Ravi, V. (2022). Blockchain-based cloud-enabled security monitoring using the Internet of Things in smart agriculture. *Future Internet, 14*(9), 250.
50. Quy, V. K., et al. (2022). IoT-enabled smart agriculture: Architecture, Applications, and challenges. *Applied Sciences, 12*(7), 3396. https://doi.org/10.3390/app12073396
51. Bera, B., Vangala, A., Das, A. K., Lorenz, P., & Khan, M. K. (2022). Private blockchain-envisioned drones-assisted authentication scheme in IoT-enabled agricultural environment. *Computer Standards & Interfaces, 80*, 103567. https://doi.org/10.1016/j.csi.2021.103567
54. Zhang, F., Zhang, W., Luo, X., Zhang, Z., Lu, Y., & Wang, B. (2022). Developing an IoT-enabled cloud management platform for agricultural machinery equipped with automatic navigation systems. *Agriculture, 12*(2), 310. https://doi.org/10.3390/agriculture12020310
55. Kashyap, P. K., Kumar, S., Jaiswal, A., Prasad, M., & Gandomi, A. H. (2021). Towards precision agriculture: IoT-enabled intelligent irrigation systems using deep learning neural network. *IEEE Sensors Journal, 21*(16), 17479–17491. https://doi.org/10.1109/JSEN.2021.3069266
56. Amiri-Zarandi, M., Hazrati Fard, M., Yousefinaghani, S., Kaviani, M., & Dara, R. (2020). A platform approach to smart farm information processing. *Agriculture, 12*(6), 838. https://doi.org/10.3390/agriculture12060838.
57. Haseeb, K., Ud Din, I., Almogren, A., & Islam, N. (2020). An energy-efficient and secure IoT-based WSN framework: An application to smart agriculture. *Sensors, 20*(7), 2081.
58. Vangala, A., Das, A. K., Chamola, V., Korotaev, V., & Rodrigues, J. J. (2022). Security in IoT-enabled smart agriculture: architecture, security solutions, and challenges. *Cluster Computing*, 1–24.
59. Nayyar, A., & Puri, V. (2016). Smart farming: IoT based smart sensors agriculture stick for live temperature and moisture monitoring using Arduino, cloud computing & solar technology. In Proceedings of the international conference on communication and computing systems (ICCCS-2016) pp. 9781315364094–121.
60. Ferrag, M. A., Shu, L., Yang, X., Derhab, A., & Maglaras, L. (2020). Security and privacy for green IoT-based agriculture: Review, blockchain solutions, and challenges. *IEEE Access, 8*, 32031–32053.

Comprehensive Study of Intelligent Transference System in a Contemporary IoT Commuting Environment

Shaurya Gupta, Sonali Vyas, and Vinod Kumar Shukla

Abstract Nowadays sequence concerning roadside cases is aggregating rapidly throughout the world considerably, which in turn initiates the need to develop polices inculcating road fortification and directives. It stands as an enormous aim for the developed nations to augment the eminence of people's life via the progress of an intellectual system of transference. VANETs, which are Ad hoc networks, are known for their efficiency in the implementation of intelligent transference systems. It will enrich highway and street safety besides providing ease to commuters. However, these sorts of expertise are quite exposed to numerous security encounters, which require to be addressed the afore-secure disposition of VANET expertise to overall working systems. Vehicular Ad Hoc Networks converses an active and dynamic system because of the presence of indefinite vehicles on the road besides data broadcast amid automobiles plus roadside units. Resourceful message communication amid the automobiles and RSUs is quite beneficial for the Intelligent Transference System. The procedures or protocols for routing are responsible for routing data over the network. A wide-ranging assessment of varied clustering algorithms with suitable cluster head selection is being discussed for effective VANET implementation along with a discourse on appropriate simulators for effective VANET implementation. Lastly, the existing challenges and their implementation in developing nations are being deliberated.

Keywords VANET · MANET · V2V · V2I · NS-3 · OpenStreetMap · Clustering algorithm

S. Gupta · S. Vyas (✉)
School of Computer Science, UPES, Dehradun, Uttarakhand, India
e-mail: vyas.sonali86@gmail.com

S. Gupta
e-mail: shaurya55@gmail.com

V. K. Shukla
Department of Engineering and Architecture, Amity University, Dubai, UAE

A. Prasad et al. (eds.), *Communication Technologies and Security Challenges in IoT*, Internet of Things, https://doi.org/10.1007/978-981-97-0052-3_15

Introduction

VANET is a wireless network for communication that uses cars to build a mobile network as mobile nodes. VANET is a dynamic sys-tem, which was developed to provide roadside assistance to moving vehicles [1]. The prime concern of this technology is to restructure the safety of the vehicle in traffic. The vehicles are equipped with sensors, which are responsible for collecting information, and this in turn is broadcasted to other vehicles and RSUs (Road Side Units) according to the necessity of the information [2]. Mobile devices create a multi-hop path to transport data packets from base to terminus nodes. Many routing protocols have been introduced that facilitate communication in "ad-hoc networks". Several challenges in the VANETs need to be addressed for effective data communication in a wireless environment. The "vehicular ad-hoc network" is essentially centred on "mobile ad-hoc network", which includes "AODV (Ad-hoc on Demand Distance Vector Routing)" [3], "DSR (Dynamic Source Routing)" [4], "TORA (Temporally-Ordered Routing Algorithm)" [5], "OLSR (Optimized Link State Routing)" [6] whereas AODV and DSR accomplish multi-hop wireless ad-hoc networks very efficiently. The chief characteristics of VANET embrace higher node mobility, rapid access and pattern movement [7], which are assisted by setting up nodes, which provide geographical data, and communicate amongst themselves to assist the vehicular network which is being showcased in Fig. 1.

This chapter starts with a basic introduction, then focuses on the motivation for research, objectives, and research methodology, and in the end summarization of the chapter is done. "Mobile ad-hoc networks" are secondary groups of "ad hoc networks" with mobile nodes, whereas "vehicle ad hoc networks" are special classes of MANETs that are distinct and expressed as self-organized, distributed networks recognized by moving vehicle nodes. The high mobility and predictable mobility pattern differ from MANETs [8]. The driving pattern of vehicles is different in terms of inter and intra-city. The recurrent partitioning of the network and infrastructure less condition of networks caused by high-speed moving vehicles create a serious connectivity problem in VANET. The study's focus is on providing an effective and optimized routing algorithm for "ad-hoc vehicle delay tolerant networks." The idea of holding and forwarding the data [9] is supported by DTN. Some of the DTN-VANET applications are as follows:

Accident Information Dissemination: In case of an accident, which is causing a traffic jam on the highways, the vehicles approaching the point of occurrence must be informed in advance and they must change to alternate routes to avoid traffic jams. Intelligent Transportation Systems (ITS) [10] is an example of such a system, which uses the Internet to Vehicle (I2V) communication system.

Location-based Services: The utility store on the road can promote their products and notify people in vehicles on the way about current offers at their stores [11].

Parking Space Reservation: drivers can reserve the parking space in advance before reaching a certain destination. GPS receivers, and collision sensors in vehicles,

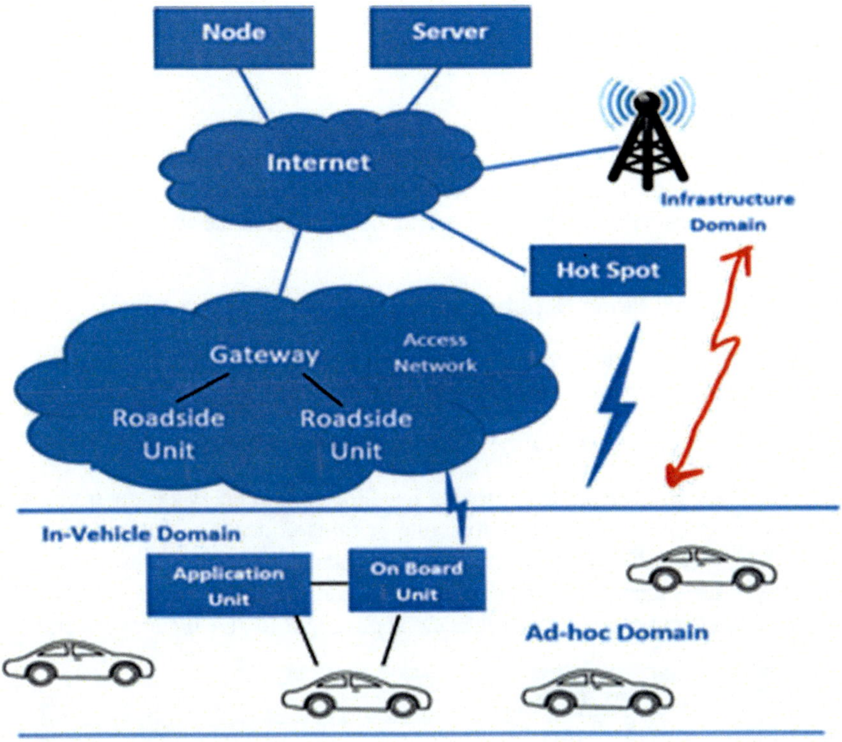

Fig. 1 Framework of VANET [1]

facilitate all these applications. There is a variety of applications in VANETs, which range from real-time communication, internet access to DTN applications [12].

Characteristics of VANET

- **Exceedingly Dynamic Topology**

It is quite critical when it comes to describing a precise formation for a VANETs setup as the nodes are in continual motion, and message communication between V2V and V2I takes place in a short period, which makes it quite difficult in describing the formation of such kinds of networks [13].

- **Inconstant Channels in Time and Frequency**

The message communication amongst vehicles is always in a state of heavily degrading reason being the variable speed of vehicles, which might be high, or low at times, besides the setting, which is facilitating the movement of vehicles, taking into consideration of certain obstacles to a wireless signal, which includes high constructions.

- **Sovereignty**

Each node of a network is quite independent when it comes to the selection of medium or transmission route in sending or receiving packets without any central control. The onboard unit beside the roadside unit is also independent in the above-said selection [14].

- **Unconstrained Power Supply**

The battery power is quite enough for supplying enough power to onboard the unit and authenticate users in a networking assignment [15].

- **Great Computational Ability**

Node on-board units are quite capable of handling extraordinary system traffic flow while considering cases of video streaming or cases of dealing with low data rates but with high priority, which is very much dependable on applications. Apart from that, they observe channeling and connection to roadside units, as devices necessitate computational power, specifically in acute applications to give higher performance.

Vanet Directing Protocols

There are quite numerous routing protocols, which are projected for competent message communication in VANET, and the prime emphasis routing protocols involve improving the enactment of VANET by minimalizing two factors i.e., delay and packet loss. A detailed description of the route protocol based on each category is provided below:

A. **Topology-Based Routing Protocol**

It is a supplementary sub separated into two categories, which are as follows:

- **Proactive Routing**: In this scheme of routing, all algorithms follow the smallest and the most optimized route [16]. The pertinent data, of nodes, are protected and pre-defined in the table and the routing table is reorganized as the network topology changes. These protocols ingest supplementary bandwidth, specifically in the areas of high traffic movement density, besides extraordinary overhead; therefore, it is not appropriate in the case of VANET. Some of such routing protocols are as follows [17]:

  a. DSDV: "Dynamic Destination Sequenced Distance Vector"
  b. FSR: "Fisheye State Routing"

- **Reactive Routing:** It includes on-demand direction-finding procedures or protocols and is called reactive routing protocol. As they are reverse in nature in terms of proactive protocol besides periodically, keep recommencing the routing table now and then. The significant purpose of such routing protocols involves generating plus upholding directions or routes of message communications [18]. The most prominent advantage of it is its adaptability to topology transformation in a highly dynamic VANET environment. In it, a flooding method determines the initial route in addition to encountering the overhead issue. Instances of reactive routing protocol include [19]:

    a. AODV: "Ad-Hoc On-Demand Distance Vector"
    b. TORA: "Temporary-Ordered Routing Algorithm"

Table 1 discusses the difference between AODV, DSDV and OLSR protocols amongst varied factors. AODV being a part of reactive routing protocol its benefit involves occupying less memory and avoiding generating traffic overheads during following an active connection. In it, the attacker might attempt to publish a path with the help of a minor interval metric which is quite different from the actual interval. A supplementary improvement version of AOVD proposed solved the above issues while making AODV more secure and protected apart from being fair in AODV data transmission with the help of multihop links [20]. Whereas, in DSR i.e. the dynamic state routing the data packets are accelerated via hop to hop mode. The act of re-moving any node from the list of reordering the arrangement of nodes results in potential hazards in terms of security. Figure 2 discourses cataloguing of routing protocols, which is completely centered on the implementation methodology.

- **Hybrid Routing:** It is an amalgamation of Proactive and Re-active routing protocols. It provides the benefit of a link-state routing communication system plus a vector-based route protocol instead of scalable and high overhead cost as compared to the route issue [21].

B. Cluster-based Routing Protocol

Table 1 Difference among "AODV" "DSDV" and "OLSR" Protocols

Factors of Protocol	AODV	DSDV	OLSR
Reactive	Applicable	Not applicable	Not applicable
Route retained	Routing table	Routing table	Routing table
QoS (Quality of service) support	Not applicable	Applicable	Not applicable
Multicast routes	Not applicable	Applicable	Not applicable
Distributed	Applicable	Applicable	Applicable
Unidirectional link	Not applicable	Applicable	Not applicable
Support multicast	Applicable	Applicable	Not applicable
Periodic broadcast	Applicable	Applicable	Applicable

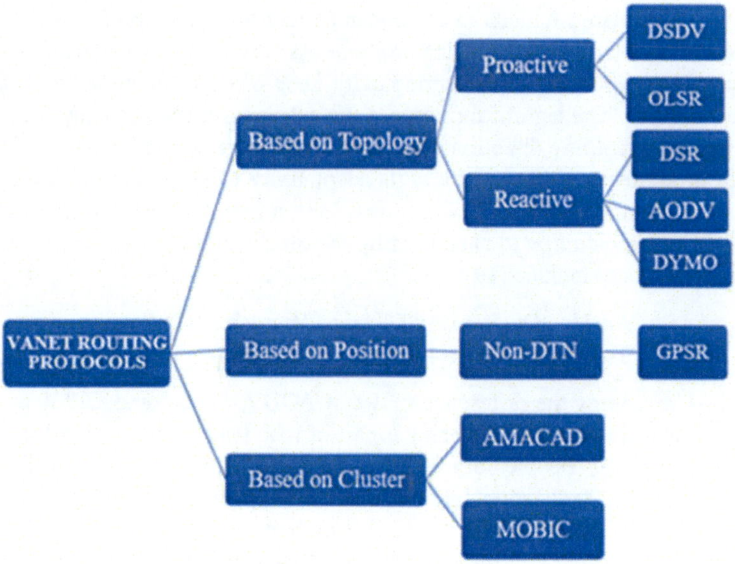

Fig. 2 Routing protocols of VANET [13]

Here, nodes are grouped all together in a cluster. The connection is initiated from a member cluster to a head cluster and thereafter from cluster heads to other cluster heads. In a cluster, similar-nature nodes are grouped under one cluster. Instances of cluster-based routing protocols are as follows:

- DMAC: "Distributed and mobility adaptive clustering algorithm".
- RMAC: "Robust mobility adaptive clustering".

The nodes, which are out of the range of the cluster head, are termed orphan nodes therefore, these kinds of nodes create a sub-cluster plus choose a cluster head which is constructed on the shorter distance amongst the entry nodes [22].

C. **Broadcast-based Routing Protocol**

It involves simultaneous data sharing on manifold nodes in VANET. The foremost restraint of it is that it works only on a smaller amount of nodes; as numerous nodes increase in a network the overall performance of the protocol drops. Some of the broadcast routing protocols are [23]:

- UMB: "Urban multi-hop broadcast protocol".
- PBSM: "Parameters broadcast in static to highly mobile".

D. **Position-based Routing Protocol**

The physical location of automobile nodes along with the neighbor nodes is evaluated with the help of the GPS position marking facility in the device and thereby facilitating position-based routing procedures. It doesn't require any direction-finding table [24]. This is a radical departure from conventional network architecture, and DTN researchers are actively designing and implementing protocols to enable message routing through networks that lack end-to-end routes. The performance of DTN routing and application protocols is vastly reliant on the primary mobility and node characteristics of a network. The primary concept of DTN is to spread da-ta/ messages from the basis to the terminus node without having comprehensive network connectivity. In this kind of network, each node stores the messages it receives from other nodes and forwards them to the next node as the connection is established. DTNs are acknowledged as Opportunistic Networks because the nodes connect and transfer messages as soon as they get the opportunity to connect [25]. Examples of non-DTN routing protocols are as follows:

- LAR: "Location Aided Routing".
- MORA: "Movement-based routing algorithm".
- VADD: "Vehicle Assisted Data Delivery".
- MaxProp.

E. **Geocast-based routing protocol:** In it, the info is broadcasted only to automobiles, which are under a definite topographical zone called Zone of Relevance (ZOR). ZOR ignores and drops all messages, which fall, outside its zonal area. Numerous kinds of Geocast routing protocols are offered by WSN. It keeps into consideration the following parameters like geometric, trajectory-based, and path-sharing [26]. Following are the examples of Geocast based routing protocol:

    - Inter-Vehicle Geocast.
    - Distributed Robust Geocast.

Routing Simulators

When considering the case of ITS which talks about numerous components clubbed together and spread over a network [27]. Taking an instance, travel statistics and transference can be grouped under a smart municipal plan. Vehicular ad-hoc networks are a sophisticated version of ITS [28]. Network Simulator stands as an open-source network simulator, which is mainly applied in the education or research field. The network presents a framework for diverse simulations; which highlights how replicas of computer network function. Figure 3 discourses the basic model of the network simulator.

Urban Mobility Simulation is an open-source, simulator that is quite accessible to every end user for modifications [29]. Apart from that, it embraces simulation packages involving modelling constituents. Dissimilar exploration areas, which can be studied using SUMO, are as follows:

- Algorithm of Traffic Light
- Route Selections

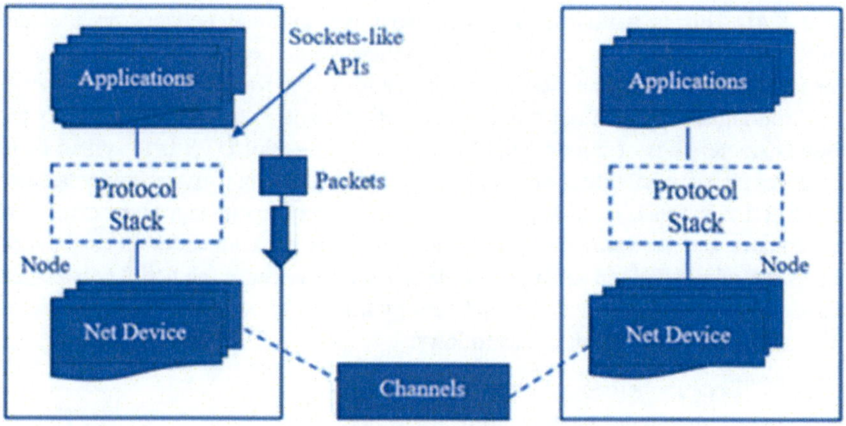

Fig. 3 Model of network simulator [29]

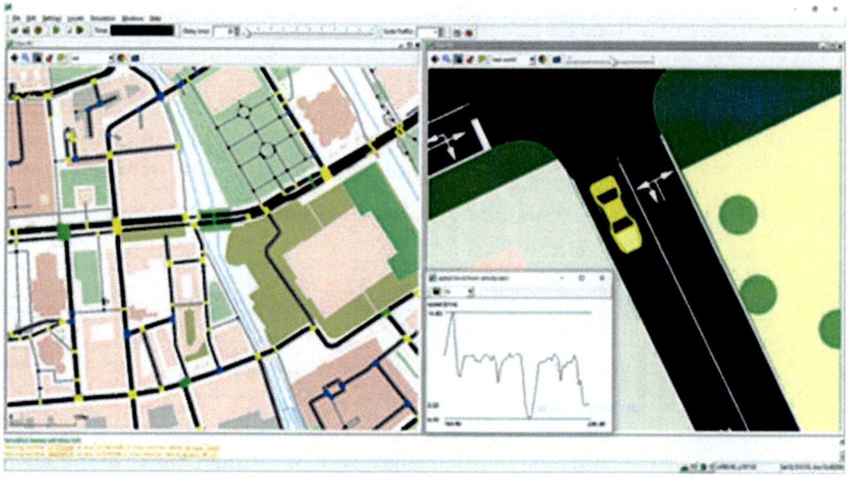

Fig. 4 Simulation of urban mobility [30]

- Vehicular interaction simulation.

SUMO simulations are centered on discrete time and space continuous vehicular motions.

GNU plot is a software based on command-line statements, which produces 2D, and 3D plotting of functions, data, and data fits and executions on all computer systems [30]. Figures 4 and 5 have showcased the simulation modes of SUMO and GNU plot simulation respectively. Moreover, the routing algorithm entails a strategy ensuring an association amongst pair of network nodes involving which are linked

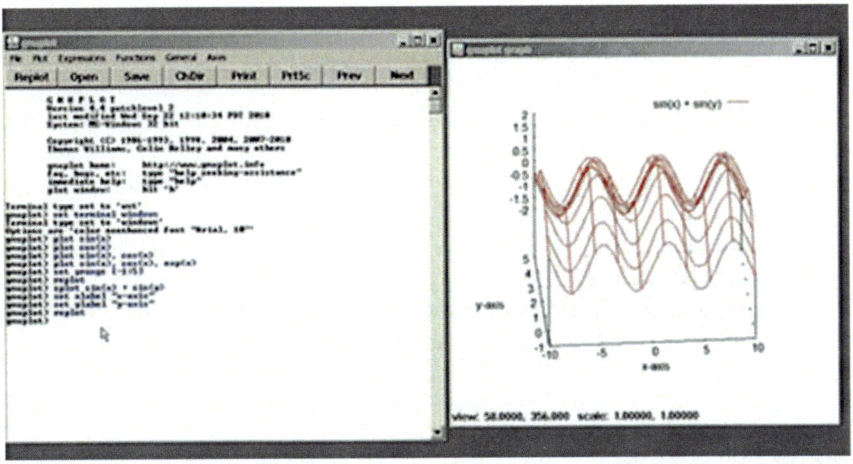

Fig. 5 GNU Plot Simulation [30]

with a network at any moment. Numerous factors, which are taken into consideration, involve:

- Numerous bonds
- Bandwidth
- Energy Constraints

The foremost aim of the study involved assessing routing protocols based on topology in the VANET area using numerous tools.

Recent Trends of Routing Protocol Simulation

Contemporarily, there are various articles and research in the VANET domain, but the maximum of works discusses algorithm strategy in addition to the discussion which is briefly regarding the simulation procedure. Authors in [31] deliberate regarding the DSDV routing protocol, which is completely created on deep learning, i.e. GD-DSDV. It discusses simulation tools, and functions but is not discussing the simulation execution procedure, here authors have premeditated a wireless Mesh network DSDV routing algorithm that supports QoS. In [32] authors have expansively associated and analyzed the presentation of each direction-finding procedure. Considering a VANET scenario, it is very challenging for scholars to choose which simulation software to be used in which scenario. In [33] authors, have discussed OPNET Modeler 14.5 which was used in simulating and improving the performance of AODV, whereas in [34] authors have deliberated about the NS-3 platform which is used in simulating cross-layer routing protocol in a metropolitan scenario. Supplementary simulation software, which was conversed in additional literature, comprised "QualNet, NS-2, loMosim" etc. Many difficulties in combining multiple software routing simulations manually whereas including the following factors:

Design of Vehicle Mobility: Node flexibility stands as the key point in VANET. The movement of the Vehicle considers the macroscopic lane topology limit, along with the characteristics of microscopic vehicle movement. For making the simulation extremely close to reality, a particular vehicle movement model generation tool is required for efficient implementation.

Simulation Process Visualization: For viewing the process of simulation in an additionally specified manner, importing varied visual tools is also necessary. Following is the list of differences between Internet of things (IOT) and traditional network safety issues is provided under this section.

Simulation Strategy

Ns-3 is a network simulation framework, built on an inclusive deliberation of simulation performance, cost, development effort and elasticity and utilized in the education and research areas, in which the communication method in a huge network area [35]. In comparison to the existing simulation software packages like OPENET, the NS-3 possess subsequent characteristics:

- Open-source and Simple
- Flexible
- Better performance
- Modular-approach

It is easy to utilize and provides a facility for micro road traffic simulation. At the macro road traffic level, speed limit, lanes, traffic lights etc. are taken into consideration, while at the micro-level, the car-drive model is considered. In vehicle motion simulation the SUMO generates data which helps in the simulation of vehicle turning, speed, overtaking, maintaining safe distance and other factors. The overall design process is standard and can be utilized for advanced algorithm verification, current algorithm assessments, and high-speed group. Initially, researchers need to incorporate a routing algorithm, and then make use of SUMO for simulating the mobility of vehicle traffic. After that, the simulation function is evaluated for setting simulation parameters. Subsequently, the data by the NS-3 simulator is generated utilizing visual tools. Lastly, AWK text is scripted for processing and analyzing the simulation log and its performance. To get close to the actual movement section, the car movement model ought to contain parameters like road lanes, joints, crossings, traffic lights, car initial point, speeding limit, etc.

IoT in VANET

With time the concept of the Internet of Things (IoT) has been introduced in vehicular systems and termed as Internet of Vehicles. It is responsible for using real-time or dynamic data communications amongst vehicles to better manage the vehicular traffic management system. Though the concept of implementing IoT in VANET is still in the implementation phase and uses the core concepts of wireless computing which itself is based on the platform of edge and fog computing. The concept of Intelligent Transportation System has significantly improved transportation, especially in urban and freeway contexts and it will be implemented more efficiently and effectively in future with the concept implementation of automotive vehicles and infrastructures. IoT networks require high availability all the time, adding some risk in the transference of data amongst varied vehicles. The acceptance of IoT in terms of vehicular networks involves the following:

- Engineering Framework
- Information Managing
- Ubiquitous Advancements in Registration

As VANETS was developed to facilitate communication amongst vehicular networks and roadside units but apart from that they are simplifying varied applications like checking traffic conditions, vehicle diagnostics and crash evasion. Figure 6 discourses varied parameters of inter-vehicular communication with associated IoT components and clubbing them altogether. When considering the concept of inter-vehicular communication and vehicle communication with IoT-enabled road sides unit all the factors like precise navigation, vehicle condition monitoring and emergency services etc. can be taken into consideration.

Figure 7 discusses varied sorts of communication amongst intra-vehicular communication covering the aspects like vehicle to vehicle communication, vehicle to infrastructure communication, vehicle to server communication, vehicle to sensor communication, vehicle to cloud server communication etc. Here also IoT plays a very crucial role in facilitating communication amongst varied components of communication considering intravehicular communication. Though internet connectivity in sparse and far-flung areas can still be a challenge for overall efficient and effective vehicular communication. Many applications are being developed for facilitating communications amongst vehicles to nearby roadside units, vehicle-to-cloud servers, vehicle to personal devices for the overall augmentation of traffic management systems.

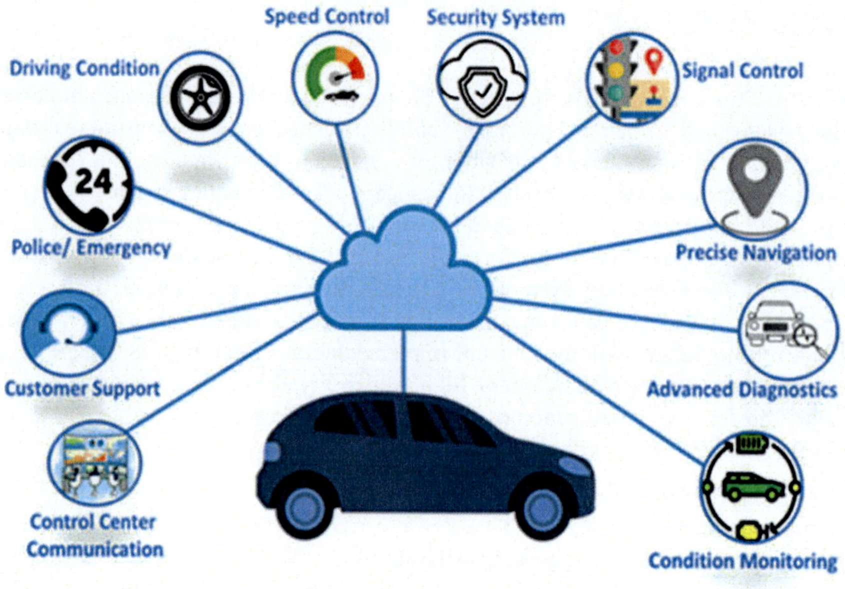

Fig. 6 IoT supportive factors in inter-vehicular communication [42]

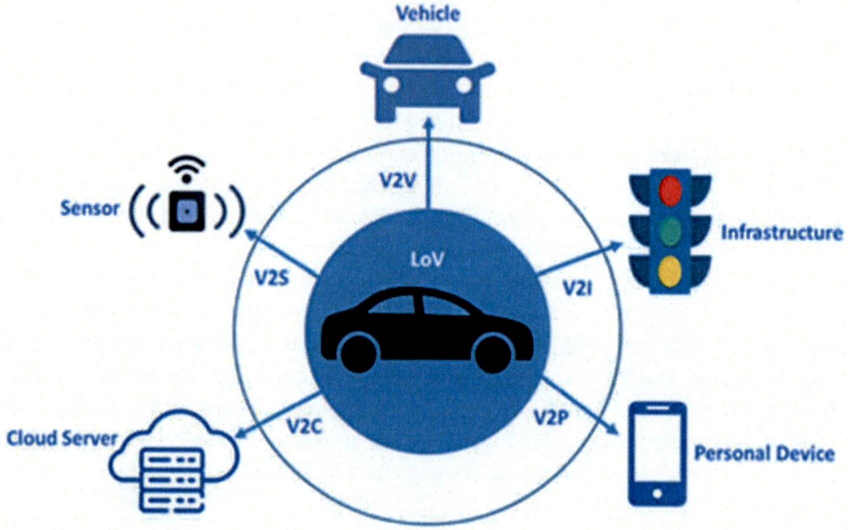

Fig. 7 IoT supportive factors in intra-vehicular communication [42]

Vanet Security Challenges

The matter of security is really important for VANETs that ensure the safety of drivers and passengers. This is the responsibility of designing key algorithms to ensure security. Challenges related to safety as set out in VANETs are accessibility, validation, veracity, privacy, denial, movement, data and location authentication [36].

Some issues are discussed below:

1. **Accessibility:** It is considered an important issue in the protection of VANETs. This certifies that all resources are permanently accessible to the network because of the risk and denial-of-service attack challenges. Cryptographic algorithms founded on trustworthiness and regulations help shield VANETs from such attacks [37].
2. **Authentication:** It allows the valid participants to log into the network later double verification. It also confirms that the contributor or the sender of the message is not a criminal. Alternatively, user privacy is maintained using pseudonyms.
3. **Data Integrity:** It guarantees that no modification is there in the actual packets of data delivered by the source. Otherwise, it is required to be safeguarded from intruders in all communication pathways. The digital signature, public key structure, and encryption method can be used to certify integrity with the sender and receiver [38].
4. **Confidentiality:** It means hiding information from intruders. In this, it is made sure that only authorized users have entree to data using cryptography. By this method, the data will stay private, while other illegal users may not be able to access the information [39].
5. **Pseudonymity**: It means hiding your true identity. The legitimate participants can use pseudonyms as a substitute for using real identities. In this way, legal entities can connect anonymously without disclosing their true identity. This guarantees privacy protection to subscribers [40].
6. **Privacy**: For VANETs, privacy means hiding driver ID and location data from other illegal users on the network.
7. **Scalability:** It means network capability in responding to changing needs. The ever-varying topology of the vehicle network is another big issue for researchers.
8. **Mobility:** It is everything on VANETs as the nodes that connect to VANTEs modify their position very rapidly and often in a network. The nature of VANETs changes with each second, and the node location changes. The mobility feature emphasizes the requirement for highly secure and flexible algorithms that maintain the quality-of-service necessities [41].
9. **Data Authentication**: It is utilized to delete malevolent messages from the network. It confirms checking the accuracy of the data and guarantees the validity of the active nodes [42].
10. **Key management**: In this, a key is used in cryptographic mechanisms while establishing connections between nodes. Key management and issuance are fixed all through the design of network security agreements.

Conclusion and Future Scope

A detailed study to understand the effectiveness of an Intelligent Transference System in a Contemporary Commuting Environment is discussed in the chapter. The various performance parameters in the network are closely monitored and analyzed with the existing benchmarking algorithms. The design and maintenance of the collection provide excellent communication to VANET and improve network stability besides choosing a headset can improve network communication performance. Nevertheless, as there is a change in technology and VANETs become added prevalent, security vulnerabilities increase rapidly, which ultimately limits the extensive use of VANETs. In this chapter, the vulnerability of VANET safety is examined. This also delivers specific layer attack categories in the stack of the VANETS protocol. Apart from that, also the chapter discusses a few countermeasures or issues which can be useful for future research and studies. One of the future challenges in the case of VANET is intermittent connectivity because of which the management and control of vehicles becomes quite difficult and there is a lot of scope in the future to optimize the intermittent connectivity amongst vehicles for efficient and effective network traffic management.

References

1. Ashokkumar, K., Baron, S., & Arshadprabhu, R. (2015). Cloud based intelligent transport system. *Procedia Computer Science, 50*, 58–63.
2. Shamsoshoara, A., & Darmani, Y. (2015). Enhanced multi-route ad hoc on-demand distance vector routing. In *2015 23rd Iranian conference on electrical engineering*. IEEE.
3. Sharma, R., & Sabharwal, S. (2013). Dynamic source routing protocol (dsr). *International Journal of Advanced Research in Computer Science and Software Engineering, 3*, 7
4. Thiagarajan, R., & Moorthi, M. (2017). Efficient routing protocols for mobile ad hoc network. In *2017 Third international conference on advances in electrical, electronics, information, communication and bio-informatics (AEEICB)*. IEEE
5. Younis, Z. A., Abdulazeez, A. M., Zeebaree, S. R. M., Zebari, R. R., & Zeebaree, D. Q. (2021). Mobile Ad Hoc network in disaster area network scenario: a review on routing protocols. *International Journal of Online and Biomedical Engineering, 17*, 3
6. Zhang, D., Zhang, T., Dong, Y., Liu, X., & Cui, Y. (2018). Novel optimized link state routing protocol based on quantum genetic strategy for mobile learning. *Journal of Network and Computer Applications, 122*, 37–49.
7. Wahid, I., Ikram, A. A., Ahmad, M., Ali, S., Ali, A. (2018). State of the art routing protocols in VANETs: A review. *Procedia Computer Science, 130*, 689–694.
8. Ahamed, A., & Hamid, V. (2018). Issues and challenges in VANET routing protocols. In: *2018 IEEE international conference on electro/information technology (EIT)*. IEEE
9. Boussoufa-Lahlah, S., Semchedine, F., & Bouallouche-Medjkoune, L. (2018). Geographic routing protocols for Vehicular Ad hoc Networks (VANETs): A survey. *Vehicular communications, 11*, 20–31.
10. Hamdi, M. M., Audah, L., Rashid, S. A., Mohammed, A. H., Alani, S., Mustafa, A. S. (2020). A review of applications, characteristics and challenges in vehicular ad hoc networks (VANETs). In *2020 international congress on human-computer interaction, optimization and robotic applications (HORA)*. IEEE.

11. Khelifi, H., Luo, S., Nour, B., Moungla, H., Faheem, Y., Hussain, R., & Ksentini, A. (2019). Named data networking in vehicular ad hoc networks: State-of-the-art and challenges. *IEEE Communications Surveys and Tutorials, 22*(1), 320–351.
12. Menon, V. (2019). Optimized opportunistic routing in highly dynamic ad hoc networks.
13. Rizwan Ghori, M., Safa Sadiq, A., & Ghani, A. (2018). VANET routing protocols: review, implementation and analysis. *Journal of Physics: Conference Series, 1049*. IOP Publishing.
14. Yasser, A., Zorkany, M., & Abdel Kader, N. (2017). VANET routing protocol for V2V implementation: A suitable solution for developing countries. *Cogent Engineering, 4.1*, 1362802
15. Jerbi, W., Guermazi, A., & Trabelsi, H. (2015). A clustering protocol for maximum coverage in large-scale wireless sensor networks. *International Journal of Business Data Communications and Networking (IJBDCN), 11*(2), 1–21.
16. Allani, S. (2018). *Data dissemination and aggregation in vehicular Adhoc network*. Diss. Université Pau et des Pays de l'Adour.
17. Liu, Y. (2021). Vanet routing protocol simulation research based on ns-3 and sumo. In: *2021 IEEE 4th International Conference on Electronics Technology (ICET)*. IEEE.
18. Houaidia, C., Idoudi, H., Van den Bossche, A. (2017). Inter-flow and intra-flow interference mitigation routing in wireless mesh networks. *Computer Networks, 120*, 141–156.
19. Kumareshan, N., et al. (2019). Performance analysis of various routing protocols for VANET environments. *International Journal of Innovative Technology and Exploring Engineering, 8*, 4381–4384.
20. Zhang, W., Xiao, X., Wang, J., & Lu, P. (2018). An improved AODV routing protocol based on social relationship mining for VANET. In *Proceedings of the 4th international conference on communication and information processing*.
21. Gawas, M. A., & Govekar, S. (2021). State-of-art and open issues of cross-layer design and QOS routing in internet of vehicles. *Wireless Personal Communications, 116*(3), 2261–2297.
22. Ahmed, A. A., & Alzahrani, A. A. (2019). A comprehensive survey on handover management for vehicular ad hoc network based on 5G mobile networks technology. *Transactions on Emerging Telecommunications Technologies, 30*(3), e3546.
23. Amina, B., & Elboukhari, M. (2018). Performance evaluation of VANETs routing protocols using SUMO and NS3. In *2018 IEEE 5th international congress on information science and technology (CiSt)*. IEEE.
24. Maygua-Marcillo, L., Luis, U. -A., & Paredes, M. P. (2019). Creating a new model in NS3 network simulator.
25. Yao, Y., Chang, X., Mišić, J., Mišić, V. B., & Li, L. (2019). BLA: Blockchain-assisted lightweight anonymous authentication for distributed vehicular fog services. *IEEE Internet of Things Journal, 6.2*, 3775–3784.
26. Azees, M., Pandi, V., & Deborah, L. J. (2016). Comprehensive survey on security services in vehicular ad-hoc networks. *IET Intelligent Transport Systems, 10*(6), 379–388
27. Hamdi, M. M., Audah, L., Abood, M. S., & Rashid, S. A. (2021). A review on various security attacks in vehicular ad hoc networks. *Bulletin of Electrical Engineering and Informatics, 10*(5), 2627–2635
28. Zhang, L., Wu, Q., Domingo-Ferrer, J., Qin, B., & Hu, C. (2016). Distributed aggregate privacy-preserving authentication in VANETs. *IEEE Transactions on Intelligent Transportation Systems, 18*(3), 516–526.
29. Ye, X., Xu, G., Cheng, X., Li, Y., & Qin, Z. (2021). Certificateless-based anonymous authentication and aggregate signature scheme for vehicular ad hoc networks. *Wireless Communications and Mobile Computing*, 1–16.
30. Janert, P. K. (2016). Gnuplot in action: understanding data with graphs. Simon and Schuster.
31. Anupriya, V., Sangavi, S., Shivani, P. C., & Dharany, P. V. (2020). An efficient approach for vehicle traffic monitoring by collaborating vehicular mobility module and network simulator 3. In 2020 4th *International conference on intelligent computing and control systems (ICICCS)*. IEEE.

32. Goswami, B., Wilson, S., Asadollahi, S., & Manuel, T. (2020). Data visualization: Experiment to impose ddos attack and its recovery on software-defined networks. *Data Visualization: Trends and Challenges Toward Multidisciplinary Perception* 141–160.
33. Sinwar, D., Sinwar, D., Sharma, N., Maakar, S. K., & Kumar, S. (2020). Analysis and comparison of ant colony optimization algorithm with DSDV, AODV, and AOMDV based on shortest path in MANET. *Journal of Information and Optimization Sciences, 41*(2), 621–632.
34. Kurniawan, A., Prima K., & Hadi, M. Z. S. (2020). Performance analysis of routing protocols AODV, OLSR and DSDV on MANET using NS3. 2020 international electronics symposium (IES). IEEE.
35. Arega, K. L., Raga, G., & Bareto, R. (2020). Survey on performance analysis of AODV, DSR and DSDV in MANET. *Computer Engineering and Intelligent Systems 11*(3), 23–32.
36. Shaban, A. M., Kurnaz, S., & Shantaf, A. M. (2020). Evaluation DSDV, AODV and OLSR routing protocols in real live by using SUMO with NS3 simulation in VANET. In *2020 International Congress on Human-Computer Interaction, Optimization and Robotic Applications (HORA)*. IEEE.
37. Bhatia, J., Dave, R., Bhayani, H., Tanwar, S., & Nayyar, A., (2020). SDN-based real-time urban traffic analysis in VANET environment. *Computer Communications, 149*, 162–175.
38. Al-Heety, O. S., Zakaria, Z., Ismail, M., Shakir, M. M., Alani, S., & Alsariera, H. (2020). A comprehensive survey: benefits, services, recent works, challenges, security, and use cases for sdn-vanet. *IEEE Access, 8*, 91028–91047.
39. Guo, R., Xu, L., Li, X., Zhang, Y., & Li, X. (2022). An efficient certificateless ring signcryption scheme with conditional privacy-preserving in VANETs. *Journal of Systems Architecture, 129*, 102633.
40. Al-Shareeda, M. A., Anbar, M., & Manickam, S., Yassin, A. A. (2020). Vppcs: Vanet-based privacy-preserving communication scheme. *IEEE Access, 8*, 150914–150928.
41. Grover, J. (2022). Security of vehicular Ad Hoc networks using blockchain: A comprehensive review. *Vehicular Communications, 34*, 100458.
42. El Zouka, H. A. (2022). An efficient and secure vehicular networks based on IoT and cloud computing. *SN Computer Science, 3*(3), 240.

Redefining Urban Development: Technology and Automation in India

Himam Saheb Shaik, Battula Harshini, and Vonteri Sanjana Reddy

Abstract Urban development plays a crucial role in shaping the future of any nation. With the rapid growth of population and urbanization, it has become imperative to redefine traditional approaches to urban development by embracing technology and automation. This chapter explores the significance of technology and automation in transforming urban development in India. It highlights the key areas where technology and automation can revolutionize urban planning, infrastructure, governance, and sustainability. Additionally, the challenges and opportunities associated with implementing technology-driven solutions are discussed, along with examples of successful initiatives in Indian cities. The chapter concludes by emphasizing the need for a comprehensive strategy that leverages technology and automation to create smart, sustainable, and inclusive urban spaces in India.

Keywords Technology · Automation · Urban development · India · Smart cities · Data analytics · Sustainable development · Geospatial technology · E-governance · Renewable energy integration

Introduction

Urban development is a critical aspect of a nation's progress, particularly in countries experiencing rapid urbanization like India. With a significant portion of the population migrating to cities, there is an urgent need to redefine conventional approaches

H. S. Shaik (✉)
Department of Mechanical Engineering, Malla Reddy Engineering College for Women, Hyderabad, India
e-mail: himam.mech@gmail.com

B. Harshini · V. S. Reddy
Faculty of Science and Technology, The ICFAI Foundation for Higher Education, Hyderabad, India
e-mail: battula.harshini19@ifheindia.org

V. S. Reddy
e-mail: vonteri.sanjanareddy19@ifheindia.org

A. Prasad et al. (eds.), *Communication Technologies and Security Challenges in IoT*, Internet of Things, https://doi.org/10.1007/978-981-97-0052-3_16

to urban development. This redefinition involves embracing technology and automation to create smart, sustainable, and inclusive urban spaces. By leveraging technological advancements, cities can overcome challenges such as overpopulation, inadequate infrastructure, resource scarcity, and environmental degradation. India is witnessing an unprecedented wave of urbanization because of the vast opportunities and improved living facilities available in the urban areas attracting the rural population to city. This rapid urbanization has put immense pressure on existing infrastructure, leading to issues such as overcrowding, inadequate housing, traffic congestion, and strain on essential services like water supply and waste management. Traditional methods of urban development struggle to keep pace with this exponential growth, necessitating a paradigm shift towards technology-driven solutions.

Problem Statement

The conventional methods of urban development in India are insufficient to address the challenges posed by rapid urbanization. Inefficient planning, substandard infrastructure, and limited resources hinder the provision of quality services to urban residents. Additionally, the lack of technology integration and automation leads to inefficiencies in governance, service delivery, and resource management. Therefore, there is a pressing need to redefine urban development by harnessing the potential of technology and automation to create sustainable, smart, and resilient cities.

Objectives of the Chapter

This chapter main aim to explore the following

- To explore the role of technology and automation in redefining urban development in India.
- To highlight the potential of technology-driven solutions in addressing urban challenges such as infrastructure, governance, and sustainability.
- To discuss the challenges and opportunities associated with the implementation of technology and automation in urban development.
- To provide examples and case studies of successful initiatives that have utilized technology and automation to transform urban spaces in India.
- To propose recommendation to policymakers, urban planners, and bureaucrats to effectively integrate technology and automation in urban development strategies. By achieving these objectives, this chapter aims to contribute to the discourse on redefining urban development in India and inspire the adoption of technology and automation as catalysts for sustainable and inclusive urban growth.

Technology and Automation in Urban Planning

Urban planning plays a pivotal role in shaping cities and ensuring their sustainable development. By integrating technology and automation into the urban planning process, cities can enhance efficiency, accuracy, and effectiveness in decision-making, leading to better outcomes for residents and the environment. The following subtopics delve into different aspects of technology and automation in urban planning.

Geospatial Technology and GIS Mapping

Geospatial technology, including Geographic Information System (GIS) mapping, offers a powerful tool for urban planners. GIS collect the spatial data, that data can be analyzed for better outputs, and visualization of spatial data, enabling planners to understand the existing urban landscape and make informed decisions. GIS mapping provides valuable insights into land use patterns, infrastructure networks, transportation systems, and environmental features. It facilitates the identification of suitable locations for new developments, the assessment of potential risks and vulnerabilities, and the optimization of resource allocation. By leveraging geospatial technology, urban planners can create comprehensive and data-driven strategies for urban development [1].

Smart City Concepts and Applications

The idea of smart-cities has gained prominence as a transformative approach to urban development. Smart cities harness expertise and automation to improve the excellence of life for residents while promoting sustainability and efficiency. Various applications of smart city technology include smart mobility systems, intelligent energy management, integrated data platforms, and digital governance solutions. These technologies enable cities to monitor and manage infrastructure in real-time, respond proactively to challenges, and provide personalized services to residents. By adopting smart city concepts and applications, urban planners can create livable, connected, and future-ready urban environments.

Exploratory Data Analysis (EDA) and Modelling for Prediction

Data predictive modeling have become indispensable tools for urban planners. By comparing high volumes of data, including demographic information, environmental factors, and socio-economic indicators, planners can gain useful insights into urban trends, anticipate future demands, and optimize resource allocation. Data analytics helps identify patterns, correlations, and anomalies, enabling evidence-based decision-making. Predictive modeling techniques, such as machine learning algorithms, can forecast future scenarios, simulate urban development patterns, and evaluate the potential impact of different policies. By harnessing the value of data predictive modeling, urban planners can make informed choices and develop sustainable and resilient cities [2, 3].

Metaverse in Urban Design

Metaverse is a major combination of AR, VR, MR and XR. The VR and AR technologies offer innovative ways to visualize and experience urban design concepts. VR allows stakeholders to immerse themselves in virtual environments, providing a realistic simulation of future urban spaces. This enables urban planners to evaluate design options, test pedestrian flow, assess the impact of buildings on the surrounding area, and gather feedback from residents. Augmented Reality overlays digital information onto the real world, providing interactive and contextual information about urban spaces. AR can assist in urban design presentations, public consultations, and decision-making processes, enhancing transparency and inclusivity. The Fig. 1 describes the use of metaverse technologies for safe navigation. By integrating VR and AR into urban design practices, planners can create more engaging, participatory, and user-centric urban spaces [4].

By embracing geospatial technology, smart city concepts, data analytics, and virtual/augmented reality, urban planners can leverage technology and automation to optimize urban planning processes, enhance decision-making, and create sustainable and people-centric cities. These tools empower planners to tackle complex urban challenges, anticipate future needs, and design urban spaces that meet the diverse needs of residents.

Transforming Urban Infrastructure

Urban infrastructure forms the backbone of cities, encompassing transportation networks, energy systems, and waste management. By harnessing technology and automation, cities can transform their infrastructure to be more intelligent, efficient,

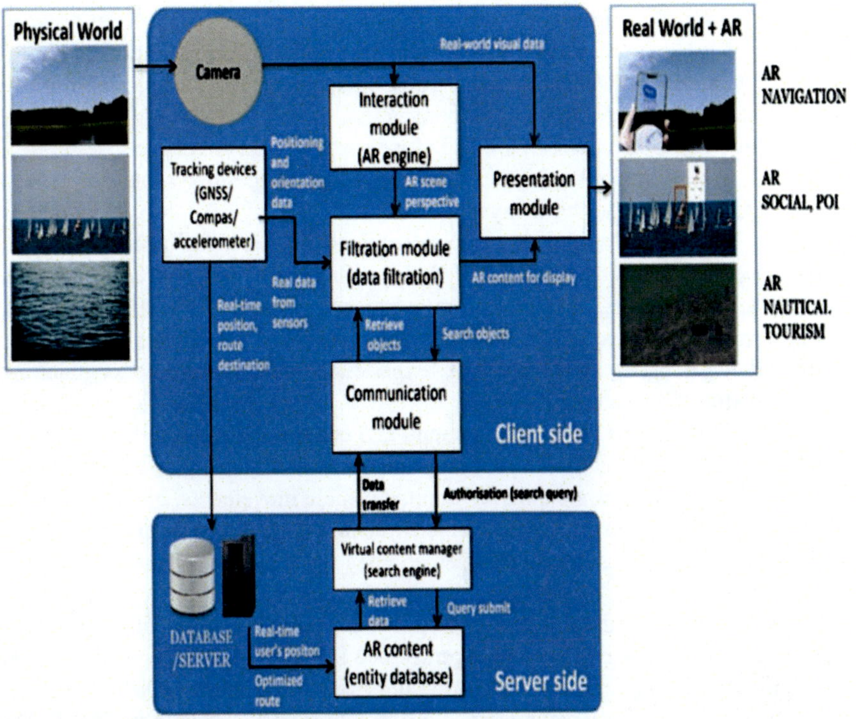

Fig. 1 Integrating AR and VR in safe navigation

and sustainable. The following subtopics explore the application of technology and automation in transforming urban infrastructure.

Intelligent Transportation Systems(ITS)

It leverage technology and automation to optimize the movement of people and goods within urban areas. ITS encompasses a range of solutions, including traffic management systems, smart parking, real-time public transportation information, and intelligent signaling. These systems use sensors, cameras, and data analytics to monitor traffic patterns, alleviate congestion, and enhance safety. ITS enables dynamic traffic management, adaptive signal control, and smart routing, resulting in reduced travel times, improved air quality, and enhanced overall transportation efficiency. By adopting ITS, cities can create seamless and sustainable transportation networks that enhance mobility and reduce environmental impact. The Key points of ITS in India as follows.

Traffic Management: ITS technologies are used to monitor and manage traffic flow, improve signal timings, and optimize the utilization of road infrastructure.

Intelligent Traffic Management Systems (ITMS) use real-time data from sensors, cameras, and other sources to provide information about traffic conditions, congestion, and incidents. This helps in better traffic management and enables authorities to take proactive measures to mitigate congestion.

Electronic Toll Collection (ETC): To reduce congestion at toll plazas and improve the overall toll collection process, Electronic Toll Collection systems have been implemented across major highways in India.

Public Transport Management: ITS is utilized in managing public transport systems such as buses and metros. It includes technologies like Automatic Vehicle Location (AVL) systems, fare collection systems, passenger information systems, and fleet management systems. These systems improve the efficiency of public transport operations, enhance passenger experience, and facilitate better planning and monitoring of services.

Vehicle Tracking and Fleet Management: ITS enables real-time tracking of vehicles through GPS technology. This is particularly useful for fleet management, logistics, and transportation companies to monitor vehicle movements, optimize routes, and ensure timely delivery of goods. Vehicle tracking systems also help in improving security, reducing theft, and providing emergency assistance in case of breakdowns or accidents.

Road Safety and Surveillance: ITS plays a crucial role in enhancing road safety through technologies like Intelligent Video Surveillance (IVS) systems, speed enforcement systems, and automated number plate recognition (ANPR) systems. These technologies help in monitoring and detecting traffic violations, identifying stolen vehicles, and ensuring compliance with traffic rules.

Information and Communication Systems: ITS facilitates the dissemination of real-time information to travelers through various channels, including Variable Message Signs (VMS), mobile applications, and online platforms. This includes providing updates on traffic conditions, route guidance, public transport schedules, and emergency alerts.

Smart Grids and Energy Management

These technologies enable efficient monitoring, and coordination of electricity generation, distribution, and consumption. Here's an overview of smart grids and energy management. Smart grid technologies enable continuous monitoring of the electrical grid's condition, including voltage levels, power quality, and line losses. This data allows operators to proactively manage grid operations, optimize power flow, and prevent blackouts or outages. Smart grids enable the creation of localized microgrids that can operate autonomously or in connection with the main grid. Microgrids increase grid resiliency by providing backup power during outages and supporting localized renewable energy generation and consumption.

Energy Management Systems (EMS)

Energy management systems are software-based platforms that enable efficient monitoring, control, and optimization of energy usage in buildings, industries, and residential sectors. EMS platforms offer the following benefits:

- Energy Monitoring and Analysis: Detailed energy analytics and reporting help identify energy-saving opportunities, benchmark performance, and track progress towards energy efficiency goals.
- Load Management and Optimization: EMS enables load forecasting, load balancing, and load shedding strategies for energy optimization and to minimize peak demand. By dynamically managing energy loads, EMS platforms help lower electricity costs and enhance grid stability.
- Automated Control and Optimization: EMS integrates with building automation systems, smart appliances, and energy storage systems to automate control functions based on energy availability, pricing signals, and user preferences. This enables smart scheduling, optimization of energy consumption, and efficient use of on-site generation or storage resources.
- Demand Side Management (DSM): EMS supports demand-side management strategies by coordinating energy usage, responding to demand response programs, and implementing energy-saving measures. DSM helps businesses and consumers reduce energy costs, participate in incentive programs, and contribute to grid stability.

Waste Management and Recycling Technologies

Effective waste management is essential for creating sustainable and livable cities. Technology and automation offer innovative solutions for waste collection, sorting, recycling, and disposal. Automated waste collection systems utilize sensors, RFID tags, and robotic arms to optimize waste collection routes, reduce collection costs, and minimize environmental impact. Smart bins equipped with sensors provide real-time data on waste levels, enabling efficient bin emptying and reducing overflow. Recycling technologies, such as advanced sorting systems and material recovery facilities, enhance the recovery of valuable resources from waste streams. Additionally, digital platforms and mobile applications facilitate citizen engagement, enabling residents to report issues, schedule pickups, and access recycling information. By leveraging waste management and recycling technologies, cities can achieve higher recycling rates, reduce landfill waste, and promote a circular economy. By implementing intelligent transportation systems, smart grids, and advanced waste management and recycling technologies, cities can transform their infrastructure into interconnected, efficient, and sustainable systems.

These technologies improve the quality of urban services, reduce resource consumption, and minimize environmental impacts. The integration of technology and automation in urban infrastructure sets the foundation for smart, resilient, and future-ready cities [5–7]. Waste management and recycling technologies play a crucial role in addressing the environmental challenges posed by increasing waste generation in India. The Indian government, along with various organizations and stakeholders, has been focusing on implementing advanced technologies and sustainable practices to manage waste effectively. Here are some key waste management and recycling technologies in India.

Waste Segregation and Processing: Waste segregation at source is an essential step in waste management. Various technologies, including manual and automated segregation systems, are employed to break up different types of waste like natural waste, recyclables (plastics, paper, metal), and non-recyclables. Segregated waste can then be processed through composting, anaerobic digestion, or recycling facilities.

Composting: Composting is widely used for the treatment of organic waste in India. Technologies like aerobic and vermin composting are employed to convert biodegradable waste into nutrient-rich compost. Large-scale composting facilities and decentralized composting units are implemented to manage organic waste from households, markets, and institutions.

Biogas Generation: Biogas plants are utilized to treat organic waste and generate biogas, Anaerobic digestion technology is employed in biogas plants to break down organic waste and produce biogas as a byproduct. Biogas plants are commonly used in rural areas and can help in dropping the dependence on traditional fuels.

Waste-to-Energy: Waste-to-Energy WtE plants are used to convert non-recyclable waste into electricity or heat energy. These plants employ technologies such as incineration, gasification, or pyrolysis to treat waste and generate energy.

E-Waste Recycling: Recycling technologies for electronic products have gained importance. E-waste recycling units use specialized techniques to extract valuable metals and components from discarded electronic devices while ensuring proper disposal of hazardous materials.

Plastic Recycling: Plastic waste recycling technologies are being promoted to reduce the environmental impact of plastic waste. Mechanical recycling processes, such as sorting, shredding, and reprocessing, are employed to convert plastic waste into granules or pellets.

Construction Waste Recycling: These are established to process and reuse construction and demolition debris. Technologies like crushing, screening, and sorting equipment are utilized to segregate and recycle materials like concrete, bricks, wood, and metal.

The Indian government has implemented various policies and regulations to promote waste management and recycling practices in the country. Initiatives like the Swachh Bharat Mission and Extended-Producer-Responsibility (EPR) regulations aim to create awareness, encourage citizen participation, and promote sustainable waste management practices.

Technology-Enabled Governance and Citizen Services

Technology plays a crucial role in transforming governance practices and enhancing citizen services in urban areas. By leveraging technology-enabled solutions, cities can streamline administrative processes, promote transparency, and foster citizen engagement. The following subtopics explore the application of technology in governance and citizen services:

E-Governance and Online Service Delivery

E-Governance involves the use of technologies to deliver government services to citizens. Online service delivery platforms provide a convenient and accessible way for residents to access various government services, such as obtaining permits, paying taxes, and accessing public records. E-Governance systems streamline administrative processes, reduce bureaucratic inefficiencies, and enhance the overall citizen experience. These platforms often include features such as online forms, digital signatures, and secure payment gateways, enabling citizens to complete transactions remotely and saving them time and effort. The architecture of the E Governance is shown in Fig. 2. By adopting e-governance and online service delivery, cities can improve administrative efficiency, reduce corruption, and enhance citizen satisfaction.

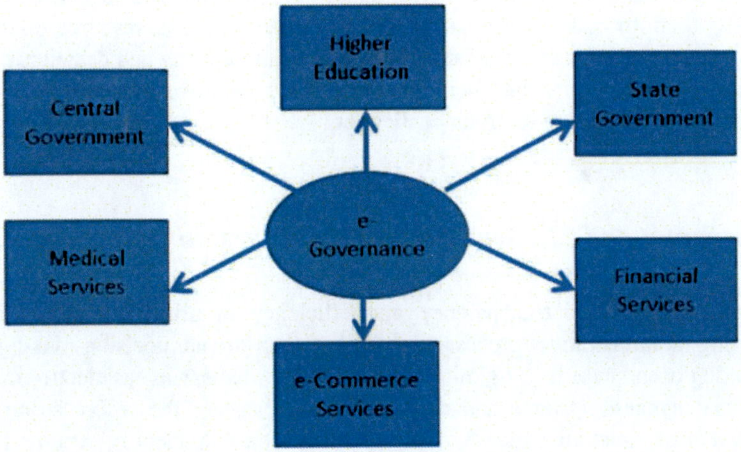

Fig. 2 E-Governance

Digital Platforms for Citizen Engagement

Digital platforms have revolutionized citizen engagement by providing avenues for participation, collaboration, and feedback. These platforms include websites, mobile applications, and social media channels that enable residents to communicate with government officials, provide input on urban policies and projects, and report issues in their neighborhoods. Digital platforms foster transparency and inclusivity by ensuring that citizens' voices are heard and considered in decision-making processes. Additionally, these platforms facilitate real-time communication between government agencies and citizens, enabling timely updates on public services, emergencies, and civic initiatives. By embracing digital platforms for citizen engagement, cities can promote participatory governance, strengthen community ties, and build trust between the government and its constituents [8, 9].

Real-Time Monitoring and Decision Support Systems

Real-time monitoring systems and decision support tools enable urban administrators to gather and analyze data in real-time, providing them with insights for effective decision-making. These systems utilize sensors, data analytics, and visualization techniques to monitor various urban parameters. Real-time monitoring helps identify trends, detect anomalies, and respond swiftly to emerging issues. Decision support systems provide decision-makers with data-driven insights and scenario analysis, enabling them to make informed choices for urban planning, resource allocation, and emergency response. By leveraging real-time monitoring and decision support systems, cities can enhance operational efficiency, improve service delivery, and ensure a timely response to urban challenges.

Transparent and Efficient Urban Administration

Technology promotes transparency and efficiency in urban administration by digitizing administrative processes, enabling data-driven decision-making, and promoting accountability. Digital record-keeping systems, such as electronic document management systems and online databases, ensure the secure storage and retrieval of administrative records. Data analytics tools help identify patterns, trends, and areas for improvement in urban administration. Automation of administrative tasks reduces paperwork, minimizes errors, and speeds up processes [10]. By embracing transparent and efficient urban administration practices, cities can enhance trust in government institutions, optimize resource utilization, and foster innovation.

By adopting technology-enabled governance practices and enhancing citizen services, cities can create more efficient, transparent, and responsive administrative systems. These initiatives empower citizens, streamline administrative processes, and strengthen the relationship between the government and its constituents. Technology plays a transformative role in improving governance and citizen services, ultimately contributing to the overall livability and well-being of urban residents.

Ensuring Sustainable Urban Development

Sustainable urban development is crucial for creating livable, resilient, and environmentally friendly cities. Technology and innovation play a significant role in achieving sustainable urban development goals. The following subtopics explore different areas where technology can ensure sustainable urban development:

Green Building Technologies

Green building technologies focus on constructing energy-efficient and environmentally friendly buildings. Technologies such as energy-efficient lighting, smart HVAC systems, and sustainable building materials help reduce energy consumption and minimize the carbon footprint of buildings. Building automation systems enable efficient management of energy usage, temperature control, and occupancy sensing, further enhancing energy efficiency. By adopting green building technologies, cities can promote energy conservation, reduce environmental impact, and create healthier indoor environments for residents.

Smart Irrigation and Green Spaces

Smart irrigation systems and green spaces contribute to sustainable urban development by optimizing water usage and enhancing urban biodiversity. Technology can support the management of green spaces by monitoring soil moisture, automating irrigation systems, and tracking plant health. By implementing smart irrigation and promoting the development of green spaces, cities can conserve water, enhance urban biodiversity for better quality of life. Smart irrigation and green spaces initiatives in India focus on leveraging technology and sustainable practices to manage water resources efficiently and promote the development and maintenance of green spaces. These initiatives aim to address water scarcity, enhance urban biodiversity, and create healthier and more livable environments. Here's an overview of smart irrigation and green spaces in India.

Smart Irrigation Systems: These systems utilizes advanced technologies to optimize water usage in agricultural fields, gardens, parks, and landscapes. These systems employ sensors, weather data, and automation to determine precise irrigation requirements and deliver water efficiently. Key features include:

Soil Moisture Sensors: The moisture levels of soil is changes with various reasons. For plants the moisture levels are most important parameter, too high or too less moisture content damages the plant growth The sensors placed in the soil transmit data to a central control system. This allows irrigation to be triggered only when necessary, preventing overwatering and water wastage.

Weather-based Irrigation Controllers: These controllers utilize weather data, such as temperature, humidity, and precipitation forecasts, to adjust irrigation schedules and amounts accordingly. They ensure that watering is adjusted based on actual environmental conditions.

Drip Irrigation and Micro-Irrigation: These techniques deliver water directly to plant roots in a targeted and efficient manner, reducing water loss due to evaporation and ensuring optimal water uptake by plants.

Remote Monitoring and Control: Smart irrigation systems can be monitored and controlled remotely through mobile applications or web-based platforms. This allows farmers and landscape managers to monitor irrigation activities, adjust settings, and troubleshoot issues from a distance.

Urban Green Spaces and Urban Forestry

Green spaces, including parks, gardens, and urban forests, are crucial for creating sustainable and healthy urban environments.

Increasing Green Cover: Efforts are made to increase the green cover in urban areas through tree plantation drives, rooftop gardens, vertical gardens, and pocket parks. These initiatives enhance urban biodiversity, mitigate heat island effects, and improve air quality.

Community Participation: Various programs encourage community involvement in developing and maintaining green spaces. This includes citizen-led initiatives, volunteer groups, and public–private partnerships for the establishment and upkeep of parks and gardens.

Biodiversity Conservation: Efforts are made to promote native plant species, preserve natural habitats within urban areas, and support urban wildlife conservation. Such initiatives contribute to ecological balance and enhance urban resilience.

Education and Awareness: Education and awareness campaigns are conducted to promote sustainable water management practices and the importance of green spaces. These initiatives aim to encourage responsible water use, educate communities on the benefits of green spaces, and foster a sense of environmental stewardship.

Smart irrigation and green spaces initiatives in India contribute to water conservation, sustainable urban development, and the well-being of communities. By adopting technology-driven irrigation practices and prioritizing the development of green spaces, India can create more livable and environmentally friendly cities.

Climate Change Adaptation and Mitigation

Cities must adapt to the challenges posed by climate change and take steps to mitigate its effects. Technology plays a crucial role in both aspects. Climate change adaptation involves implementing measures to withstand and recover from climate-related events. This includes early warning systems for natural disasters, real-time monitoring of weather patterns, and resilient infrastructure design. Mitigation efforts focus on reducing greenhouse gas emissions and minimizing the impact on the environment. Technologies such as electric vehicles, renewable energy generation, waste management systems, and carbon capture technologies contribute to climate change mitigation. By prioritizing climate change adaptation and mitigation, cities can enhance their resilience, protect vulnerable populations, and contribute to global efforts to combat climate change. By incorporating renewable energy integration, green building technologies, smart irrigation and green spaces, and climate change adaptation and mitigation strategies, cities can ensure sustainable urban development. These technologies and practices promote resource efficiency, environmental protection, and resilience in the face of climate challenges. Sustainable urban development not only benefits the environment but also improves the quality of life for residents, creating vibrant and sustainable cities for future generations.

Challenges and Opportunities

Implementing technology and automation in urban development comes with its own set of challenges and opportunities. Addressing these challenges and leveraging the opportunities is crucial for successful implementation. The following subtopics delve into the ultimatums, barriers and opportunity linked with these technologies and automation in urban development are as follows.

Technological Barriers and Infrastructure Gaps

One of the primary challenges is the presence of technological barriers and infrastructure gaps. Not all cities have access to advanced technologies or the necessary infrastructure to support their implementation. Limited internet connectivity, inadequate power supply, and outdated infrastructure can hinder the adoption of technology-enabled solutions. Overcoming these barriers requires investments in improving digital infrastructure, expanding internet connectivity, and upgrading existing systems. Collaboration between government entities, private sector partners, and technology providers is essential to bridge the technological divide and ensure equitable access to technology-enabled urban development.

Data Privacy and Security Concerns

As cities collect and analyze vast amounts of information to make fast and effective decisions and service delivery, data privacy and security concerns become prominent. Safeguarding sensitive information, protecting citizen privacy, and preventing data breaches are critical considerations. Cities must establish robust data protection protocols, comply with relevant data privacy regulations, and implement secure data storage and transmission mechanisms. Building trust among citizens regarding data handling practices is essential to ensure their active participation and engagement in technology-enabled urban initiatives. Investing in cybersecurity measures, data encryption, and regular audits can help address data privacy and security concerns and ensure responsible data usage [11].

Capacity Building and Skill Development

Training programs, workshops, and educational initiatives can help build expertise in areas such as data analytics, GIS mapping, smart infrastructure management, and cyber security. Collaborations between educational institutions, government agencies, and industry stakeholders can facilitate the development of relevant curricula, certification programs, and practical training opportunities. By investing in capacity building and skill development, cities can foster a technologically competent workforce capable of driving innovation and managing technology-enabled urban projects.

Collaboration and Stakeholder Engagement

Technology-enabled urban development requires collaboration and engagement among many stakeholders, Which includes public agencies, private agencies, community organizations, and citizens. Effective collaboration ensures that diverse perspectives are considered, and solutions are co-created to address urban challenges. Stakeholder engagement mechanisms, such as public consultations, participatory planning processes, and community feedback processes, foster a sense of ownership and shared responsibility. Building strong partnerships and fostering a collaborative ecosystem can help overcome resistance to change, address conflicting interests, and harness collective intelligence for urban development. Successful collaboration and stakeholder engagement can lead to more inclusive, sustainable, and contextually relevant technology-enabled solutions. By acknowledging and addressing the challenges related to technological barriers, data privacy and security, capacity building, and collaboration, cities can unlock the opportunities presented by technology and automation in urban development. These challenges can be seen as opportunities for innovation, learning, and creating resilient and future-ready cities. By proactively addressing these challenges and leveraging the opportunities, cities can pave the way for sustainable, inclusive, and technologically advanced urban environments.

Success Stories in Indian Cities

Several Indian cities have successfully implemented technology and automation solutions to address urban challenges and to increase the eminence of residence quality life. The following subtopics highlight some success stories in Indian cities:

Surat: Intelligent Traffic Management System

Surat, a city in Gujarat, implemented an Intelligent Traffic Management System (ITMS) to tackle traffic blocking and to improve road safety. The ITMS utilizes advanced technologies such as adaptive traffic signal control, surveillance cameras, and real-time data analysis to monitor traffic flow, identify congestion hotspots, and optimize signal timings. The system uses artificial intelligence algorithms to dynamically adjust signal timings based on traffic conditions, thereby reducing travel time and minimizing traffic congestion. Surat's ITMS has significantly improved traffic management, leading to smoother traffic, less fuel consumption.

Pune: Integrated Command and Control Center (ICCC)

Pune, a city in Maharashtra, established an ICCC to enhance urban governance and service delivery. The ICCC integrates various urban systems, including traffic management, public transportation, water supply, solid waste management, and emergency response, into a centralized platform. The center enables real-time monitoring, data analysis, and decision-making for efficient management of city services. Pune's ICCC has improved coordination among government departments, facilitated data-driven decision-making, and enhanced the overall quality of urban services for residents.

Bangalore: Smart Waste Management System

Bengaluru, the capital city of Karnataka, implemented a smart waste management system to address the city's waste management challenges effectively. The system incorporates smart bins equipped with sensors that monitor waste levels and provide real-time data on bin status. This data is analyzed to optimize waste collection routes, reduce collection costs, and minimize environmental impact. Additionally, citizens can use mobile applications to report waste-related issues and track waste collection schedules. Bengaluru's smart waste management system has resulted in improved waste collection efficiency, reduced littering, and enhanced cleanliness in the city[12].

Indore: Smart Parking Solutions

Indore, a city in Madhya Pradesh, implemented smart parking solutions to address parking challenges and optimize the utilization of parking spaces. The city introduced a mobile application that provides real-time information on available parking spaces, parking rates, and navigation to parking locations. Sensors installed in parking lots detect vehicle occupancy, enabling users to find vacant parking spots efficiently. The smart parking system reduces the time spent searching for parking, minimizes traffic congestion, and encourages efficient use of parking spaces. Indore's smart parking solutions have significantly improved the parking experience for residents and visitors, making parking more convenient and reducing traffic congestion. These success stories demonstrate the positive impact of technology and automation in Indian cities. By leveraging innovative solutions and embracing smart city concepts, these cities have overcome urban challenges, improved service delivery, and enhanced the overall quality of life for residents. The success of these initiatives highlights the potential of technology-enabled urban development in creating sustainable, efficient, and livable cities in India [13–15].

Path Towards Smart and Sustainable Urban Development

To pave the path towards smart and sustainable urban development in India, a holistic approach is needed. Embrace renewable energy integration, green building technologies, smart irrigation, and climate change adaptation and mitigation measures to ensure long-term sustainability and resilience. Strengthen e-governance initiatives and online service delivery mechanisms to enhance transparency, efficiency, and accessibility of government services. Utilize real-time monitoring and decision support systems to enable data-driven governance and optimize resource allocation. Foster a culture of innovation, experimentation, and continuous learning to adapt to evolving technologies and address emerging urban challenges. By following this path, Indian cities can strive towards becoming smarter, more sustainable, and inclusive urban spaces that prioritize the well-being and quality of life for their residents.

Conclusions

The integration of technology and automation presents immense opportunities for redefining urban development in India. By leveraging these technologies, addressing challenges, and embracing sustainable practices, cities can create smart, livable, and resilient environments that cater to the needs of present and future generations. Through collaborative efforts and a shared vision, India can embark on a transformative journey towards smart and sustainable urban development, ensuring a prosperous and sustainable future for its citizens. The following subtopics summarize the key findings, provide recommendations for future implementation, and outline the path towards smart and sustainable urban development.

Throughout this chapter, several key findings have emerged regarding the use of technology and automation in urban development. Geospatial technology and GIS mapping play a vital role in urban planning, facilitating data-driven decision-making and spatial analysis. Smart city concepts and applications encompass a wide range of initiatives, including smart transportation, energy management, and citizen engagement, aimed at improving urban livability and sustainability. Data analytics and predictive modeling enable cities to extract insights from vast datasets, leading to informed decision-making and proactive urban management. Virtual reality and augmented reality technologies are increasingly being used in urban design, enabling immersive experiences and enhancing stakeholder engagement. Additionally, the chapter has highlighted success stories in Indian cities, including Surat's Intelligent Traffic Management System, Pune's Integrated Command and Control Center, Bangalore's Smart Waste Management System, and Indore's Smart Parking Solutions. These success stories demonstrate the tangible benefits of implementing technology-enabled solutions in addressing urban challenges and improving urban services.

Recommendations for Future Implementation

To further harness the potential of technology and automation in urban development, the following recommendations are put forth. Invest in digital infrastructure and address technological barriers to ensure equitable access to technology-enabled solutions across all cities.

Strengthen data privacy and security measures to protect citizen information and build trust in the use of data for urban development purposes. Prioritize capacity building to equip urban professionals with expertise needed to utilize technology in their work. Foster collaboration and stakeholder engagement, involving citizens, government entities, private sector partners, and community organizations in the co-creation and implementation of technology-enabled urban projects.

References

1. Ali, E. (2020). *Geographic information system (GIS): Definition, development, applications and components*. Department of Geography, Ananda Chandra College.
2. Kumar, A. (2019). Beyond technical smartness: Rethinking the development and implementation of sociotechnical smart grids in India. *Energy Research and Social Science, 49*, 158–168.
3. Batty, M. (2013). Big data, smart cities and city planning. *Dialogues in human geography, 3*(3), 274–279.
4. Schrom-Feiertag, H., Stubenschrott M., Regal, G., Matyus, T., & Seer, S. (2020). An interactive and responsive virtual reality environment for participatory urban planning.
5. Yu, L., Yu, T., Wu, Y., & Wu, G. (2020). Rethinking the identification of urban centers from the perspective of function distribution: A framework based on point-of-interest data. *Sustainability, 12*(4), 1543.
6. Reymond, P., Chandragiri, R., & Ulrich, L. (2020). Governance arrangements for the scaling up of small-scale wastewater treatment and reuse systems–Lessons from India. *Frontiers in Environmental Science, 8*, 72.
7. Yang, C. (2020). The transformation of foreign investment-induced 'exo (genous)-urbanisation'amidst industrial restructuring in the Pearl River Delta, China. *Urban Studies, 57*(3), 618–635.
8. Höffken, S., Streich, B. (2013). Mobile participation: Citizen Engagement in urban planning via smartphones. 199–225. https://doi.org/10.4018/978-1-4666-4169-3.ch011
9. Wiig, A., Karvonen, A., McFarlane, C., & Rutherford, J. (2022). From the guest editors splintering urbanism at 20: Mapping trajectories of research on urban infrastructures. *Journal of Urban Technology, 29*(1), 1–11.
10. Morozov, E., & Bria, F. (2018). Rethinking the smart city. *Democratizing Urban Technology, 2*.
11. Verhoeven, G., Wild, B., Schlegel, J., Wieser, M., Pfeifer, N., Wogrin, S., & Watzinger, A. (2022). Project INDIGO: document, disseminate & analyse a graffiti-scape. In 9th International Workshop 3D-ARCH" 3D Virtual Reconstruction and Visualization of Complex Architectures"(3D-ARCH 2022) (vol. 46, pp. 513–520). ISPRS.
12. Hayat, P. (2023). Integration of advanced technologies in urban waste management. *Advancements in Urban Environmental Studies: Application of Geospatial Technology and Artificial Intelligence in Urban Studies* (pp. 397–418). Springer International Publishing.
13. Khan, N., Shahid, Z., Alam, M., Abu Bakar Sajak, A., Mazliham, M., Khan, T., & Rizvi, S. (2022). Energy management systems using smart grids: An exhaustive parametric comprehensive analysis of existing trends, significance, opportunities, and challenges. *International Transactions on Electrical Energy Systems*, 1-38. https://doi.org/10.1155/2022/3358795
14. Gaur, R. C. (2003). Rethinking the Indian digital divide: The present state of digitization in Indian management libraries. *The International Information and Library Review, 35*(2–4), 189–203.
15. Allam, Z., Bibri, S. E., Jones, D. S., Chabaud, D., & Moreno, C. (2022). Unpacking the '15-minute city'via 6G, IoT, and digital twins: Towards a new narrative for increasing urban efficiency, resilience, and sustainability. *Sensors, 22*(4), 1369.

Trust-Free Homes: The Zero-Trust Paradigm in a Smart Home Setting

Aditya Damodhar Dhanapal, S. M. Anantha Ramanujan, and V. Jeyalakshmi

Abstract The Internet of Things (IoT) is one of the novel concepts which has taken the world by storm. Although the technology holds many advantages, the security concerns in these systems make them double-edged swords. Zero-Trust has been introduced as the answer to present-day cyber threats in the security industry. The segmentation of access rights, even within internal nodes of the system, is a vital feature of this model. This article describes the implementation of a Zero-Trust System to analyse its capabilities in IoT networks. Implementing Zero-Trust without affecting the regular functioning of IoT devices at scale is a challenge. The implementation was carried out by simulation of IoT devices' behaviour using Microsoft Azure's simulation capabilities to model the network, followed by a Python implementation of a very small IoT network. This article is aimed to serve as a guide to implementing a Zero-Trust network model.

Keywords Internet of Things (IoT) · Zero-Trust · Secure architecture

Introduction

Internet-of-Things (IoT) is one of the critical concepts leading into Industry 4.0. It is the integration of all devices over a single network, setting up a continuous stream of data all devices on the network can access. All this data is freely accessible and is stored on low-power, limited-capability IoT devices which affect security. The state of the cyber-security landscape is such that the more integrated and exposed a system

A. D. Dhanapal (✉) · S. M. A. Ramanujan · V. Jeyalakshmi
Department of Electronics and Communication, College of Engineering Guindy, Anna University, Chennai, Tamilnadu, India
e-mail: aditya.damodhar.d@gmail.com

S. M. A. Ramanujan
e-mail: ananthramanujan@gmail.com

V. Jeyalakshmi
e-mail: jpjeya@annauniv.edu

A. Prasad et al. (eds.), *Communication Technologies and Security Challenges in IoT*, Internet of Things, https://doi.org/10.1007/978-981-97-0052-3_17

is, the easier it is for malicious actors to exploit the same. IoT is particularly at risk due to the low grade of security in IoT devices.

The Zero-Trust Paradigm is a promising new idea among the many methods proposed and used in the industry to secure the same. Its novelty is that, unlike traditional perimeter-based methods, Zero-Trust preaches complete isolation and segmentation, even within a network. It aims to secure all data and functionality of the network and its nodes by verifying and encrypting all data and communications and granting only required privileges post-authentication on-demand.

Zero-Trust functions on the motto, "Never Trust, Always Verify"—this policy of segmentation and verification safeguards sensitive data from the threat of compromised internal nodes. The segmentation localises the compromise to a specific area. The Intrusion Detection and Prevention System (IDPS) notifies the administrator about the security breach and takes further preventive measures to stop data leakage.

Zero-Trust has the edge over perimeter-based security because of the human factor in security breaches. Weak passwords, unchanged router or server default passwords and accidental introduction of malware into internal nodes are possible with human intervention. Human intervention is required for smooth functioning and the repair and recovery of networks; thus, setting external perimeters alone compromises the network's overall security.

Despite the apparent advantages of adopting Zero-Trust, there need to be more codified guidelines for adoption. Various organisations have multiple implementations, tailoring the guidelines to match their requirements. The need for better standardisation and investment has significantly impacted adoption.

This article describes an attempt to codify guidelines and present a working example for small-scale implementations of Zero-Trust security in networks. A home automation network was redesigned based on the Zero-Trust guidelines and implemented using two Raspberry Pi 4 units. The security was tested against attack vectors from the OWASP top ten vulnerabilities and demonstrated to be secure. This work was intended to be a basic representative implementation and is unsuitable for real-time deployment. Future constructions based on this work for deployment in high-speed, low-power environments are also described in this article.

This work is structured as follows: It starts with an overview of the references and a detailed explanation of the Zero-Trust model. This is followed by the description of the implemented Zero-Trust based IoT network and the simulation platforms and hardware usage. The article concludes with an illustration of security features' functioning when one device is compromised.

Literary Presence of Zero-Trust

The Zero-Trust Paradigm emerged from the industry as a policy for securing organisations' internal networks. It has multiple different proprietary implementations by companies offering security as a Service. Academic contributions to the field generally involve architectural propositions inspired by the traditional security model and economic analysis of a Zero-Trust implementation.

Kindervag, in his pioneering work [1] on Zero-Trust, illustrates the prime focus of a Zero-Trust network architecture. The work differentiates the traditional architecture, where security is an overlay, from the Zero-Trust architecture envisioned in 2010, where security is an integral component of the architecture's design.

Friedman [2], and the first two chapters of Gilman and Barth [3] provide detailed insight into the Zero-Trust paradigm. They cohesively explain the failure of the promise of traditional security and the assurances of Zero-Trust. Home and Nair [4] provides insight into designing systems with Zero-Trust in mind. Michael et al. [5] provides some in-depth criticism of the Zero-Trust paradigm and is noted as essential for the industry to move further into an era of secure computing.

Buck et al. [6] and He et al. [7] provide a comprehensive overview of the current trends and research interests in Zero-Trust security. They evaluate the authentication, control and trust evaluation mechanisms and summarise their main advantages and problems. They also look at grey literature to gain insight into the industry's view of the Zero-Trust paradigm.

Palmo et al. by [8] provides insight into Software-Defined Perimeters (SDP), a security model built with Zero-Trust in mind. They outlined the disadvantages of existing SDN models and proposed scalable SDN architectures which satisfy modern network requirements. Qazi [9] reviews the impact of Zero-Trust architectures on APIs, which were intuitively useful in modelling communications in this project.

Ward and Beyer [10] describes Google's rendition of a Zero-Trust system in BeyondCorp. Google's new approach to enterprise security is thoroughly documented and available for reference for the design and construction of secure systems.

Mahajan et al. [11] by provides insight into the basics of IoT and the necessary description of a simple IoT network envisioned for this project. It describes the components required and the working of an IoT-based Smart Refrigerator. We have taken inspiration from the same to build the physical IoT network on which the constructed Zero-Trust model is deployed.

The Zero-Trust Paradigm

All traditional security models are based on trust levels. Some networks, users and devices are considered trustworthy. Stringent firewalls, rigorous safelists and constant monitoring of data and communications establish the trust boundary. This security model has been relatively successful, barring the case of the malicious insider. In this scenario, a "trustworthy" user, device or network is the source of malicious activity. In this case, the premise of the security model fails, and the malicious agent can compromise the network.

The Zero-Trust model's security improvements result from changing the model of trust. It applies the same conditions and regulations to a node from an internal system as to an external one. This encompasses encryption, verification and authentication of all devices and providing access to network resources and data on a need-only basis. Zero-Trust does not mean not trusting the employees of an organisation; instead, it applies to the data and communications being transferred around in the organisation's network.

One of the advantages of the Zero-Trust model is that it is platform agnostic. The same principles can be applied to all platforms and networks, achieving the expected result. It is also highly modular; newer network segments can be integrated with older segments effectively. The architecture of the Zero-Trust system is inside-out, i.e. smaller segments are wholly secured prior to integrating all segments. Once the segments are integrated, the connections between segments are secured as well.

It must be noted that the commonly used security measures, namely firewalls, VPNs, safelists, access control mechanisms, intrusion prevention systems, cryptography, and so on, are still relevant in this model; instead, their deployment method is altered. They are used extensively in the boundaries between segments for encrypting all network communications and providing access to resources on a need-only basis. The process of a single connection in a Zero-Trust environment is shown in Fig. 1. Authentication of identity, integrity verification, and encryption of all communications is implicit in a Zero-trust system. Apart from all this, all data is logged and stored for future reference.

Securing IoT

With the Zero-Trust Principles in mind, a secure IoT network was designed and simulated. A physical IoT set-up was constructed post-simulation, and the zero-trust system was implemented. A home automation network was chosen due to an intuitive knowledge of functionality instead of an industrial IoT network.

Zero-Trust involves securing the system by design rather than as an add-on layer. The illustrated system architecture in Fig. 2 represents the same. Segmentation was implemented by the system's design, using three different servers to perform the function of a regular home automation server. The communications between the servers were heavily restricted, and the allowed communication occurred only post-authentication. All messages were encrypted using RSA, and the capability to perform other communications was removed.

Traditionally, home automation systems are controlled by a central server, either set locally or in the cloud. The sensors collect and send data to this Server, which logs the data and communications, and decides based on the same. It then instructs

Fig. 1 Sample Zero-Trust process

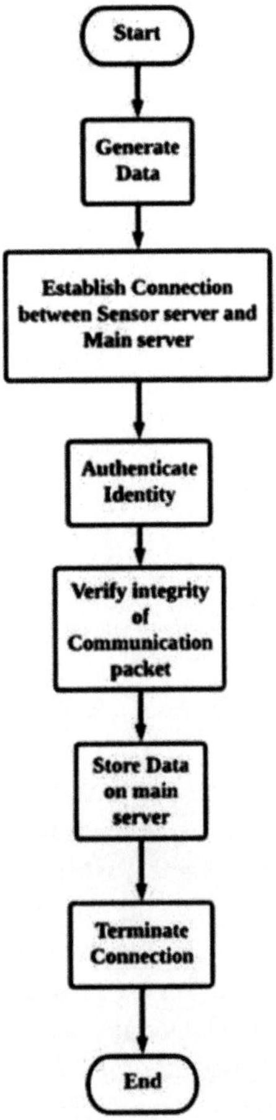

the actuators to perform necessary actions. This architecture trusts all nodes on the network. Thus, the compromise of the internet module of the edge devices exposes the Main Server to attack, thus resulting in the loss of personal data. This work segmented the functions of the Main Server into three smaller servers to minimise the risk associated with connecting to vulnerable IoT devices.

This could be modelled as the Main Server flanked by two gateway servers which double up as the central hubs to two unique star networks. The Central Sensor Hub (CSH) and the Central Actuator Hub (CAH) control the star networks. They are active star networks, and the Central Hubs regulate the data flow.

The entire system was built using the Python programming language. Despite the obvious speed advantage, Python was chosen over C because of the vast support for Python in libraries and the overall code simplicity. The representative nature of the system, rather than a robust, fully functional implementation, allowed us to simplify the code as much as possible.

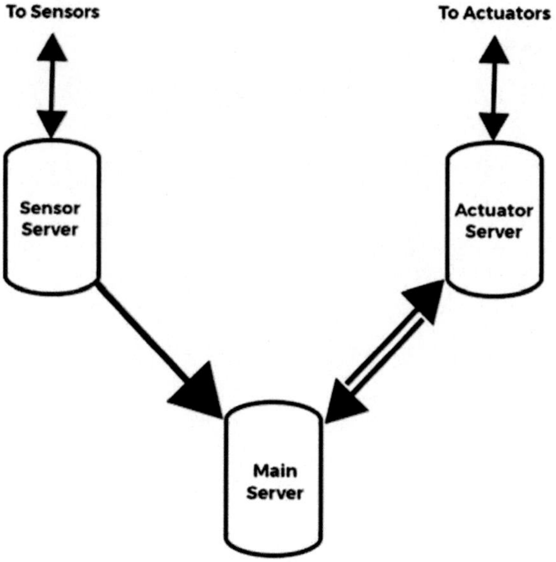

Fig. 2 IoT system designed with Zero-Trust principles in mind

The usage of Python allowed us to select from the extensive collection of libraries. Thus came the following design choice, to use existing IoT protocols like Bluetooth and Zigbee, which have extensive Python support, or to use the Socket library and complete the project using socket programming. Socket programming allowed more basic TCP connections between devices, and using a cloud EC2 instance for the Main Server provided further impetus to use Sockets. Another feature of Sockets requiring one party to initiate the connection and the other to accept it synchronously provided a realistic environment. If the second party does not accept the connection request, the socket stays open until closed, thus vulnerable to masquerading attacks. The Zero-Trust system implements an authentication mechanism to counter such attacks.

Central Sensor Hub (CSH)

The Central Sensor Hub (CSH) directly interfaced with the sensors to accumulate real-time data and sent it to the Main Server. In the simulated network, Microsoft Azure's IoT Central was used to simulate readings of sensors. This data was loaded into the Server, and the output of real-time sensors was simulated. During simulations, an Amazon Web Services (AWS) Electronic Compute Cloud (EC2) functioned as the CSH. Figure 3 shows the simulated CSH.

In the implemented network, a Raspberry Pi 4 was used to act as the CSH and the real-time output of sensors was collected. The collected data was sent to the Main Server after a predetermined time interval.

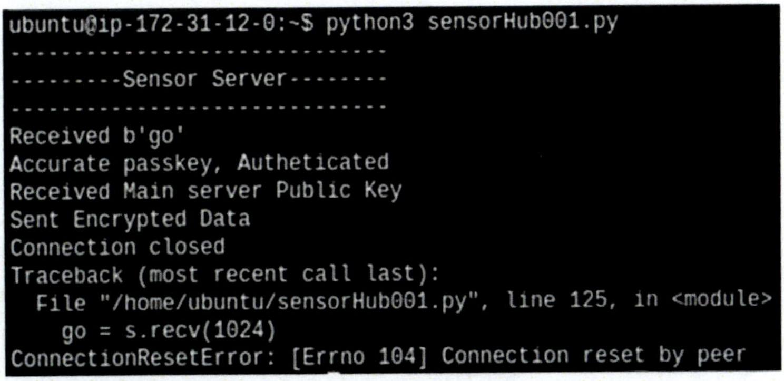

Fig. 3 Simulated CSH working

The Main Server

The Main Server was the heart of the Smart-Home System. All communications to and from the edge devices terminated there. It performed the function of encrypting and logging all data and controlled the connection and termination with the other two hubs.

Despite its control role, it did not make any decisions regarding actuation. It only provided the CAH with the data necessary to make the decisions. It must be noted that this Server refrained from sending sensitive data to the CSH and sent data from logs to the CAH upon an authenticated request's arrival. Even then, the data required for the current computation is only shared.

An AWS EC2 instance was used to function as the Main Server. An EC2 instance was used over pre-existing IoT interfaces in the cloud for the modularity of running sockets on the EC2 instance. Figure 4 shows the AWS instance in action.

Central Actuator Hub (CAH)

The Central Actuator Hub (CAH) controlled the actuators based on data from the Main Server. The CAH functioned by requesting the data needed to compute the output state of a specific actuator from the Main Server and made state decisions based on the received data. At a given time, only one data type was present in the CAH. An AWS EC2 instance, shown in Fig. 5, acted as the CAH during simulations.

In the implemented network, a Raspberry Pi 4 was used to act as the CAH and actuators were connected. The data was sent from the Main Server after a fixed interval, after which actuation occurred.

Fig. 4 Main server working

Simulations

Azure IoT Central was used to simulate the data of sensors. It allowed the user to create a sensor template and specify its functionality, i.e. the types of data it must output. The data of the same type were randomly output at one reading per minute. These values were downloaded in a CSV format and randomly accessed by a Python script running on the EC2 instance functioning as the CSH. Figure 6 shows the device templates created.

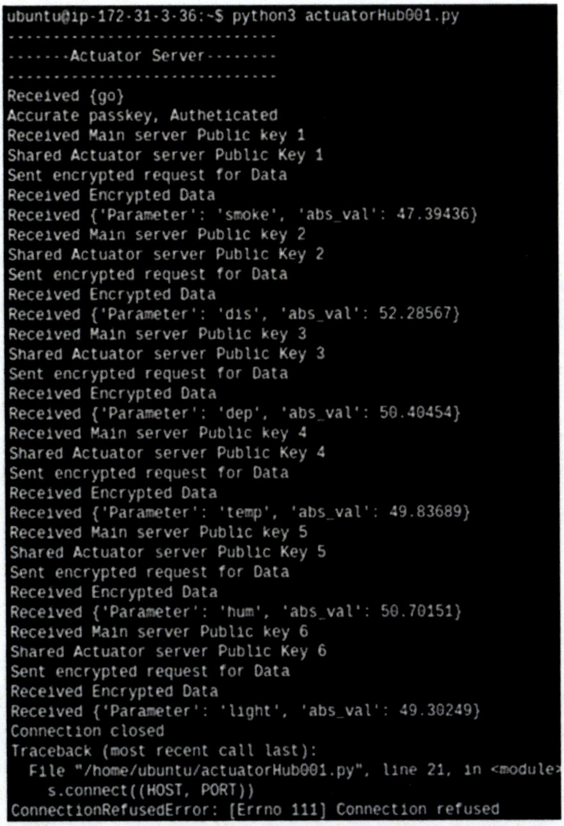

Fig. 5 CAH working

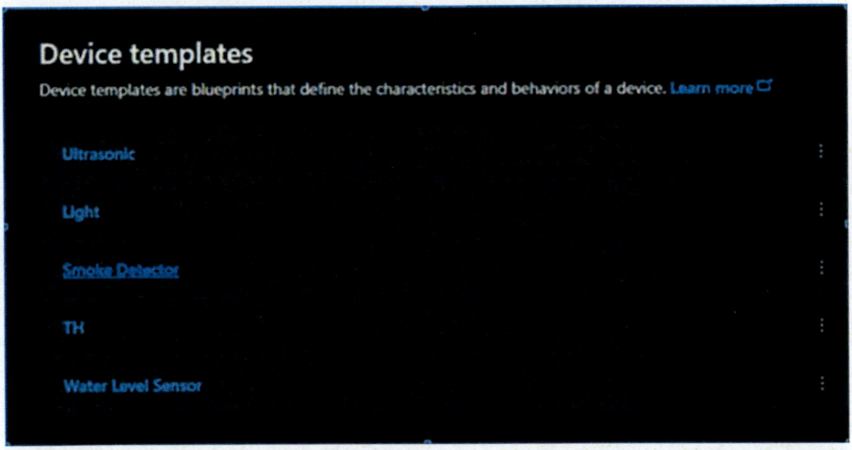

Fig. 6 Device templates on Azure IoT central

Physical Network Implementation

The Physical IoT Network had two Raspberry Pi 4 s acting as the CSH and CAH. The sensors were meant to be deployed in a fridge to make it an intelligent appliance. An HC-SR04 ultrasonic distance sensing module measured the distance to the closest obstacle. Based on a set threshold, the user was alerted to the availability of food items. This set-up was meant to be deployed near an egg tray. The CAH controlled two red and green LEDs and a buzzer to alert the user of egg availability. A periodic flashing of the Green LED indicated the eggs' availability, and the flashing of the red LED and the buzzer noise indicated the non-availability of eggs.

The CSH also interfaced with a DHT-11 Temperature and Humidity sensor, which measured the ambient temperature inside the fridge. Depending on the ambient temperature, the cooling intensity was changed by controlling a Servo motor connected to the CAH. The input of the DHT-11 controlled the angle of the Servo motor's arm. The above figures show the physical implementation of the IoT network. Figure 7 shows the CSH, and Fig. 8 shows the CAH before integrating the LEDs.

Security of Implementation

This network was meant to provide complete isolation and localisation of damage during a security incident. To verify the same, the physical network was tested, assuming the existence of the following two known security vulnerabilities in the CSH. The vulnerabilities are inspired by the web security vulnerabilities defined in OWASP [12].

The first vulnerability was inspired by A2:2017-Broken Authentication [12]. It is the lack of efficient measures to prevent the breach of authentication. Due to the unchanged default login credentials of the Raspberry Pi acting as the CSH (Username: pi, Password: default), an adversary could remotely access it. This allowed her to

Fig. 7 CSH set-up

Fig. 8 CAH set-up

view all processes running on the Server, kill them and even begin new processes, as shown in Figs. 9 and 10.

Despite these privileges, the adversary could not connect to the Main Server to request data as she was unaware of the authentication mechanism, as shown in Fig. 11. If the adversary has not entirely compromised the CSH, this prevents any further connection from the CSH to the Main Server. In the event of a complete failure, the adversary could view the code being run on the Server.

Here, for the sake of demonstration, we introduced another vulnerability inspired by A3:2017—Sensitive Data Exposure [12]. This includes sensitive information

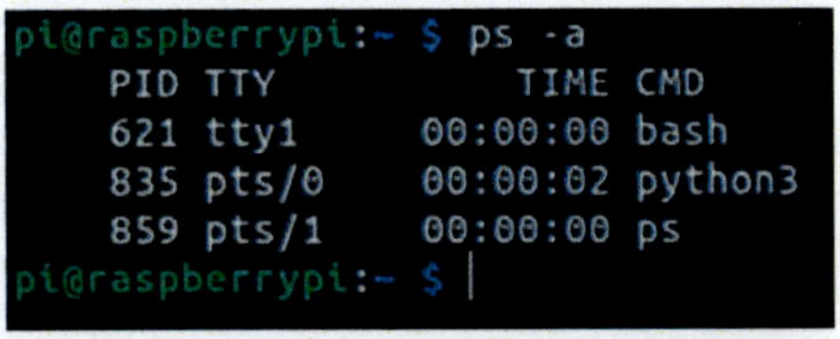

Fig. 9 Adversary viewing all running processes on the server

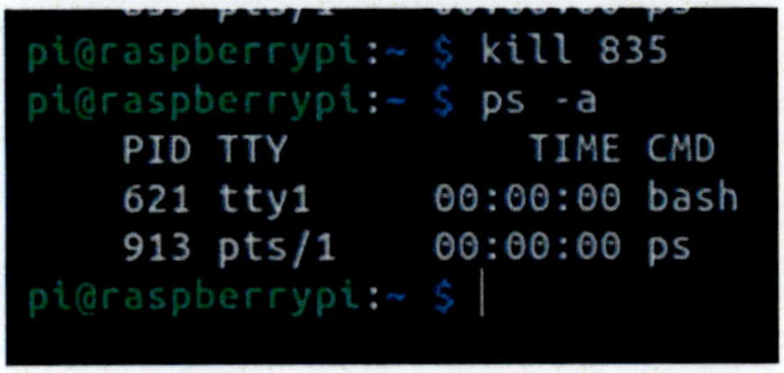

Fig. 10 Adversary killing running processes on the server

```
pi@raspberrypi:~ $ python3 FYP/adversary.py
Sending request for Logs Data
Received Data: b'no go'
pi@raspberrypi:~ $
```

Fig. 11 Adversary unable to connect to main server

```
while(True):

    s = socket.socket(socket.AF_INET, socket.SOCK_STREAM)
    s.connect((HOST, PORT))
    go = s.recv(1024)
    print(f'Received {go}')
    password = b'12345'
    pass_hash = hashlib.sha256(password).hexdigest()
    s.sendall(pass_hash.encode())
```

Fig. 12 Weak authentication keys stored in plaintext

stored in cleartext, which the adversary observed. Despite the available authentication keys (as shown in Fig. 12), the adversary could not retrieve any stored data from the Main Server. This is due to hard-coded segmentation, where the Main Server lacked the ability to send messages to the CSH. This demonstrated the primary security benefits of segmentation and authentication proposed by the Zero-Trust paradigm.

Keeping in line with the Zero-Trust Principles, all communications between the nodes occurred only after authentication. All nodes encrypted the communications with freshly generated RSA keys, one for each message, to ensure security. The Main Server logged the data which was encrypted prior to logging.

Suggested Improvements

With the most elementary requirements of a Zero-Trust based network satisfied, this project currently implements complete identity control and access management by design. It can be improved by using certain other security mechanisms described below.

Intrusion Detection and Prevention Systems

Intrusion Detection and Prevention Systems (IDPS) are physical or software entities that detect and prevent access to an intruder on a computer device or network. An Intrusion is defined by Azeez et al. [13] as a criminal act committed against an information system, e.g., computer system, network or web infrastructure, such that security on the system is breached or compromised, thereby putting it in an insecure state which allows for unauthorised access to the data being hosted by the system. IDPS are classified as Network-Based IDPS, Wireless IDPS, Network Behaviour Analysis and Host-Based IDPS, each with advantages and disadvantages. In general, all IDPS must be fast and accurate in detecting intrusions, log the security incidents for reference and gather information on the host, its OS and the network characteristics to determine susceptibility to various attacks.

Demilitarised Zones

A Demilitarised Zone (DMZ) is an extra layer of protection flanked by two firewalls, ensuring that all network traffic going into a system in the DMZ is inspected by firewalls. Rather than running firewalls on the system, a DMZ provides effective isolation from the unprotected LAN network by running dedicated firewalls outside the system to be protected.

Firewall

A Firewall is a prevalent security measure in modern-day networks. It provides security by filtering traffic. Mukkamala and Rajendran [14] describes the various types of Firewalls and their techniques to filter traffic. Broadly classified into two types, host-based and network- based, they can further be classified into the following based on their working methodologies:

Packet-Filtering Firewalls create checkpoints on network nodes and check the contents of a packet. The packet is dropped or forwarded depending on the rules based on IP address, port number and other information. Circuit-Level gateways, on the other hand, do not inspect the packet contents but instead drop or forward packets based on the TCP Handshake.

Stateful Inspection firewalls use packet inspection and the TCP handshake to provide greater security. They use session tables to maintain a flow of the state of connections, whose processing is essential for the accuracy of these Firewalls. Application Level Firewalls operate at an application level and use proxy servers to deliver filtered packets between the network and the source.

Next-Generation Firewalls combine modern methods like Deep packet inspection, Intrusion Prevention Systems, Bandwidth management, URL filtering, Malware detection, and older packet filtering and TCP handshakes based firewalls. These provide a higher level of security with advanced filtering capabilities. Cloud Firewalls are implemented with Cloud Solutions.

Antivirus

Hacking was initiated with IBM's experiment in 1967, allowing a group of students to explore their new system. With the development of ARPANET, cyber security became a noteworthy term, with the first virus developed called Creeper. The Reaper was developed to follow the Creeper and delete it from systems, becoming the first documented Antivirus. Commercial Antivirus was first released in 1987, and their usage has been instrumental in securing Cybersystems. They are diverse software to combat all types of viruses. Their techniques range from deep inspection to signature matching.

Along with the aforementioned security measures and a lot more classical security measures like access-control lists, there are multiple modern security measures described by Aslan et al. [15], like using machine learning, deep learning, cloud platforms, big data, and blockchain in detecting malware, intrusion detection, spam identification, DNS attack classification, fraud detection, recognising hidden channels, and distinguishing advanced persistent threats. These measures will further strengthen a Zero-Trust system and is an exciting research direction for the future of Cybersecurity.

Conclusion

The Internet is insecure by nature, with security generally an add-on feature. The Zero-Trust methodology suggests securing networks by design and implementing security as a part of the architecture. This article showed one such implementation, whose security was demonstrated to be sound against two primary attack vectors. The basics of Zero-Trust, specifically Segmentation and Authentication was, shown to be implemented.

The implementation functioned as expected both in terms of speed and security. It was slower than optimal for deployment in a real-life scenario but its latency was heavily affected by the internet speed. As we were relying on wireless connections, the speed difference in low signal strength areas was considerable from high signal strength areas. On the security front, the devices proved successful in defense against attacks which would otherwise be possible without segmentation. The inherent lack of communication capabilities between servers, except permitted hard-coded messages, prevented information leakage. Thus, the implementation is judged to be satisfactory as a reference model.

As mentioned, this project was implemented in Python, focusing on simplicity. A C language implementation would be much faster and can even be deployed on real-time medical and industrial systems. Faster and more robust implementations while not compromising security would be the ideal scenario arising from the Zero-Trust philosophy.

Acknowledgements This work was carried out as part of the Project Phase 1 for final-year students at the Department of Electronics and Communication Engineering at the College of Engineering, Guindy, Anna University. We thank the HoD and the staff for providing the necessary resources for this project. We also thank Dr. Jeyalakshmi V for her guidance, whose insights and technical know-how were instrumental in the fruition of this project. The source code to our project is available on GitHub: https://github.com/AdityaDamodhar25/SImple-Zero-Trust.

References

1. Kindervag, J. (2012). Build security into your network's DNA: The Zero-Trust network architecture. Available via Forrester https://www.virtualstarmedia.com/downloads/Forrester_zero_trust_DNA.pdf
2. Friedman, J. (2020). Definitive guide to zero trust security. Available via Colortokens. https://colortokens.com/ebook/definitive-guide-zero-trust-security/
3. Gilman, E., & Barth, D. (2017). Zero trust networks. Available via Wordpress. https://itjumpstart.files.wordpress.com/2019/02/zerotrust.pdf
4. Home, D., & Nair, S. (2021). Introducing zero trust by design: principles and practice beyond the zero trust hype. In *Advances in security, networks, and internet of things*. Springer. Available via Researchgate. https://www.researchgate.net/publication/354054404_Introducing_Zero_Trust_by_Design_Principles_and_Practice_Beyond_the_Zero_Trust_Hype
5. Michael, J. B., Dinolt, G. C., Cohen, F. B., & Wijesekera, D. (2022). Can you trust zero trust? *Computer*. https://doi.org/10.1109/MC.2022.3178813
6. Buck, C., Olenberger, C., Schweizer, A., Völter lter, F., & Eymann, T. (2021). Never trust, always verify: A multivocal literature review on current knowledge and research gaps of zero-trust. *Computers & Security*. https://doi.org/10.1016/j.cose.2021.102436
7. He, Y., Huang, D., Chen, L., Ni, Y., & Ma, X. (2022). A survey on zero trust architecture: Challenges and future trends. *Wireless Communications and Mobile Computing*. https://doi.org/10.1155/2022/6476274
8. Palmo, Y., Tanimoto, S., Sato, H., & Kanai, A. (2021). A consideration of scalability for software defined perimeter based on the zero-trust model. In *10th International Congress on Advanced Applied Informatics (IIAI-AAI)*. Niigata, Japan, pp. 717–724. https://doi.org/10.1109/IIAI-AAI53430.2021.00127
9. Qazi, F. A. (2022). Study of zero trust architecture for applications and network security. In *IEEE 19th International Conference on Smart Communities: Improving Quality of Life Using ICT, IoT and AI (HONET)*. https://doi.org/10.1109/HONET56683.2022.10019186
10. Ward, R., & Beyer, B. (2014). BeyondCorp: A new approach to enterprise security *39*(6), 6–11.
11. Mahajan, M. P., Nikam, R. R., Patil, V. P., & Dond, R. D. (2017). Smart refrigerator using IOT. *IJLERA Writing, 2*(3), 86–91.
12. OWASP. (2017). OWASP Top Ten 2017. Available via OWASP. https://owasp.org/www-project-top-ten/2017/Top_10
13. Azeez, N. A., Bada, T. M., Misra, S., Adewumi, A., Vyver, C. V. D., & Ahuja, R. (2020). Intrusion detection and prevention systems: an updated review. In: Data management, analytics and innovation. Advances in Intelligent Systems and Computing. https://doi.org/10.1007/978-981-32-9949-8_48
14. Mukkamala, P. P., & Rajendran, S. (2020). A survey on the different firewall technologies. *International Journal of Engineering Applied Sciences and Technology*. https://doi.org/10.33564/IJEAST.2020.v05i01.059
15. Aslan, O., Aktuğ, S. S., Ozkan-Okay, M., Yilmaz, A. A., & Akin, E. (2023). A comprehensive review of cyber security vulnerabilities, threats, attacks, and solutions. *Electronics*. https://doi.org/10.3390/electronics12061333

Security Challenges of IoT-Enabled Vehicular Communications and Their Countermeasures

Nishan Rai, A. R. Badrinath, Abhishek Kamath, Veerishetty Arun Kumar, and Rathishchandra R. Gatti

Abstract Vehicular Communication Networks are the recent intensification of data-based, interconnected communication between computers, network entities, and sensory systems that have created a new path to analyse, predict, and prevent potential road misbehaviour by utilising real-time information transfer and environmental simulation. The first part of this book chapter examines the hidden and pervasive potential of IoT in-vehicle networking in terms of real-time traffic forecasting, traffic welfare, and traffic efficiency. The first part also provides an overview of the many cutting-edge technologies, their functionality, structure, laws now being used in vehicular networking for vehicle CPS, and their difficulties. The widespread deployment of IoT-enabled vehicular communication infrastructure also brings various security concerns and challenges that could any breach or attack could significantly harm human life, physical assets, and the environment. Hence, in the second part of the paper, we will examine the various security issues and threats in vehicular communications and the various countermeasures proposed to address them. The overall objective of this chapter is to provide a comprehensive overview of the state of the art in this area and to identify challenges and solutions for securing current and future IoT-enabled vehicular communications.

Keywords Internet of Things · Vehicular communication networks · VANETs · Cyber-Security · Intelligent transportation systems

N. Rai
Toyota Industries Engine India Private Limited, Bengaluru, India

A. R. Badrinath
Ace Micromatic Machine Tools Private Limited, Bengaluru, India

A. Kamath
Cognizant Technologies India Private Limited, Bengaluru, India

V. A. Kumar
SRS Decors, Faridabad, India

R. R. Gatti (✉)
Department of Mechanical Engineering and Robotics, Sahyadri College of Engineering & Management, Bengaluru, Karnataka, India
e-mail: gattirathish@gmail.com

A. Prasad et al. (eds.), *Communication Technologies and Security Challenges in IoT*, Internet of Things, https://doi.org/10.1007/978-981-97-0052-3_18

Introduction

In the 1970s, the dawn of vehicular communication began to lay the foundations of ITS. Countries such as the United States and Japan began their work on projects like Electronic Route Guidance System (ERGS), Comprehensive Automobile Traffic System (CACS) and Vehicle Information and Communication System (VICS). Real-time data tracking, analysis and networking have experienced widespread advancements leading up to the modern-day ubiquitous communication systems that have resulted in the development of Intelligent Transport Systems (ITS) [1]. Rapid urbanisation and increased dependency on vehicles as a mode of commutation have led to unhealthy traffic pollution, traffic congestion, increased road fatality rates and inefficient traffic flow. For such problems, one can envisage an Online Ad-hoc Distributed Simulation (OADS), which uses a dynamic reservoir of self-governing simulations in a constantly coordinating, networked, real-time simulated environment, where they interact with one other and with live data [2]. Modern Vehicles are already complex computing systems, with various sensing units on board, each committed to a different aspect of the vehicle's performance parameters. They also enable intelligent vehicular networking through new wireless communication devices with processing, computing, and sensing functionalities [3, 4]. One of the most common approaches in Intelligent Vehicular Communication is testing the system on a transportation network with arbitrary network segmentation and varying geographical distributions of driver (client) locations under diverse traffic demand conditions in an attempt to simulate and cover all the various plausible scenarios that might occur [5]. The term Inter-Vehicle Communication (IVC) refers to the communication or exchange of data between drivers or vehicles within the ITS region, along with Dedicated Short-Range Communications (DSRC) such as infrastructure to vehicle communications, Wide Area Wireless Communications (WAWC) and Wireless Communications (WC). It enables an operator or a driver to communicate with the other participating drivers and permits the exchange of required information [6].

Architecture of Vehicular Communication

Vehicle networking is a crucial component of any ITS. It allows data to be exchanged between vehicles and other networks and aid in real-time simulations to anticipate potential collisions, find optimal routes, and solve other driver assistance technologies. This part will review the essential technologies in a vehicular communication system and their applications and security implications. Vehicle-to-vehicle, vehicle-to-infrastructure, and vehicle-to-pedestrian interactions are the three most common vehicle communication modes, as shown in Fig. 1.

Vehicle to Vehicle (V2V) communications are carried out if two vehicles are in each other's vicinity. This model of the network is used with ad-hoc and Inter-Vehicle Communications. It employs DSRC frequencies in the transmission of data.

Fig. 1 Vehicle networking architecture

Multi-hop communication technology sends data between vehicles separated by a significant distance [1, 7].

Vehicle to Infrastructure (V2I) technology is a networking architecture that allows data to be exchanged between vehicles and a set of devices along a path. It is a wireless and bidirectional information exchange system that utilises an ad-hoc network, often known as a vehicle ad-hoc Network or VANET, to enable simple data transmission between infrastructure equipment and automobiles. It also employs DSRC technology [1]. Vehicle to Pedestrian (V2P) allows a vehicle to communicate with a pedestrian or a group of pedestrians close to it. It negotiates the information supplied by both the pedestrian and the vehicle and then utilise its real-time simulation technique to determine the outcome, allowing the pedestrian to cross safely by providing an oncoming vehicle/pedestrian alert to both the vehicle and the pedestrian. It aims to ensure traffic safety, allow for practical traffic commuting, reduce congestion, and determine the best possible scenarios. GPS Fleet tracking is a widely used efficient technology that warns the vehicle of possible collisions and blind spots by capturing real-time data through dash cameras or interior recorders. The assistance provided to disabled or visually impaired persons is one of the most significant elements. Real-time monitoring and simulation assist people with suitable warnings and supply essential information when crossing roadways or travelling along the roadside [8].

Deployed Vehicular Technologies

In the United States, many programmes such as Crash Avoidance Metrics Partners (CAMP), The California Partners for Advanced Transit and Highways (PATH), Cooperative Intersection Collision Avoidance System (CICAS) and US IntelliDrive are being developed and deployed. The WAVE Protocol standard, which IEEE standardises, was utilised in these projects. The Intelligent Transportation Society of America (ITSA) was approached in the late 1990s to assist the US Department of Transportation with implementing the Intelligent vehicle systems programme, which intended to improve traffic safety and minimise accidents [9]. Furthermore, a design known as the national intelligent system architecture has been established, which may be used to implement already available services such as the ITS.

CACS was one of the most critical initiatives carried out in Japan in the 1970s; it encompassed the same technological objectives as previously described, such as traffic welfare. This can be accomplished by linking the communications networks to a central processing facility. A CACS prototype test system was created in Tokyo to demonstrate this model. The test included 85 critical crossroads, 103 roadside devices networked into a central system and even 1330 onboard units. The CACS project's advanced technology was applied to the transportation system. Later, a CACS-based communication system was extended along prime roadways in Tsukuba Expo85, 60 kms from downtown Tokyo [6, 8].

The system contained eight roadside units and 150 vehicles with onboard units, with a two-way inductive transmission unit with a transmission speed of 9.6 kbps, and roughly 600 bits and data was collected from all units on the road and processed at the control facility. Car manufacturers also focused on R&D activities on various Driver Information Systems. Much such electronic equipment which could be operated within the vehicle became popular. NAVICOM, DRIVE GUIDE, and GYROCATOR were all introduced to the market. GYRO or Magnetic sensors, display units, and microprocessors are all part of these devices [8].

Dedicated Short Range Communication is a wireless communication technology based on the IEEE 802.11p standard that allows for robust security and fast, uninterrupted communication between automobiles and the neighbouring infrastructure without cellular infrastructure. IEEE 802.11p is a revision to the IEEE 802.11 standard that specifies upgrades for ITS implementations. The Intelligent Transportation System of America (ITSA) petitioned the board to allocate 75 MHz spectrum in the 5.9 GHz band for ITS, specifically for DSRC. The board embraced a report and order establishing licensing and service standards for Dedicated Short-Range Communications (DSRC) service in the 5.85–5.925 GHz band in the Intelligent Transportation System Radio service [4, 5]. Onboard Units (OBUs) and Road Side Units (RSUs)

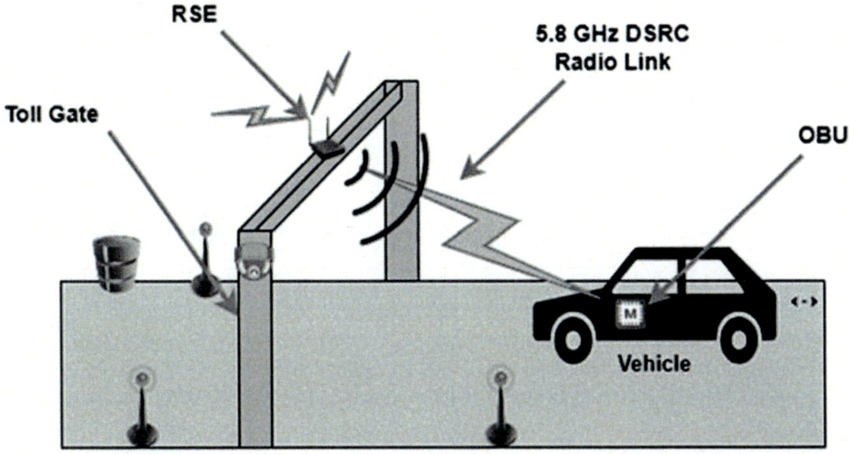

Fig. 2 DSRC system [11]

are part of the DSRC system, as shown in Fig. 2. An OBU is a transmitter usually installed on the vehicle but may also be a portable unit in specific cases. An RSU is a transceiver positioned by the side of a road or a pedestrian walkway. The RSU can be installed on a vehicle or carried by hand, but it can only function when stationary. The RSU transmits data to OBUs in its communication zone or exchanges data with them [7, 9, 10].

Navigation is guiding a vehicle from its initial position to its goal. It is also the process of choosing the optimum route to a given location. This can significantly assist visually impaired people who cannot navigate themselves as efficiently throughout the journey. Also, newcomers or tourists who have ventured onto the land and are unfamiliar with the roadways are much more likely to use navigation systems. This is an area where navigation systems play an essential role [12]. One such previously established navigation system is known as NAVICOM. Toyota introduced NAVICOM, a type of navigation system, in the 1980s. It can continuously deliver data on the direction and straight-line percentage to the location with the help of the vehicle speed sensor and geomagnetic sensor when confronting unfamiliar road conditions. Then the distance to the location and the proportion of the distance already covered is displayed. NAVICOM was where it all began regarding a functional navigation system [8, 12]. A Navigation System Consists of a Microcontroller, GPS Unit, A/D and signal conditioning unit, proximity unit, and mapping unit, as shown in Fig. 3. The GPS unit receives the GPS satellite signals, while the maps are fetched by the mapping unit from the updated Map database. The vehicle heading direction, distance and location are captured by the heading sensors, distance sensors and location sensors, respectively. The user input is usually a touch screen, and the user output is a LED display processed by input and output units.

Additional navigation systems that are significantly more competent than the one discussed have been introduced—introducing a vast feature set, increased efficiency,

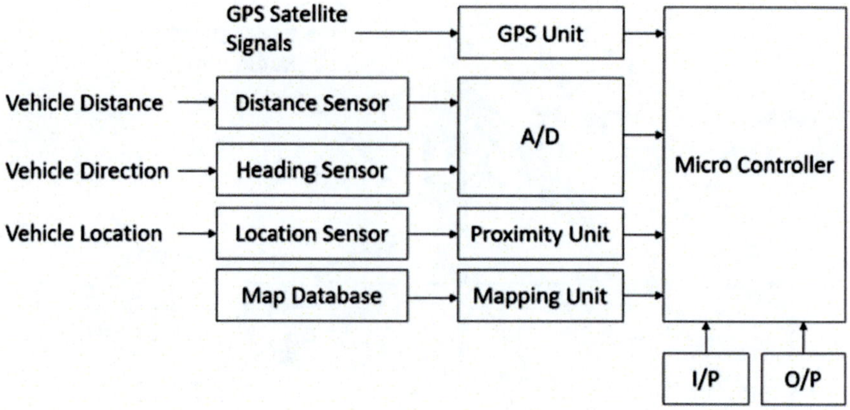

Fig. 3 Block diagram of a navigation system [13]

and real-time audio functionality to assist the driver. Venturing into artificial intelligence, many different systems are being worked upon, and some even already implemented into the commercial market have proven successful.

Applications of Vehicular Networking

Vehicular Networking applications are designed to deliver integrated solutions like the shortest path connecting the start and the end location based on information collected by data collection technologies, considering various aspects, for example, time, distance, and power usage, in a data-great travel atmosphere. These apps can assist transportation control facilities in tracking and controlling transportation setup performance by altering traffic signals, continuously regulating transit functions, or sending emergency maintenance services.

When a car communicates its geographical data as determined by GPS and some other vehicle gets the information and compares it to its data of position, the driver can discover the existence of another vehicle that would otherwise be difficult to detect. This could assist a driver who is having trouble assessing the conditions before him, maybe due to technical difficulties or weather conditions, and this could help prevent many fatalities on the road [5, 6].

An advanced automated highway system, described as dynamic routing and scheduling, consists of autonomous cars over a series of lanes within a small inter-vehicular distance. A connection exists between all the autonomous vehicles in the vicinity, allowing them to undertake safe, regulated overtaking, passing-by or efficient lane transitions [1, 4]. The European Union, in the mid-2010s, approved legislation mandating that all new vehicles be equipped with the e-Call feature. In an accident, the in-vehicle e-Call is made manually by the vehicle passengers or automatically by in-vehicle sensors activating. Whenever the in-vehicle e-Call device is enabled, it will initiate an emergency call conveying audio and data to the nearest emergency location. The car occupant can converse with a certified e-Call operator through voice call. A minimal data set will be delivered to the e-Call operator handling the voice call simultaneously. This is another example of self-driving vehicle-to-infrastructure technology that comes in handy when the driver cannot contact for assistance.

Such other applications of vehicle networking technologies may be explored indefinitely, providing aid to various aspects of our society and, as a result, enabling a more efficient and effective way of transportation. These systems' development and successful implementation demonstrate the viability and promise of inter-vehicle communications and automated vehicle control technology [6, 14].

Security Implications

Concerns regarding data authenticity arise from the effects and trustworthiness of a system that gathers and shares information across distinct units. Any entity might misuse the information to gather or regulate data about other participating vehicles. This could lead to a perilous situation because that individual could control other vehicles' operations. As we have seen, inaccurate information leads to disastrous outcomes. A user might spread false information to gain the upper hand, such as when a vehicle communicates that a particular path is congested or has been in an accident, causing others to avoid that road and giving the misuser a less-congested journey. Some may even feed false information into the data communication channel between connected vehicles for enjoyment to commit evil deeds that could lead to high-level traffic accidents and cause loss of lives. A hacker can even force the road's safety features to be activated by manipulating the system, which might result in significant damage and unnecessary congestion [15].

A robust, difficult-to-bypass security system and an appropriate alerting mechanism to indicate the existence of a foreign entity within the system can help mitigate such security consequences. The sender being decisively acknowledged as a trusted source is vital to a dedicated communication channel. Spammers and users who send false information into the network could be detected using a mechanism similar to a mobile call spam alert approach; they could be identified by multiple vehicles and tagged as "Spam", causing the system to disregard further updates from such individuals [16].

Currently, several cooperatives are operating on Vehicle Ad-hoc Network (VANET). VANET is an applied concept of Mobile Ad-hoc Network (MANET) that is applied to vehicular networks. MANET involves the spontaneous establishment of a wireless connection between moving devices. This is applied to the field of automobiles by VANET to produce vehicular ad hoc networks to ensure safety and security regulations. VANET also considers other design requirements of projects such as the Crash Avoidance Metrics Partnership (CAMP), Vehicle Safety Communications project, the SeVeCom project, the Embedded Security for Cars (ESCAR)

Consortium, the Vehicle Infrastructure Integration (VII) project. A particular challenge in guarding VANETs is providing sender authentication during inter-vehicle communication. Such authentication is tricky because vehicles may engage with one another for the first time, and link-layer losses may impact various broadcast receivers to varying degrees [17].

State of the Art Vehicular Communication Technologies

Various vehicle networking techniques are being introduced based on their functionality, range, implementation, reliability, efficiency, and relevance in the current transportation environment. Some known vehicular communication technologies are VANET, mm-Wave, SEVECOM, CAMP/VSC-2, IEEE Wave, Geo-Net and C2C-CC. This section discusses some brief information about the technologies mentioned above.

VANET: VANETs are a subset of conventional Mobile Ad-hoc Networks (MANETs), following upon similar technologies but for autonomous vehicles. Vehicles are integrated with advanced onboard electronics. While travelling, they communicate via Vehicle-to-Vehicle, Vehicle-to-Infrastructure, and vehicle-to-pedestrian communications [2]. VANET refers to a real-time ad-hoc network that develops among moving vehicles on the road. High mobility, a rapidly evolving structure, and momentary, one-time contacts are all characteristics of VANETs. Vehicles can communicate beforehand to surrounding vehicles about diversions, traffic accidents, and congestion to help relieve traffic congestion near the concerned locations. Vehicles having a VANET system can connect Online to get real-time news, traffic, and weather updates via applications [14].

The onboard vehicle radar might detect traffic blocks and automatically slow the vehicle down. Sensing units are employed in various accident warning systems to identify whether the vehicle safety systems have functioned accordingly. This data is subsequently communicated via V2V or V2I inside the vehicle network. VANETs use various technologies, primarily as facilitators of Intelligent Transportation Systems such as GSM, Wi-MAX restricted Wi-Fi, UMTS and Wireless Connectivity in Vehicular Environments. Wireless Access in Vehicular Environments (WAVE) is a unique and specific technology for this application type. The last several advancements in ITS have resulted in the developing of the new Dedicated Short Range Communication (DSRC) protocol, which uses WAVE to provide high-speed, reduced latency V2V and V2I connectivity [2].

As previously mentioned, DSRC is a networking system that operates in the 5.850–5.925 GHz range for civilian security and privacy implications. V2V and V2I wireless communication received the above frequency spectrum from the Federal Communications Commission. Onboard and roadside units may construct VANETs using the assigned frequency and previously invented services, allowing nodes to connect wirelessly without a gateway node. The transmission is mainly in the 5.850 GHz-5.925 GHz bandwidth, which allows for a 70 MHz band [18].

The employment of a GPS within the vehicle allows the vehicle's location to be determined. Each vehicle in the VANET must know its specific position to transmit correct data to its peers. The accuracy of a localisation service can be improved by combining a GPS with an Inertial Navigation System (INS). Additionally, supplemented GPS technologies such as differential GPS are frequently utilised to improve accuracy. Ultimately, while the main objective of VANET is to sanction safety functionalities, implementing non-security features like entertainment devices is estimated to increase the number of vehicles equipped with onboard wireless devices, resulting in high commercial potential. The goal of comfort and multimedia apps is to deliver information and entertainment to the passengers, in effect making their journey more worthwhile [9, 14, 17].

mm-Wave: The millimetre wave (mm-Wave) spectrum is already used in vehicle radars. For fully linked vehicles, mm-Wave technology is a practical addition to DSRC. The incorporation of mm-Wave gives vehicles and infrastructure access to high-bandwidth communication channels, enabling high Giga-Bit-Per-Second (GBPS) data rates and achieving unfiltered sensor data interchange. Multiple mm-Wave transceivers should be used for mm-Wave V2X communications to avoid mm-Wave signals being blocked by neighbouring vehicles or even people [10, 19]. To utilise data at a high transfer rate from nearby vehicles, many mm-Wave transceivers on a vehicle will likely be linked to the vehicle's central controller through a Controller Area Network (CAN) bus or a more robust onboard network system that can handle high transfer rates. Millimetre Wave frequency might be used in 5G cellular networks to enable GBPS data speeds. 5G communication networks may allow mm-Wave V2X communications. In this case, 5G base stations will act as Infrastructure for V2I communications, and 5G device-to-device mode will support V2V communications. Rather than depending on 5G networks or existing Wi-Fi standards, a new standard with a specialised spectrum for mm-Wave V2X (X = Vehicles, Infrastructure or Pedestrians) communications, akin to DSRC, might be developed. A distinct mm-Wave V2X spectrum will allow greater flexibility in optimising vehicle communication systems [10].

SeVeCom: Secular Vehicular Communication (SeVeCom) is a fully operational project incorporating the essential security measures for vehicular networking. To facilitate the security and privacy required in properly operating various vehicular communications like V2X. It is necessary to enhance several main features, such as authentication, availability, and privacy, to establish these networks' security architecture and provide a blueprint for integrating security systems into this network architecture. For successful implementation, it is vital to concentrate on operational factors such as network range, capital, and system trustworthiness [15, 16].

SeVeCom conducted a thorough requirements study based on these factors, analysing general application features and security needs for various VC implementations. Focusing on the hierarchy of various types of information or warnings based on their importance is a critical feature. For example, an obstacle warning may be considered a low priority compared to a collision warning message, ensuring that the network channel is entirely secure and that no false information is flowing [15].

Hardware Security Modules (HSMs) are integrated into Onboard and RSUs, aiming to safeguard V2X connection by storing and protecting critical data. This data primarily consists of private keys used to generate signatures. If modules interfered with any intervention to obtain private keys, the unit's security system would guarantee that the critical information was deleted, limiting the attacker from acquiring them. The HSM automatically executes cryptic operations to guarantee that critical information does not leak from the secure HSM network. Without the HSM, this essential information might be compromised [15, 20].

IEEE WAVE: An ITS technical framework for systems integration was designed based on the IEEE Wireless Access for Vehicular Environment (IEEE WAVE) standard for safe V2V and V2I wireless communication. It specifies the structures and processing of secure messages. It also defines the circumstances under which encrypted message exchanges should be executed and how such communications should be processed based on their intended purpose [1, 9].

It also establishes networking and transport layer features for safe WAVE data exchange, including addressing and routing. It also specifies WAVE Short Messages, a WAVE-specific counterpart to IPv6 applications that may be efficiently utilised. WAVE Services creates and administers the implementation of the Provider Service Identifier (PSID) number, which allows an RSU to broadcast information to cars. PSID applications also create a subnet of PSID identifiers that will be transmitted to the ITS Network [1]. The IEEE 1609 series uses the PSID as a 4-byte numeric string to determine a specific application or service provider. The PSID is made available to potential consumers of an application or service within a secured WSA frame [1, 9].

C2C-CC: The Car-to-Car Communication Consortium (C2C-CC) is an organisation to study interactions between vehicles and cars and infrastructures, also referred to as C2X. The C2C-CC attempts to standardise wireless communications interfaces and protocols between automobiles and their surroundings so that vehicles from various manufacturers may connect with roadside equipment [18, 21].

Spontaneously, C2C Communications interconnects individual information scattered over several cars, allowing vehicles to work together. The widely established VANET, as discussed earlier, also functions as a sensor, expanding the vehicles' spectrum of perception to areas that the driver and onboard sensory systems would typically miss. In the Cooperative Forward Collision Warning System (CFCWS) of C2C-CC, equipped cars autonomously transmit vital information like location, velocity, and direction. Each car analyses its very own driver's activities as well as the location and behaviour of all other adjacent vehicles in order to forecast a potential rear-end collision. The car informs the driver through a display or audio when it senses a dangerously close scenario. As a result, the operator will have sufficient time to act and avoid the collision. Also, the Hazardous Location Vehicle to Vehicle Warning System (HLV2V-WS) uses the vehicular network to relay information about unsafe road conditions. All of which contribute to the system's overall performance [21].

Along with safety and security features, this system also provides infotainment and internet applications used by the driver to ensure and enhance the journey. A network connection lets the vehicle utilise various standard internet protocol services. A multi-hop link to an RSU as an interconnection point is constructed and maintained. Since the multi-hop link is invisible to the protocol stack's upper levels, practically any IP-based protocol and service may be installed in automobiles. This case demonstrates the advantages of providing the automobile or driver with the ability to obtain any resources on the web [14, 21]. However, if a high degree of dependability and security can be guaranteed, the C2C-CC and its applications will be successful and embraced by users.

Key Elements of IoT-Enabled Vehicular Communication

IoT-enabled vehicular communication involves the integration of various technologies, such as sensors, cameras, GPS, wireless networks, and cloud computing, to provide a seamless and connected experience. The following are some of the key components of IoT-enabled vehicular communication.

Sensors and Actuators: Sensors are essential for IoT-enabled vehicular communication systems as they capture data about the vehicle's status, environment, and performance. Examples of sensors used in vehicular communication include temperature sensors, light sensors, speed sensors, and accelerometer sensors. Actuators are used to control the vehicle's actions, such as braking, acceleration, and steering, based on the data captured by sensors [22].

Communication networks: Communication networks connect the various components of the IoT-enabled vehicular communication system. These networks can include cellular networks, Wi-Fi, Bluetooth, and dedicated short-range communication (DSRC) networks. DSRC networks are specifically intended for V2V and V2I communication and operate at a frequency of 5.9 GHz [23].

Cloud Computing: Cloud computing provides a scalable and secure platform for storing and processing the massive amounts of data generated by IoT-enabled vehicular communication systems. The cloud can also deliver real-time and on-demand services to vehicles, such as traffic updates, weather information, and entertainment services [24].

Data Analytics: Data analytics extract valuable insights from the data generated by IoT-enabled vehicular communication systems. These insights can improve vehicle performance, optimise traffic flow, and enhance passenger experience. TrasoNET is an integrated network framework that explores data analytics and networking to give real-time intelligent transportation services to linked automobiles. Two components form TrasoNET. The first analyses local traffic, user preferences, service applications, and network circumstances to direct vehicles to the right access networks [25].

Security: Security is a critical component of IoT-enabled vehicular communication systems. Due to the sensitive nature of the data generated by these systems, security measures must be implemented to prevent unauthorised access, tampering, and cyber-attacks. Some security measures used in IoT-enabled vehicular communication systems include encryption, authentication, access control, and intrusion detection

[26]. IoT-enabled vehicular communication is an emerging technology that has the potential to transform the automotive industry. Integrating various technologies, such as sensors, communication networks, cloud computing, and data analytics, enables real-time and intelligent transportation services. However, ensuring the security and privacy of these systems is essential to ensure safe and reliable vehicle operations.

Need for Security in Vehicular Communication Systems

Several recent cyber-attacks on vehicular communication systems have highlighted the security vulnerabilities of these systems. Security researchers remotely operated a Jeep Cherokee in 2015 by exploiting a vehicle's infotainment system flaw [27]. They seized control of the vehicle's brakes, steering, and transmission, resulting in a 1.4 million vehicle recall. In 2016, a team of Chinese researchers demonstrated a successful hack on a Tesla Model S vehicle. They were able to manipulate the car's braking systems, locking mechanisms, and dashboard computer remotely by exploiting a flaw in the controller area network (CAN) bus [28]. In 2016, a security researcher was able to hack into a Nissan Leaf's mobile app and gain control of the car's heating and air conditioning system. The vulnerability was related to the app's lack of encryption [29]. In 2018, Fiat Chrysler recalled 4.8 million vehicles due to a vulnerability in their U-Connect system. The vulnerability allowed attackers to take control of the car's engine, brakes remotely, and other critical systems [30].

In 2018, an autonomous Uber car crashed into a pedestrian, leading to the first death caused by an autonomous vehicle. The crash was attributed to a software bug that failed to identify the pedestrian [31]. These attacks demonstrate the potential consequences of security vulnerabilities in vehicular communication systems. They also emphasise the importance of comprehensive security precautions to safeguard these systems and assure the safety and security of drivers and passengers.

The lack of cyber-security in vehicular communications can have significant economic implications. The following are some of the economic impacts of cyber-security breaches in vehicular communication systems.

Loss of trust: Cyber-security breaches in vehicular communication systems can lead to losing trust among consumers and stakeholders. This can result in declining sales and revenue for automotive manufacturers, suppliers, and service providers.

Product recalls: Cyber-security breaches can lead to product recalls, which can be expensive and time-consuming for automotive manufacturers. Product recalls can also damage the reputation of the manufacturer and lead to a loss of consumer confidence.

Litigation costs: Cyber-security breaches in vehicular communication systems can lead to litigation costs, including legal fees, settlements, and damages. These costs can be substantial and can significantly impact the profitability and financial stability of automotive manufacturers, suppliers, and service providers.

Business disruption: Cyber-security breaches in vehicular communication systems can lead to business disruption, including service outages and production delays. This can result in lost productivity, revenue, and increased costs.

Regulatory fines: Cyber-security breaches in vehicular communication systems can result in regulatory fines and penalties. This can result in significant financial costs and damage to the reputation of the manufacturer.

Increased insurance premiums: Cyber-security breaches can increase the cost of insurance premiums for automotive manufacturers, suppliers, and service providers. This can result in higher operating costs and reduced profitability.

In conclusion, the lack of Cybersecurity in vehicular communication systems can have significant economic implications. The costs of Cybersecurity breaches can be substantial and can impact the financial stability, reputation, and profitability of automotive manufacturers, suppliers, and service providers. It is, therefore, crucial to implement robust Cybersecurity measures to ensure the safety and security of drivers and passengers and the economic viability of the automotive industry.

Security Challenges and Their Countermeasures in IoT-Enabled Vehicular Communication

The security challenges of IoT-enabled vehicular communications are complex and diverse, and they can compromise the safety and reliability of the entire transportation system. The following are some of the key security challenges of IoT-enabled vehicular communication.

Device authentication and authorization: IoT-enabled vehicular communication systems involve integrating multiple devices and services, and verifying the identity and authorisation of these devices can be challenging. Attackers can exploit this vulnerability to gain unauthorised access to the system and launch cyber-attacks. Countermeasures include implementing strong authentication mechanisms, such as multi-factor authentication, to ensure the identity and authorisation of devices. In one reported research, using the Smooth Projective Hash Function (SPHF), a (group) two-factor authenticated key exchange (AKE) with low-interactive communication rounds and security validated via the Bellare-Pointcheval-Rogaway (BPR) model is developed [32].

Encryption: Encryption is essential to protect the privacy and confidentiality of data generated by IoT-enabled vehicular communication systems. However, implementing encryption in a real-time and dynamic vehicular communication environment can be challenging. Countermeasures include using robust encryption algorithms and keys to secure data and implementing a secure key management system. Recently, A method for encrypting information known as Efficient Algorithm for Secure Transmission (EAST) blends data encryption with steganographic methods with a greater time efficiency of 0.86 ms, avalanche effect of 58.81%, and PSNR of 78.58% [33].

Anti-tampering measures: Attackers can physically tamper with IoT-enabled vehicular communication systems, such as replacing legitimate devices with malicious ones or intercepting and modifying data during transmission. Countermeasures include implementing anti-tampering mechanisms, such as tamper-resistant hardware and software, such as PKI technology [34], and monitoring systems to detect and alert any tampering attempts.

Intrusion detection: Intrusion detection is essential to detect and respond to cyberattacks on IoT-enabled vehicular communication systems. Countermeasures include implementing intrusion detection systems that can detect anomalies in the network traffic and system behaviour and providing real-time alerts to the system administrators. One such method proposed for intrusion detection was using an unsupervised Kohonen Self-Organizing Map to attack intrusion messages sent over the CAN bus [35].

GPS spoofing and jamming: GPS (Global Positioning System) is a critical component of IoT-enabled vehicular communication, and its disruption can cause significant problems for the vehicle's operation. GPS spoofing involves deceiving the GPS receiver by sending false location information, while GPS jamming involves blocking the GPS signal. Countermeasures include anti-spoofing and anti-jamming techniques, such as signal filtering, power monitoring, and cross-checking of multiple sources of location information and implementing advanced GPS technologies, such as DGPS (Differential GPS) [36] and RTK (Real-Time Kinematic) [37].

Privacy concerns: IoT-enabled vehicular communication systems generate a vast amount of data that can be used to track and monitor the activities of drivers and passengers. This can raise privacy concerns, and attackers can exploit this vulnerability to access sensitive information. Countermeasures include implementing data minimisation techniques, such as limiting the collection and storage of personal information and implementing privacy-preserving protocols, such as differential privacy [38].

Denial of Service attacks: Denial of Service (DoS) attacks can interfere with the regular operation of IoT-enabled vehicular communication systems, leading to service outages and safety concerns. Countermeasures include implementing DoS protection mechanisms, such as rate limiting, throttling, and load balancing [39].

Driver monitoring: Monitoring driver behaviour is essential to ensure safe and reliable vehicle operations. However, monitoring drivers can raise privacy concerns and compromise the system's security if the monitoring systems are not adequately secured. Countermeasures include implementing secure driver monitoring systems that use encryption, authentication, and access control mechanisms to protect driver data [40, 41].

Lack of standardization: The lack of standardisation in IoT-enabled vehicular communications can create security vulnerabilities. One countermeasure is to adopt industry-wide standards and guidelines, such as the ISO/SAE 21,434 standard for Cybersecurity engineering of road vehicles [42].

Vehicle-to-vehicle (V2V) and Vehicle-to-Infrastructure (V2I) communication security: V2V and V2I communication is an essential aspect of IoT-enabled vehicular communication, but it can be vulnerable to attacks, such as eavesdropping and man-in-the-middle attacks. One countermeasure is using secure communication protocols, such as TLS (Transport Layer Security), to ensure the communication is encrypted and authenticated [43].

Human Factors: Humans are also a security challenge in IoT-enabled vehicular communication. For instance, drivers can be distracted or compromised, resulting in unsafe driving behaviours. Countermeasures include implementing driver monitoring systems, such as cameras and sensors, to detect unsafe driving behaviours and provide alerts to the driver [44].

Supply Chain Security: The IoT ecosystem involves multiple vendors and suppliers, increasing the supply chain's complexity. Attackers can exploit vulnerabilities in the supply chain to compromise IoT-enabled vehicular communication systems [45]. Countermeasures include conducting security audits on third-party vendors and suppliers and implementing security controls, such as access controls, to limit the attack surface.

Data Integrity: Ensuring the accuracy and consistency of data generated by IoT-enabled vehicular communication systems is critical to maintaining safe and reliable vehicle operations [46]. Countermeasures include implementing data validation controls, such as checksums, to verify that the data is accurate and has not been modified.

Insider threats: Internal threats like employees or contractors accessing the IoT-enabled vehicular communication systems can pose a security risk. Countermeasures include implementing access controls and monitoring tools to detect unauthorised activities and provide alerts.

Cybersecurity Education: Education and training are essential in mitigating IoT-enabled vehicular communication security challenges. Countermeasures include providing cyber security training to employees and drivers to raise awareness of security risks and best practices for maintaining secure operations.

Conclusion

Regarding execution and dependability, vehicular communication or networking is a highly significant and challenging undertaking continually redesigned and adjusted as new technologies are introduced. It is clear that technology has much potential in terms of assisting upcoming intelligent transportation systems. V2V networking offers undeniable qualities that make the development of next-generation intelligent transportation systems possible at a much faster rate. Maintaining the security system and establishing its trustworthiness with the help of interdisciplinary techniques is an important area to focus on so that future vehicle manufacturers widely implement these technologies in their products as an essential feature, as well as demonstrate that drivers and passengers can be confident enough to accept the guidance and assistance provided by these intelligent services and gain their trust. Vehicle communication technology is, without a doubt, among the most significant technological advancements in safe and efficient transportation in the modern world.

IoT-enabled vehicular communication systems pose significant security challenges that require a multi-layered approach to address them. Implementing a comprehensive security framework, including strong authentication, encryption, anti-tampering measures, intrusion detection, GPS anti-spoofing and anti-jamming techniques, privacy-preserving protocols, DoS protection, and secure driver monitoring systems, can help mitigate these challenges and improve the overall safety and security of the transportation system.

References

1. Karagiannis, G., et al. (2011). Vehicular networking: A survey and tutorial on requirements, architectures, challenges, standards and solutions. *IEEE Communications Surveys & Tutorials, 13*(4), 584–616.
2. Suh, W., Hunter, M. P., & Fujimoto, R. (2014). Ad hoc distributed simulation for transportation system monitoring and near-term prediction. *Simulation Modelling Practice and Theory, 41*, 1–14.
3. Papadimitratos, P., La Fortelle, A., Svensson, K., Brignolo, R., & Cosenza, S. (2009). Vehicular communication systems: Enabling technologies, applications, and future outlook. *Intelligent Transport Communication Magazine*, 84–95.
4. Wang, S. Y., & Chou, C. L. (2009). NCTUns tool for wireless vehicular communication network researches. *Simulation Modelling Practice and Theory, 17*(7), 1211–1226.
5. Kato, S., Tsugawa, S., Tokuda, K., Matsui, T., & Fujii, H. (2002). Vehicle control algorithms for cooperative driving with automated vehicles and intervehicle communications. *IEEE Transactions on Intelligent Transportation Systems, 3*(3), 155–161.
6. Tsugawa, S. (2002). Inter-vehicle communications and their applications to intelligent vehicles: An overview. In *Intelligent vehicle symposium, 2002*. IEEE (Vol. 2, pp. 564–569).
7. Demba, A., & Möller, D. P. F. (2018). Vehicle-to-vehicle communication technology. In *IEEE international conference on electro/information technology (EIT)* (pp. 459–464).
8. Kawashima, H. (1990). Japanese perspective of driver information systems. *Transportation (Amst), 17*, 263–284.
9. Weil, T. (2009). Service management for ITS using WAVE (1609.3) networking. In *2009 IEEE globecom workshops* (pp. 1–6).
10. Choi, J., Va, V., Gonzalez-Prelcic, N., Daniels, R., Bhat, C. R., & Heath, R. W. (2016). Millimeter-wave vehicular communication to support massive automotive sensing. *IEEE Communications Magazine, 54*(12), 160–167.
11. Ali, I., et al. (2020). A highly reliable, 5.8 GHz DSRC wake-up receiver with an intelligent digital controller for an ETC system. *Sensors, 20*, 4012. https://doi.org/10.3390/s20144012
12. Claussen, H. (1991). Vehicle navigation systems. In *Modern cartography series* (vol. 1, pp. 225–235). Elsevier.
13. CVEL. (2023). Navigation systems. Clemson University Vehicular Electronics Laboratory. Retrieved from 05 March 2023. https://cecas.clemson.edu/cvel/auto/systems/navigation.html
14. Vegni, A. M., Biagi, M., & Cusani, R. (2013). Smart vehicles, technologies and main applications in vehicular ad hoc networks. *Vehicular Technologies and Applications*, 3–20.
15. Kargl, F., et al. (2008). Secure vehicular communication systems: Implementation, performance, and research challenges. *IEEE Communications Magazine, 46*(11), 110–118.

16. Kosch, T., Kulp, I., Bechler, M., Strassberger, M., Weyl, B., & Lasowski, R. (2009). Communication architecture for cooperative systems in Europe. *IEEE Communications Magazine, 47*(5), 116–125.
17. Sarika, S., Pravin, A., Vijayakumar, A., & Selvamani, K. (2016). Security issues in mobile ad hoc networks. *Procedia Computer Science, 92*, 329–335.
18. Eichler, S., Schroth, C., & Eberspächer, J. (2006). Car-to-car communication
19. Becker, J. C. (2000). Fusion of heterogeneous sensors for the guidance of an autonomous vehicle. In *Proceedings of the third international conference on information fusion* (vol. 2, pp. WED5–11).
20. Leinmüller, T., et al. (2006). Sevecom-secure vehicle communication. In *IST mobile and wireless communication summit*, no. POST_TALK.
21. Baldessari, R., et al. (2007). Car-2-car communication consortium-manifesto.
22. Pendor, R. B., & Tasgaonkar, P. P. (2016). An IoT framework for intelligent vehicle monitoring system. In *International conference on communication and signal processing (ICCSP)* (pp. 1694–1696).
23. Cheng, L., Henty, B. E., Stancil, D. D., Bai, F., & Mudalige, P. (2007). Mobile vehicle-to-vehicle narrow-band channel measurement and characterisation of the 5.9 GHz dedicated short-range communication (DSRC) frequency band. *IEEE Journal on Selected Areas in Communications, 25*(8), 1501–1516.
24. Whaiduzzaman, M., Sookhak, M., Gani, A., & Buyya, R. (2014). A survey on vehicular cloud computing. *Journal of Network and Computer Applications, 40*, 325–344.
25. Chen, C., Luan, T. H., Guan, X., Lu, N., & Liu, Y. (2017). Connected vehicular transportation: Data analytics and traffic-dependent networking. *IEEE Vehicular Technology Magazine, 12*(3), 42–54. https://doi.org/10.1109/MVT.2016.2645318
26. Ahmad, F., Adnane, A., & Franqueira, V. N. L. (2016). A systematic approach for cyber security in vehicular networks. *Journal of Computing and Communication, 4*(16), 38–62.
27. Drozhzhin, A. (2015). Black Hat USA 2015: The full story of how that Jeep was hacked. Kaspersky Daily. https://www.kaspersky.com/blog/blackhat-jeep-cherokee-hack-explained/9493/
28. Pauli, D. (2016). Hackers hijack Tesla Model S from afar, while the cars are moving. The Register. https://www.theregister.com/2016/09/20/tesla_model_s_hijacked_remotely/
29. Micro, T. (2016). Nissan leaf can be hacked via mobile app and web browser. https://www.trendmicro.com/vinfo/fr/security/news/internet-of-things/nissan-leaf-can-be-hacked-via-mobile-app-and-web-browser
30. Brewster, T. (2015). Chrysler recalls 1.4 million cars after jeep vulnerability exposed. Forbes. https://www.forbes.com/sites/thomasbrewster/2015/07/24/chrysler-recall-exploit/?sh=37e07c411b70
31. Wakabayashi, D. (2018). Self-driving uber car kills pedestrian in Arizona, where robots roam. NY Times. https://www.nytimes.com/2018/03/19/technology/uber-driverless-fatality.html
32. Li, Z., Wang, M., Sharma, V., & Gope, P. (2022). Sustainable and round-optimised group authenticated key exchange in vehicle communication. *IEEE Transactions on Intelligent Transportation Systems*, 1–14. https://doi.org/10.1109/TITS.2022.3169182
33. Rathore, M. S., et al. (2022). A novel trust-based security and privacy model for internet of vehicles using encryption and steganography. *Computers & Electrical Engineering, 102*, 108205. https://doi.org/10.1016/j.compeleceng.2022.108205
34. Wang, F., Xu, Y., Zhang, H., Zhang, Y., & Zhu, L. (2016). 2FLIP: A two-factor lightweight privacy-preserving authentication scheme for VANET. *IEEE Transactions on Vehicular Technology, 65*(2), 896–911. https://doi.org/10.1109/TVT.2015.2402166
35. Barletta, V. S., Caivano, D., Nannavecchia, A., & Scalera, M. (2020). Intrusion detection for in-vehicle communication networks: An unsupervised Kohonen SOM approach. *Future Internet, 12*(7), 119.
36. Wen, H., Huang, P. Y. R., Dyer, J., Archinal, A., & Fagan, J. (2005). Countermeasures for GPS signal spoofing. In *Proceedings of the 18th international technical meeting of the Satellite division of the Institute of Navigation (ION GNSS 2005)* (pp. 1285–1290).

37. Feng, Y., & Wang, J. (2008). GPS RTK performance characteristics and analysis. *Positioning, 1*(13).
38. Zhao, P., Zhang, G., Wan, S., Liu, G., & Umer, T. (2020). A survey of local differential privacy for securing internet of vehicles. *The Journal of Supercomputing, 76*, 8391–8412.
39. Loukas, G., & Öke, G. (2010). Protection against denial of service attacks: A survey. *The Computer Journal, 53*(7), 1020–1037.
40. Levshun, D., Chechulin, A., & Kotenko, I. (2022). Security and privacy analysis of smartphone-based driver monitoring systems from the developer's point of view. *Sensors, 22*(13), 5063.
41. Alsibai, M. H., & Siang, H. M. (2015). A smart driver monitoring system using android application and embedded system. In *2015 IEEE international conference on control system, computing and engineering (ICCSCE)* (pp. 242–247).
42. Macher, G., Schmittner, C., Veledar, O., & Brenner, E. (2020). ISO/SAE DIS 21434 automotive cybersecurity standard-in a nutshell. In *Computer safety, reliability, and security. SAFECOMP 2020 Workshops: DECSoS 2020, DepDevOps 2020, USDAI 2020, and WAISE 2020*, Lisbon, Portugal, 15 September 2020, Proceedings vol. 39, pp. 123–135.
43. Van Der Heijden, R. (2010). Security architectures in V2V and V2I communication. In *Proceedings of the 20th student conference IT* (pp. 1–10).
44. Zhang, H., Zhang, Z., & Liang, J. (2022). Dynamic driving intention recognition of vehicles with different driving styles of surrounding vehicles. *IET Intelligent Transport Systems, 16*(5), 571–585.
45. Moghadasi, N., et al. (2022). Trust and security of electric vehicle-to-grid systems and hardware supply chains. *Reliability Engineering and System Safety, 225*, 108565. https://doi.org/10.1016/j.ress.2022.108565
46. Lin, J., Yu, W., Zhang, N., Yang, X., & Ge, L. (2018). Data integrity attacks against dynamic route guidance in transportation-based cyber-physical systems: Modeling, analysis, and defense. *IEEE Transactions on Vehicular Technology, 67*(9), 8738–8753.

Industrial IoT Security Infrastructures and Threats

Daniel Dauda Wisdom, Olufunke Rebecca Vincent, Kingsley Igulu, Eneh Agozie Hyacinth, Arinze Uchechukwu Christian, Odunayo Esther Oduntan, and Adamu Ganya Hauni

Abstract The term "Industrial Internet of Things" (IIoT) is used to describe the network of sensors, industrial devices, and systems with the internet to enhance automation, efficiency, and optimization of industrial processes. IIoT has brought about significant improvements in industrial automation, operational efficiency, and cost reduction. However, the IIoT is vulnerable to cyber-attacks, which can cause significant disruptions to critical infrastructure, leading to financial losses and loss of life. This has made security a critical concern in IIoT systems. Hence, this chapter discusses Industrial IoT Security Threats and Counter Measures. In the chapter,

D. D. Wisdom (✉) · O. R. Vincent
Department of Computer Science, Chrisland University Abeokuta, Abeokuta 110104, Ogun State, Nigeria
e-mail: ddaniel@chrislanduniversity.edu.ng

O. R. Vincent
e-mail: rebecca@gmail.com; vincentor@funaab.edu.ng

O. R. Vincent
Department of Computer Science, Federal University of Agriculture, Abeokuta 111101, Ogun State, Nigeria

K. Igulu
Department of Computer Science, Ignatius Ajuru University, Port Harcourt, River State, Nigeria
e-mail: igulu.kingsley@iaue.edu.ng

E. A. Hyacinth · A. U. Christian · O. E. Oduntan
Department of Computer Science, University of Nigeria Nsukka (UNN), Enugu State, Nigeria
e-mail: agozieeneh@gmail.com; agozie.eneh@unn.edu.ng

A. U. Christian
e-mail: arinzechukwu@gmail.com; uchechukwu.arinze@nnra.gov.ng

O. E. Oduntan
e-mail: eoduntan@chrislanduniversity.edu.ng

A. G. Hauni
Department of Mathematics, Usmanu Danfodio University Sokoto (UDUS), Sokoto, Sokoto State, Nigeria
e-mail: adamuganya@gmail.com

A. Prasad et al. (eds.), *Communication Technologies and Security Challenges in IoT*, Internet of Things, https://doi.org/10.1007/978-981-97-0052-3_19

essential ways to mitigate cyber-attacks on IIoT systems is presented and a comprehensive review and security strategy that includes risk assessment, threat modeling, and vulnerability testing as well as the proposed counter measure model. In the proposed model, the strategy incorporates security protocols and standards. Such as encryption, access control, and authentication, in a bid to guarantee the integrity, confidentiality, and availability of the IIoT system. Additionally, network segmentation is also introduced, which involves dividing the IIoT network into smaller, isolated segments that can be secured independently mitigating cyber-attack. The proposed model limits the impact of a potential breach and allows for more granular control over access to different parts of the IIoT system. In addition, the proposed model implemented an intrusion detection and prevention systems (IDPS) which helps in detecting and mitigating cyber-attacks by monitoring traffic, identifying potential threats, and blocking malicious activity. The proposed model is also useful in creating awareness about the various cyber-attacks; as well as presents essential countermeasures against cyber-attacks. Finally, the proposed scheme mitigates the risk of cyber-attacks on IIoT systems while ensuring the safety and reliable operations of critical infrastructure respectively.

Keywords IIoT · Cyber-attack · Threats · Counter measures and infrastructure

Introduction

The concept "Industrial Internet of Things," or IIoT for short, describes the process of connecting sensors, devices, and other types of industrial systems to the internet. This gives these systems the ability to collect and share data in real time. The IIoT is also known as Industry 4.0 due to the fact that it is the fourth industrial revolution and is altering conventional manufacturing processes and general industrial operations [1]. The major objective of the IIoT is to improve operational efficiency, hence reducing costs, increasing production, and improving overall product quality. By connecting industrial systems, IIoT enables businesses to collect data on their operations, identify inefficiencies, and optimize processes. IIoT can also help businesses gain insights into their supply chain, allowing them to make better decisions and improve their competitive advantage. Included within its scope are manufacturing equipment, supply chain management systems, energy management systems, and building automation systems [2]. These systems can connect to the internet, collect data, and send it to the cloud for analysis thanks to sensors, actuators, and other devices. IIoT relies on a range of technologies to for collection of data and analysis, including machine learning, artificial intelligence (AI), cloud computing, and big data analytics. These technologies make it possible for businesses to analyze vast amounts of data in real time, recognize patterns and trends, and base their decisions on the data analyzed. One of the key advantages of IIoT is that it enables businesses to move from reactive to proactive maintenance. By collecting data on equipment performance and using predictive analytics, businesses can identify potential issues before they become major problems [3]. This strategy can aid firms in reducing downtime, enhancing equipment dependability, and extending the life of their assets. IIoT also has implications for safety and environmental monitoring. By integrating safety systems and sensors, IIoT can help identify potential safety hazards and alert workers in real-time. Similarly, by monitoring environmental conditions, IIoT can help businesses optimize energy consumption, reduce waste, and meet regulatory requirements. Industrial Internet of Things (IIoT) is the basis of how IoT technologies are applied in industrial and manufacturing environments, including factories, power plants, and oil rigs [4]. These systems aim to boost productivity, cut expenses, and streamline operations, but they also present new security risks. More specifically, the manufacturing and production sectors have undergone a revolution thanks to the IIoT, which has increased automation, efficiency, and productivity. To protect sensitive data and critical infrastructure, it is necessary to address the security risks that the integration of IoT devices and systems also brings about. IIoT security threats also include potential flaws and dangers related to the use of connected devices, sensors, and other industrial systems in an industrial setting [5]. These threats include cyber-attacks, data breaches, physical security breaches, and insider threats. One existing problem in Industrial IIoT security is the vulnerability of IIoT devices and systems to cyber-attacks, which can have serious consequences on industrial operations and public safety. Hackers can easily gain access to private information, take control of industrial processes, and even cause physical damage to equipment by taking advantage of flaws in the design of many IIoT devices [6]. Vital solution to this problem is to implement robust security measures that can protect IIoT devices and systems against cyber threats. This includes: Device Authentication: Implementing device authentication can prevent unauthorized access to IIoT devices and systems. This can be achieved by using secure passwords, encryption, and digital certificates. Data Encryption: Data encryption can protect sensitive information from unauthorized access by encrypting data in transit and at rest. Network Segmentation: Network segmentation can help to prevent hackers from accessing critical systems and data by segmenting the network into smaller, isolated segments. Regular Software Updates: Regular software updates can help to patch vulnerabilities and improve the security of IIoT devices and systems [7–10]. Threat Monitoring: Implementing threat monitoring can help to identify and react to cyber-attacks in real-time. Thus, Industrial IoT Security Threats and Counter Measures scheme is proposed. The proposed scheme presents some of the most common IIoT security threats and explores potential countermeasures to mitigate these risks. The various level layers are shown in Fig. 1 [11]. The rest of the paper is structured as follows: This section follows introduction, sections "Related Works", "IIoT Security threats", "IIoT Architecture" and "Conclusion" research study.

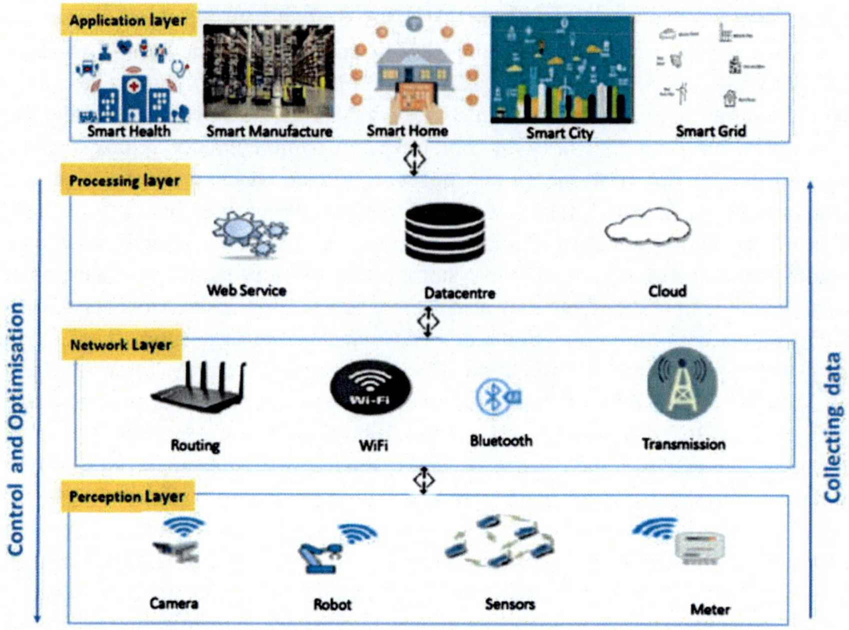

Fig. 1 A comprehensive IoT architecture with level layers

Related Works

Internet of Things (IoT) more essentially, Industrial Internet of Things (IIoT). It's a vital field of study in the digital era. Previous studies focused on IIoT system integrity first, and then on IoT system integrity. This paper provides a comprehensive study of IIoT with different goals than the previous ones. With the increasing use of IoT in industries, it's now possible to connect anything at any time and place in the industry system context. Studies have shown that privacy and security issues are significant challenges for industrial IoT. In addition, Vehicles, home appliances, smart grids, aerospace, and defense are just some of the many uses for embedded electronics. These devices need to be secured from time with an efficient encryption that meets various resource constraints respectively. This section present related research works as follows: In the study referenced as [12], various lightweight algorithm schemes suitable for embedded systems are presented to provide both integrity and confidentiality. In this study, various lightweight authenticated encryption algorithms are compared and contrasted in terms of their effects on performance metrics such as throughput, latency, and energy usage. The establishment of keys is one of the countermeasures that is contemplated as a possible option for ensuring the safety and confidentiality of IIoT systems.

The study presented in [13] provides an in-depth examination of both physical-layer and higher-layer for establishing symmetric and asymmetric keys techniques.

While the authors stress the need to develop cross-layer key-establishment protocols for IIoT. They look into how cross-layer design allows IIoT gadgets to set up communication keys independent of a trusted third party or any presumptions about the safe and secure transfer of confidential information. IIoT systems were the focus of a comprehensive study [14], which zeroed in on the challenges connected with ensuring their security and privacy. Security threats, attack vectors, and the need for additional security were all factors. The authors propose a variety of solutions for securing industrial IoT, including protection design, integrity testing, secure device and sensor management, and the technology that make it possible can be found in Reference [15] The authors demonstrate how the integration of these technologies might alter traditional industries such as healthcare, manufacturing, smart grids, and transportation by highlighting a number of different domains in which the industrial internet can be utilized. In their paper, which is cited in [16], the authors provide an exhaustive survey of Supervisory Control and Data Acquisition (SCADA) system architectures. After that, we'll look at how vulnerable SCADA systems are and what we can do to make them safer. This is followed by an analysis of attacks on SCADA systems as well as the security needs for SCADA systems. The authors carry out a thorough investigation into the various intrusion detection methods that can be used to SCADA.

Reviews on IoT System Integrity

A thorough examination of the Internet of Things' (IoT) current security flaws and vulnerabilities is given in reference number [17]. The design, communication, and application of the IoT are all covered in this analysis. The authors offer some ideas for potential remedies to the issues and risks brought on by the internet of things environment. Similarly, [18], looks at the current initiatives being taken to secure Internet of Things apps within smart cities and focuses on security and privacy issues related to those applications. Critical security challenges are identified and categorized using a methodical methodology in [19]. Internet of Things (IoT) security roadmap, Data privacy, trust, identification, and authentication are some of the key issues. The authors of [20], provide a thorough analysis that concentrates on privacy issues in IoT and covers both potential issues and potential solutions. The Internet of Things' characteristics are examined in [21], which also discusses DoS and DDoS attacks on IoT networks, particularly those that target the MAC layer. The mechanisms used to mitigate these attacks are also examined by [21], such as intrusion detection systems (IDSs).

Smart devices that make up the family of Internet of Things present unique security challenges, which are outlined by the authors of [22, 23]. They continue by discussing many different security protocols like Datagram Transport Layer Security (DTLS), Host Identity Protocol (HIP). Internet Protocol Security (IPSec). Elliptic Curve Digital Signature Algorithm (ECDSA), Rivest-Shamir-Adleman (RSA), Advanced Encryption Standard (AES), and Tiny Encryption Algorithm (TEA) are some of the cryptographic methods they investigated for use in the IoT. The effect that cryptographic algorithms have had on the IoT is discussed in [24]. Lightweight algorithms that have been implemented on different kinds of software and hardware tools are compared in reference [25], with a focus on memory and processing capa. The authors classified lightweight algorithms into hash functions, stream ciphers, and block ciphers based on their function. According to their findings, lightweight algorithms are crucial for boosting IoT security while simultaneously decreasing device power consumption and memory usage.

In-depth analysis of potential preventative measures for Internet of Things threats is provided in the article [26]. The survey analyzes the risks posed to each layer separately with the aid of a model-based risk assessment strategy. While [27] addresses security issues related to Internet of Things protocols, including the Message Authentication Code (MAC) and physical layer communication drawbacks, the study focuses on flaws in communication protocols. The communication taking place at the physical and MAC layer levels is the focus of the latter study.

References [28, 29] investigates Internet of Things security risks and categorized them based on the specific threats each model presents. This paper also highlights security solutions for IoT communication protocols and suitable protection tailored to address the technological heterogeneity and limited resources of the IoT paradigm.

Despite the fact that the preceding sections presented studies covering a wide range of security problems, attacks, and layers, these studies have limitation and do not offer a comprehensive analysis of the numerous security solutions available today. Thus, there is still a need for a comprehensive study that covers the vast majority of the currently available Internet of Things and Industrial Internet of Things security attacks, solutions, and countermeasures. This paper aims to close that knowledge gap by providing an in-depth analysis of the current state of IoT and IIoT security threats, solutions, and countermeasures.

IIoT Security Threats

The IIoT is revolutionizing the way industries operate by enabling machines, devices, sensors, and other equipment to connect and communicate with each other over a network. While IIoT has the potential to enhance efficiency, productivity, and overall business operations, it also exposes industrial systems to a variety of security threats [21, 30]. In this response, we will explain some of the most common IIoT security threats.

1. Cyberattacks and data breaches: Since IIoT devices are frequently connected to the internet, they are susceptible to malware, hacking, and other cyberattacks. Attackers can take advantage of security vulnerabilities in IIoT devices to steal confidential information, halt operations, and physically damage equipment.
2. Malwares: Malicious software programs, commonly referred to as malware, are typically intended to damage, disrupt, or obtain illegal access to a computer system. IIoT devices are not immune to malware attacks, and if an attacker manages to inject malware into an IIoT device, they can potentially gain control

of the device and use it to launch further attacks [31, 32]. IIoT systems are vulnerable to malware attacks, which can cause significant damage. For example, an attacker could use malware to modify a factory's control system, causing it to malfunction and potentially cause physical harm. More so, Malware and ransomware attacks can be devastating to IIoT systems, as they can quickly spread throughout a network, disrupting operations and compromising sensitive data. Malware can be introduced into an IIoT system through a variety of vectors, including email phishing attacks, malicious downloads, and infected USB devices [33]. Ransomware attacks, where an attacker demands payment in exchange for releasing control of a system or restoring access to data, can be particularly damaging to IIoT systems, as they can cause significant downtime and financial losses.

3. Physical Attacks: Physical attacks on IIoT systems can occur in a number of ways, such as tampering with devices, stealing sensitive data, or disrupting operations. Physical attacks can be particularly challenging to defend against, as they often require physical access to the system or device [34]. IIoT devices that control critical infrastructure such as power grids, transportation systems, and water treatment plants can be targeted by physical attacks. Attackers can damage or destroy the devices, causing widespread disruption and even endangering human life.
4. Denial-of-service attacks: DoS attacks, or denial-of-service attacks, are a type of cyberattack in which the target network is flooded with so much traffic that it either crashes or is rendered inaccessible to legitimate users. IIoT devices that are connected to the internet can be targeted by DoS attacks, disrupting operations and causing financial losses [35]. DoS attacks are a common threat to IIoT systems. These attacks may also overload the system with traffic, rendering it unusable. For example, an attacker could launch a DoS attack against a factory's control system, causing it to shut down and halt production. DoS attacks are also common threat to IIoT systems, where attackers flood a network with traffic or requests, causing it to become overloaded and unresponsive. These attacks can be launched using botnets or other malicious tools, and can cause significant disruption to IIoT systems, leading to downtime, lost revenue, and damage to the organization's reputation [36].
5. Supply chain attacks: The purpose of supply chain attacks is to implant malware or other malicious code into IIoT devices before they reach the end user. Attacks of this caliber can be challenging to identify and have negative effects on the entire supply chain.
6. Lack of security updates and patches: Many IIoT devices are designed to operate for long periods of time without being updated or patched. This can leave them vulnerable to known security vulnerabilities that have been patched in more recent versions of the device or software [37].
7. Inadequate authentication and access control: IIoT devices often have weak authentication and access control mechanisms, making it easy for attackers to gain access to them. Attackers can use stolen or guessed credentials to gain access to IIoT devices, allowing them to cause damage or steal data.

8. Unauthorized access: One of the most significant IIoT security threats is unauthorized access. This can happen if an attacker breaks into a system by using a weak point in the code or the physical components [38]. For example, an attacker could use a phishing attack to trick an employee into giving up their login credentials or exploit vulnerability in an unsecured network. More-so, unauthorized access can occur through a variety of means, including exploiting weak passwords, using stolen credentials in the system's software or firmware. A hacker who gains access to an IIoT system may be able to compromise or disrupt its operations, steal confidential information, or carry out attacks on other connected devices or networks.
9. Data breaches: A data breach occurs when an adversary obtains access to sensitive information, such as production processes or consumer data, and uses that information for their own malicious purposes. This can result in monetary loss, damage to reputation, and fines imposed by regulatory agencies. An adversary, for instance, could get access to information regarding a customer's energy consumption by exploiting a vulnerability in a smart meter system [39].
10. "Man in the Middle" (MitM) attacks happen when a fraudster is able to monitor or even alter the ongoing communication between two computers or other electronic gadgets [40]. These attacks provide the adversary the ability to listen in on private conversations without being discovered. Attacks utilizing a man in the middle technique can be carried out in various ways, such as by taking advantage of loopholes in communication protocols, connecting to rogue access points, or infiltrating a device or network.

Hence, IIoT security threats are numerous and diverse. To mitigate these threats, Strong authentication and access control mechanisms, regular security updates and patches, and network segmentation to isolate critical systems from potentially compromised devices are all essential components of a comprehensive security strategy. Additionally, organizations must adopt a proactive approach to security, continuously monitoring for threats and actively responding to any incidents that arise.

IIoT Countermeasures

Securing IIoT systems requires a comprehensive approach that incorporates a range of countermeasures, including secure network architecture, authentication and access controls, encryption, patching and updates, monitoring and logging, incident response planning, physical security, and training and awareness. By implementing these measures, organizations can help ensure that IIoT systems are secure and can deliver the benefits of increased automation and efficiency without compromising the safety or security of critical systems [41]. IIoT which describes the use of connected devices and sensors in industrial environments to collect data and automate processes. While IIoT has the potential to revolutionize industrial operations, it also poses unique security challenges due to the scale and complexity of the systems involved. Countermeasures are therefore necessary to protect against potential threats and attacks. Here are some of the countermeasures that can be employed to secure

IIoT systems: Secure network architecture: Secure network architecture is critical to ensuring the security of IIoT systems [42]. Network segmentation, firewalls, and access controls can help limit the potential attack surface and prevent unauthorized access to critical systems. Authentication and access controls: Robust authentication and access controls are necessary to ensure that only authorized personnel can access IIoT systems. This can include measures such as two-factor authentication, strong passwords, and access controls based on job function [43]. Encryption: Encryption is a critical component of IIoT security. Data in transit should be encrypted to prevent interception and data at rest should be encrypted to prevent unauthorized access. Patching and updates: Regular patching and updates to IIoT devices and systems are necessary to address security vulnerabilities and ensure that systems are up-to-date with the newest security patches. Monitoring and logging: Monitoring and logging are necessary to detect and respond to potential security threats. This can include measures such as intrusion detection systems, network traffic analysis, and event logging [44, 45]. Incident response planning: To ensure that IIoT systems can be secured rapidly and efficiently in the event of a security incident, a well-defined incident response plan is required. Procedures for reporting incidents, establishing response teams, and establishing channels of communication are all part of this. Protecting IIoT systems from theft, vandalism, or sabotage that involves the use of physical force requires the use of physical security measures like access controls, video surveillance, and security personnel. Education and understanding: If you want to make sure your staff is prepared to deal with any security issues that may arise, training and awareness programmes are essential. Training for responding to security incidents, phishing attacks, and other similar activities fall under this category [46]. To be more precise and emphatic we have itemize some counter measures that will be of great importance to data integrity as follows: Network segmentation: which involves dividing a network into smaller, more manageable subnetworks. This helps to prevent unauthorized access and contain any security breaches. For example, a factory could segment its network into separate zones for production, administration, and maintenance. Network segmentation may also involve dividing an IIoT system into smaller, isolated subnetworks, each with its own security measures and access controls. This can help prevent the spread of malware or other threats throughout the network, as well as limit the impact of a potential attack [47].

Strong authentication: Strong authentication involves using biometric or multi-factor authentication (MFA) to confirm a user's identity. This helps to prevent unauthorized access, even if an attacker has stolen login credentials. For example, a factory could require employees to use MFA to access its control systems.

Encryption: Encryption involves encoding data so that it can only be read by authorized users. This helps to prevent data breaches and protect sensitive information. For example, a power plant could encrypt its control system data to prevent attackers from gaining access. Encryption can help protect sensitive data as it is transmitted across an IIoT network [48]. End-to-end encryption can ensure that data is only accessible to authorized parties, even if it is intercepted during transmission.

Regular patching and updates: Regular patching and updates help to prevent vulnerabilities in software and hardware from being exploited. IIoT systems should be regularly updated with the latest security patches to prevent attacks. For example, a smart meter system should be updated regularly to prevent vulnerabilities from being exploited [49].

Monitoring and alerting: Monitoring and alerting involves using tools to detect and alert security teams to any suspicious activity. This helps to prevent attacks from going undetected and limits the damage that an attacker can cause. For example, a factory could use intrusion detection systems to monitor its control systems for any suspicious activity [11].

Strong Authentication: Strong authentication procedures, such as two-factor authentication, are one method that can assist in preventing illegal access to IIoT systems. Both complicated passwords and passwords that are changed frequently should be used, and access controls should be examined and modified on a regular basis.

Intrusion Detection and Prevention: Intrusion prevention and detection systems are useful for finding and preventing attacks on an IIoT network. Attempts to exploit vulnerabilities or access sensitive data are two examples of behaviors that these systems can be set up to detect [50].

Firmware and software updates should be applied routinely to fix security flaws and foil attacks. It is important for businesses to keep track of all of their IIoT equipment.

Finally, IIoT security threats are real growing concerns. However, by implementing the right countermeasures, organizations can mitigate these risks and protect their critical systems and data.

IIoT Security Threats and Countermeasures

Cyber-attacks: Cyberattacks can take many forms, such as malware, ransomware, phishing, and denial-of-service (DoS) attacks, among others. These attacks can disrupt the operation of connected industrial systems and cause serious damage to equipment and facilities [51]. Countermeasure: To mitigate cyber-attacks, IIoT systems need to be protected by robust cybersecurity protocols, such as intrusion detection systems, firewalls, and encryption. Additionally, regular software updates and patches should be implemented to address any known vulnerabilities. Data breaches: IIoT systems generate a significant amount of data, which if compromised, can result in serious consequences, including loss of intellectual property, confidential information, and financial losses. Countermeasure: To mitigate data breaches, IIoT systems should implement strong data encryption, access controls, and authentication mechanisms. Also, data should be backed up regularly and stored securely to prevent unauthorized access [45, 52].

Physical security breaches: Physical security breaches refer to unauthorized access to industrial systems, which can result in equipment damage, theft, or sabotage. Countermeasure: Physical security measures such as cameras, biometric scanners, and access control systems can help prevent physical security breaches. Additionally, monitoring and tracking of all access to critical systems and assets can provide an early warning system in case of any suspicious activity [11].

Insider threats: The term "insider threat" refers to any activity, deliberate or not, taken by an employee, contractor, or vendor that compromises the security of an industrial internet of things (IIoT) system. Countermeasure: Access control measures should be implemented in IIoT systems to ensure that only those workers and contractors who need access to sensitive data and systems are granted that access. Additionally, continuous monitoring of employee behavior and activity can help detect any suspicious behavior and provide an early warning system.

Supply chain attacks: Supply chain attacks refer to attacks that target the suppliers of IIoT systems, compromising their security and leading to a cascading effect on the entire IIoT ecosystem [53]. Countermeasure: It is essential for industrial Internet of Things (IIoT) systems to perform due diligence on all of their suppliers, which includes verifying their security protocols and conducting regular security audits. This is a crucial component of their protection against attacks on the supply chain. In addition, contingency measures should be developed in case there are breaks in the supply chain.

Finally, mitigating IIoT security threats requires a comprehensive approach that involves a combination of cyber security, physical security, access control, and continuous monitoring. It is critical to implement robust security protocols, conduct regular security audits and risk assessments, and have contingency plans in place to deal with any potential security breaches. By taking these measures, IIoT systems can help ensure the security and integrity of their operations and protect themselves against potential threats.

IoT Architecture

Multiple layers make up the Internet of Things architecture, and these layers are differentiated from one another according to the devices and functions they support [54]. Although there is some debate on the number of layers, the majority of scholars are in agreement that there are three fundamental layers: The importance of learning about the architecture of the Internet of Things is explored in this study. Researchers disagree on how many levels make up the IoT, but they do agree that there are at least three: the Perception, Network, and Application layers. Due to the inherent security threats, individual layers require their own layer of protection. An overview of these three tiers and how they encompass the rest of the IoT architecture layers is provided in this article (IoT). In Fig. 1, the technologies and gadgets that make up the Internet of Things are shown in relation to the three-layer architecture, and in Fig. 2, the framework of the three-layer design is shown to include all of the other layers. You can find both numbers down below.

Perception Layer

The Internet of Things' perception layer, also referred to as the sensors layer, is in charge of using sensors to collect data from the environment. This layer's main duty is to use sensors to detect and collect data, which it then processes before sending

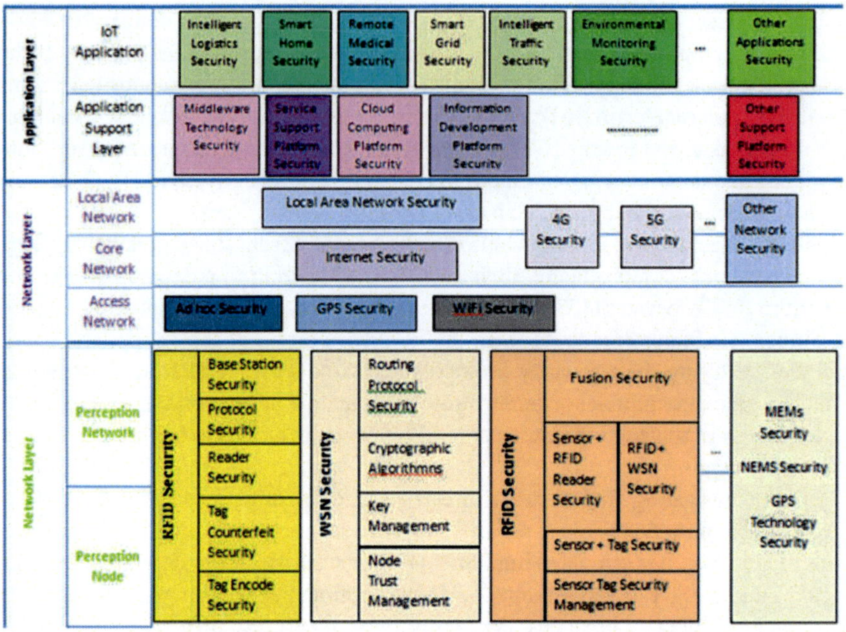

Fig. 2 Proposed comprehensive IoT architecture, digital trends

to the network layer. Additionally, this layer may enable IoT nodes to cooperate with one another in local and short-range networks [55]. The Internet of Things' perception layer is vulnerable to security flaws like interference signals, which could lower the potency and capabilities of wireless connection. IoT devices are vulnerable to attacks due to a variety of factors, such as their energy-constrained components and their active participation in outdoor environments, both of which expose them to physical assault.

Node Tampering

Supposing an attacker obtains physical access to Internet of Things nodes, they will be able to do damage by either completely changing the node or a portion of its hardware, or by employing electronic methods in order to access important information. Cryptographic keys and routing tables are examples of sensitive information that is susceptible to modification. To put it another way, an attacker who has physical access to Internet of Things nodes can either alter portions of the device or the entire device itself, or they can employ electronic ways to gain sensitive information including cryptographic keys or routing tables. This has the potential to cause harm to the nodes, which in turn could jeopardize their functionality or security [17].

RF Interference on RFIDs

By creating and broadcasting unstable signals with RFIDs, an attacker is able to carry out a denial of service (DoS) assault on Internet of Things devices. These signals are able to interfere with RFID signals on other devices, which in turn disrupts communication in the internet of things [18]. Threat actors can launch a denial of service attack by utilizing RFIDs to generate and transmit unstable signals that disrupt the communication of other Internet of Things devices, which can then lead to the loss of service. This prevents Internet of Things devices from working as intended and preventing them from communicating with one another.

Node Jamming on WSNs

Similar to the RF Interference attack, the WSN Interference attack is meant to interfere with Wireless Sensor Networks (WSNs). The attacker in this type of attack interferes with the wireless sensor nodes' radio frequencies, causing signal jamming and hindering the nodes' ability to communicate with one another. If the Internet of Things (IoT) nodes' signals can be disrupted, the malicious actor can effectively halt the services provided by the IoT system [17]. To disrupt Wireless Sensor Networks, an attack known as WSN Interference can disrupt the radio frequencies used by wireless sensor nodes. As a result, the Internet of Things system may grind to a halt as signal jamming prevents nodes from exchanging data with one another.

Malicious Node Injection/Fake Node

An attacker can infiltrate an Internet of Things (IoT) system by inserting a malicious or fake node between two or more nodes, which is known as a "Man-in-the-Middle" (MitM) attack. If the insertion is successful, they will be able to control all data flow operations to and from the IoT system and even introduce malicious data into the network [19].

Sleep Deprivation Attack

The majority of IoT nodes get their power from user-replaceable batteries and are pre-set to hibernate in order to extend the battery's lifespan. Nevertheless, the Sleep Deprivation attack involves continuously reawakening the nodes. This leads to a rise in the amount of power that must be consumed, which, in the end, causes the nodes to shut down [20, 56]. The Sleep Deprivation assault is an attack that upsets Internet of Things nodes by continuously waking them up from their usual naps. This causes a spike in the amount of power that is consumed by the nodes, which ultimately leads to the nodes being shut down. Because Internet of Things nodes are frequently powered by rechargeable batteries and are intended to conserve battery life through nap schedules, this assault presents a particularly difficult challenge.

Malicious Code Injection

The Code Injection attack is an attack where an adversary gains illegal access to an IoT system by injecting malicious or fake code into an IoT node. By doing so, the attacker can take control of the node and manipulate its functionality to their advantage [21]. The Code Injection attack is an attack where an attacker gains unauthorized access to an IoT system by injecting malicious or fake code into an IoT node. This allows the attacker to take control of the node and manipulate its behavior for their own purposes.

Social Engineering/Physical Attack

One type of cyberattack is known as "Physical Tampering," in which an attacker physically interacts with an IoT system in an effort to steal sensitive data or otherwise affect how the system's users experience it [57]. In order to disrupt services or access private information, an attacker can resort to physical interaction with an Internet of Things (IoT) system. The attacker gains physical access to the target Internet of Things (IoT) device or system and then attempts to compromise it.

The Side-Channel attack is a type of attack that targets encryption systems in Internet of Things devices by taking advantage of data from side channels. The amount of power used, the amount of time spent, and electromagnetic radiations from sensor nodes are a few examples of side channel information. The confidentiality of the system is at risk and might even be compromised when the sensitive data stored on Internet of Things devices is altered [40]. More clarifications on various physical attacks are discussed in [28, 31–34]

Network Layer

The network layer of the IoT is responsible for distributing and routing data across IoT nodes in the network. Some examples of devices that employ cutting-edge technologies like WiFi, LTE, Bluetooth, 4G, and 5G in order to function and provide a wide range of network services include Internet gateways, routing devices, and switching devices. Some examples of these methods are shown in Fig. 2. In the Internet of Things (IoT), gateways filter, aggregate, and transmit data to and from various IoT notes [58]. DoS attacks, as was previously mentioned, can also target the network layer of Internet of Things systems. In addition to denial-of-service attacks, the attacker can use traffic analysis, conversation listening, and passive monitoring to exploit network layer vulnerabilities related to confidentiality and privacy [59]. These types of attacks are quite likely to happen due to the remote access capabilities and the frequent data exchange between devices. At the network layer, the "Man in the Middle" attack, which can be detected by listening to conversations, is a major source of worry. The integrity of the encrypted communication channel will be totally destroyed if the private data kept on the devices is listened to. A secure key exchange system is necessary for the Internet of Things in order to prevent identity theft and unauthorized intruding.

Due to the fact that it is machines talking to machines, Internet of Things communication is distinct from Internet communication. This diversity, however, creates a compatibility problem, making it tough to use preexisting network protocols and developing ineffective defense mechanisms. Because of the interconnected nature of modern technology, attackers can easily gain access to previously inaccessible data about their targets [60]. It is critical to safeguard not only the network itself but also the devices that make up the Internet of Things. Security-sensitive gadgets need the ability to monitor their environment and report any suspicious activity to the user. To accomplish this, reliable security protocols and programs must be created [61].

Traffic Analysis Attacks

An RFID Data Sniffing attack is a type of attack in which an attacker intercepts and captures data that is being transmitted wirelessly between RFID devices with the intention of stealing confidential information or gaining unauthorized access to the system. This type of attack is also known as an RFID Data Sniffing attack. This can be accomplished with the help of a number of different apps, such as those for port scanning or packet sniffing [17]. The RFID Data Sniffing attack is an attack where an attacker intercepts and captures wireless data being transmitted between RFID devices. These data may contain sensitive information, and in order to obtain it, the attacker may make use of a variety of tools and apps, such as port scanning or packet sniffing software. Due to the fact that it is a wireless technology, RFID is susceptible to these kinds of assaults; therefore, it is imperative that suitable security precautions be taken in order to protect against them.

RFID Spoofing

The technique known as RFID Spoofing is a form of attack in which an adversary falsifies or spoofs an RFID signal in order to obtain data transfer information from RFID tags. The attacker can then use this information to either steal confidential information or obtain unauthorized access to the system [36]. An RFID Spoofing attack is one in which an adversary generates a spoofed RFID signal in order to steal information from RFID tags. This information may comprise data transfer information, which can be exploited to obtain unwanted access to the system or steal confidential information. Also included in this information may be data. The adversary is able to accomplish this goal by employing specialized tools and methods to imitate the behavior of a genuine RFID device; as a result, they are able to fool the system into granting access or releasing information. It is essential to protect oneself from attacks of this nature by putting in place the appropriate security precautions [62].

RFID Cloning

Data from one legitimate RFID tag is copied to another tag using an unauthorized tag, a form of attack known as "cloning." The duplicate tag is identical to the original except that it lacks the unique identifier that was used to identify the original. As a result, an adversary can forge a tag and use it to gain access to a system or commit other forms of malicious activities [28]. An RFID clone is a fake RFID tag made by hackers using the same technology as the original. The duplicate has the same information as the original but a different ID, making them hard to distinguish. The imposter can then use the counterfeit tag to access restricted areas or engage in other unlawful behavior. Cloned objects can be read and copied with RFID tag readers and writers. After the information has been copied, it can be written to a new tag. Keep in mind that if you get caught cloning, you could face serious consequences. Use of encryption and authentication mechanisms to safeguard RFID tags from cloning attacks is strongly suggested. Tracking and monitoring tag usage for signs of anomalous behavior is also crucial [63].

RFID Unauthorized Access

RFID systems may lack proper validation mechanisms which can make RFID tags accessible to anyone. As a consequence of this, anyone is able to read the data stored in the RFID nodes, edit it, or even erase it [28]. This indicates that there is a potential for unauthorized users to get access to the system, which could result in the modification of sensitive data. Because RFID systems do not have validation processes, they are susceptible to assaults. This makes it much simpler for malicious actors to obtain access to the system and change the data contained within it. Because of this, the RFID system needs to have the appropriate validation processes put into place so that only authorized personnel may use it [64].

Sinkhole Attack

When a sinkhole attack is launched, all data from Wireless Sensor Networks (WSNs) is funneled into a single location. The term "sinkhole attack" describes this specific form of attack. When this kind of attack is carried out, the data privacy is at risk because all data packets are sent to the sinkhole instead of the intended node [35]. Data packets destined for the sinkhole can be intercepted and tampered with by the attacker, potentially leading to data breaches and the loss of sensitive information. The sinkhole attack is a form of Denial of Service (DoS) technique in which an attacker prevents data from reaching its intended destination by directing it to a single-entry point. The entire loss of data or even the modification of critical information might come from this, which poses major security consequences. Therefore, effective security measures must be put in place to identify and forestall sinkhole vulnerabilities [65].

Passive Man in the Middle Attack

Passive man-in-the-middle (PMITM) attack refers to the particular kind of attack that enables a fraudster to intercept a conversation taking place between two nodes in an Internet of Things (IoT) network, restricting or modifying the flow of data without the communicating nodes' knowledge, and then passively monitoring the transit of the data. The attacker does not need to inject a malicious node into the network, but rather relies on the communication protocols of the IoT system to carry out the attack [28]. To carry out the PMITM attack, the attacker needs to be in close proximity to the communication path of the nodes. Once the attacker intercepts the communication, they can manipulate the data transmitted between the nodes, either to modify it, delete it, or redirect it to a different location. This attack has the potential to result in a breach of the data's confidentiality and privacy while it is being transported between the nodes. An important aspect of the PMITM attack is that it requires an attacker to be physically present between two nodes in an Internet of Things network in order to be able to intercept and manipulate data that is being transmitted between the nodes. The attacker does not inject any malicious nodes but instead relies on the communication protocols of the IoT system to carry out the attack [66].

Denial of Service

An attacker can cause a disruption in an IoT network by sending too much data traffic, which overwhelms the capacity of the network. This is known as overburdening, or bombarding, the network. When this happens, the network becomes unavailable for legitimate users who are trying to use the network for useful services [36]. In other words, a threat actor can flood an IoT network with too much data traffic, which causes the network to crash. This means that the network cannot be used for its intended purposes by legitimate users.

Routing Information Attacks

An attacker can manipulate the routing information in an IoT network by using techniques such as spoofing or replaying routing information. This can cause the routing information to be spread incorrectly throughout the network, which can lead to a number of problems [37]. For example, the network may form routing loops, which means that packets of data are sent in circles and never reach their intended destination. Additionally, false error messages may be distributed, causing unnecessary disruption to the network. The network may also become divided, with certain parts of the network becoming isolated from others. The attacker may even be able to alter the source routes of the network, either shortening or extending them as they see fit. Finally, the network may experience an overload of traffic, as the attacker is able to redirect network traffic to a single point, causing that part of the network to become overwhelmed and unable to function properly.

Sybil Attack

Sybil attack, is a kind of attack that occurs when a malicious node (known as a Sybil node) claims to be multiple IoT nodes within a network, effectively taking over the identities of those nodes. This allows the attacker to have multiple identities within the network, which can be used to distribute false information or disrupt network processes [38]. By claiming to be multiple IoT nodes, the Sybil node is able to send false information throughout the network, which can cause confusion and disruption. For example, the Sybil node could send fake sensor readings to other nodes within the network, which could lead to incorrect decisions being made based on that false data. Another potential consequence of a Sybil attack is the disruption of the WSN (Wireless Sensor Network) election process. In a WSN, nodes are often elected to perform specific tasks or act as leaders within the network. If a Sybil node is able to disrupt this process by taking over multiple identities and skewing the results of the election, it can cause chaos within the network and potentially lead to a breakdown in communication and coordination. Thus, a Sybil attack allows a single malicious node to take over multiple entities within an IoT network, which can be used to distribute false information and disrupt network processes, potentially causing significant damage to the entire functionality of the network [67].

Heterogeneity problem

The non-homogeneous nature of IoT devices presents one of the most significant challenges posed by these devices. This indicates that there are a wide variety of distinct types of devices, each of which has its own set of particular requirements and standards. As a result, ensuring compatibility and interoperability between these devices can be very challenging, especially at the network layer [39]. The network layer of an IoT system is manages the communication between devices, as well as routing data and ensuring that it is delivered to the correct destination. However, because of a variety of devices in an IoT network, each with its own unique communication protocols and data formats, it can be difficult to ensure that they can all communicate with each other effectively. For example, some IoT devices may use Wi-Fi to communicate, while others may use Bluetooth or cellular networks. Some devices may use different encryption protocols or data formats, which can make it difficult to ensure that they can all work together seamlessly. As a result, ensuring compatibility and interoperability of the network layer of an IoT system is a complex and challenging task. It requires careful planning and consideration of varying devices that form the IoT system, as well as the communication protocols and data formats that they support. Only by addressing these challenges can we ensure that IoT devices can communicate with each other effectively and work together to achieve their intended goals [68].

Network Congestion Problem

In IoT systems, there is often a huge volume of sensor data present, and in order to verify that the data is secure and trustworthy, a large number of authentications are required. This can create a congestion problem, as the authentication process can slow down the flow of data within the network [39]. To solve this problem, a feasible device authentication mechanism can be used. This mechanism ensures that only trusted devices are allowed to send data within the network, which can help to reduce the congestion caused by the authentication process. Once the data has been authenticated and processed at the network layer, it is then passed on to the application layer for further processing. The application layer is responsible for analyzing the data and using it to make decisions or take actions within the IoT system. Overall, managing the flow of data within an IoT system is a complex process that involves multiple layers of processing and authentication. By using a feasible device authentication mechanism and carefully managing the flow of data between layers, it is possible to reduce congestion and ensure that the system operates smoothly and efficiently [69].

Application Layer

Application layer of IoT ensure data integrity, authenticity, and confidentiality. This layer enables the creation of smart environments which is the primary goal of IoT. Ensuring security at the application layer is crucial in IoT systems [70]. However, there are many issues related to this layer, such as the lack of standard and global policy enforcement, which can make it challenging to develop applications efficiently. Integration can be challenging for a number of reasons, one of which is that different apps utilize a variety of validation methods, each of which has the potential to have an effect on data privacy, confidentiality, and identity authentication. The distribution of data between IoT devices has the potential to impose additional burden on application processing, which may reduce the availability of services. It is important to consider how different users will interact with one another, how much data may be shared among users, and who will be in charge of managing the apps when developing Internet of Things (IoT) applications. Additionally, users must be in charge of what data they disclose and be aware of how, by whom, and when data will be used. The following are the main points of attack at the application layer:

Phishing Attacks

A threat actor can compromise an IoT system in many ways, including through the transmission of malicious emails or through phishing attacks. The objective of these kinds of attacks is to acquire the authentic users' login credentials in order to gain access to the IoT system [40]. When a user opens an infected email, malicious software or other potentially harmful software may be revealed. If the user does not remove the software, the attacker will be able to infect their device and gain access to the system. In contrast, phishing attacks may involve sending emails or other messages that appear to have come from a trusted source in an effort to trick the user into revealing sensitive information. An attacker can gain unauthorized access to a device connected to the internet of things by obtaining the user's credentials and using them to log in (IoT). As a result, criminals could potentially gain access to sensitive information or take control of IoT devices, wreaking havoc on the system or causing it to malfunction. Users must be aware of the dangers and take measures to safeguard their login information if attacks of this type are to be avoided. This includes being wary of emails or messages from unknown or suspicious sources, as well as making use of strong passwords and not sharing them with others. Multi-factor authentication and encryption are two examples of security measures that should be built into IoT systems from the start to keep sensitive user information safe from prying eyes.

Malicious Active X Scripts

A threat actor could target an Internet of Things user with an Active X script in order to access that user's system without authorization. Sending the user an infected email or message that includes the malicious Active X script is one way to achieve this. After receiving the malicious email or message, the user will then be prompted to run the script. If the user executes the script, it may cause the system to crash completely or cause data to be stolen [24]. Small programs known as Active X scripts can be downloaded and run by a web browser. They are frequently used to improve websites' functionality, but they can also be maliciously used to access a user's system. The script can run commands on the user's device after it has been executed, such as installing malware or stealing data. Users should exercise caution when downloading and running scripts from the internet in order to avoid such attacks. They should refrain from opening emails or messages from unidentified or dubious sources and only download and execute scripts from reliable sources. IoT systems should also be built with security in mind, including safeguards like firewalls and antivirus software to stop the installation of malware and other harmful programs.

Malwares attack

Malware can be deployed at the application layer of an IoT system to gain unauthorized access, steal data, or disrupt service. Trojan horses, worms, and viruses are among the most pernicious forms of malware used in attacks today [41]. A Trojan horse is a type of malware that appears to be a trusted program or file but, behind the scenes, steals information or gains access to a computer without the owner's knowledge. Once the malicious user has installed the Trojan horse, they can take complete control of the system, access all of its data, and cause extensive damage to it. Unlike viruses, worms can infect a computer without the help of another program. As an alternative, they can propagate throughout a system by taking advantage of loopholes in the system's software or operating system. A worm's ability to replicate and spread to other systems means it can cause significant disruption and harm when it infects a single system.

Malware like viruses can also be used to compromise IoT devices. They can infect files or programs and then spread to other computers, where they can do harm or steal private information. IoT systems are especially vulnerable to viruses because, like worms, they can spread through a network. Designing Internet of Things systems with security in mind is crucial for preventing these kinds of attacks. The installation of malware and other harmful software can be avoided with the help of preventative measures like antivirus software, firewalls, and routine software updates. Users should also regularly back up their data to prevent data loss in the event of an attack, and they should exercise caution when downloading and opening files or programs from unknown or suspicious sources.

Distributed Denial of Service

The application layer is the primary focus of Denial of Service (DoS) and Distributed Denial of Service (DDoS) attacks on Internet of Things (IoT) networks. The attack is executed by penetrating the IoT network and taking advantage of all its users. The attacker then launches a denial-of-service (DoS) or distributed denial-of-service (DDoS) attack, which floods the system with requests and traffic and prevents legitimate users from reaching the application layer [43]. The attacker exploits a vulnerability in the application layer to gain access to data stores and perform other malicious actions. This is particularly dangerous because it can result in the theft of private information, including financial and personal data. The attacker can then use this data to commit crimes like stealing someone's identity or blackmailing them for money.

Strong security measures at the application layer of IoT systems are essential for protecting against this kind of attack. Intrusion prevention systems, firewalls, and routine software updates are all examples of what this category encompasses. Users should also be educated on how to spot and avoid the phishing attacks that are increasingly being used to breach IoT security. Protecting vital information and infrastructure from DOS/DDoS attacks is possible with these precautions in place.

Finally, Table1 provides a thorough examination of the proposed architecture's layer-by-layer examination of various attacks at each layer, along with possible solutions.

IoT Security Issues

To ensure the security of the internet of things, it is essential to achieve the core security goals of confidentiality, integrity, and availability, which apply to all forms of computer and network communication (IoT). However, the Internet of Things faces a number of component and device limitations, resource constraints, as well as its own heterogeneous and ubiquitous nature, all of which present additional security-related challenges. This section is divided into two parts: the first part covers the essential security requirements for IoT, and the second half covers the security issues unique to each IoT layer.

The General Security Features of IoT

According to [87], there are two main types of risks associated with the IoT, and they are technological and security. Both are related to the need to incorporate various security ideas and features into a network in order to ensure its safety, the former being caused by the sheer number and variety of IoT devices and the latter by the necessity of doing so. Both groups contain examples of the problematic situations listed in Table 1. Security measures should be built into all Internet of Things devices and hubs from the design phase all the way through their use [57]. Authorizing software on IoT devices, authenticating devices before data transmission, and implementing firewalls to filter packets destined for devices are just a few of the mechanisms that must be used to achieve security. Communication in the IoT must be secure between people, software, processes, and things, so security principles must be adhered to.

Confidentiality

Humans, machines, and services, both inside and outside of the network, all need to be accounted for when considering data security and authorized access in the IoT. It's crucial that sensors don't share their data with nearby nodes [88], and that users have full visibility into the data management mechanisms that have been put in place. Throughout the procedure, data security must be ensured by the responsible parties and process [61].

Integrity

In order to guarantee the accuracy of the data shared in the IoT, it is crucial to confirm that it is coming from the intended sender. Furthermore, it is crucial to make sure that the data was not altered during transmission as a result of interference or other problems [89]. At this point, the quality known as "integrity" begins to work. Firewalls and protocols are frequently used in data traffic management. This is done in order to maintain end-to-end security for IoT connections. However, these

Table 1 A comprehensive layer-wise attacks on IoT and proposed solutions

Types of attack	The specific layer	The proposed achievement	Security problems	The affected layer	Proposed Scheme counter measure	Similar schemes counter measures	The paper
Node tampering RF interference on RFID's Node jamming Malicious node injection/ Fake node Sleep deprivation Malicious code injection Frequency jamming Side channel attack	Perception layer	Protecting the area around nodes from potential physical threats for data confidentiality at every node, strong authentication and access control mechanisms are required. Lightweight encryption protocols tailored to low-power mobile devices are a welcome development	Trust	Perception Layer	Trust management	[54, 71, 72] [73, 74]	[17, 75] [18] [17] [19, 39] [20, 21] [76, 77] [21, 78] [79–81] [30, 39]

(continued)

Table 1 (continued)

Types of attack	The specific layer	The proposed achievement	Security problems	The affected layer	Proposed Scheme counter measure	Similar schemes counter measures	The paper
Traffic analysis attack RFID spoofing RFID cloning RFID unauthorized access Routing attacks Sinkhole attack Man in the middle attack Sybil attack Denial of service attack Congestion and problem of heterogeneity	Network layer	To improve the efficiency of IoT systems, a secure communication protocol must be provided against all possible attacks	Heterogeneity	Network layer	Heterogeneity management	[26, 70, 82] [73, 74]	[17] [18] [28] [28] [37, 83] [84] [35, 52] [85] [28] [38, 86] [36] [39]
Phishing attack Malicious active-X scripts Malwares	Application layer	The advent of safe application code, Creating elaborate, login keys In order to provide reliable methods of access control, Databases are being watched in real time, Anti-malware software update verification checks	Authentication and Access control	Application layer	Authentication control	[23, 47, 48] [51] [52] [53]	[40] [24] [24, 41]

mechanisms might not be able to fully ensure the integrity of endpoints in the Internet of Things due to the low computing capacity of IoT nodes, which cannot support these processes efficiently (IoT).

Availability

The primary goal of the IoT is to link together a vast ecosystem of intelligent gadgets [90]. All data should be readily available to IoT users at all times. However, in order to meet the demands of the IoT, not only data but also devices and services need to be easily accessible and available when needed.

Authentication

It is crucial for all objects to be able to correctly identify and authenticate one another in the Internet of Things. However, this can be difficult due to the interconnected nature of the many devices, services, people, processing units, and service providers that make up the Internet of Things (IoT). Having the ability to correctly identify and authenticate other objects is fundamental. It may also be important for objects to interact with one another [91] under certain conditions. As a result, it's crucial that every IoT contact be equipped with a system for mutual authentication and verification.

Lightweight Solutions

Although IoT may place certain constraints and characteristics on each of the security goals discussed above, they are not unique to IoT. Confidentiality, integrity, availability, and authentication are fundermental to computer and network security. But because of the limited processing and power of IoT devices, a novel security feature has been developed: the lightweight solution [75]. In the context of designing and implementing authentication or encryption protocols for the data and devices that make up the Internet of Things, it is not a goal in and of itself but rather a constraint that must be taken into consideration. These approaches must be compatible with the limited capabilities of IoT devices because they are intended to run on those devices.

Heterogeneity

A wide range of entities, each with its own distinct set of capabilities, complexities, and providers, are interconnected in the IoT. Additionally, these electronic devices come in a variety of release versions and were produced on various dates. They each had a specific goal in mind when they were developed, and they each use different technological interfaces and bitrates. As a result, protocols must be created in a way that allows them to function with a wide range of equipment and situations [57, 60, 91]. By connecting existing technologies, people, and other networks, the Internet of Things (IoT) seeks to build a network made up of a wide range of various things and networks. Another challenge presented by the Internet of Things is the dynamic nature of the environment, where devices may connect to a completely different group of devices at various times (IoT). To ensure the security of the system, an appropriate cryptographic system, key management, and security protocols are needed.

Policies

Establishing policies and standards is necessary for the IoT's data to be managed, protected, and transmitted effectively [76]. However, merely having these regulations in place is not enough. Additionally, there needs to be a mechanism in place to ensure that the standards are being followed and enforced by all parties. For each service that makes up the IoT, it is advised to set up and clearly identify Service Level Objectives (SLO). Since the Internet of Things is a complex and ever-changing network, traditional computer and network security measures may not be adequate to protect it. Putting these ideas into practice will improve users' levels of trust in the Internet of Things, which will ultimately result in the IoT's expansion and scalability.

Key Management Systems

The exchange of encryption materials between devices and sensors is necessary to keep data private in the IoT. All frameworks need a lightweight key management system to accomplish this [77]. Trust between devices and key distribution should be established with minimal interference to device functionality using this system.

Security Challenges in Each Layer

Any part of the IoT infrastructure can be compromised by an attack. Insider attacks are just one example of these kinds of attacks that can come from within an organization or from the outside [59]. While an active attack would immediately shut down the service, a passive attack would simply monitor data from the IoT network. Denial of Service (DoS) attacks, which render a device, resource, or network inaccessible to authorized users, are a threat to all layers of the Internet of Things (IoT). The security issues at each layer are listed in Table 1, along with a detailed analysis of the issue.

Perception Layer

The perception layer of the Internet of Things is subject to three primary security vulnerabilities. The first one has to do with the intensity of wireless signals that are used to communicate data between sensor nodes. These signals, which can be degraded by interference, are utilized to transmit data. The second concern is the potential for physical attacks on Internet of Things (IoT) sensors and devices. These attacks can happen because IoT sensors and devices often operate in external and outdoor environments, which makes it easier for attackers to tamper with the hardware components of the devices. IoT nodes are regularly moved to new places, which offers a third and final barrier to network security [78]. The ever-changing topology of today's networks contributes to this issue. Sensors and RFIDs are the backbone of the Internet of Things' perception layer, but they're especially vulnerable to attacks due to their poor resilience against data loss, high power consumption, and lack of processing power.

There are many potential security flaws in the Internet of Things' perception layer. The Replay Attack is an example of a threat that could undermine the layer's security. It can be accomplished by forging, replaying, or otherwise manipulating the unique identifiers of IoT devices. The Timing Attack is a further threat to the privacy of the information because it allows the adversary to try to figure out the encryption key by timing how long it takes to encrypt the data. Node Capture is an example of an attack that could compromise privacy. The attacker gains access to the node and then steals its data and information. There is a potential threat to data confidentiality at this layer due to the introduction of a new node into the network, which could lead to the transmission of malicious data. The system may have to expend more energy because the nodes are unable to enter sleep mode as a result of such an attack [88].

The effects of these vulnerabilities can be mitigated [12] through the use of point-to-point or end-to-end encryption, authentication (to verify the sender's genuine identity), and access control. Additional safety measures and preventative strategies to address these issues are outlined in section "IIoT Architecture".

Authentication Measures

The author of the paper [13] presented a hashing and feature extraction-based mutual authentication strategy for the Internet of Things. The scheme's primary concern is authentication during the process by which the platform distributes data to terminal nodes. Because it is lightweight and provides the needed irreversibility, the method of feature extraction is suited for use with the Internet of Things. Nevertheless, despite the fact that the plan appears to have potential in theory, there is no proof of concept to back it in practice.

Reference [76] presents in their study an additional approach for ID authentication at sensor nodes in the Internet of Things (IoT). Their one-time, one-cipher technique is based on a request-reply system, and it requires all communicating parties to have a pre-shared matrix in order to function. A random coordinate is created and utilized as the key coordinate; this random coordinate is what is then passed between the parties in place of the actual key. The key, combined with the key position, the device ID, and the timestamp, are used to encrypt the messages. Validating timestamps allows two devices to connect with one another and determine whether or not to continue with the session. Even if this encryption can be utilized in Internet of Things scenarios with a lower level of sensitivity, the pre-shared matrix installation needs to be safe in order for it to be employed on a broad scale.

In order to ensure the safety of the internet of things (IoT), Mahalle et al. [87] devised an integrated protocol called IACAC that combines identity authentication and capability-based access control. This protocol uses a public key approach that is compatible with existing access technologies such as Bluetooth, 4G, WiMax, and Wi-Fi in order to circumvent the limitations posed by lightweight, mobile, and computationally limited Internet of Things (IoT) devices. These limitations can be caused by a lack of processing power. The IACAC model protects against man-in-the-middle attacks by including a timestamp in the authentication message sent between devices. This model operates in three stages as follows: generating a secret key using the Elliptical Curve Cryptography-Diffie Hellman algorithm (ECCDH), establishing identities using one-way and mutual authentication protocols, and implementing access control by storing a capability with access rights, device identifier, and a random number in each IoT device. These steps are all part of the Elliptical Curve Cryptography-Diffie Hellman algorithm. The IACAC model restricts access to only one user ID at a time, which reduces the likelihood of a denial-of-service (DoS) attack occurring. However, the model does not totally prohibit such attacks. In addition, a lightweight authentication mechanism was created in [15] to safeguard RFID tags against attacks that involve the sniffing and programming of EPCs. These types of attacks are designed to steal information from the tags. This protocol enables RFID readers and tagged goods to authenticate each other without imposing a considerable additional burden on the devices themselves by adding unnecessary overhead.

Proposed Scheme

Industrial IoT Security Threats and Counter Measures is proposed which aims to address the critical security concerns of IIoT systems. IIoT is the interconnection of industrial devices, sensors, and systems with the internet, which enables greater automation, efficiency, and optimization of industrial processes. However, the IIoT is vulnerable to cyber-attacks, which can cause significant disruptions to critical infrastructure, leading to financial losses as well as loss of life.

To mitigate the risks of cyber-attacks, the proposed model includes a comprehensive review and security strategy that includes risk assessment, threat modeling, and vulnerability testing, as well as a counter measure model. The proposed model incorporates security protocols and standards, such as encryption, access control, and authentication, to ensure the confidentiality, integrity, and availability of the IIoT system.

One of the key elements of the proposed model is network segmentation, which involves dividing the IIoT network into smaller, isolated segments that can be secured independently, thereby mitigating cyber-attacks. This approach limits the impact of a potential breach and allows for more granular control over access to different parts of the IIoT system.

Furthermore, the proposed model also implements an intrusion detection and prevention system (IDPS) to help detect and mitigate cyber-attacks by monitoring network traffic, identifying potential threats, and blocking malicious activity. This system can alert administrators of any suspicious activity and provide them with the necessary information to take corrective action.

The proposed strategy also includes disseminating information about cyberattacks and outlining necessary defenses. As a result, businesses will be able to foresee potential threats to their IIoT systems and take preventative measures.

Network segmentation, security protocols and standards, and intrusion detection and prevention systems are all part of the comprehensive security strategy proposed for the Industrial Internet of Things (IIoT). The model effectively reduces vulnerabilities to cyberattacks on IIoT systems, allowing vital infrastructures to continue running without interruption (Fig. 3).

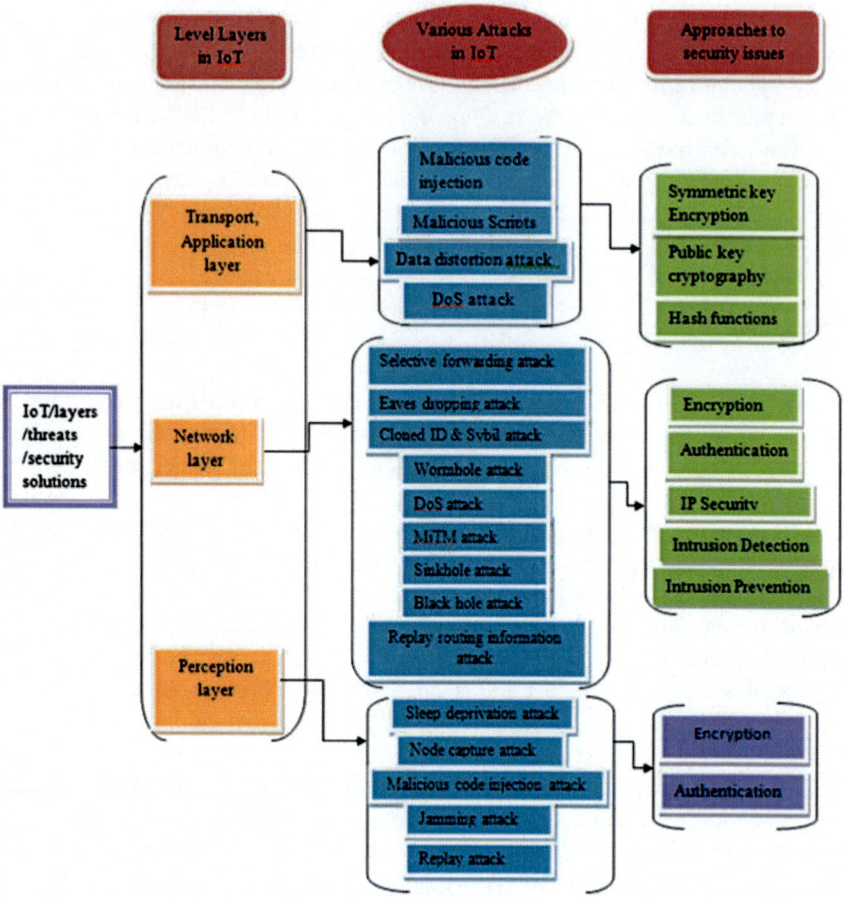

Fig. 3 A proposed taxonomy of IoT layers, security threats and solutions

Conclusion

The Internet of Industrial Things (IIoT) is revolutionizing the industrial landscape by facilitating real-time data collection and analysis, process optimization, and increased productivity. The IIoT, however, has implications for many different kinds of industrial systems and technologies, and this will continue to spur development and shifts in the market in the coming years. The paper's authors recognized the IIoT's susceptibility to cyberattacks and proposed solutions to that problem in the form of threats to and countermeasures for that network's security. Risk assessment, threat modeling, and vulnerability testing are all covered, and a model is proposed that takes advantage of security protocols and standards like encryption, access control, and authentication. For critical infrastructure to run securely and reliably, the model also includes network segmentation and intrusion detection and prevention systems (IDPS). Similarly, the proposed scheme raises awareness of various cyberattacks to critical infrastructure and industries and presents essential countermeasures to defend against them.

References

1. Alcaraz Calero, J. M., & Liotta, A. (2016). Security and privacy in 5G networks: Challenges and solutions. *IEEE Communications Magazine, 54*(2), 80–87.
2. Chen, X., Yang, X., & Wang, Y. (2019). 5G security: Current status and future challenges. *IEEE Network, 33*(6), 176–181.
3. He, H., & Wu, D. (2020). 5G security and privacy: A survey. *IEEE Internet of Things Journal, 7*(9), 8259–8280.
4. Li, Y., Zhou, Z., & Zhang, Q. (2021). A survey on 5G security: Challenges and solutions. *Future Generation Computer Systems, 119*, 339–354.
5. Saeed, S., & Huang, X. (2021). 5G security: A survey of current trends and future research directions. *IEEE Communications Surveys & Tutorials, 23*(1), 45–71.
6. Khan, A. A., Amin, M. B., & Hamza, A. (2017). Security and privacy in 5G cellular networks: A survey. *IEEE Access, 5*, 26587–26609.
7. Wang, P., Wu, J., & Li, X. (2020). A survey on 5G security and privacy issues. *Journal of Network and Computer Applications, 150*, 102527.
8. Zhang, X., Liu, Y., & Cheng, L. (2018). Security and privacy in 5G networks: Challenges and opportunities. *IEEE Wireless Communications, 25*(1), 120–127.
9. Zhang, H., Fang, Y., & Zhang, H. (2022). A comprehensive survey of 5G network security: Threats, defenses, and challenges. *IEEE Journal on Selected Areas in Communications, 40*(3), 516–532.
10. Zhou, K., Zhang, Q., & Liu, Y. (2019). 5G security: A survey and future research directions. *Science China Information Sciences, 62*(3), 031101.
11. Abosat, N., Al-Rubaye, S., Inalhan, G., & Emmanouilidis, C. (2021). Internet of Things for system integrity: A comprehensive survey on security, attacks and countermeasures for industrial applications. *IEEE Access, 9*, 12325–12339. https://doi.org/10.1109/ACCESS.2021.3056161
12. Alsubaei, F., Abuhussein, A., & Shivai, S. (2019). Ontology-based security recommendation for the internet of medical things. *IEEE, 7*, 48948–48960.
13. Silvai, J. D. C., Rodrigues, J. J. P. C., Saleem, K., Kozlov, S. A., & Rabelo, R. A. L. (2019). M4DN.IoT-A networks and devices management platform for Internet of Things. *IEEE Access, 7*, 53305–53313.
14. Sharma, V., Choudary, G., Ko, Y., & You, I. (2018). Behavior and vulnerability assessment of drones-enabled Industrial Internet of Things (IIoT). *IEEE, 6*, 43368–43383.
15. Chen, L., Thombre, S., Järvinen, K., Lohan, E. S., et al. (2017). Robustness, security and privacy in location-based services for future IoT: A survey. *IEEE Access, 5*, 8956–8977.
16. Zhao, K., & Ge, L. (2013). A survey on the internet of things security. In *Int'l Conference on computational intelligence and security (CIS)* (pp. 663–667).
17. Uke, S. N., Mahajan, A. R., & Thool, R. C. (2013). UML modeling of physical and data link layer security attacks in WSN. *International Journal of Computer Applications, 70*(11).
18. Li, H., Chen, Y., & He, Z. (2012). The survey of RFID attacks and defenses. In *8th international conference on IEEE wireless communications, networking and mobile computing (WiCOM)* (pp. 1–4). IEEE.
19. Kandah, F., Singh, Y., Zhang, W., & Wang, C. (2013). Mitigating colluding injected attack using monitoring verification in mobile ad-hoc networks. *Security and Communication Networks, 6*(12), 1468–1479.

20. Brun, O., Yin, Y., & Gelenbe, E. (2018). Deep learning with dense random neural network for detecting attacks against IoT-connected home environments. *Procedia Computer Science, 134*, 458–463.
21. Farooq, M. U., Waseem, M., Khairi, A., & Mazhar, S. (2015). A critical analysis on the security concerns of Internet of Things (IoT). *International Journal of Computer Applications, 111*(7), 1–6.
22. Andrea, I., Chrysostomou, C., & Hadjichristofi, G. (2015). Internet of Things: Security vulnerabilities and challenges. In *2015 IEEE symposium on computers and communication (ISCC)* (pp. 180–187). IEEE.
23. Mahalle, P. N., Anggorojati, B., Prasad, N. R., & Prasad, R. (2013). Identity authentication and capability based access control (IACAC) for the internet of things. *Journal of Cyber Security and Mobility, 1*(4), 309–348.
24. Tobias, H., Huebscher, M., & Randles, M. (2011). Security challenges in the IP-based Internet of Things. *Wireless Personal Communications, 61*(3), 527–542.
25. Mahmoud, R., Yousuf, T., Aloul, F., & Zualkernan, I. (2015). Internet of things (IoT) security: Current status, challenges and prospective measures. In *2015 10th International conference for internet technology and secured transactions (ICITST)* (pp. 336–341). IEEE.
26. Leo, M., Battisti, F., Carli, M., & Neri, A. (2014). A federated architecture approach for Internet of Things security. In *2014 Euro med telco conference (EMTC)* (pp. 1–5). IEEE.
27. Matharu, G. S., Upadhyay, P., & Chaudhary, L. (2014). The Internet of Things: Challenges & security issues. In *2014 International conference on emerging technologies (ICET)* (pp. 54–59). IEEE.
28. Andrea, I., Chrysostomou, C., & Hadjichristofi, G. (2015). Internet of Things security vulnerabilities and challenges. In *2015 IEEE symposium computer and communication (ISCC)* (pp. 180–187). Larnaca.
29. Roman, R., Zhou, J., & Lopez, J. (2013). On the feature and challenges of security and privacy in distributed Internet of Things. *Computer Networks, 57*(10), 2266–2279.
30. Zulkifli, M. Z. W. M., & Mohd, Z. W. (2008). Attack on cryptography. *Computer Security, 12*(5), 33–45.
31. Uke, S. N., Mahajan, A. R., & Thool, R. C. (2013). UML modeling of physical and data link layer.
32. Li, H., Chen, Y., & He, Z. (2012). The survey of RFID attacks and defenses. In *8th international conference on IEEE wireless communications, networking and mobile Computing (WiCOM)* (pp. 1–4).
33. Kandah, F., Singh, Y., Zhang, W., & Wang, C. (2013). Mitigating colluding injected attack using monitoring verification in mobile ad-hoc networks. *Security and Communication Networks, 6*(10), 1295–1304. https://doi.org/10.1002/sec.748
34. Liu, Z., Longa, P., Pereira, G., Reparaz, O., & Seo, H. (2018). FourQ on embedded devices with strong countermeasures against side-channel attacks. *IEEE Transactions on Dependable and Secure Computing, 15*(1), 95–107. https://doi.org/10.1109/TDSC.2016.2594438
35. Abdullah, M. I., Rahman, M., & Roy, M. C. (2015). Detecting sinkhole attacks in wireless sensor network using hop count. *International Journal of Computer Network and Information Security, 7*(11), 50–56. https://doi.org/10.5815/ijcnis.2015.11.06
36. Wahid, A., & Kumar, P. (2015). A survey on attacks, challenges and security mechanism in wireless sensor network. *JIRST-International Journal for Research in Science & Technology, 1*(8), 189–196.
37. Dvir, A., Holczer, T., & Buttyan, L. (2011). VeRA-version number and rank authentication in RPL. In *2011 IEEE eighth international conference on mobile Ad-Hoc and sensor systems* (pp. 709–714). https://doi.org/10.1109/MASS.2011.47
38. Zhang, K., Liang, X., Lu, R., & Shen, X. (2014). Sybil attacks and their defenses in the internet of things. *IEEE Internet of Things Journal, 1*(4), 372–383. https://doi.org/10.1109/JIOT.2014.2331212
39. Zhao, K., & Ge, L. (2013). A survey on the Internet of Things security. In *Computational intelligence and security-ninth international IEEE conference* (pp. 663–667). https://doi.org/10.1109/CIS.2013.124

80. Vadlamani, S., Eksioglu, B., Medal, H., & Nandi, A. (2016). Jamming attacks on wireless networks: A taxonomic survey. *International Journal of Production Economics, 172*, 76–94.
81. Guan, Y., & Ge, X. (2017). Distributed attack detection and secure estimation of networked cyber-physical systems against false data injection attacks and jamming attacks. *IEEE Transactions on Signal and Information Processing over Networks, 4*(1), 48–59.
82. Anggorojati, B., Mahalle, P. N., Prasad, N. R., & Prasad, R. (2012). Capability base access control delegation model on the federated IoT network. In *International symposium on wireless personal multimedia communication (WPMC)* (pp. 604–608).
83. Le, A., Loo, J., Lasebae, A., Vinel, A., Chen, Y., & Chai, M. (2013). The impact of rank attack on network topology of routing protocol for low-power and lossy networks. *IEEE Sensors Journal, 13*(10), 3685–3692.
84. Wu, D., & Hu, G. (2008). Research and improve on secure routing protocols in wireless sensor networks. In *4th IEEE international conference on circuits and systems for communications, 2008 (ICCSC 2008)* (pp. 853–856). IEEE.
85. Cao, Q., & Yang, X. (2013). SybilFence: Improving social-graph-based sybil defenses with user negative feedback. arXiv preprint: 1304–3819.
86. Newsome, J., Shi, E., Song, D., & Perrig, A. (2004). The sybil attack in sensor networks: analysis & defenses. In *Proceedings of the 3rd international symposium on Information processing in sensor networks* (pp. 259–268). ACM.
87. Bouij-Pasquier, I., Ait Ouahman, A., Abou El Kalam, A., & Ouabida de Montfort, M. (2015). SmartOrBAC security and privacy in the Internet of Things. In *2015 IEEE/ACS 12th International conference of computer systems and applications (AICCSA)* (pp. 1–8).
88. Ammar, M., Russello, G., & Crispo, B. (2018). Internet of Things: A survey on the security of IoT frameworks. *Journal of Information Security and Applications, 38*, 8.
89. Castrucci, M., Neri, A., Caldeira, F., Aubert, J., Khadraoui, D., Aubigny, M., et al. (2012). Design and implementation of a mediation system enabling secure communication among critical Infrastructures. *International Journal of Critical Infrastructure Protection, 5*(2), 86–97. https://doi.org/10.1016/j.ijcip.2012.06.001
90. Duan, X., & Wang, X. (2015). Authentication handover and privacy protection in 5G HetNets using software-defined networking. *IEEE Communications Magazine, 53*(4), 28–35. https://doi.org/10.1109/MCOM.2015.7081089
91. Khemissa, H., & Tandjaoui, D. (2016). A novel lightweight authentication scheme for heterogeneous wireless sensor networks in the context of internet of things. In *2016 Wireless telecommunications symposium (WTS)* (pp. 1–6).
92. Tsiknas, K., Taketzis, D., Demertzis, K., & Skianis, C. (2021). Cyber threats to industrial IoT: A survey on attacks and countermeasures. *Sensors, 21*(6), 2076. https://doi.org/10.3390/s21062076
93. Dustdar, S., & Rosenberg, F. (2007). A survey on context-aware systems. *International Journal of ad Hoc and ubiquitous Computing, 2*(4), 263–277.
94. Neisse, R., Steri, G., Nai Fovino, I., & Baldini, G. (2015). SecKit: A model-based security toolkit for the Internet of Things. *Computers and Security, 54*, 60.
95. Atzori, L., Iera, A., Morabito, G., & Nitti, M. (2012). The social internet of things (SIOT) when social networks meet the Internet of Things: Concept, architecture and network characterization. *Computer Networks, 56*(16), 3594–3608.
96. Li, J. H. (2018). Cyber security meets artificial intelligence: A survey. *Frontiers of Information Technology & Electronic Engineering, 19*(12), 1462–1474.
97. Sharmila, S., & Umamaheswari, G. (2011). Detection of sinkhole attack in wireless sensor networks using message digest algorithms. In *2011 International conference on process automation, control and computing* (pp. 1–6). IEEE.
98. Farooq, M., Waseem, M., Khairi, A., & Mazhar, S. (2015). A critical analysis on the security concerns of Internet of Things (IoT). *Perception*, 111.
99. Roman, R., Zhou, J., & Lopez, J. (2013). On the features and challenges of security and privacy in distributed Internet of Things. *Computer Networks, 57*(10), 2266–2279.

100. Mashal, I., Alsaryrah, O., Chung, T.-Y., Yang, C.-Z., Kuo, W.-H., & Agrawal, D. P. (2015). Choices for interaction with things on internet and underlying issues. *Ad Hoc Networks, 28*, 68–90.
101. Al-Fuqaha, A., Guizani, M., Mohammadi, M., Aledhari, M., & Ayyash, M. (2015). Internet of Things: A survey on enabling technologies, protocols, and applications. *IEEE Communications Surveys & Tutorials, 17*(4), 2347–2376.

Industrial IOT: Security Threats and Counter Measures

S. C. Vetrivel, R. Maheswari, and T. P. Saravanan

Abstract The rapid proliferation of Industrial Internet of Things (IIoT) technologies has brought forth numerous advancements, along with significant security challenges. This chapter comprehensively examines the security threats faced by industrial IoT systems and presents effective countermeasures to mitigate these risks. By exploring the unique characteristics of IIoT environments, such as interconnectedness, heterogeneity, and resource constraints, this study identifies potential vulnerabilities that can be exploited by adversaries. Various attack vectors, including unauthorized access, data tampering, and device spoofing, are analyzed in depth, highlighting their potential impact on industrial operations. To address these threats, a multi-layered security framework is proposed, encompassing preventive, detective, and responsive measures. The preventive measures focus on secure network design, robust authentication and access control mechanisms, and secure software and firmware development practices. Detection mechanisms involving anomaly detection, intrusion detection systems, and machine learning algorithms are discussed to promptly identify and respond to potential attacks. Additionally, this paper emphasizes the importance of implementing incident response plans, data backup strategies, and security awareness training to minimize the impact of successful attacks. Furthermore, the role of encryption, secure communication protocols, and hardware security modules in enhancing the security posture of industrial IoT systems is explored. Finally, challenges related to the adoption of security measures, such as compatibility, scalability, and cost, are addressed, along with potential future research directions. This comprehensive analysis and proposed countermeasures aim to assist industrial organizations, researchers, and policymakers in establishing robust security frameworks to safeguard critical infrastructure in the era of Industrial IoT.

S. C. Vetrivel (✉) · R. Maheswari · T. P. Saravanan
Department of Management Studies, Kongu Engineering College, Perundurai, Erode 638 060, Tamilnadu, India
e-mail: scvetrivel@gmail.com

R. Maheswari
e-mail: mahyas2000@gmail.com

T. P. Saravanan
e-mail: vanakamtps@gmail.com

A. Prasad et al. (eds.), *Communication Technologies and Security Challenges in IoT*, Internet of Things, https://doi.org/10.1007/978-981-97-0052-3_20

Keywords Industrial IoT · Security threats · Data confidentiality · Cybersecurity · Threat modeling · Risk mitigation

Introduction

This is an era of the Internet of Things (IoT). A universal network of physical things is referred to as the IoT (Viriyasitavat et al. 2019) [1]. The dawn of the Industrial Internet of Things (IIoT) has enabled a new level of connectivity and automation in industrial settings, with the potential to improve productivity, efficiency, and operational optimization. The IIoT relates to all networked sensors, devices, and other equipment used in industrial applications such as manufacturing and energy conservation. The use of automation is made possible to a larger extent by this intricate web of services (Tsiknas et al. 2021) [2]. Despite the advantages, this vast network of connected devices and systems brings forth a range of security risks that must be addressed to protect the confidentiality, availability, and integrity of industrial systems. This paper aims to explore the security threats that arise in the context of Industrial IoT deployments and to propose effective countermeasures to mitigate these risks. The security landscape in industrial environments differs significantly from traditional IT networks, as it encompasses a wide range of critical infrastructure sectors such as manufacturing, energy, transportation, and healthcare. Consequently, the potential consequences of security breaches in these sectors are not only financial but can also pose significant physical and safety risks. This paper also provides an overview of the Industrial IoT architecture, highlighting the key components and their interdependencies. Understanding this architecture is essential for comprehending the potential attack vectors and vulnerabilities that may be exploited by malicious actors. We delve into the unique characteristics of IIoT systems, including resource-constrained devices, heterogeneous protocols, and the convergence of operational technology (OT) and information technology (IT) networks.

The subsequent sections discuss the prominent security threats faced by Industrial IoT deployments. It presents a taxonomy of threats categorized into various dimensions, such as unauthorized access, data breaches, device tampering, network attacks, and insider threats. Each category is explored in detail, illustrating real-world examples and their potential impact on industrial systems and processes. By providing a comprehensive understanding of these threats, organizations can prioritize their security efforts and allocate appropriate resources. IIoT taxonomy is not only feasible but also essential for the development of both individual systems and a true cross-industry IIoT (Schneider et al. 2017) [3]. In response to the identified threats, the paper then proposes a set of countermeasures and best practices to enhance the security posture of Industrial IoT environments. These countermeasures encompass a multi-layered approach, addressing both technical and organizational aspects. Technical measures include secure device provisioning, network segmentation, encryption, authentication, anomaly detection, and secure update mechanisms.

Organizational measures emphasize the importance of security awareness training, incident response planning, supply chain security, and regulatory compliance.

Furthermore, the paper discusses emerging technologies and standards that can aid in bolstering Industrial IoT security. These include the adoption of blockchain for secure data integrity and supply chain traceability, leveraging machine learning and artificial intelligence for threat detection and response, and embracing industry-specific security frameworks and guidelines. Machine Learning methods have been made to reduce potential risks in IIoT networks (Vargas et al. 2021) [4]. The Industrial IoT has tremendous potential to revolutionize industries, but its widespread adoption necessitates a robust security framework. This paper serves as a comprehensive guide for understanding the security threats inherent in Industrial IoT deployments and provides a roadmap for implementing effective countermeasures. By adopting these recommendations, organizations can mitigate risks, protect critical infrastructure, and build resilient and secure Industrial IoT ecosystems.

Background and Motivation

The swift augmentation and adoption of the Industrial Internet of Things (IIoT) have revolutionized various industries by enabling the seamless integration of physical devices, sensors, and networks. The IIoT offers tremendous opportunities for increased efficiency, automation, and data-driven decision-making in industrial settings such as manufacturing, energy, transportation, and healthcare. However, this connectivity also introduces significant security challenges that must be addressed to ensure the integrity, confidentiality, and availability of critical industrial systems.

The motivation behind the full paper on "Industrial IoT Security Threats and Countermeasures" is to comprehensively analyze the security threats associated with IIoT deployments and propose effective countermeasures. The paper aims to raise awareness among industry professionals, researchers, and policymakers about the potential risks and vulnerabilities in industrial environments and provide actionable recommendations to mitigate them.

1. **Identification and Classification of Threats:** The paper aims to identify and classify the security threats that can affect IIoT systems. This involves analyzing various attack vectors, such as unauthorized access, data breaches, tampering, denial of service, and supply chain attacks. By understanding the nature and characteristics of these threats, the paper lays the foundation for developing targeted countermeasures.
2. **Vulnerability Analysis:** The paper conducts a comprehensive analysis of vulnerabilities specific to IIoT systems. It examines potential weaknesses in the hardware components, software platforms, communication protocols, and network infrastructure that are commonly used in industrial environments. Understanding these vulnerabilities is crucial for devising effective security measures.

3. **Impact Assessment:** The paper investigates the potential consequences and impact of successful attacks on IIoT systems. This includes analyzing the potential damage to critical infrastructure, financial losses, safety risks, and potential threats to human life. By highlighting the potential ramifications of security breaches, the paper emphasizes the urgent need for robust security measures.
4. **Countermeasures and Best Practices:** The paper proposes a range of countermeasures and best practices to enhance the security of IIoT deployments. This includes recommendations for secure design principles, network segmentation, encryption, authentication mechanisms, anomaly detection, intrusion prevention systems, and incident response protocols. The proposed countermeasures are practical and can be implemented by industrial organizations to mitigate the identified threats effectively.
5. **Regulatory and Policy Considerations:** The paper also addresses the regulatory and policy implications surrounding IIoT security. It examines existing standards and frameworks, such as NIST Cybersecurity Framework and IEC 62443, and discusses the need for industry-wide collaboration, information sharing, and regulatory measures to address security challenges effectively.

Problem Statement and Study Objectives

(a) *To identify and analyze the key security threats faced by Industrial IoT systems*

This objective aims to explore the various security vulnerabilities and risks that arise due to the integration of IoT devices, networks, and data in industrial environments. The study will investigate potential attack vectors such as unauthorized access, data tampering, physical manipulation, and network-based attacks, along with their potential impact on industrial operations and critical infrastructure.

(b) *To evaluate existing security measures and their effectiveness in mitigating Industrial IoT threats*

This objective focuses on assessing the current state of security measures implemented in Industrial IoT systems. The study will analyze encryption protocols, access control mechanisms, intrusion detection systems, and other security practices commonly employed in industrial settings. By evaluating the strengths and weaknesses of these measures, the research aims to identify gaps and limitations that need to be addressed for better protection against emerging threats.

(c) *To propose comprehensive countermeasures for securing Industrial IoT systems*

Building upon the findings from the previous objectives, this objective aims to develop a set of practical and effective countermeasures tailored to the unique requirements and challenges of Industrial IoT environments. The proposed countermeasures may include security frameworks, authentication mechanisms, anomaly detection systems, secure communication protocols, and guidelines for the secure deployment and configuration of IoT devices.

(d) *To validate the proposed countermeasures through simulations and real-world experiments*

This objective involves implementing the proposed countermeasures in a simulated Industrial IoT environment and conducting experiments to evaluate their efficacy. By simulating various attack scenarios and analyzing the system's response, the study aims to assess the effectiveness of countermeasures in detecting, preventing, and mitigating security threats. Real-world experiments may also be conducted in collaboration with industrial partners to validate the countermeasures in operational environments.

(e) *To provide recommendations and guidelines for securing Industrial IoT systems*

Based on the research findings and results, this objective aims to provide practical recommendations, best practices, and guidelines for organizations and industrial stakeholders to enhance the security of their IoT systems. The study will highlight the critical areas of focus, necessary security controls, and ongoing monitoring and maintenance practices needed to ensure a robust and resilient Industrial IoT infrastructure.

Fundamentals of Industrial IoT

Jaidka (2020) [5]explored the transformation of IoT to IIoT. The Industrial Internet of Things (IIoT) refers to the extension of IoT technologies and concepts to industrial settings, enabling connectivity and data exchange between various devices and systems. It involves the integration of physical machinery, sensors, actuators, and other devices with networked systems to create a connected infrastructure for industrial operations. The key elements of Industrial IoT include:

1. Connectivity: IIoT relies on robust network infrastructure to connect devices and enable seamless communication and data transfer. This can include wired or wireless networks, such as Ethernet, Wi-Fi, or cellular networks.
2. Sensors and Actuators: Industrial devices are equipped with sensors to collect data from the physical environment, and actuators to control physical processes or machinery based on the received data. These components enable real-time monitoring and control of industrial operations.
3. Data Analytics: The collected data is processed and analyzed using various techniques, such as machine learning and artificial intelligence, to derive insights and make informed decisions. For example, Li et al. (2021) [6] found a surge in the application of ML-based techniques for the improvement of various IoT applications.
4. Cloud Computing: IIoT systems often leverage cloud-based platforms to store and process large volumes of data, provide scalable infrastructure, and enable remote access and management.

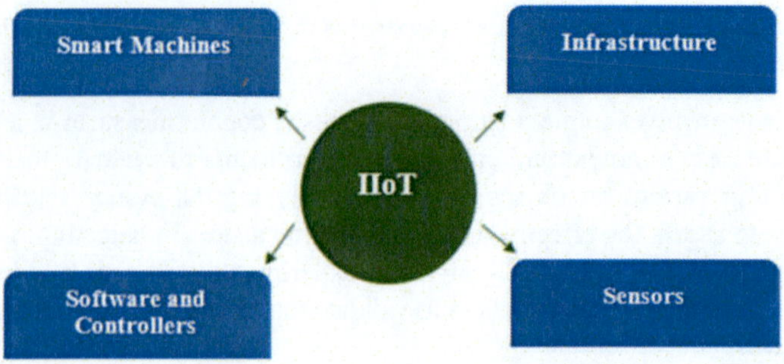

Fig. 1 Components of II_OT network. *Source* https://www.omega.com/en-us/resources/iiot-components, accessed on 5th July 2023

IIoT is one such technology, which is well-known among business insiders but virtually unheard of by the average public. It is made of some important components. Figure 1 shows the Key components of Industrial IoT.

Security Threats and Countermeasures

The implementation of IIoT introduces new security challenges due to the increased connectivity and integration of various devices and systems. All industrial systems must be highly attentive to the safety and security of IIOT devices to prevent damage to the assets and employees (Karmakar et al. 2020) [7]. Some common security threats in the context of Industrial IoT include:

1. Unauthorized Access: Unauthorized individuals or malicious entities may attempt to gain unauthorized access to industrial systems, devices, or data, potentially causing disruptions or data breaches. Robust authentication mechanisms, access controls, and encryption techniques can help mitigate this threat.
2. Data Integrity and Privacy: Ensuring the integrity and privacy of data transmitted and stored within IIoT systems is crucial. Encryption, secure protocols, data validation mechanisms, and secure storage practices are essential to protect sensitive information.
3. Denial-of-Service (DoS) Attacks: Attackers may try to overwhelm IIoT systems or specific devices with a high volume of requests, causing service disruptions. Implementing intrusion detection and prevention systems, traffic filtering, and redundancy measures can help mitigate DoS attacks.
4. Device Compromises: IIoT devices can be vulnerable to exploitation, where attackers may compromise their firmware or software to gain control or disrupt their functionality. Regular software updates, strong device authentication, and secure device management practices can reduce the risk of device compromises.

5. Supply Chain Risks: The complex supply chains involved in IIoT deployments can introduce security vulnerabilities. Verifying the security practices of suppliers, implementing secure communication channels, and conducting thorough security assessments are essential to mitigate supply chain risks.

The countermeasures to address these threats involve a combination of technical and organizational measures, such as:

1. Implementing strong access controls and authentication mechanisms for devices and users.
2. Encrypting data during transmission and storage.
3. Regularly updating and patching firmware and software on IIoT devices.
4. Deploying firewalls, intrusion detection systems, and security monitoring tools.
5. Conducting security assessments and penetration testing to identify vulnerabilities.
6. Training employees and stakeholders about security best practices.
7. Establishing incident response plans to address and mitigate security incidents.

Key Components and Architecture Requirements

The Industrial Internet of Things (IIoT) has emerged as a transformative force in various industries, revolutionizing processes and enabling new levels of automation and efficiency. However, the rapid proliferation of interconnected devices in industrial environments has also introduced significant security challenges. Litoussi (2020) [8] outlined the three basic layers of the IoT architecture, as well as many attack types and examples for each layer.Attackers find IIoT endpoint devices to be a desirable target (Zhou et al. 2018) [9]. The key components and architecture necessary to address the security threats faced by industrial IoT deployments are;

Key Components of Industrial IoT Security Architecture

1. Device-Level Security: The foundation of industrial IoT security lies in ensuring the integrity, confidentiality, and availability of individual devices. Key components in device-level security include secure boot mechanisms, tamper-resistant hardware, robust authentication protocols, and secure firmware updates. Additionally, implementing strong access controls, encryption algorithms, and device-level monitoring capabilities is essential for protecting against unauthorized access and data breaches.
2. Network Security: Securing the communication channels within the IIoT network is crucial to prevent unauthorized access, eavesdropping, and data manipulation. Key components of network security include secure protocols, such as Transport Layer Security (TLS), virtual private networks (VPNs), and secure authentication mechanisms like digital certificates. Implementing intrusion detection

and prevention systems (IDPS) and robust firewalls help mitigate network-based attacks and enforce network segmentation for improved security.

3. Data Security: Protecting the confidentiality, integrity, and availability of data generated and transmitted by IIoT devices is paramount. Key components of data security include encryption techniques, secure data storage, and secure data transmission protocols. Implementing data access controls, data encryption at rest and in transit, and data anonymization techniques helps safeguard sensitive information from unauthorized disclosure or tampering.
4. Cloud and Edge Security: As IIoT deployments rely heavily on cloud and edge computing, securing these platforms is critical. Key components of cloud and edge security include robust identity and access management (IAM) systems, secure APIs, secure containerization, and security monitoring and analytics. Implementing encryption, strong authentication, and secure data storage techniques within the cloud and edge infrastructure helps protect against attacks targeting these components.

Architecture for Industrial IoT Security

1. Centralized Security Management: An effective architecture for industrial IoT security requires a centralized management system that provides visibility, control, and monitoring of the entire IIoT infrastructure. This component enables security administrators to implement security policies, monitor devices and network traffic, detect anomalies, and respond to security incidents promptly. It should integrate with existing security information and event management (SIEM) systems for comprehensive threat detection and analysis.
2. Threat Intelligence and Analytics: Incorporating threat intelligence and analytics into the security architecture allows organizations to stay ahead of emerging threats. This component involves collecting and analyzing data from various sources, including security logs, network traffic, and threat intelligence feeds. By leveraging machine learning and artificial intelligence techniques, it becomes possible to detect anomalies, identify potential security breaches, and proactively respond to emerging threats.
3. Security Incident Response: An efficient incident response mechanism is a vital component of the security architecture. It includes predefined incident response plans, incident reporting, and escalation procedures, and a dedicated team responsible for handling security incidents. Automation tools and playbooks can be employed to facilitate rapid response and minimize the impact of security breaches, reducing downtime and ensuring business continuity.
4. Continuous Monitoring and Updates: Continuous monitoring of the IIoT environment is essential for identifying security gaps, detecting anomalies, and

ensuring ongoing protection. Security architecture should incorporate mechanisms for real-time monitoring, log analysis, and vulnerability scanning. Furthermore, regular security updates and patches should be deployed to address newly discovered vulnerabilities and maintain a robust security posture.

Importance of Security in Industrial IoT

1. Protecting Critical Infrastructure: Industrial IoT involves the integration of physical systems and industrial processes with the digital world, making it essential to secure critical infrastructure. Breaches or attacks on industrial systems can have severe consequences, including disruption of services, financial loss, and even physical harm. Studies such as Yussoff et al. (2012) [10] reviewed attacks on trusted platforms and proposed a summary of a framework for that.
2. Safeguarding Sensitive Data: Industrial IoT environments generate and transmit vast amounts of data. This data often includes sensitive information such as proprietary designs, trade secrets, customer data, and operational details. Protecting this data from unauthorized access, modification, or theft is crucial to maintaining business integrity and competitive advantage.
3. Ensuring Operational Continuity: Industrial IoT deployments are designed to enhance efficiency, productivity, and automation. However, without robust security measures, these systems are susceptible to disruptions, malfunctions, or cyberattacks. Ensuring the integrity, availability, and reliability of Industrial IoT devices and networks is essential to maintain operational continuity.
4. Mitigating Financial Losses: Industrial IoT security breaches can lead to significant financial losses. These may include costs associated with system downtime, equipment damage or replacement, regulatory fines, legal liabilities, reputation damage, and the expenses required to remediate the security breach.
5. Preserving Worker Safety: Industrial IoT devices often interact with workers and physical equipment in industrial settings. Compromised security in IoT systems can pose risks to worker safety, resulting in accidents, injuries, or even fatalities. By implementing robust security measures, potential hazards can be mitigated and worker safety can be ensured.
6. Upholding Regulatory Compliance: Many industries, such as healthcare, energy, and manufacturing, are subject to strict regulations and compliance requirements. Industrial IoT security plays a crucial role in meeting these standards, protecting sensitive data, and avoiding penalties or legal consequences associated with non-compliance.

The Common Security Threats and Countermeasures Are

1. Unauthorized Access: Unauthorized access to Industrial IoT devices or networks can lead to data breaches, disruption of operations, or unauthorized control of

critical infrastructure. Countermeasures include strong authentication mechanisms, access controls, and encryption protocols to ensure that only authorized individuals or devices can access IoT systems.

2. Data Integrity and Privacy: Industrial IoT generates a vast amount of data that needs to be protected from unauthorized modification, tampering, or exposure. Encryption, secure data storage, and secure communication protocols help safeguard data integrity and privacy.
3. Device Compromise: Industrial IoT devices can be compromised and used as entry points into the network or as platforms for launching attacks. Implementing secure boot processes, regular software updates, and endpoint security measures help mitigate the risk of device compromise.
4. Denial-of-Service (DoS) Attacks: DoS attacks aim to disrupt the availability and functionality of Industrial IoT systems by overwhelming them with excessive traffic or requests. Implementing traffic monitoring, rate limiting, and anomaly detection techniques can help detect and mitigate DoS attacks.
5. Supply Chain Vulnerabilities: Industrial IoT security can be compromised through vulnerabilities in the supply chain, such as the inclusion of malicious components or insecure software/firmware. Ensuring supply chain integrity, performing security audits, and vetting suppliers can help mitigate these risks.
6. Lack of Standardized Security Practices: The absence of standardized security practices in Industrial IoT deployments can lead to inconsistencies and vulnerabilities.

Security Challenges in Industrial IoT

As the application of IIoT is multifold, its security issues in various sectors like manufacturing, medicine, automobiles, etc. are also increasing. Gaurav et al. (2022) [11] underlined the significance of establishing the appropriate security measures to increase the Internet of Medical Things(IoMT) resistance to cyberattacks. Bakhshi (2018) [12] divided threats based on security flaws and attacks after addressing potential security issues with the Cisco and Azure architecture models. Figure 2 depicts the prominent challenges in employing IIoT.

Overview of Security Threats and Vulnerabilities

1. Unauthorized Access and Data Breaches: One of the primary concerns in IIoT security is unauthorized access to sensitive data or control systems. Attackers may exploit vulnerabilities in network infrastructure, devices, or software to gain unauthorized access and compromise critical information.

Fig. 2 Challenges in employing IIoT. *Source* https://www.iotforall.com/the-challenges-of-iiot-and-how-to-solve-them, accessed on 5th July 2023

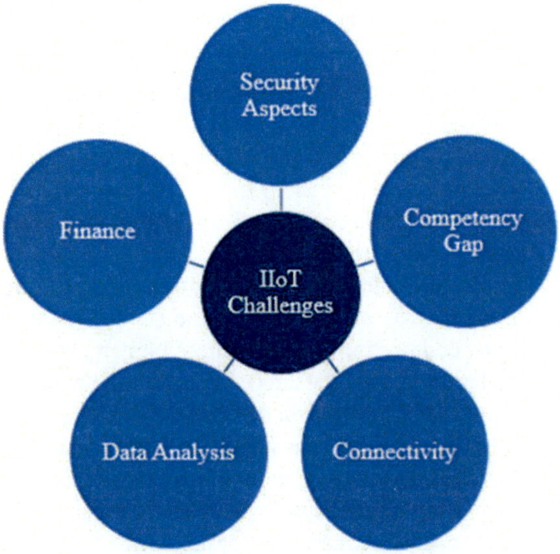

2. Malware and Ransomware Attacks: IIoT devices can be targeted by malware, including viruses, worms, and ransomware. Malicious software can disrupt operations, compromise data integrity, and even hold systems hostage in exchange for ransom payments.
3. Denial-of-Service (DoS) and Distributed Denial-of-Service (DDoS) Attacks: These attacks aim to disrupt or incapacitate IIoT systems by overwhelming them with a flood of network traffic. DoS and DDoS attacks can lead to service interruptions, rendering devices or systems unavailable to legitimate users.
4. Physical Attacks: IIoT devices located in industrial environments are susceptible to physical tampering or theft. Attackers may physically compromise or manipulate devices, sensors, or communication links to gain control or extract sensitive information.
5. Insecure Network Communication: IIoT devices often rely on wireless communication protocols, such as Wi-Fi, Bluetooth, or cellular networks. Inadequate encryption, weak authentication, or improper configuration of network protocols can expose data to interception or unauthorized access.
6. Lack of Security Updates and Patch Management: Failure to apply security updates and patches to IIoT devices and systems leaves them vulnerable to known exploits. Regular updates are crucial to addressing newly discovered vulnerabilities and strengthening overall security.
7. Insider Threats: Insider attacks involve individuals with authorized access to IIoT systems who intentionally or unintentionally compromise security. This may include employees, contractors, or other trusted individuals who misuse their privileges or unknowingly introduce risks.

Physical Security Considerations

1. Access control: Implement strict access controls to physical areas where IoT devices are deployed. This includes using mechanisms such as secure locks, key cards, biometric authentication, and surveillance systems to restrict unauthorized access.
2. Secure physical location: Ensure that IoT devices are deployed in secure physical locations that are protected against unauthorized access, theft, vandalism, and environmental hazards. Consider factors such as physical barriers, locked cabinets, secure data centers, and controlled environments.
3. Tamper detection and prevention: Implement tamper-resistant measures to detect and prevent physical tampering with IoT devices. This can include using tamper-evident seals, intrusion detection systems, and physical sensors that can trigger alerts or shut down devices if tampering is detected.
4. Physical device protection: Safeguard IoT devices from physical damage, theft, and unauthorized modifications. This can involve measures such as protective enclosures, hardened casings, anti-tamper coatings, and secure mounting mechanisms.
5. Surveillance and monitoring: Deploy surveillance cameras, motion sensors, and other monitoring systems to detect suspicious activities around IoT infrastructure. Regularly review and analyze the surveillance data to identify potential security breaches or vulnerabilities.
6. Supply chain security: Ensure the security of the supply chain for IoT devices, including the physical transportation and storage of devices. Implement measures to verify the authenticity and integrity of devices during procurement, shipping, and installation.
7. Disaster recovery and continuity planning: Develop robust disaster recovery and business continuity plans to address physical threats such as natural disasters, power outages, or physical damage to IoT infrastructure. This includes backup power systems, redundancy measures, and off-site data backups.
8. Employee awareness and training: Educate employees and stakeholders about the importance of physical security in Industrial IoT environments. Conduct training programs to raise awareness about security policies, procedures, and best practices to prevent physical breaches.

Communication Security Issues

Communication security issues in Industrial IoT generally revolve around protecting the confidentiality, integrity, and availability of data transmitted between devices, systems, and networks. Here are some common communication security issues in IIoT:

1. Unauthorized access: Unauthorized individuals or malicious entities may attempt to gain unauthorized access to IIoT networks and devices, potentially compromising sensitive data or taking control of critical infrastructure.
2. Data interception: IIoT communication channels may be vulnerable to eavesdropping, where attackers intercept and capture data transmitted between devices. This can lead to the exposure of sensitive information and potentially enable further attacks.
3. Data tampering: Attackers may attempt to modify or tamper with the data transmitted between IIoT devices, leading to potentially severe consequences such as incorrect decisions, equipment malfunction, or safety hazards.
4. Man-in-the-middle attacks: In this type of attack, an adversary positions themselves between communicating IIoT devices and intercepts or modifies the data transmitted. This allows them to manipulate the communication flow, potentially leading to unauthorized access or data manipulation.
5. Denial-of-Service (DoS) attacks: Attackers may launch DoS attacks against IIoT networks, flooding them with a high volume of requests or malicious traffic, causing service disruptions or system unavailability.
6. Insecure protocols: The use of insecure communication protocols or outdated encryption algorithms in IIoT deployments can expose vulnerabilities and make it easier for attackers to compromise communication channels.
7. Lack of device authentication: Failure to properly authenticate IIoT devices can allow unauthorized devices to gain access to the network, compromising its security.

Countermeasures to address these communication security issues typically involve a combination of technical and procedural measures. Some common countermeasures include:

1. Encryption: Implementing strong encryption mechanisms to secure data transmission and protect against eavesdropping and tampering.
2. Authentication and access control: Enforcing strict authentication mechanisms to verify the identity of IIoT devices and users, and implementing access control policies to restrict unauthorized access.
3. Secure communication protocols: Using secure and robust communication protocols that have been specifically designed for IIoT environments, taking into account the unique security requirements of industrial systems.
4. Intrusion detection and prevention systems: Deploy systems that can monitor network traffic and detect potential intrusions or abnormal behavior, enabling timely response and mitigation of security incidents.
5. Network segmentation: Dividing IIoT networks into separate segments or zones to limit the potential impact of a security breach and prevent lateral movement by attackers.
6. Security updates and patch management: Regularly update IIoT devices and systems with the latest security patches and firmware updates to address known vulnerabilities.

7. Employee training and awareness: Educating employees about security best practices, including safe communication practices and the potential risks associated with IIoT deployments.

Authentication and Access Control Challenges

1. Secure Device Authentication: Authenticating IoT devices securely is crucial in industrial IoT systems. Ensuring that only authorized devices can connect to the network and communicate with other devices is a challenge due to the large number of devices and potential vulnerabilities in their authentication mechanisms.
2. Access Control Complexity: Industrial IoT environments typically involve a complex network of devices, gateways, and systems. Implementing and managing access control policies across such a diverse and dynamic ecosystem can be challenging. Ensuring that the right individuals or entities have appropriate access rights to devices and data is crucial for maintaining security.
3. Scalability: Industrial IoT systems often consist of a large number of devices spread across multiple locations. Ensuring that authentication and access control mechanisms can scale effectively to handle the growing number of devices and users is a challenge.
4. Interoperability: Industrial IoT systems often involve devices and technologies from different vendors. Ensuring interoperability while maintaining security can be challenging, particularly when different devices and systems may have varying authentication and access control mechanisms.
5. Secure Communication: Industrial IoT systems rely on secure communication channels to transmit data between devices and backend systems. Ensuring the confidentiality, integrity, and authenticity of data exchanged between devices is a significant challenge, as these systems may be susceptible to eavesdropping, tampering, or impersonation attacks.

Data Integrity and Confidentiality Concerns

In general, data integrity and confidentiality are crucial concerns in any IoT system, especially in industrial settings where sensitive information and critical operations are involved. Here are some common data integrity and confidentiality concerns related to Industrial IoT security:

1. Data Integrity:
    - (i) Tampering: Unauthorized modification of data during transmission or storage can compromise the integrity of the information.
    - (ii) Data corruption: Errors or failures in the storage, processing, or transmission of data can lead to data corruption and loss of integrity.

(iii) Replay attacks: Malicious actors may capture and replay legitimate data, leading to unauthorized actions or manipulation of the system.

2. Data Confidentiality:

(i) Unauthorized access: Confidential data should only be accessible to authorized individuals or systems. Unauthorized access can lead to data breaches and compromise sensitive information.
(ii) Eavesdropping: Interception of data during transmission can result in unauthorized parties gaining access to sensitive information.
(iii) Insider threats: Trusted individuals within the organization may misuse their privileges to access and disclose confidential data.

The Countermeasures to Address These Concerns Can Include

1. Data Integrity Countermeasures:

(i) Cryptographic techniques: Implementing encryption and digital signatures can ensure the integrity and authenticity of data.
(ii) Message authentication codes (MAC): Using MACs can verify the integrity of messages and detect tampering attempts.
(iii) Error detection and correction codes: Employing techniques like checksums or cyclic redundancy checks (CRC) can identify and correct data corruption.

2. Data Confidentiality Countermeasures:

(i) Access control and authentication: Implementing strong access controls and authentication mechanisms can ensure that only authorized individuals can access sensitive data.
(ii) Encryption: Encrypting data during storage and transmission can protect it from unauthorized access.
(iii) Secure communication protocols: Using secure protocols like Transport Layer Security (TLS) can protect data from eavesdropping attacks.

Attack Vectors in Industrial IoT

Some of the Common Attack Vectors in Industrial IoT Security Include

1. Unauthorized Access: Attackers may attempt to gain unauthorized access to IIoT devices, networks, or systems, either by exploiting vulnerabilities in software or by using default or weak credentials.

2. Denial of Service (DoS) and Distributed Denial of Service (DDoS): Attackers overload the targeted IIoT systems with a flood of requests, causing service disruptions or complete system shutdowns. For example, in 2016, one of the biggest Distributed Denial-of-Service (DDoS) attacks was carried out on service provider Dyn using an IoT botnet. This resulted in the internet being down and affecting the companies like CNN, Twitter, Netflix, and Reddit.
3. Malware and Ransomware: IIoT devices can be infected with malware, including ransomware, which can encrypt critical data or systems until a ransom is paid. Example: McDonald's 500 GB data has been stolen in a ransomware attack.
4. Man-in-the-Middle (MitM) Attacks: Attackers intercept and alter communication between IIoT devices, leading to unauthorized access, data manipulation, or eavesdropping.
5. Physical Attacks: Physical tampering with IIoT devices or infrastructure, such as stealing or manipulating devices, can compromise the security and integrity of the entire system.
6. Data Manipulation and Spoofing: Attackers can manipulate or falsify IIoT data, leading to erroneous decisions, system malfunctions, or safety hazards.
7. Supply Chain Attacks: Malicious actors may compromise the security of IIoT devices or components during the manufacturing, distribution, or installation process, leading to backdoors or vulnerabilities.*Example:* Data breach and Data hack happened in *Target*, one of the largest US retailers. Millions of customers' payment cards became prone to fraud due to the malware in the POS system.

Different Countermeasures Can Be Used to Address These Attack Routes and Improve IIoT Security, Such As

1. Strong Authentication and Access Controls: Implementing robust authentication mechanisms and access controls, such as multi-factor authentication, can help prevent unauthorized access.
2. Encryption and Secure Communication: Encrypting IIoT data in transit and at rest, as well as implementing secure communication protocols, can protect against data interception and manipulation.
3. Intrusion Detection and Prevention Systems (IDPS): Employing IDPS solutions can help detect and mitigate potential attacks, providing early warning and response capabilities.
4. Regular Patching and Updates: Keeping IIoT devices and software up to date with the latest security patches and updates can address known vulnerabilities.
5. Network Segmentation and Firewalls: Segregating IIoT networks into separate zones and implementing firewalls can limit the impact of potential breaches and contain attacks within specific segments.
6. Security Monitoring and Incident Response: Implementing robust monitoring solutions and establishing incident response procedures enable timely detection and response to security incidents.

7. Employee Education and Awareness: Providing training and awareness programs to employees helps prevent social engineering attacks and ensures best security practices are followed.

Device-Level Attacks (e.g., Firmware Attacks, Physical Tampering)

Device-level attacks are a major threat to industrial IoT systems. These attacks can include firmware attacks, physical tampering, and malicious code injection. Firmware attacks can involve attackers gaining access to the underlying firmware of an IoT device and manipulating it to gain access to privileged functions. Physical tampering can involve physical access to an IoT device, allowing an attacker to gain access to the device and its associated data. Malicious code injection is the practice of inserting malicious code into the firmware of an IoT device. This malicious code can be used to gain access to privileged functions or to manipulate data stored on the device. All of these attacks can be mitigated by using secure firmware updates, physical device locks, and code verification techniques.

Firmware Attacks These are one of the most prominent security threats to industrial IoT systems. Such attacks leverage malicious code or commands embedded in the firmware of the devices that form the IIoT system. Attackers can exploit firmware vulnerabilities to gain access to devices or modify the system's functionality, thereby compromising data and disrupting operations.

Physical Tampering is another type of attack that can be used to compromise industrial IoT systems. In physical tampering attacks, an attacker gains physical access to a device or system and modifies its hardware or firmware to gain access to the data or to modify the system's functionality. This type of attack is particularly dangerous because the attacker can directly modify the system's hardware, making it difficult to detect and mitigate.

Network-Level Attacks These can be used to disrupt the communication between IoT devices and the network. These attacks can include denial of service (DoS) attacks, man-in-the-middle attacks, and distributed denial of service (DDoS) attacks. Additionally, network-level attacks can be used to gain access to confidential data stored on the device. Application-Level Attacks: Application-level attacks are targeted at the software running on the device. These attacks can include buffer overflows, SQL injection, and malicious code injection. These attacks can be used to gain access to confidential data or modify the functionality of the device. Device-Level Attacks: Device-level attacks target the hardware of the device. These attacks can include reverse engineering, side-channel analysis, and fault injection attacks. These attacks can be used to gain access to confidential data or modify the hardware of the device.

Software Vulnerabilities and Malware Threats

Software vulnerabilities and malware threats are a major concern when it comes to the security of Industrial IoT (IIoT) systems. As IIoT systems become increasingly interconnected and rely on a variety of operating systems, software, and applications, they are exposed to a wide range of malicious attacks. Malware such as viruses, worms, Trojan horses, and ransomware can cause serious damage to IIoTsystems, leading to data loss, downtime, and even financial losses. In addition to malware, malicious actors can exploit software vulnerabilities to gain unauthorized access to sensitive information, alter or delete data, or even hijack entire IIoT systems. Some of the common malware threats include:

1. Ransomware: Ransomware is a type of malware that encrypts a user's data and then demands a ransom payment to access the data.
2. Spyware: Spyware is a type of malicious software that can track a user's activities, collect personal information, or even take control of the system.
3. Botnets: Botnets are networks of computers that have been infected with malicious software and can be used to launch distributed denial-of-service (DDoS) attacks or to spread malware.

It is necessary to identify appropriate security services at various levels of the Industrial Internet of Things (IIoT) stack and implement the necessary mechanisms to safeguard the industrial space from bad actors to be able to defend the entire application space from malicious attacks (Ledwaba and Hancke 2021) [13]. To mitigate the threat of malicious software, organizations should regularly patch their software and keep up to date with security updates. Furthermore, organizations should regularly monitor their networks for malicious activity and respond quickly to any suspicious activity.

1. Unpatched Software: Unpatched software can be vulnerable to attackers who can use these vulnerabilities to gain access to a system or take control of the industrial IoT device.
2. Weak Authentication: Weak authentication methods, such as passwords or authentication tokens, can allow attackers to gain access to the system.
3. Backdoors: Backdoors can be used by attackers to gain access to the system, bypassing authentication methods.

Social Engineering and Human Factors

Social engineering and human factors are often overlooked when discussing industrial IoT security threats. However, humans are often the weakest link in the security chain. Hackers may use social engineering tactics such as phishing, baiting, or even physical access to gain access to a system and its data. Additionally, unauthorized personnel may access an IoT device due to weak authentication methods or a lack of personnel training. To counter these threats, organizations should implement strict authentication methods such as two-factor authentication and regularly train their personnel on security best practices. Furthermore, organizations should implement procedures to detect and respond to any unauthorized access attempts.

Risk Assessment and Threat Modeling

Importance of Risk Assessment in Industrial IoT

Risk assessment is an essential component of any security system for Industrial IoT. Risk assessment helps to identify and analyze potential threats and vulnerabilities that can lead to damage or harm to the system. It also helps to prioritize risks and allocate resources to address them. By recognizing the potential risks in advance, organizations can take preventive measures to avoid or mitigate them. Risk assessment also helps organizations identify their security objectives and develop appropriate countermeasures. Additionally, it enables organizations to monitor and review their security systems regularly to ensure continued effectiveness. With this information, organizations can also develop contingency plans to respond to any security-related incidents. Therefore, risk assessment is a key component of any Industrial IoT security system.

Threat Modeling Techniques and Methodologies

Threat modeling techniques and methodologies of risk assessment in Industrial IoT are becoming increasingly important as organizations strive to protect their devices and data from malicious actors. Threat modeling helps organizations identify potential threats, assess their risk, and then develop countermeasures to protect their systems. It involves analyzing the system to identify its vulnerabilities and then developing countermeasures to mitigate the risks associated with those vulnerabilities. Risk assessment methods such as the Risk Assessment Matrix (RAM), the NIST Risk Management Framework (RMF), and the Common Vulnerability Scoring System (CVSS) can be used to identify, prioritize, and mitigate potential threats. Additionally, organizations should also consider implementing strong authentication and authorization protocols, intrusion detection and prevention systems, and regular security audits to ensure the safety of their Industrial IoT systems.

Identifying Critical Assets and Potential Vulnerabilities

Industrial Internet of Things (IoT) is a rapidly growing technology that is revolutionizing the way industrial processes are managed and monitored. However, as with any technology, security threats and vulnerabilities exist that can put an organization's assets and data at risk. Organizations must identify their critical assets and potential vulnerabilities to ensure they are protected from the latest cyber threats. Organizations must first identify the critical assets in their industrial IoT system. These assets can include devices, networks, and data. Once the assets have been identified, organizations must assess their current security posture and identify any potential vulnerabilities. This can include outdated software, unsecured networks, weak passwords, and a lack of encryption. By identifying these potential vulnerabilities, organizations can put measures in place to protect their assets from potential threats. Organizations must also be aware of the latest cyber threats and understand how they can be used to target their assets. This includes understanding the malicious actors and the techniques they use, such as malware, phishing, and ransomware. By understanding the threats and the techniques used, organizations can develop countermeasures that can help protect their assets from potential attacks. By identifying their critical assets and potential vulnerabilities, organizations can ensure their industrial IoT system is secure and protected from the latest cyber threats. Using two case studies, Wurm et al. (2016) [14] showed that commercial and industrial IoT equipment are susceptible to IoT-specific assaults. Lee (2020) [15] evaluated cyber risk management frameworks and IoT cybersecurity technologies and created a linear programming model to assist in resource allocation choices. Through proper security measures, organizations can ensure their assets are safe and secure. The critical assets of an organization include Industrial control systems (ICS), Industrial Internet of Things (IIoT) devices, Cloud computing services, Network infrastructure, etc. The potential vulnerabilities of these critical assets happen mainly due to Unsecured wireless networks, Unpatched firmware, Unencrypted data, Poorly configured firewalls, Lack of authentication and authorization, Insufficient logging and monitoring, Inadequate physical security measures, Malware and ransomware attacks, etc. According to Gupta et al. (2023) [16] whenever digital transformation happens, it offers opportunities and challenges.

Impact Analysis and Risk Prioritization

The impact analysis can be summarized as follows:

1. Increased Awareness: The paper contributes to raising awareness about the security challenges specific to IIoT environments. By highlighting the potential risks and vulnerabilities, it helps industrial organizations and stakeholders recognize the importance of robust security measures for IIoT deployments.

2. Risk Mitigation: The research identifies and analyzes various security threats in industrial IoT systems, such as unauthorized access, data breaches, device tampering, and supply chain attacks. It offers countermeasures and best practices to mitigate these risks, providing valuable guidance to organizations implementing or operating IIoT solutions.
3. Operational Resilience: By addressing security concerns, the paper aids in enhancing the resilience of industrial operations. It emphasizes the significance of protecting critical infrastructure, manufacturing processes, and sensitive data from cyber threats, thereby ensuring uninterrupted and secure industrial processes.
4. Standardization and Regulatory Efforts: The paper's insights contribute to ongoing standardization and regulatory initiatives for IIoT security. It informs policymakers, industry associations, and regulatory bodies about the existing and emerging threats in the industrial IoT landscape, enabling them to develop appropriate guidelines and regulations to safeguard industrial systems.

To prioritize the risks discussed in the paper, several factors should be considered:

1. Potential Impact: Evaluate the potential impact of each identified threat on the availability, integrity, and confidentiality of IIoT systems. Assess the severity of consequences, such as operational disruptions, financial losses, safety hazards, or reputational damage.
2. Likelihood of Occurrence: Assess the likelihood of each threat occurring based on the probability of vulnerabilities being exploited, threat actors' motivations, and historical data on similar incidents. Consider the risk landscape specific to the organization's industry and geographical location.
3. Existing Safeguards: Examine the effectiveness of existing security measures and controls in addressing each threat. Identify any gaps or weaknesses that require immediate attention.
4. Vulnerability Severity: Determine the severity of vulnerabilities associated with each threat. Assess whether they are common, easily exploitable, or require advanced techniques, as well as the potential for widespread impact.
5. Business Impact: Consider the impact of each threat on the organization's core business processes, critical infrastructure, sensitive data, and compliance requirements. Assess the potential disruption to operations, customer trust, and legal/regulatory consequences.

By evaluating these factors, an organization can prioritize risks based on their potential impact and likelihood, allowing for the allocation of appropriate resources and the implementation of effective countermeasures to mitigate the most critical threats.

Conclusion

In conclusion, the rapid growth of Industrial Internet of Things (IIoT) technology has brought numerous benefits to industrial sectors, enhancing efficiency, productivity, and connectivity. However, it has also introduced significant security threats that must be addressed to ensure the continued success and integrity of industrial operations. This paper has provided an extensive analysis of various security threats facing IIoT systems, including device vulnerabilities, network breaches, data privacy concerns, and potential attacks on critical infrastructure. Moreover, it has explored a range of countermeasures and best practices to mitigate these threats, such as robust authentication protocols, secure communication channels, regular system updates, and employee training on cybersecurity awareness. The implementation of comprehensive security strategies, involving a combination of technological solutions, industry standards, and organizational policies, is crucial for protecting IIoT ecosystems from evolving threats. By proactively addressing these challenges, industries can embrace the full potential of IIoT while safeguarding their operations, assets, and sensitive data. Future research and collaborative efforts are needed to continually adapt and enhance security measures in response to emerging threats in the dynamic landscape of Industrial IoT.

References

1. Viriyasitavat, W., Anuphaptrirong, T., & Hoonsopon, D. (2019). When blockchain meets the Internet of Things: Characteristics, challenges, and business opportunities. *Journal of Industrial Information Integration, 15*, 21–28.
2. Tsiknas, K., Taketzis, D., Demertzis, K., & Skianis, C. (2021). Cyber threats to industrial IoT: A survey on attacks and countermeasures. *IoT, 2*(1), 163–186.
3. Schneider, S. (2017). The industrial Internet of Things (iiot) applications and taxonomy. In *Internet of Things and Data Analytics Handbook* (pp. 41–81).
4. Vargas, H., Lozano-Garzon, C., Montoya, G. A., & Donoso, Y. (2021). Detection of security attacks in industrial IoT networks: A blockchain and machine learning approach. *Electronics, 10*(21), 2662.
5. Jaidka, H., Sharma, N., & Singh, R. (2020). Evolution of IoT to IIot: Applications & challenges. In *Proceedings of the international conference on innovative computing & communications (ICICC)*.
6. Li, W., Chai, Y., Khan, F., Jan, S. R. U., Verma, S., Menon, V. G., & Li, X. (2021). A comprehensive survey on machine learning-based big data analytics for IoT-enabled smart healthcare system. *Mobile Networks and Applications, 26*, 234–252.
7. Karmakar, A., Dey, N., Baral, T., Chowdhury, M., & Rehan, M. (2019). Industrial Internet of things: A review. In *International conference on opto-electronics and applied optics (optronix)* (pp. 1–6). IEEE.
8. Litoussi, M., Kannouf, N., El Makkaoui, K., Ezzati, A., & Fartitchou, M. (2020). IoT security: Challenges and countermeasures. *Procedia Computer Science, 177*, 503–508.
9. Zhou, L., Yeh, K. H., Hancke, G., Liu, Z., & Su, C. (2018). Security and privacy for the industrial Internet of things: An overview of approaches to safeguarding endpoints. *IEEE Signal Processing Magazine, 35*(5), 76–87.

10. Yussoff, Y. M., Hashim, H., Rosli, R., & Baba, M. D. (2012). A review of physical attacks and trusted platforms in wireless sensor networks. *Procedia Engineering, 41*, 580–587.
11. Gaurav, A., Psannis, K., & Peraković, D. (2022). Security of cloud-based medical internet of things (miots): A survey. *International Journal of Software Science and Computational Intelligence (IJSSCI), 14*(1), 1–16.
12. Bakhshi, Z., Balador, A., & Mustafa, J. (2018). Industrial IoT security threats and concerns by considering Cisco and Microsoft IoT reference models. In *IEEE wireless communications and networking conference workshops (WCNCW)* (pp. 173–178). IEEE.
13. Ledwaba, L. P., & Hancke, G. P. (2021). Security challenges for industrial IoT. In *Wireless networks and industrial IoT: Applications, challenges and enablers* (pp. 193–206). Springer, Cham.
14. Wurm, J., Hoang, K., Arias, O., Sadeghi, A. R., &Jin, Y. (2016). Security analysis on consumer and industrial IoT devices. In *21st Asia and South Pacific design automation conference (ASP-DAC)* (pp. 519–524). IEEE.
15. Lee, I. (2020). Internet of Things (IoT) cybersecurity: Literature review and IoT cyber risk management. *Future internet, 12*(9), 157.
16. Gupta, S., Kapoor, M., & Sharma, H. K. (2023). Industry 5.0 and the collaborative approach of internet of things with artificial intelligence. In *Advanced research and real-world applications of industry 5.0* (pp. 167–177). IGI Global.

Internet-of-Things Enabled Smart Health Monitoring System Using AutoAI: A Graphical Tool of IBM Watson Studio

Yunika Kadayat, Sachin Sharma, Piyush Agarwal, and Seshadri Mohan

Abstract The use of internet-of-things (IoT) enabled smart health monitoring system to improve healthcare service quality has become inescapable due to the industry's growing data volume. Although big data processing technologies and platforms have been included into preventing emergencies. An IoT-enabled architecture is proposed to produce real-time warnings for healthcare systems. This proposed method is including IBM Watson studio and AutoAI services to produce a real-time warning and precise predictions of patient health. People's health awareness has effectively increased along with the current socioeconomic level, and self-personal health management is receiving an increasing amount of attention. Personal health management systems usage can produce the best outcomes in health management and, as a result, improve the protection of people's health. The personal health management system is examined and explored using IoT and big data technology so that its features can be more effectively utilized. It primarily implements administrator and user functions while also achieving the functionalities of accessing epidemic-related data and maintaining personal health information. Viewing epidemic information on a mobile device is convenient and easy for people who are managing their health. Experiments show that the system is highly practicable, highly secure, highly scalable, and has good maintainability.

Keywords Big data · AutoAI · IBM Watson · Web · Personal health systems

Y. Kadayat · S. Sharma (✉) · P. Agarwal
Graphic Era Deemed to be University, Dehradun, Uttarakhand, India
e-mail: sachin.cse@geu.ac.in

Y. Kadayat
e-mail: kadayatyunika747@gmail.com

P. Agarwal
e-mail: piyush221292@gmail.com

S. Mohan
University of Arkansas at Little Rock, Little Rock, AR, USA
e-mail: sxmohan@ualr.edu

A. Prasad et al. (eds.), *Communication Technologies and Security Challenges in IoT*, Internet of Things, https://doi.org/10.1007/978-981-97-0052-3_21

Introduction

According to HealthCatalyst2, over 400 TB of data generated through IoT sensors is stored by healthcare organizations with more than 1,000 employees. Healthcare is undoubtedly a high-data volume industry given this transactional data in the healthcare sector is subject to frequent modification. Every day, claims are paid, electronic health records (EHRs) are created using abstracted patient data, and real-time diagnostic test results are recorded electronically. The assertion that medical data complies with high-velocity requirements is supported by each of these characteristics.

The proportion of the population that is older, sometimes known as the ageing population, is a global phenomena [1] that has a significant impact on healthcare ssystem. Relying on such conventional procedures in such a situation could lower the quality of life for millions of people. To address this issue, numerous healthcare systems have been created. The standard procedure is to periodically upload medical information to an automated system that will monitor the patient's health in real time, including body temperature, blood glucose levels, blood pressure, heart rate, and ECG signals. These systems continuously assess data, allowing them to provide prompt assistance when necessary. Automated health monitoring employs proactive measures to lower hospitalisation costs and enhance healthcare services by cutting down on consultation and waiting times. IoT sensors and other sources provide enormous volumes of data for monitoring healthcare, and these data include things like biomedical imaging, and patient health monitoring. The amount and diversity of healthcare data are enormous, and it also moves quickly. Big data techniques, therefore have great potential for improving the effectiveness of healthcare systems. This chapter aims to provide an architecture for IoT-enabled healthcare monitoring applications with the capability of data storage, processing, and analysis of real-time healthcare data. This study highlights how big data analytics may be used in the healthcare industry to unearth meaningful information from extremely valuable data.

The AutoAI graphical tool in Watson Studio examines your data to identify the most effective data transformations, algorithms, and parameter settings for your predictive modeling challenge. You can choose from a list of model candidate pipelines that AutoAI presents to you as the results. Necessary service Machine learning using Watson IBM Studio. You can create and deploy a machine learning model with complex training features and no coding using AutoAI. Most of the work is done for you by the tool. The healthcare information that is now available in India is independently maintained by hospitals and other institutions and is not easily accessible in an organized, knowledgeable manner. This significantly hampers healthcare providers' efforts to raise standards of performance. In this essay, we examine the problem of combining information from diverse sources in real-time, which has the potential to save lives. Additionally, India's low per capita income and low doctor-to-patient ratio drive up medical costs, making it harder for patients, particularly those living in rural regions, to afford quality health care. The focus is on finding ways to bridge the gap between patients and doctors and how to treat patients more cheaply. The development of a mobile/web application that allows patients to transmit their symptomatic queries to doctors via a server is the main emphasis of this chapter.

Literature Survey

Lower communication overhead is accomplished with self-certified public key cryptography based on elliptical curves. It allows for reciprocal authentication that is anonymous [2]. User authentication is performed jointly by two devices on the user side and a server. however, neither device is capable of completing the authentication entirely. The two-level security mechanism raises the bar for security. Without managing the device, the attacker won't be able to break the encryption even if they compromise one device [3]. A Verifiable Three-factor remote user authentication that is secure and effective allows for anonymity for the user [4]. the strengthening of the server authorization procedure through the application of threshold cryptography [5]. In the healthcare industry, it can stop well-known assaults such as the man in the middle, guessing, and stolen smart card, among others. IoT (Internet of Things) in the healthcare sector. Stage improvement has received increased attention because of Medicine 4.0, both in terms of programming and equipment. Because of this idea, medical care IoT (H-IoT) frameworks have been developed [6].

Background

An Overview of Big Data Approaches

Due to its superior ability to handle large volumes of data as compared to conventional technologies, big data technologies have drawn a lot of perspectives. Big data frameworks support structured, semi-structured, and unstructured data, and they also offer certain additional features. A better decision-making process is enabled by these characteristics, which include tools for large data mining and the building of predictive models. Streamline and batch processing are the two methods that were used in the proposed method. When there is no restriction on the response time, in that situation batch processing can be used otherwise stream processing. In the case of data processing, all data must be ingested before processing in batch mode at a predetermined time. MapReduce is an extensively used solution in the field of batch computing [7]. Many aspects of batch processing frameworks include the capacity to do several complicated computations and access all data; its delay is measured in minutes or more.

Evaluation of Bioinformatics

Real-world applications frequently generate large data with redundant characteristics and missing values, which complicates data analysis since it takes a lot of processing effort. For a single data point or a window of recent data, quick and simple computation processing is applied in stream processing. These calculations finish in a matter of seconds. Methods for mining big data streams, such as classification, by quickly extracting the most pertinent information, clustering, and frequent pattern mining reduce the computing effort. Techniques based on data enable summarization to choose a portion of the continuous flock rather than the full dataset of processing of streaming data [8]. Both the aggregation method and the summary data structure approach are database-based techniques. Task-based methodologies upgrade current practices or designs. To speed up computations when processing data streams, new techniques must be developed. They can be divided into three groups: approximation algorithms to produce outputs with acceptable error margins, and sliding windows for analysis of recent data. Modern healthcare analytics require big data techniques because they enable the instantaneous extraction of pertinent data from a vast volume of patient data [9]. Consequently, when the prediction algorithm spots potential issues, alarms are issued. This procedure aids in avoiding health emergencies, decision-making of medical professionals on illness diagnosis, and offers recommendations for specialized therapy.

Proposed System

The proposed approach is prepared to deal with a range of circumstances of early disease diagnosis. The scenario that follows serves as an illustration of how data management functions in this architecture. Figure 1 depicts the proposed system's layered architecture. Figure 2 shows the suggested system in detail.

Data Acquisition Layer

Several different sorts of data are produced when a patient's health state is continually monitored. Medical data can comprise organized information like conventional Electronic Health Records (EHRs), semi-structured information like logs created by various medical equipment, and unstructured information like biomedical images.

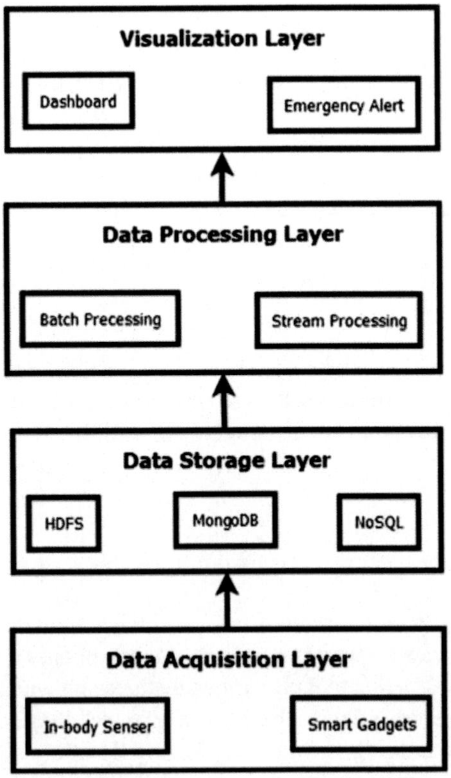

Fig. 1 Layered proposed architecture

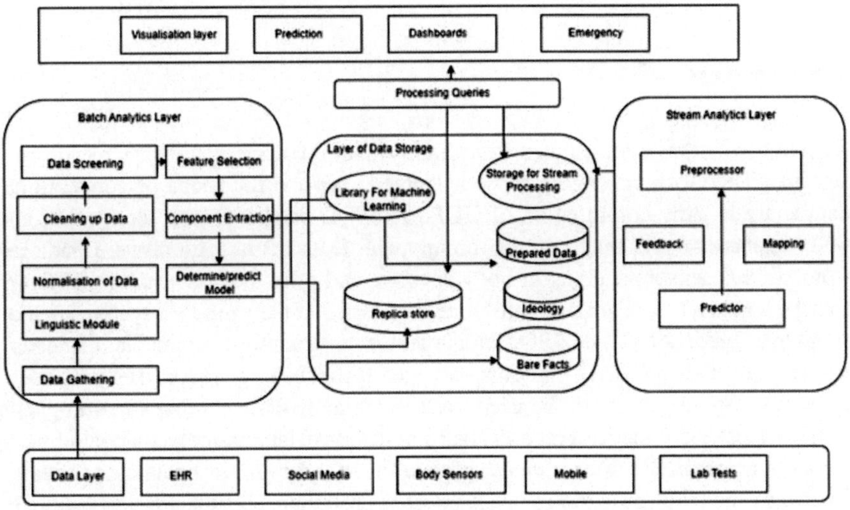

Fig. 2 Proposed system

Sensing Data

Systems for monitoring heath use a variety of sensors. Patients' living spaces could be crammed with gadgets like pressure sensors, microphones, and surveillance cameras to assure effective health monitoring. As a result, the volume of data created by health monitoring systems tends to expand dramatically, necessitating the use of advanced techniques during the processing stage.

Social Media Data

It must be gathered to do social network analysis. The following stage is obtaining information that may influence healthcare prediction. Social network data has characterized by ambiguity, making it dangerous to utilize it to develop prediction models.

Mobile Phone

Now, mobile phones have a variety of sensors, including cameras, accelerometers, satellite positioning systems, and many others. Mobile phones are the perfect devices for collecting health data due to their various features and widespread use, enabling the development of several effective healthcare apps including heart frequency monitoring, child nutrition, and pregnancy monitoring. Reading the data from various formats of healthcare sensors is the goal of the data acquisition phase. After that, the data goes via a semantic module before being normalized.

Data Storage

Selection of storage techniques have highly data-intensive healthcare monitoring systems. Therefore, managing those systems is beyond the scope of conventional data analysis. Any combination of HDFS, NoSQL databases like MongoDB, and SQL databases could make up this component. Data of the patient were gathered from various sensors such as in-body sensors, and other body sensors in different smart gadgets and categorized into three categories in the proposed system include semi-structured (XML and JSON documents), unstructured (biomedical images), and structured data (EHR). The raw data stored for the target databases will house these data. Stream temp will be used to store social media and other streaming data [10]. Clinical trials, medical records, radiological data, laboratory testing details, and payment notes. EHRs offer a wealth of data for analytics in the healthcare industry. EHR also makes it possible for data interchange among healthcare professionals. Biomedical imaging is thought to be an effective technique for treating patients and diagnosing ailments. To assist doctors, make appropriate choices, analyzing these kinds of photos is difficult since they contain noisy data that must be eliminated.

Data Processing Layer

When dealing with such a vast number of healthcare data, choosing the most important attributes becomes essential [11]. Several approaches have been put forth in this context to handle this problem, particularly when managing thousands of features. The feature extraction method, on the other hand, is an alternative strategy that entails extracting fewer attributes than the original ones. A statistical tool store is needed to apply feature selection and extraction algorithms.

Data Preparation

Processing raw data devoid of preparatory procedures could call for additional computational resources that are too expensive in the context of big data [12]. To generate precise prediction models and increase the dependability of data mining approaches. The data cleaning and data filtering are the two stages that makeup data preparation. It's crucial to identify missing values in healthcare data. Making mistakes when filling in missing variables might lower the quality of the information collected and produce false results. In the healthcare industry, processing missing data should be done with the utmost accuracy since making the wrong choice might have catastrophic repercussions. The Expectation Maximization (EM) method and the multiple Imputation algorithm are only two examples of the numerous potent algorithms used in data mining to handle missing values.

Noise Treatment

There are two main approaches to handle noisy data in general. The first method includes applying data-cleaning techniques to correct noisy numbers; these methods are difficult to employ and are only applied when there is minimal to no noise. The patient's health parameters are irregularly captured and stored in the server throughout their hospital stay, which is the primary reason of the health monitoring system's data shortage. On the other side, working with unprocessed social media data makes data preparation more difficult.

Predictive Model Design

Building a model that can make predictions for fresh observations based on historical data is the goal of this component. The accuracy of a particular prediction model is used to gauge its quality. These models are created using resources offered by the recommended architecture's statistics and machine learning store. Batch processing results will be kept in a model store.

- **Batch Processing Layer**: Through many stages, batch computing is applied to the extracted data from the prepared data storage.
- **Stream Processing Layer**: Data synchronization, adaptive pre-processing, and adaptive prediction modules make up the stream data analysis layer.

  Data Synchronization: The data synchronisation module's goal is to ensure that data is handled in accordance with time standards and in the right order. Additionally, the data synchronization mechanism eliminates inconsistent measurements and deals with missing values. Setting thresholds on the incoming parameters allows for the detection of incorrect values.
  Adaptive Learning: The creation of adaptive systems was prompted by the requirement to adjust in the face of data changes, and adaptability is a crucial component of a big data system's long-term success. Pre-processing is a part of the adaptive system and not an activity carried out in a standalone fashion. Due to this, the adaptive system is considered the component of the prediction process, which is composed of the adaptive pre-processor and the adaptive predictor.
  Adaptive Pre-processor: The data flow that is entered is separated into time windows to begin processing activities. To divide data streams into overlapping subsets of tuples in this situation, the sliding window approach is used (each tuple is included in several windows). Every measure's average for a specified timeframe is calculated, and it is then contrasted with a user-defined threshold. A particular average will be stored when an emergency warning is issued if its value reaches an upsetting predefined threshold; otherwise, it will only be stored. Following the termination of comparisons with threshold values, the knowledge extraction stage picks out pertinent characteristics to be sent to the adaptive predictor portion.
  Adaptive Predictor: Predictors need to update in response to changes in the data to maintain a particular degree of accuracy. If not, data evolution will only cause them to lose reliability over time. Therefore, the predictive model should have the capability to generate the prediction and take the new data point at the same movement. Through that link, the predictor informs the pre-processor if an update is necessary or not, and the pre-processing unit responds by sending the raw data it requires via a predetermined mapping function. Stream processing results storage will receive the output of the stream processing.

Query Processor

It merges the results of both the stream processor and the batch processor and analyzes the health status of the patient.

Visualization Layer

The analytics layer generates a variety of outputs, including reports on predictive decision-making and health monitoring. To prevent the emergency to arise and better decision-making, the analytical results must be displayed in the visualized form. Visual representation of the results better understanding of the health of the patient and also patient can identify the irregular behavior of health.

Implementation Process for Detecting Emergency Cases

We want to identify any possible risks to patients using this prototype technology. The emergency detection module that appears in the visualization layer of the suggested architecture is to be implemented by AutoAI services and IBM Watson. Spark is used to take input in the batch processing layer, from the data acquisition layer. The user-specified regular period is followed by the execution of batch jobs. AutoAI is used to analyze real-time data streams. Medical sources such as sensors share the information with the AutoAI to identify the problematic circumstances based on the threshold defined by the user. The IBM Watson cloud receives warnings from spark streaming, which are then utilized to inform clinicians about crises. The parameters radius, texture, perimeter, area smoothness, concavity, and compactness are used to train a machine learning model that we are creating. The model is trained utilising AutoAI services on the IBM Watson cloud and may be utilised in web or mobile applications. All of these factors are taken into account along with mean, standard error, and overall values. After analysing the data, we collect data from patients here. The suggested system's implementation process is depicted in Fig. 3.

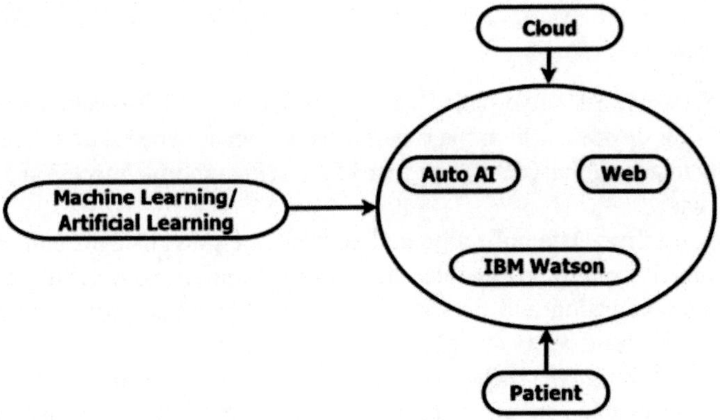

Fig. 3 Process of proposed system

We create an account on IBM CLOUD and then registered over there. Further, we create a cloud object storage service then create a Watson studio platform then activate machine learning service es. The most important part here is to build the model. For this, first, we created a project in Watson studio and added an AutoAI experiment to the project. Datasets are ready for further processing and selection of AutoAI pipeline which has ranked 1. Then deploy and test the model in Watson studio. similarly, for application building, we create node red services and manage the palette and install dashboard nodes. Then build UI with node-red. We have created an input form to collect the various input parameter and sent them to the IBM AutoAI service through the HTTP request once the machine learning its algorithm and then gives the output but the prediction is either M or B if the prediction is M, it is said to be that cell are very dangerous and can lead to cancer. Similarly, if the prediction came out to be B then it is said to be cells are safe from cancer. The application of machine learning techniques to healthcare analysis is increasing. The two elements of diagnosis that are unquestionably most important are the doctor's experience and the study of the patient's clinical data. Most likely medical errors can be avoided by utilising classification algorithms, which also allow for a quicker and more thorough assessment of healthcare data. Making the best treatment decisions for a patient requires accurate and fast cancer prognosis. Table 1 illustrates the comparative analysis of proposed methodology with existing methodologies. Figure 4 Illustrates the selection of Best Pipeline. Figure 5 describes the pipeline leaderboard. Figure 6 describes the deployment result. Figure 7 illustrates the Node RED. Figure 8 describes the output of proposed system.

Challenges

IoT Enabled Smart Health Monitoring System using AutoAI, a graphical tool of IBM Watson Studio, has several challenges that need to be addressed. Some of the challenges are (Fig. 9):

- *Data Security*: Sensitive patient data is collected by IoT-enabled smart health monitoring devices, which must be secured against cyberattacks. The system's security must be robust enough to guard against unauthorised access and guarantee data privacy.
- *Interoperability*: Data collection and analysis are done using a number of platforms and devices by IoT-enabled smart health monitoring systems. To guarantee seamless connection and data sharing, it is essential to provide compatibility between various devices and platforms.
- *Technical Knowledge*: Technical competence is necessary for creating and maintaining IoT-enabled smart health monitoring systems employing AutoAI, which might not be present in all healthcare settings. The management and upkeep of these systems may require healthcare providers to invest in training their staff.

Table 1 Comparative analysis of proposed methodology with existing methodologies

References	Methodologies	IoT	AI	Big data	Accuracy (%)
Boland et al. [1]	Structural health monitoring	Remote patient monitoring	Virtual health assistants	No	91
Bates et al. [13]	Gaussian mixture model	NO	Automation of redundant healthcare tasks	No	96.67
Forrest et al. [14]	Computational intelligence techniques	NO	Management of medical records	No	91.33
Andreu-Perez et al. [15]	Knowledge-driven techniques, fuzzy logic, ontologies	Depression and mood monitoring	Improved healthcare acess	No	93.44
Nunes et al. [7]	Computational intelligence techniques, decision tree, clustering	Heart-rate monitoring	Accurate cancer diagnosis	No	94
Proposed internet-of-things enabled smart health monitoring system using AutoAI: a graphical tool of IBM watson studio	machine learning model	IoT sensors	AutoAI services	Yes	97.66

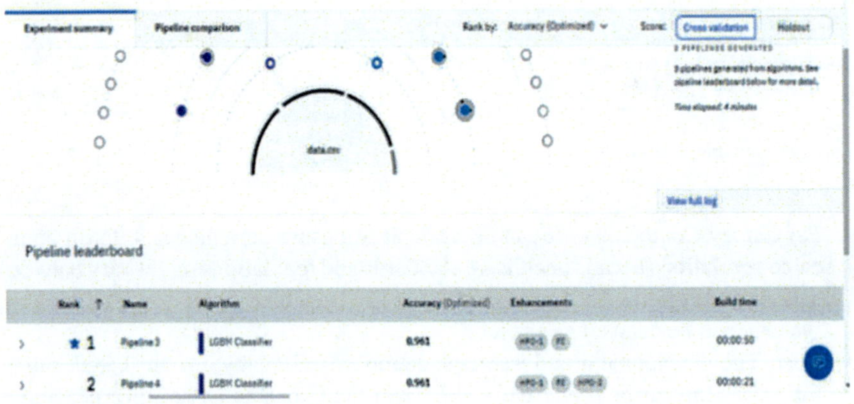

Fig. 4 Selection of best pipeline

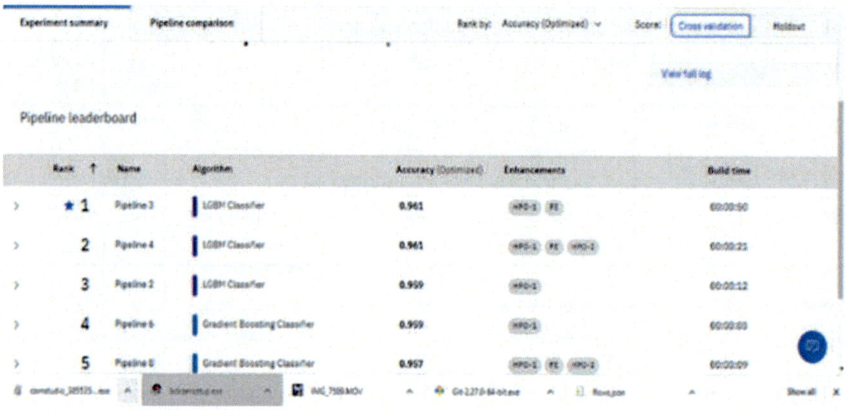

Fig. 5 Pipeline leaderboard

Fig. 6 Deployment result

- *Regulatory Compliance*: It can be difficult to ensure compliance with all applicable regulations becaus healthcare standards and regulations might vary between nations and regions. The implementation of IoT-enabled smart health monitoring systems can be delayed as a result.
- *Cost*: The development and implementation of IoT-enabled smart health monitoring systems can be expensive, making it difficult for smaller healthcare facilities to embrace these systems.
- *Ethical Issues*: The adoption of Internet of Things (IoT)-enabled smart health monitoring devices presents ethical issues with patient permission, data privacy, and transparency. These issues must be rectified in order to ensure that patients are appropriately informed about how their data is being used and to promote trust in these systems.

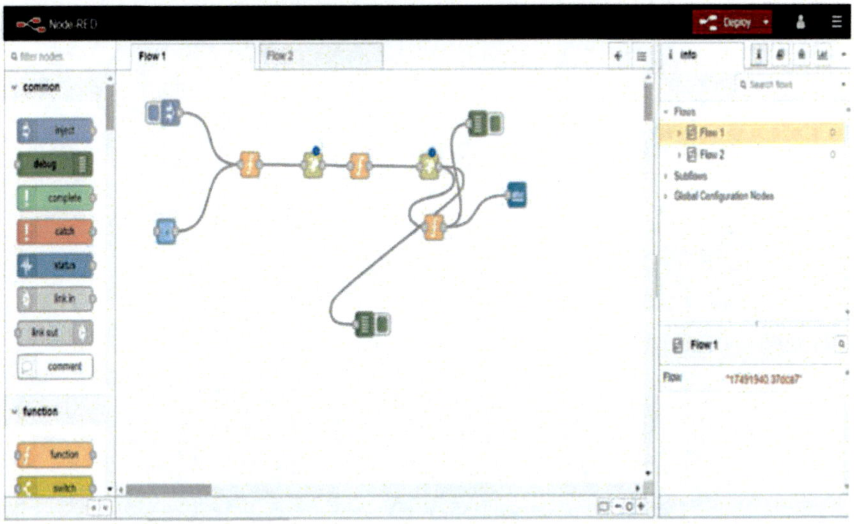

Fig. 7 Node RED

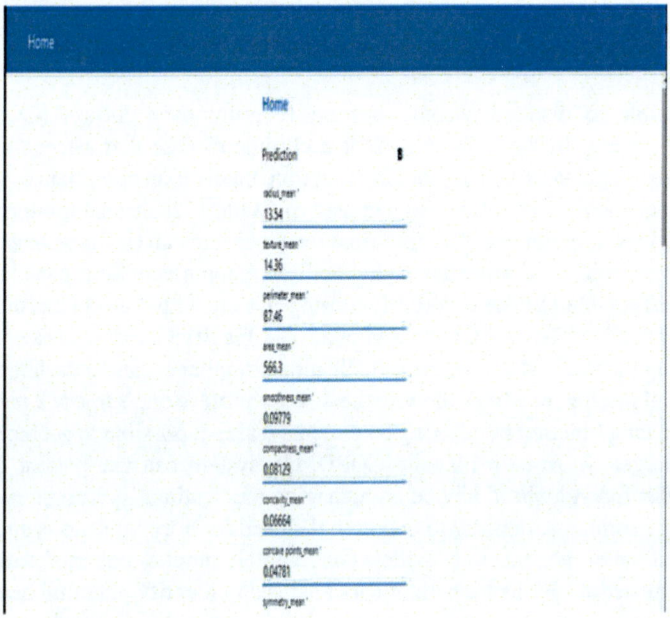

Fig. 8 Output of proposed system

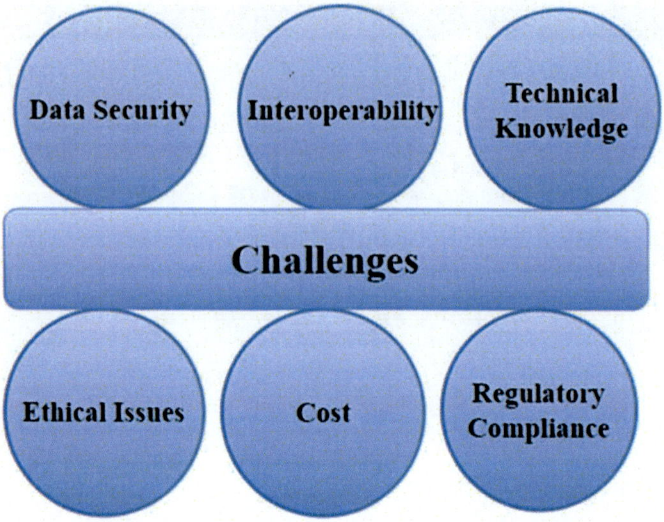

Fig. 9 Challenges of proposed system

Case Study

It is possible to manage chronic obstructive pulmonary disease (COPD) with proposed system. In the USA, COPD is a chronic disease that affects millions of individuals. Regular monitoring of lung function, medication compliance, and symptoms is necessary for effective management of COPD. Traditional monitoring techniques, however, might not offer real-time feedback and can be time-consuming. By offering real-time monitoring and individualised recommendations based on patient data, an IoT-enabled Smart Health Monitoring System with AutoAI can aid in better managing COPD. The system consists of a wearable device that assesses lung function, a drug dispenser that monitors medication compliance, and a dashboard driven by artificial intelligence (AI) allowing healthcare professionals to track patient data. Real-time lung function monitoring by patients makes it possible to act more quickly in an emergency. To help manage COPD, the system can use patient data analysis to offer individualised advice on managing medications, quitting smoking, and managing symptoms. In order to decrease the need for hospitalisations and improve patient outcomes, healthcare providers can remotely monitor patients' data and take appropriate action. By using patient data analysis to identify potential health issues before they arise, the system enables healthcare professionals to take preventative steps and enhance patient outcomes.

The "Wellness Hub" created by Panasonic Corporation is an additional example of proposed system that may be used in Japan. The method was created to support elderly people in maintaining an active lifestyle and tracking their health. The Wellness Hub is made up of a variety of sensors and gadgets that gather information on body weight, blood pressure, sleep habits, and other things. Wirelessly transmitting this data to a cloud-based data platform allows for analysis and sharing with medical staff, family members, and carers. IoT and artificial intelligence (AI) technologies are combined by the system to offer users individualised health coaching and support. For instance, the system can assess a user's level of physical activity and make suggestions on how to raise it, or it can identify abnormal blood pressure readings and notify a healthcare professional. In Japan, the Wellness Hub has been set up in a number of places, including private homes, assisted living facilities, and elder care facilities. Users of the system say they feel more supported in their daily activities and like they have greater control over their health. The Wellness Hub is a great illustration of how IoT and AI technology may be used to benefit people's health and quality of life, especially the ageing population.

The "MyActive and Healthy Ageing" (MAHA) project is one case study of proposed system in Europe. The MAHA project, which is supported by the European Union, intends to create a smart home system that can enable senior citizens to maintain their independence for longer. The MAHA system tracks the health and happiness of elderly individuals in their homes using a variety of IoT devices. The system has sensors for air quality, temperature, humidity, and motion detection. It also has wearable sensors for blood pressure, heart rate, and other vital indications. The MAHA system analyses the sensor data to look for any potential dangers or health issues. For instance, the system might alert a senior's doctor or family members to check on them if it notices that they have been less active than usual. Similar to this, if the system notices that the home's air quality is bad, it might advise the user to open a window or switch on a fan to increase ventilation. The MAHA system also comes with a virtual assistant that may offer the older person individualised health advice and direction. The virtual assistant can offer advice on how to eat well or suggest exercises to help someone remember to take their medication. Italy, Portugal, Spain, and other nations in Europe are now testing the MAHA initiative. The MAHA system is intended to assist seniors maintain their independence for longer, lessen the strain on healthcare systems, and enhance seniors' overall quality of life, according to the project partners.

The "Smart Stethoscope" created by the Commonwealth Scientific and Industrial Research Organisation (CSIRO) is a case study of proposed system in Australia. An IoT-enabled medical tool called the Smart Stethoscope aids physicians in the diagnosis and monitoring of lung diseases like pneumonia and asthma. The device has a microphone and other sensors that can record lung sounds and analyse them using machine learning methods. A mobile app that links to the Smart Stethoscope enables clinicians to remotely record and examine patients' lung sounds. The app can offer immediate feedback on the recording's quality and make suggestions for changes to enhance the diagnosis. A patient's digital health record can be made using the information gathered by the Smart Stethoscope. To guarantee continuity of care and enhance patient outcomes, this record can be shared with additional healthcare professionals. Several facilities in Australia, notably the Royal Children's Hospital in Melbourne, are now testing the Smart Stethoscope. The purpose of the experiment is to assess how well the tool diagnoses and tracks children's lung diseases. Particularly in remote or impoverished locations where access to medical specialists may be constrained, the Smart Stethoscope has the potential to revolutionise how clinicians diagnose and manage lung problems. The gadget may also enable earlier diagnosis and treatment of lung problems, which could improve patient outcomes and lessen the need for invasive treatments like X-rays.

The "Smart Hospital" project at the Seoul National University Bundang Hospital in South Korea is a case study of proposed system. The Smart Hospital initiative makes use of IoT technologies to raise patient satisfaction, lower medical expenses, and improve patient outcomes. To track medical equipment, manage hospital operations, and monitor patient health, the hospital has integrated a variety of IoT devices and sensors. The "smart bed" system is a crucial element of the Smart Hospital idea. The patient's vital signs, such as heart rate, blood pressure, and breathing rate, can be tracked by sensors on the smart bed. The bed can also recognise when a patient tries to leave their bed and, if necessary, warn medical staff. Because the hospital's electronic medical record system is linked to the smart bed system, doctors and nurses may access real-time patient data from their mobile devices. Informed decisions may be made rapidly, increasing patient outcomes and lowering the possibility of medical errors. The "smart pharmacy" system is yet another crucial element of the Smart Hospital initiative. To make sure that medical supplies and equipment are always available when needed, the system tracks the usage and position of IoT sensors. In order to avoid supply shortages, the system can also recognise when a medication is getting low and immediately restock it. Both patient outcomes and hospital costs have improved thanks to the Smart Hospital programme. Since starting the Smart Hospital project, the hospital has recorded a 25% decrease in patient readmissions and a 15% decrease in prescription mistakes. The Smart Hospital project exemplifies how IoT technology may be used to change the way healthcare is delivered, enhancing patient outcomes and cutting costs. To make sure that medical supplies and equipment are always available when needed, the system tracks the usage and position of IoT sensors. In order to avoid supply shortages, the system can also recognise when a medication is getting low and immediately restock it. Both patient outcomes and hospital costs have improved thanks to the Smart Hospital programme. Since starting the Smart Hospital project, the hospital has recorded a 25% decrease in patient readmissions and a 15% decrease in prescription mistakes. The Smart Hospital project exemplifies how IoT technology may be used to change the way healthcare is delivered, enhancing patient outcomes and cutting costs. The suggestions for nutrition are further tailored using this data to the unique requirements of the person. With features like barcode scanning to instantly enter food consumption and a virtual coach to offer support and encouragement, the Smart Personalised Nutrition app is made to be user-friendly and simple to use. By encouraging healthy eating practises and lowering the risk of chronic diseases, the Smart Personalised Nutrition project has the potential to enhance public health outcomes. Toronto and Vancouver are two of the Canadian cities where the initiative is now being tested. The Smart Personalised Nutrition project shows how IoT technology may be used to deliver personalised and efficient health interventions, empowering people to take control of their health and make educated nutrition decisions.

The "Swasth" project created by the Indian Institute of Technology (IIT) Madras is a case study of proposed system in India. An IoT-based healthcare platform called the Swasth project offers remote medical services to underprivileged communities in India. The platform is made up of a variety of Internet of Things (IoT) gadgets and sensors, such as wearable sensors and mobile health clinics, that can track and identify a number of medical issues. Additionally, the Swasth platform uses machine learning algorithms to offer individuals personalised healthcare suggestions based on their health data. With capabilities like text messaging and video conferencing to connect patients with healthcare practitioners, the platform is made to be accessible and user-friendly. The "Mobile Health Clinic" is a crucial part of the Swasth platform and is outfitted with a variety of Internet of Things sensors and gadgets that can detect and treat a variety of medical ailments. The clinic's team consists of skilled medical specialists who, if required, can conduct remote consultations and send patients to hospitals. For underserved populations in India, the Swasth platform has been beneficial in improving healthcare results. In both urban and rural parts of India, the platform has been used to give healthcare services to more than 50,000 people. The Swasth project shows how Internet of Things (IoT) technology can be utilised to enhance access to healthcare services in low-resource settings, lowering healthcare inequities, and improving health outcomes for disadvantaged groups.

Future Perspective

IoT Enabled Smart Health Monitoring System has the potential to completely transform the healthcare sector. This system can be made even more successful and efficient at improving patient outcomes with the aid of AutoAI, a graphical tool of IBM Watson Studio.

Future prospects for IoT-enabled smart health monitoring systems that employ AI include (Fig. 10):

- *Personalised Medicine*: Doctors may gather and analyse a lot of data about their patients with the aid of IoT-enabled smart health monitoring equipment and AutoAI. As a result, they will be able to offer the patient personalised treatment based on their particular health profile.
- *Remote Monitoring*: With the help of IoT-enabled smart health monitoring devices, doctors may keep an eye on their patients' health problems from a distance. In an emergency, this will enable doctors to act fast, improving patient outcomes.
- *Predictive Analytics*: By analysing data gathered by IoT-enabled smart health monitoring systems, AutoAI can identify potential health issues before they arise. This will enable medical professionals to take preventative steps, lowering the likelihood of health issues.
- *Increased Patient Engagement*: By giving patients real-time feedback on their health state, IoT-enabled smart health monitoring devices can help patients stay

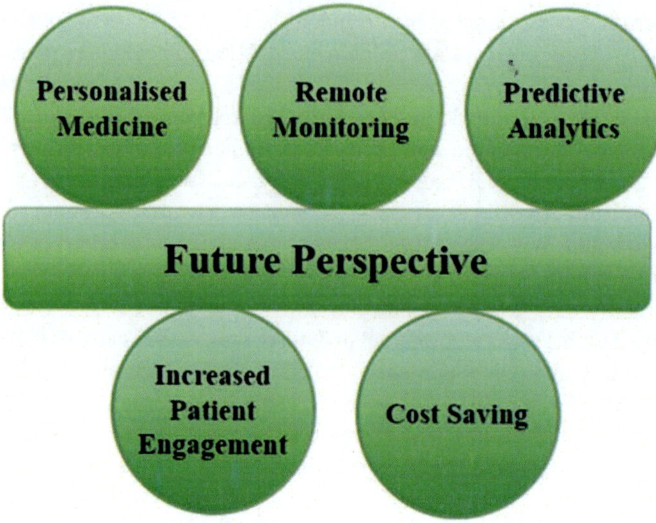

Fig. 10 Future perspective of proposed system

involved in their health. Patients' quality of life can be improved by using cellphones or other gadgets to check their health problems.

- *Cost Savings*: By enhancing patient outcomes and lowering the need for hospitalisations, IoT-enabled smart health monitoring systems can aid in lowering healthcare expenditures. Both patients and healthcare professionals will gain from this.

Conclusion

The medical solution is offered by the Internet of Things-enabled healthcare monitoring system. A healthcare monitoring architecture that can handle the massive amounts of data produced in real-time by diverse medical sources is offered. Using batch computing and stream computing, the suggested approach can forecast patient health status with a visual display and 97.66% accuracy. The solution's prototype implementation based on AutoAI and IBM Watson has then been suggested to detect worrying situations and issue real-time warnings. A potent tool for enhancing outcomes is the IoT enabled smart health monitoring system integrating AutoAI and IBM Watson Studio. This system gathers real-time patient data using sensors and other IoT devices, then analyses it with cutting-edge algorithms and machine learning approaches. This system's use of AutoAI enables it to automatically produce prediction models and other insights, which can aid medical personnel in making better decisions and offering patients better treatment. Additionally, non-technical people can easily engage with the system and obtain insights into patient data thanks to the graphical user interface offered by IBM Watson Studio. By increasing the precision and effectiveness of patient treatment, an Internet of Things (IoT) enabled smart health monitoring system has the potential to revolutionise healthcare. It has the potential to turn into a crucial tool for medical professionals all over the world with further development and improvement.

References

1. Boland, M. R., et al. (2013). Discovering medical conditionsassociated with periodontitis using linked electronic health records. *Journal of Clinical Periodontology, 40*, 474–482.
2. Brüssow, H. (2020). The novel coronavirus—a snapshot of current knowledge. *Microbial Biotechnology, 13*, 607–612.
3. Dini, G., et al. (2019). Big Data in occupational medicine: The convergence of-omics sciences, participatory research and e-health. *Medicina del Lavoro, 110*, 102–114.
4. Bragazzi, N. L., et al. (2019). SleepOMICS: How big data can revolutionize sleep science. *International Journal of Environmental Research and Public Health, 16*, 291.
5. Kataka, E., et al. (2020). Edgetic perturbation signatures represent known and novel cancer biomarkers. *Scientific Reports, 10*(1), 1–16.
6. Shynu et al. (2020). A fuzzy based data perturbation technique for privacy preserved data mining. In *2020 International Conference on Emerging Trends in Information Technology and Engineering (icETITE)* (pp. 1–4). IEEE.
7. Nunes, M., et al. (2015). Evaluating patient-derived colorectal cancer xenograftsas preclinical models by comparison with patient clinical data. *Cancer Research, 75*, 1560–1566.
8. Pandey, C., et al. (2022). data analysis and modeling of body sensor network in healthcare application. In *2022 6th International Conference on Electronics, Communication and Aerospace Technology* (pp. 590–596). IEEE.
9. Gupta, N., et al. (2022) A network data analytic technique in a 5G-IoT-based smart healthcare system using machine learning. In *5G Internet of Things and Changing Standards for Computing and Electronic Systems* (pp. 81–93). IGI Global.
10. Gupta, I., et al. (2022). Augmented reality based human-machine interfaces in healthcare environment: benefits, challenges, and future trends. In 2022 International Conference on Wireless Communications Signal Processing and Networking (WiSPNET) (pp. 251–257). IEEE.
11. Pandey, C., et al. (2021). Body sensor network architectures in healthcare internet-of-things (HIoT): A survey. In *2021 6th International Conference on Communication and Electronics Systems (ICCES)* (pp. 494–499). IEEE.
12. Pandey, C., et al. (2021). DME: An efficient encryption technique for body sensor network in healthcare internet of things (HIoT). In *2021 2nd International Conference on Intelligent Engineering and Management (ICIEM)* (pp. 80–85). IEEE.
13. Bates, D. W., et al. (2014). Big data in health care: Using analytics to identify and manage highrisk and high-cost patients. *Health Affairs, 33*, 1123–1131.
14. Forrest, G. N., et al. (2014). Use of electronic health records and clinical decisionsupport systems for antimicrobial stewardship. *Clinical Infectious Diseases, 59*, 122–133.
15. Andreu-Perez, J., et al. (2015). From wearable sensors to smart implants—Towards pervasive and personalised healthcare. IEEE Transactions on Biomedical Engineering 1–13. Submitted for publication.

Phase Noise Performance of MIMO—GFDM Systems for Millimeter Wave 5G Technology

Udayakumar Easwaran and V. Krishnaveni

Abstract Millimeter waves are a type of high-frequency electromagnetic waves that lie in the frequency scope of 30 GHz to 300 GHz. This huge bandwidth of millimeter waves, which enables the rapid transmission of enormous volumes of data, is one of their key advantages. Moreover, millimeter waves are better suited for outdoor communication where line-of-sight is accessible because they can pass through structures like buildings and trees less effectively than lower-frequency waves. To increase spectrum efficiency and handle numerous devices, 5G communication systems use the multiple access technology known as GFDM (Generalized Frequency Division Multiplexing). This chapter focuses on investigating the phase noise performance of MIMO-GFDM systems in millimeter wave 5G networks. It addresses the challenges posed by conventional channel estimations and interference caused by non-orthogonal subcarrier waveforms. To address these challenges, the proposed complex training sequence decomposition (CTSD) algorithm is introduced as an effective approach for channel estimation in MIMO-GFDM systems. The evaluation of bit error rate (BER) and symbol error in the presence of phase noise and noise further contributes to understanding system performance. However, it is important to acknowledge the limitations of ICI self-cancellation techniques in fading channels and the complexities associated with receiver design in GFDM systems. These aspects should be considered when analyzing the overall system performance. Overall, this research sheds light on the impact of phase noise and presents relevant solutions for synchronization and channel estimation in GFDM-based 5G wireless communications. The study's findings contribute to the understanding of phase noise effects and their implications for MIMO-GFDM systems, thus enhancing the development of efficient communication systems for millimeter wave 5G networks.

U. Easwaran (✉)
Department of ECE, KIT-Kalaignar Karunanidhi Institute of Technology, Coimbatore, Tamilnadu, India
e-mail: udayakumar.sujith@gmail.com

V. Krishnaveni
Department of ECE, PSG College of Technology, Coimbatore, Tamilnadu, India

A. Prasad et al. (eds.), *Communication Technologies and Security Challenges in IoT*, Internet of Things, https://doi.org/10.1007/978-981-97-0052-3_22

Keywords GFDM · Phase noise · Millimeter wave · ICI · 5G · CTSD · OFDM · LTE

Introduction

Cellular networks' fifth generation (5G) is anticipated to be able to meet the demands of rising data traffic, a sizable number of devices, and reduced latency, frame design, various access methods, waveform numerology, and scheduling technique are only a few of the technological components that have undergone modification. Each user equipment (UE) in an LTE uplink sends the base station (BS) a scheduling request before waiting for the BS to release a radio reference, resulting in significant network overhead and latency. These proposed techniques allow UEs to communicate data without the scheduling message & grant activity by requiring each user to transmit their data packet immediately. Short packets are sent by UE using a grant-free transmission mechanism in an arrive-and-go fashion. Different propagation delays and each user's mobility in the uplink might be a serious issue, resulting in asynchronous scenarios. Particularly, faulty synchronization drastically reduces the dependability of connection. The interference signaling brought on by faulty synchronization should be reduced to allow reliable communication. As a result, when the targeted & other user's offsets are innovated, the receiver must remove interference [1].

Due to the explosive growth in smartphone use over the past few years and the resulting increase in demand, mobile internet has become a widely accessible and reasonably priced service for many individuals. This trend opens up new uses for wireless cellular systems that weren't previously possible, along with higher data speeds and better coverage. One of these that sticks out is the Internet of Things (IoT). In the concept of an IoT, a wide range of machines and things will use the Internet as a communication infrastructure in addition to people [2]. The number of subscribers for the cellular systems of the next generation will be significantly higher if it is assumed that each person would own many IoT-enabled gadgets.

This will bring forth a wide range of new requirements, such as those for quality of service, mobility, latency, energy efficiency, and data rates concerning inexpensive powered tendencies. Cognitive radio is a different strategy that aims to meet the aforementioned characteristics in a spectrally flexible manner. One specific objective is to employ dynamic spectrum accession to utilize spectrum references that, despite being allotted to a specific service, aren't being used at the moment in a specific place. The capacity to transfer narrow-band signaling with little out-of-band emission that can be dispersed over a wide frequency range is required for this application field [3].

Mobile communication has become a vital part of daily living in the modern world. To provide mobile consumers with a high data rate and wide coverage, the current mobile communication networks have been developed. OFDM is the least widely used technique of the fourth generation due to its excellent endurance and low complexity. Many technologies, including WLAN, IEEE 802.16, LTE, and digital video transmission, make extensive use of this technique. However, obstacles for OFDM in upcoming fifth-generation (5G) request premise include machine-type communication, tactile internet & IoT application scenarios. Furthermore, due to its orthogonal sub-carrier, OFDM is low-cost compared to MTC because it demands minimal power consumption [4].

The cyclic prefix per symbol in the OFDM signaling and short bursts of data with low spectral efficiency are required for tactile internet's low latency requirements. Additionally, IoT intent become more favorite with 5G, allowing for the connection of an increasing number of items to the internet, when this is now not the case. Spectrum fragmentation is a solution; for example, spectrum gaps in the UHF TV bands are a classic example of spectrum fragmentation brought on by the digital dividend. However, because the OFDM signaling cannot escape significant interference, more out-of-band OFDM emissions have trouble transmission in the disconnected spectrum [5].

Beyond 5G wireless networks are being developed to considerably better the quantity and quality of wireless service. These networks will provide huge machine-type communications (MTC), improved mobile broadband ultra-reliable and low-latency data transmission [6]. This is required to meet the extremely high needs of newly developed Internet-enabled use instances & employ, like virtual and augmented reality, ultra-high quality video streaming. Modern society now relies heavily on mobile communications. The initial cellular systems offered simple yet ground-breaking voice transmission. Instead of being tied to specific areas, communication began to become more personal.

To improve system capacity, device battery life, and service quality, the second-generation digitalized voice. Additionally, the Short Message Service was established, completely altering how people communicate today. The third generation allowed for data speeds that were not too far down those of wired statements at the time and enabled mobile Internet access [7]. The emergence of smartphones with significant amounts of storage and processing power, high-definition screens and cameras, as well as social meshwork that allowed users to transition from existing media, devour to content creators, has pushed the fourth generation toward even higher throughput. The development of mobile communication has prioritized improving throughput since the second generation.

Related Works

Orthogonality must be strictly maintained to use the commonly used wireless OFDMA, which is a technique. Due to the strict closed-loop frequency & timing correction used in uplink OFDMA, there is an increase in latency between the mobile transmitter & the base station. The uplink transmission method utilized in LTE, single carrier (SC-FDMA), is substantially more susceptible to timing faults than OFDMA. OFDMA or SC-FDMA is suitable for relaxed synchronization since a waveform needs to be durable against faulty synchronization. The out-of-band (OOB) emission is produced by the FBMC and is immune to synchronization mistakes in the uplink, which is one of the possible waveforms. But it is useless for low latency communications because of its large filter lifetime, which is four times the symbol period and short burst transmission. Some other potential waveform for low latency communication theory is GFDM, which may be ideal because it uses a circular filter to shape the pulse and only uses one cyclic prefix (CP) for many sub-symbols. However, because of its circular filter, GFDM exhibits greater OOB emission than FBMC as well as intrinsic interference [8].

The IEEE 802.11 wireless network technologies have advanced steadily. They are becoming more and more popular in daily life for both personal and professional use since they are more flexible, affordable, and simple to set up than wired networks. They must do this to meet the increasing supply of data flow for heterogeneous work. The IEEE 802.11ac physical layer employs the OFDM waveform and high throughputs [9]. The subcarriers around the borders of each band must be disabled due to its high OOB [10] emission, which lowers the spectral efficiency. To achieve improved spectrum efficiency, this chapter looks into the PHY's waveform component for high throughput communication.

The OFDM technique is a favored choice for modern wireless standards due to its many benefits. By divisional a single broad data rate stream into various concurrent low-level data rate streams that are broadcast on distinct sub-carriers, OFDM, like all multi-carrier systems, welfare from the ability to exploit frequency diversity. By permitting the individual sub-carriers to be conceived of as frequency flat when paired with a cyclic prefix, the method provides a straightforward single-tap equalization. Additionally, the orthogonality of the sub-carriers allows for a fast Fourier transform (FFT)-based transmitter and receiver implementation that is effective and low-complexity. The design, however, also demonstrates some negative characteristics that prevent it from meeting a number of the prerequisites outlined earlier [11]. Because OFDM's potent OOB radiation can provide non-negligible interference in the overlay system employ, extra filters must be used to obtain the proper spectral mask.

To prevent the orthogonality from being compromised, complex synchronization mechanisms are required due to the extreme sensitivity of OFDM to carrier frequency offset. The cyclic prefix approach, which is last but not least, is an essential overhead that can decrease the system's total energy efficiency. Due to the superposition of several sub-carriers, the approach also has a high PAPR, which, depending on the application, may increase the need for amplifiers. Thus, cutting-edge transmission methods are researched. This field has recently seen the emergence of the filtered multi-carrier transmitting method, which was well-known even before OFDM became extensively utilized [12]. GFDM, a cutting-edge method of flexible multi-carrier transmission that provides more degrees of freedom than conventional OFDM, is one of them.

OFDM is widely used as the physical layer due to its resiliency against frequency-selective channels and ease of installation interface in modern digital communication systems. However, OFDM has certain significant flaws that make it doubtful for 5G networks & other future wireless systems. The broad out-of-band release of OFDM signals, in particular, provide a challenge for the application of this technology in the context of dynamic spectrum allocation and fragmented spectrum [13]. OFDM is not the best option for 5G networks, hence other waveforms are being researched for next-generation regulation. It would be very practical if the methods created for OFDM could somehow be extended to other waveforms. Typically, synchronization methods for OFDM do not take into account the preamble's OOB impacts.

However, generalized frequency division multiplexing must handle this crucial issue. (GFDM). The examination of a GFDM synchronization solution that doesn't increase OOB emission & has execution comparable to a mixture made for OFDM constitutes the main contribution of this chapter. Fortunately, we may rely on carefully thought-out receiving strategies to reduce the interference, for example, it has been effectively proved shown for a variety of applications [14]. By modulation & multiple access, and the rationale behind using well-designed reception mechanisms in next-generation networks. For completeness' sake, interested readers may also consult for additional information on frequency-domain reception. By sending and receiving numerous sub-streams concurrently through a MIMO design, spatial division multiplexing (SDM) devices can achieve high data transmission rates in the spatial domain. To resolve this problem and improve the efficiency of the V-BLAST detection approach. Performance has improved as a result of repeating this iterative technique. While waiting, an under-determined MIMO scheme should be considered in scenarios with a lot of Systems like IoT applications and extremely dense networks [15].

A novel iterative detection algorithm that, after lowering the CAI using soft symbol estimations, makes use of the residual interference's non-circular nature to enhance MIMO-GFDM system performance. It is a general form of OFDM that retains the majority of the benefits of OFDM while addressing its drawbacks. It is discussed how employing the GFDM system, which emits very little out-of-band radiation, can reduce the drawbacks associated with carrier aggregation in an OFDM system. Moreover, GFDM eliminates prototype filter transient intervals and therefore delay by using circular filtering rather than the linear filtering used in OFDM [16]. As a result, it can support low-latency applications like M2M and the Internet of Things (IoT). The inter-carrier interference (ICI) is also limited by using a well-designed prototype filter to filter the subcarriers, making it resistant to loss of orthogonality.

GFDM-MIMO System

A flexible non-orthogonal waveform called GFDM has been suggested for the 5G PHY layer. It has several benefits over OFDM, chief among them being reduced out-of-band emission, enhanced multi-path performance, and increased spectrum efficiency. Due to the high cost of its implementation in terms of processing resources, its application is less tempting for devices with limited resources, such as Wireless Sensor Networks & Internet of Things (IoT) devices. Due to its adaptability, GFDM can be utilized with both single-carrier (SC-FDE) and filtered bank multi-carrier (FBMC) frequency domain equalization. The core of GFDM is per-subcarrier modulation, where each subcarrier is modulated separately and independently utilizing a variety of symbols [17].

A prototype filter that has been circularly displacement in both the time & frequency domains is then used to filter the subcarrier. The goal of this filtering procedure is to decrease the out-of-band (OOB) remnants, which will enable the fragmented & dynamic spectrum assignment without seriously interfering with the GFDM system per other users. A filtering procedure might eliminate both ISI & ICI between subcarriers. The GFDM transmitter is

$$y[n] = \sum_{k=0}^{K\_1} \sum_{m=0}^{M-1} P_k[m] g_{tx_{k,m}}[n] \tag{1}$$

$$g_{tx_{k,m}}[n] = g_{tx}\ [n - mk]\ \text{mod}\ N]\, e^{\frac{-j2\pi kn}{K}} \tag{2}$$

$P_k[m]$ is Pulse shaped data, Sub-carriers at time m, and index k.

Each entry of $d_k[m]$ is pulse shaped sub-symbol, where M sub-symbols & K sub-carriers, y[n] is transmitted signal and n = {0,1,2, …., N − 1}.

The Total number of symbols is defined as N = KM.

The new scenarios anticipated for the 5G networks place demands on the networks, and GFDM provides a flexible solution to meet those demands. In this method, a symbol made up of several sub-carriers & symbols is used to convey a data block, with each sub-carrier having a transmission filter that is pulse-shaped. The impulse outcome for each sub-symbol is practical by circular convolution to prevent using additive samples for the ramp up & ramp down of the filter response [18]. Tail biting is another name for this strategy. To give the system more flexibility, several pulse forms can be utilized as prototype filters. In addition to reducing OOB emissions, sub-carrier filtering may also cause inter-carrier interference & inter-symbol interference. Using successive interference cancellation (SIC) in conjunction with zero forcing or a matched filter to receive signals can lessen the determinant of self-generated interference & achieve error rate execution comparable to OFDM. The Fast Fourier transform (FFT)-based implementation strategies demonstrate that GFDM transceivers may be built effectively using current technology. The channel is modeled by a linear system as:

$$x[n] = s[n] * g[n] + w[n]^2 \tag{3}$$

The sent signal s[n] is passed through the frequency flat fading channel with the response of g[n], "*" denoting their convolution, and the AWGN channel w[n] with the help of x[n], where x[n] is the received signal.

Estimating symbol time offset (STO) & carrier frequency offset (CFO) is one of the major difficulties in the multi-carrier receiving chain [19]. Data-aided techniques enable the use of both auto-correlation, cross-correlation characteristics, and one-shot synchronization can be used as one of the synchronization mechanisms for OFDM, realized in bit pipe & burst communications. It was advised to use shorter data symbols to target a wider CFO estimation range and a sequence made up of two continual OFDM symbols was originally put out to investigate a significant auto-correlation belongings for CFO approximation. This concept is solved by creating a single OFDM symbol made up of two repeating components.

A coarse STO can also be obtained, however, the plateau effect origin by the usage of cyclic prefix (CP) and cyclic suffix (CS) decreases the accuracy of the technique and introduces ambiguity. To eliminate plateau effects, a proposal is made to use further segregation, i.e., a moving average along the CP & CS length. However, the metric's pyramidal structure results in mistakes in STO estimation in noisy environments. The integrated auto-correlation approach is united with a cross-correlation after the CFO has been corrected to create the best impulsive time metric. The algorithm investigates these techniques to produce a reliable CFO estimation [20]. The Flow chart of GFDM-MIMO systems is shown in Fig. 1.

Researchers have recently paid a large deal of care to GFDM as a result of its benefits in addressing the requirements of 5G wireless networks. KM samples with M timeslots are carried by each of the K sub-carriers in a GFDM symbol. These properties may be modified to suit the application's requirements. Therefore, GFDM can engineer the time–frequency structure to correspond to the right context [21]. Circular convolution is used to independently filter each subcarrier in GFDM. Due to its extremely low OOB emission, GFDM can be utilized for irregular and flexible spectrum allocation. The GFDM uses a single CP for a whole block made up of aggregate sub-symbols. As a result, frequency domain equalization (FDE) is made possible and spectral efficiency is enhanced. As a result, it is simple to alter the GFDM's flexible properties to meet the new demands.

A crucial technique for faster data speeds and greater spectrum efficiency is MIMO transmission [22]. To increase spectrum efficiency and accommodate numerous devices, 5G communication systems use the multiple access technology known as FDM. Data is modulated using GFDM onto several sub-carriers that are organized into rectangular-shaped symbols. Although the sub-carriers are always orthogonal within a symbol, this is not always the case between symbols.

With an appropriate transmit filter, these zero-filled sequences are filtered,

$$g_{k,m}[n] = g[(n - mK)modN]exp\left(-j2\pi\frac{k}{K}n\right) \tag{4}$$

While k, m, and n are subcarrier, subsymbol, and time sample indices, g[n] denotes the impulse response of a prototype filter with N samples.

To handle the large data rates and low latency needs of 5G applications, GFDM can be deployed in both sub-6 GHz and millimeter wave frequency bands. For enabling high data rates and a large number of devices in high-density settings, GFDM offers

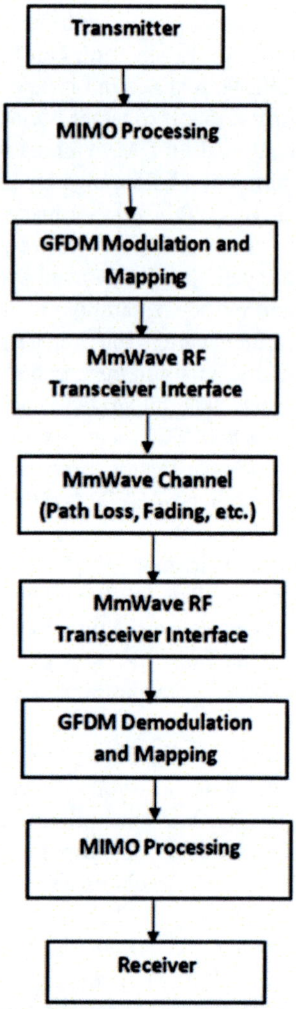

Fig. 1 Flowchart of GFDM-MIMO systems

various advantages over conventional OFDM systems, including decreased inter-carrier interference and greater spectral efficiency. In millimeter wave communications, GFDM is a common modulation technique. Radio frequencies between 30 to 300 gigahertz (GHz), or 1 to 10 mm in wavelength, are referred to as millimeter waves.

Transmitter block processes all necessary digital signal processing steps, including encoding, filtering, and shaping, on the input data. The data is subsequently transformed into GFDM symbols. The Multiple-input multiple-output (MIMO) approaches, where multiple antennas are employed at the transmitter & receiver to improve system performance, are handled by the MIMO Processing block. It entails procedures like beamforming, spatial multiplexing, and precoding. The GFDM

Modulation and mapping block generate the GFDM waveform by mapping the GFDM symbols onto subcarriers, applying the required modulation scheme (such as QAM or PSK), then shaping and filtering the pulses [23].

The millimeter wave RF transceiver, which consists of mixers, local oscillators, amplifiers, and filters, is interfaced with using the millimeter wave RF transceiver interface. For transmission and reception, it transforms the baseband GFDM signal into the millimeter wave frequency range [24]. The millimeter wave propagation channel, which has traits including path loss, fading, and interference, is represented by the millimeter wave channel. The sent signals experience these effects as they travel across the wireless medium.

Demodulation of the received GFDM waveform is carried out by GFDM Demodulation and Mapping, which then extracts the GFDM symbols. To extract the transmitted data, it uses inverse modulation and mapping procedures. To lessen the effects of channel distortions and improve the quality of the received signal, MIMO Processing employs MIMO processing techniques such as channel estimation, equalization, and combining. To extract the original data from the GFDM symbols, the receiver block conducts any required signal processing tasks, such as decoding, error correction, and demodulation.

Multi-carrier modulation with excellent spectral efficiency and resistance to frequency-selective fading is known as GFDM. It functions by breaking up the available spectrum into various sub-carriers, each of which carries a different portion of the data. Because of the frequency and temporal overlap of these sub-carriers, the spectrum can be used effectively. It is now possible to achieve high data speeds, low latency, and support for a large number of linked devices in millimeter wave communications by using GFDM. Due to these features, millimeter wave GFDM is a viable technology for future wireless communication networks, high-speed internet access, and wireless backhaul. For ensuring dependable and high-quality communication, phase noise reduction in millimeter wave GFDM systems is essential. The block diagram of MIMO-GFDM systems [25] is shown in Fig. 2. The guard band bandwidths and the quantity of guard band subcarriers [26] for GFDM are

$$B_g = 157 * \frac{2640MHz}{512} \tag{4}$$

$$N_{gGFDM} = \frac{B_g}{2640MHz/K} \tag{5}$$

where B_g and N_g, GFDM are the guard band's bandwidths and the number of subcarriers in the guard band for GFDM, respectively. There are many synchronization strategies for GFDM, in comparison to the enormous number that are accessible for OFDM. The rectangular pulse form that creates non-negligible OOB emissions & low decaying spectrum side-lobes is taken into account by all of the aforementioned approaches [27].

The transmit signal's OOB emission is dominated by a programmable pulse shaping filter that is applied to each sub-carrier in GFDM. The introduction of a

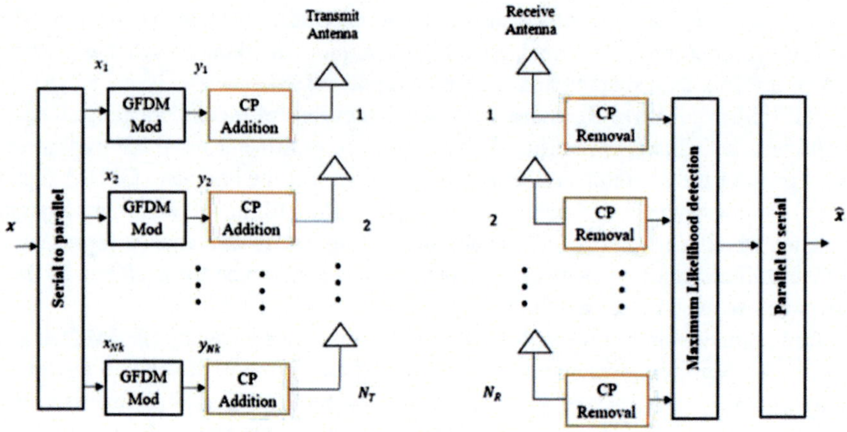

Fig. 2 Block diagram of MIMO-GFDM Systems

two-dimensional data structure allows for the organization of time slots into different sub-carriers and data symbols into blocks. A programmable parameter called block size permits the use of longer filters or the modification of the total amount of sub-carriers. Tail-biting digital filters are used to process these blocks, maintaining their circular characteristics in both the time domain and the frequency domain. Related to OFDM, GFDM can use a cyclic prefix to prevent ISI in a multipath channel. FBMC is an alternative strategy to GFDM that may effectively reduce side lobes in the broadcast signal [28]. A polyphase network is used to implement the pulse shaping filter there. To further prevent inter-carrier-interference (ICI) between nearby subcarriers, offset QAM modulation is used. The method for preventing inter-symbol interference does away with the idea of cyclic prefix and instead depends on a per-subcarrier equalization.

The Phase term is given by

$$\phi_{k,n} = \frac{\pi}{2}(n+k) - nk \tag{6}$$

When the phase term $\phi_{k,n} = 0$, the imaginary and represented as

$$r_{k,n} = i_{k,n}\left(a_{k,n} + jw_{k,n}\right) + N_{k,n} \tag{7}$$

Hence, $w_{k,n}$ is a real-valued function.

The modern waveform known as GFDM [29] can be customized to handle different use cases. The M sub-symbols & K sub-carriers make up the time–frequency grid that GFDM uses to arrange the data symbols in a circular prototype filter for each sub-carrier. Other waveforms, such as OFDM [30] & Single Carrier SC-FDMA as special examples, can be covered by GFDM with ease. The sub-carrier filtering can manage the (PAPR), lower OOB emissions, and dynamically allocate spectrum.

With its block-based architecture, GFDM can use various OFDM-formulated solutions, such as the idea of a cyclic prefix (CP) to keep inter-frame interference. To effectively resist the accord of multipath channels before the demodulation process, frequency-domain equalization can be used. These characteristics allow GFDM to meet the needs of 5G networks.

This makes detection in MIMO-GFDM tough because, when MIMO scenarios are taken into account, the nullification of different types of interference, such as inter-antenna interference (IAI) paired with ICI & ISI, gets over a significant challenge. To address this problem, several MIMO-GFDM equalization & perception techniques have been investigated. The symbol error rate of linear equalizers for MIMO-GFDM, such as zero-forcing (ZF) & minimal mean square error, is relatively high (SER). You may broadly divide existing MIMO-GFDM detection techniques into iterative and non-iterative approaches. Non-iterative receivers have an unreasonably high complexity that makes it impossible to eliminate ICI, ISI, and IAI. Contrarily, iterative receivers offer a manageable level of complexity with each iteration.

However, the necessary iterations could result in processor lag, making them unsuitable for applications requiring crucial timing [31]. This explores a grade of ICI loose prototype filters that have a low level of complexity and latency when used in MIMO-GFDM. A case in point is the previously published but little-used Dirichlet filter [32]. GFDM purposefully permits non-orthogonal sub-carriers, although this leads to reciprocal interference. GFDM enables a trade-off between this interference and several advantages. They consist of low-level out-of-band emissions, good spectral efficiency, latency, low PAPR, digital implementation of filter banks, low OOB emissions, and low power consumption. Owing to these benefits, GFDM has lately undergone substantial research for optical networks, the Internet of Things, space–time coding systems, cognitive radio networks, and filter designs [33]. Moreover, GFDM enables the improvement of pulse-shaping filters to improve desired execution metrics. GFDM thus offers a feasible substitute that might meet the performance goals.

To study channel estimation for GFDM systems, complete frequency synchronization was assumed. The interference was pre-canceled at the transmitter in the scattered-pilot-assisted channel estimate method that was provided. Yet, frequency selective fading could hurt its execution. It has been incontestable that LS channel estimation with a rectangular-pattern pilot performs better than LS channel approximation with a block-pattern pilot. However, because of ICI and ISI, they had a large training overhead and were prejudiced. The interference-free pilot-aided channel estimate was made possible by modifying the prototype filters, although this only applied to particular GFDM systems [34].

The fresh waveform celebrated as GFDM has a low OOB and good spectral efficiency. Because of its adaptability, it can be set up to operate by IEEE 802.11ac. GFDM's reduced OOB emission makes it possible to use fewer guard subcarriers than OFDM. By using these subcarriers for data transmission, throughput can be increased. Moreover, GFDM can be set up to operate with numerous data symbols per subcarrier and on a symbol basis. We can conserve time and resources by protecting many data symbols per subcarrier with a single cyclic prefix (CP). In this instance, a GFDM sign is created by fusing two CP-OFDM symbols. The kind that looks into the IEEE 802.11ac waveform aspect for better PHY performance [35].

Recently, GFDM has been presented as a solution to issues in 5G application scenarios. High bandwidth efficiency is achieved by GFDM by dividing the high-rate input data can be processed in parallel using several overlapping spectra [36]. The low-rate data stream and sending them simultaneously over several overlapping spectrums. Because each GFDM block contains numerous subcarriers and sub-symbols, GFDM is a flexible modulation technique [37]. GFDM signals exhibit good transmission efficiency when employed in low-latency applications like tactile internet because they have a lower CP than OFDM signals. Moreover, the peak-to-average power ratio of the GFDM signal is lower than that of the OFDM signal, making it an appropriate waveform for MTC transmission. The pulse-shaping filter architecture, which is also the foundation of the GFDM signal, minimizes OOB emissions, making it possible to use fragmented spectrum for IoT & dynamic spectrum access for cognitive radio without significantly interfering with already-existing services or other users.

In-depth research is being done on GFDM, a well-known non-orthogonal multiplexing technology, to satisfy the demands of the future era. By fully using frequency diversity, the MIMO-GFDM communication scheme can outperform MIMO-OFDM systems [38]. The cyclic prefix (CP), which is also utilized in OFDM, is used in GFDM to reduce inter-symbol interference. Due to its lower out-of-band radiation, GFDM is also reasoned for the cognitive radio application (OOB). However, a self-induced interference appeared as a result of the pulse-shaping sub-carriers and non-orthogonal structure of the GFDM system. Therefore, it was essential to create a receiver that can filter out this interference impact. For research purposes, only AWGN channels were used.

GFDM is a state-of-the-art non-orthogonal multi-carrier access technique. A simple GFDM transceiver architecture in which this research used for simulation. The demodulated GFDM sub-carriers at the receiver's self-interference and inter-antenna interference (IAI) cause substantial BER performance reduction when V-BLAST is applied to MIMO-GFDM. The idea of spatial modulation (SM) involves choosing active transmit antenna indexes for various information symbols by a mapping table that has been created by the user. The procedure table states that each symbol is sent via an antenna that is switched on specifically for that symbol at that precise moment. At that moment in the symbol, all other antennas are silent. Inter-antenna interference (IAI), a problem that limits the BER execution of the MIMO scheme under low-level SNR conditions, is avoided by this method.

The design of the transmitter & receiver, the computation of transmission property, and the production of non-orthogonal sub-carriers are the main constituent of the GFDM system. In this study, we show how to use a pulse-shaping filter that removes out-of-band (OOB) emissions while producing non-orthogonal radiation [39]. GFDM offers a great lot of design flexibility for both the transmitter and the receiver thanks to its adaptive transceiver architecture. GFDM can be applied in a variety of situations, and the amount used can be adjusted to meet the necessitate of different work. Due to the availability of numerous low-complexity GFDM transceiver implementations, they are commercially viable. Several algorithms can be used to further reduce complexity while keeping acceptable performance in the GFDM receiver. GFDM's performance in real-world applications is impacted by a few less-than-ideal problems. Yet, because of its adaptable structure, GFDM can use a mixture of schemes to aid in the assessment and compensation of the negative effects brought on by these problems.

The good qualities of GFDM make it appropriate for a wide range of use. The high spectrum efficiency of cognitive & full-duplex radio, which help solve the lack of bandwidth resources, makes them of tremendous interest among all possible applications. As a result, full-duplex radio and cognitive radio based on GFDM are seen as appealing approaches to spectrum management in next-generation wireless networks. Pulse shaping filters are used in GFDM, a non-orthogonal multicarrier method, to cut down on Out of band radiation [40]. Comparing the GFDM block to the OFDM block, less Cyclic Prefix is added to the GFDM area. The GFDM system's total spectrum efficiency is increased as a result. The fact that GFDM offers a greater data throughput with little latency is another one of its key advantages. A down complexity frequency division equalizing method is made possible at the receiver by the CP appended to the GFDM block.

Phase noise

Phase noise is described as the noise that results from a signal's quick, random phase fluctuations. Phase jitter, a type of temporal domain instability, is what causes these irrational variations. Phase jitter causes noise to spread out on both sides of a signal, which is known as phase noise. The single side-band phase noise is a common way to represent phase noise in radio receiver applications. Single side-band phase noise is defined as phase noise that is distributed equally on both sides of the carrier but is only captured on one side.

In the frequency domain, phase noise is a sort of specification that indicates short-term frequency stability. Phase noise is a critical characteristic for many oscillators and signal sources since it regulates a wide range of performance variables. There are two ways to look at the same parameter: phase noise and phase jitter. The Phase jitter investigates the variations in phase upon the signal, whereas phase noise examines the signal spectrum, or in the frequency domain. Radio systems are impacted by phase noise on local oscillators or other signals by, among other things, shortening the distance between points on PSK and QAM constellation diagrams [37]. Due to the way, it affects the performance of several aspects of radio communications systems, it is an essential component in many standards. Phase noise, which can distort and degrade the received signal, is the term for the random fluctuations in the broadcast signal's phase. Some methods for lowering phase noise in millimeter wave GFDM systems are as follows:

High-Quality Oscillators

The local oscillators employed for frequency generation have a significant impact on the phase noise in a GFDM system. Phase noise in the system can be greatly reduced by using high-quality oscillators with low phase noise characteristics. This can entail making a careful choice of oscillator parts, like voltage-controlled oscillators (VCOs) or crystal oscillators, that perform well in terms of low phase noise at millimeter wave frequencies.

Phase-Locked Loop (PLL)

In GFDM systems, using a phase-locked loop can reduce phase noise. Phase noise is decreased by the PLL because it fixes the local oscillator's phase to a reliable reference signal. The design of the loop filter and the loop bandwidth should be optimized to efficiently reduce noise while preserving stability.

Frequency Multipliers

To produce larger carrier frequencies in millimeter wave GFDM systems, frequency multiplication is frequently used. But frequency multipliers might also contribute more phase noise. Phase noise caused by frequency multiplication stages can be reduced by using low-noise frequency multipliers and careful design considerations. Phase noise can be reduced by adding the right filters to the system, which can also assist reduce background noise and undesirable signals. Low-pass filters can lessen high-frequency noise components whereas bandpass filters can help remove out-of-band noise. To balance noise suppression and signal quality, filter parameters must be carefully designed and optimized [24].

Signal processing methods

DSP methods for digital signal processing can be used to lessen the effects of phase noise. These methods include adaptive equalization, improved receiver algorithms that can partially correct phase noise-induced distortions, and algorithms for phase noise estimation and compensation. It is crucial to keep in mind that decreasing phase noise in millimeter wave GFDM systems frequently calls for a mix of careful component selection, hardware design considerations, and signal processing approaches. The particular strategy will depend on the system needs, the resources at hand, and the intended performance trade-offs.

Complex Training Sequence Decomposition (CTSD)

In millimeter wave systems, the "complex training sequence decomposition" algorithm is a method for lowering phase noise. It seeks to calculate and make up for the phase noise distortion the gearbox introduces [14]. The complex training sequence decomposition algorithm's general operation is described as follows:

Training Sequence

The signal being delivered has a predetermined and well-known training sequence. The characteristics of this training sequence should enable precise assessment of the phase noise.

Channel Estimation

The receiver receives the signal, together with the training sequence. This signal is received by the receiver, which utilizes it to calculate the effects of phase noise and the channel characteristics. Separating estimated channel effects from phase noise effects is the goal of the complex training sequence decomposition approach. The program attempts to detect the phase noise components and their properties by examining the received training sequence.

Phase Noise Compensation

The algorithm can use compensation approaches to reduce phase noise distortion once the phase noise components have been discovered. To make up for phase rotations, or to match the received signal with the planned phase, compensation may be necessary.

Data Recovery

The received signal is subsequently processed to recover the transmitted data once phase noise has been taken into account. To increase the precision of data recovery, the compensated signal is demodulated, decoded, and any required error correction methods are used.

Phase noise basics

The presence of phase jitter, phase noise is the noise spectrum that can be seen diffusing out on each side of a signal. For RF designers, phase noise is very important. Phase noise that extends on either side of the primary desired carrier is a sign of phase jitter. The level often decreases the more away from the carrier it is offset. Typical free-running oscillator phase noise spectrum. The phase noise spectrum or chart changes throughout the loop bandwidth due to the way frequency synthesizers work, but it eventually decreases in intensity as the carrier's offset increases.

Phase noise in signal sources

The noise spectrum that results from phase jitter, which develops as an issue of the signal's arbitrary phase fluctuations, is referred to as phase noise. The noise is caused by ordinary circuit noise, which appears as phase jitter. This can also be viewed as random frequency changes because phase and frequency are intimately related to one another. While jitter and fluctuations can happen over many periods, noise in signal origin can be seen in different ways.

Phase noise versus amplitude noise

There are many ways to introduce phase noise into circuits, particularly when frequency synthesizers are included. Nevertheless, "thermal" and "flicker" or 1/f noise is the source of phase noise in oscillators. The amplitude noise components are typically roughly 20 dB lower than the phase noise components since the least oscillators function in saturation. It indicates that amplitude noise is frequently disregarded because phase noise predominates. The amplitude elements should not be overlooked because they need to be taken into account in some concerns, even though this assumption is accurate for the majority of them.

Grandness of phase noise in communications networks

Phase jitter or noise is especially important since it gathers the signal quality and raises the communications link's mistake rate. Technically, erroneous phase modulation is much more significant than amplitude modulation in practice. This is partially due to the usage of angle modulation, which is more susceptible to phase noise, in the majority of radio links today. That also happens because the amplitude noise content in complicated signal sources is substantially weaker than the phase noise content.

Results and Discussion

In MIMO-GFDM systems, traditional channel estimations often face challenges due to the high costs associated with training, especially when dealing with multiple transmit antennas. To address these issues, we propose a new type of training sequence in our research. These sequences are generated by combining two identical zero-correlation zone sequences, where the auto and cross-correlations are zero within a specific time-shift window around the in-phase position. These challenges serve as the motivation for our study. We introduce the complex training sequence decomposition (CTSD) technique to simplify the reconstruction of the complex-field orthogonality of MIMO-GFDM signals. This approach is particularly useful as only real-valued symbols can be transmitted in the MIMO-GFDM scheme. Our simulations demonstrate that the suggested CTSD is a viable preamble-based MIMO-GFDM system's effective channel estimation method. The Phase noise versus subcarrier is shown in Fig. 3.

The Intersymbol & intercarrier interference are caused by the non-orthogonal subcarrier waveforms produced by the pulse-shaping filter architecture. However, the direct issue with the GFDM signal is that it is excitable to CFO, which is brought on by the incompatibility of local oscillators with Doppler effects in mobile networks. Synchronizing the carrier frequencies of the sent signal & the received signal is the direct solution to CFO. Thus, it is required to establish an estimate of CFO before using that estimate to adjust the carrier frequency offset. The BER_Original versus BER_N_0 Noise is shown in Table 1.

The acting of ICI nullification uses training sub-carriers to carry out CFO estimates in fast multi-path fading channels under the premise of linearity of channel variance within one GFDM block. Many schemes do not use pilot sub-carriers or activity symbols. The ICI self-cancellation has received the most attention from these strategies. These ICI self-reduction techniques allow ICI to be unreleased during reception without the need for additional symbols or equalization. The technique cannot perform well in a fading channel because the coherence detection process's ability to reduce the bit error rate may be constrained. The Comparison of symbol error with phase noise is shown in Fig. 4.

This statistic displays the ICI & ISI present by a specific prototype filter under ideal synchronization settings. To fully take advantage of the benefits of the GFDM

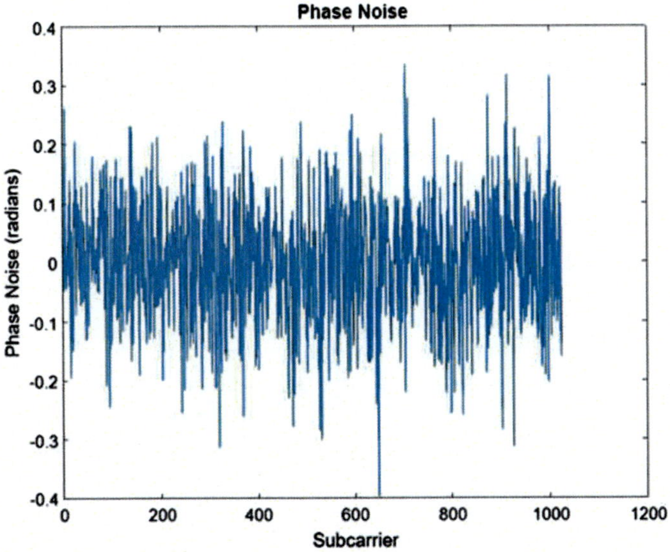

Fig. 3 Phase noise versus subcarrier

Table 1 BER_Original versus BER_N_0 Noise

BER_Original	BER_N_0 Noise
0.99974	0
1.0566	0
0.96732	0
1.0121	0
0.98842	0
1.0286	0
1.0187	0

waveform, the receiver's complexity needs to be addressed. Advances on this subject have been published by several researchers. The authors proposed a low-complexity design using the frequency sparsity of particular pulse types. Based on the block diagonal characteristic of the GFDM transmitter matrix, the authors suggest simplifying the transmitter & receiver. The Comparison of symbol error versus phase noise is shown in Table 2.

The Frequency Selective Channel only addresses the MF & ZF receivers that include the majority of current wireless communication systems. The authors, once more assuming FSC, also take advantage of the block circularity property to reduce the complexity of their MMSE receiver to O(N2). GFDM, an enlarged variant of OFDM, is a potential waveform option for 5G wireless communications. However, because GFDM's non-orthogonal prototype filter causes ICI & ISI, which make

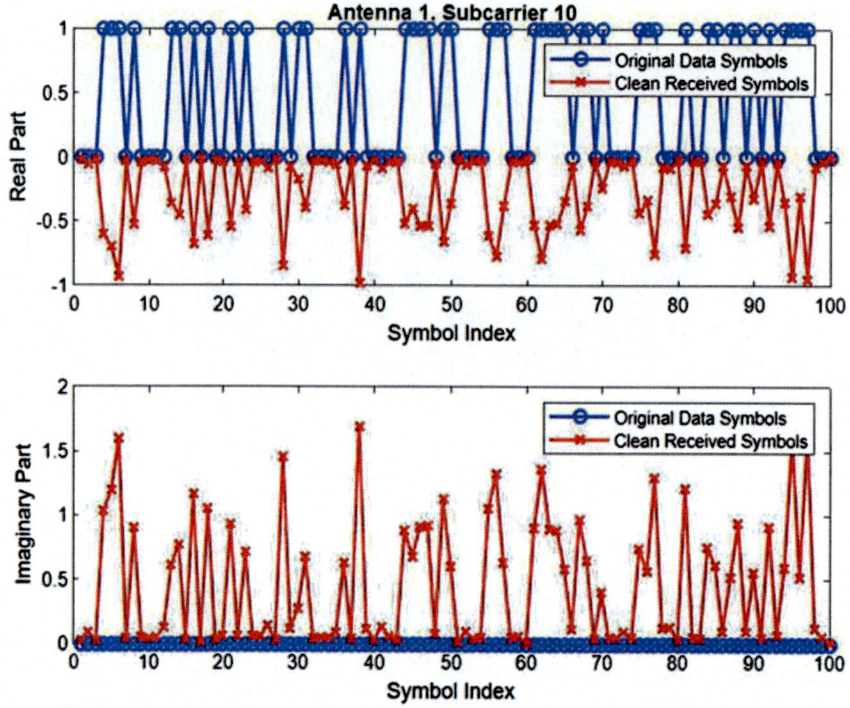

Fig. 4 Comparison of symbol error with phase noise

Table 2 Comparison of symbol error versus phase noise

Symbol index	Phase noise (degrees)
1	50.922
2	102.4
3	101.85
4	−145.46
5	−116.5
6	−105.07
7	−144.86
8	−109.27
9	32.618

channel estimation & synchronization more difficult than in OFDM, it is not possible to directly apply the solutions for OFDM to GFDM.

The Limitations and potential implementation challenges of the proposed complex training sequence decomposition (CTSD) algorithm and MIMO-GFDM systems for millimeter wave 5G systems include:

Algorithm complexity: In MIMO-GFDM systems, the CTSD algorithm may contribute to the complexity of the receiver design. It can be difficult to implement the algorithm effectively and guarantee real-time functioning, especially in devices or systems with limited resources.

Computing requirements: When dealing with large-scale MIMO systems or high-density scenarios, the CTSD method may need significant computer resources, such as processing speed and memory. It can be difficult to implement the algorithm on hardware platforms with constrained resources.

Sensitivity to channel changes: The performance of the CTSD algorithm may be affected by changes in the channel, especially in fading channels. In rapidly changing channel conditions, the accuracy of channel estimation may deteriorate, resulting in lower system performance.

Practical implementation issues: There may be issues with implementing the CTSD algorithm in actual millimeter wave 5G systems. Multiple base stations may need to work together in synchronization and coordination, and the presence of interference from nearby cells may reduce the accuracy of channel estimation.

Interference and non-orthogonal subcarrier waveforms: Non-orthogonal subcarrier waveforms can produce interference in GFDM systems, which makes them vulnerable to this problem. It might be difficult to reduce interference and guarantee dependable communication in high-density settings. These interference problems should be successfully handled by the CTSD algorithm.

Trade-off between complexity & performance: The complexity of the CTSD algorithm and the performance improvements may be subject to trade-offs. The accuracy of channel estimates may be improved by making the algorithm more complex, but this may also result in more computational work and implementation difficulties.

Generalizability and compatibility: It is necessary to assess the efficiency and performance of the proposed CTSD algorithm in a variety of conditions, including variable channel conditions, system configurations, and deployment scenarios. It can be difficult to guarantee the algorithm's generalization and interoperability with various millimeter wave 5G systems.

Practical deployment factors: The suggested CTSD algorithm must be compatible with the 5G ecosystem's current standards and infrastructure. Implementation difficulties may arise when ensuring seamless compatibility and backward compatibility with other communication system parts.

Limitations of the hardware: The use of MIMO-GFDM systems for millimeter wave 5G may necessitate specialist hardware, such as many antennas and high-frequency components, which can be expensive and difficult to integrate into real-world devices or base stations.

Conclusion

GFDM, as an advanced waveform for wireless communication systems, offers several advantages such as improved pulse shaping and reduced out-of-band emissions. The data is modulated using GFDM onto a collection of sub-carriers that are organized into rectangular-shaped symbols. GFDM is a viable multiple access strategy for 5G communication systems because it can offer great spectrum efficiency, influence for a more number of devices, & increased performance in high-density situations, but it also has downsides including phase noise and interference. However, the non-orthogonality of most GFDM prototype filters introduces challenges related to inter-symbol interference (ISI) & inter-carrier interference (ICI) between subcarriers. To address the impact of phase noise in millimeter wave systems, the complex training sequence decomposition (CTSD) method is proposed, which effectively mitigates phase noise degradation and enhances data recovery and communication efficiency. It is important to recognize that the specific implementation details and variations of the CTSD technique may vary depending on the system requirements and the unique properties of phase noise in the millimeter wave system being considered. Nonetheless, given the abundant bandwidth available in the millimeter wave frequency range, GFDM proves to be a valuable solution for millimeter wave applications. This research chapter provides a clear explanation of how millimeter waves can be utilized for transmitting data, making GFDM an appropriate choice for future high-capacity communication systems like 5G by reducing the adverse effects of phase noise.

Funding The authors have not disclosed any funding.

Data Availability Not applicable.

Declarations

Conflict of Interest No conflicts of interest.

References

1. Lim, B., & Ko, Y. (2019). Multiuser interference cancellation for GFDM with timing and frequency offsets. *IEEE Transactions on Communications, 67*(6), 4337–4349.
2. Liu, Y., Zhu, X., Lim, E. G., Jiang, Y., & Huang, Y. (2019). Robust semi-blind estimation of channel and CFO for GFDM systems. In *ICC 2019—IEEE international conference on communications (ICC)* (pp. 1–7). Shanghai, China.
3. Michailow, N., Gaspar, I., Krone, S., Lentmaier, M., Fettweis, G. (2012). Generalized frequency division multiplexing: Analysis of an alternative multi-carrier technique for next generation cellular systems. In *2012 International symposium on wireless communication systems (ISWCS)* (pp. 171–175). Paris.
4. Tai, C. L., Wang, T. H., Huang, Y. H. (2020). An overview of generalized frequency division multiplexing (GFDM), pp.1–14.

5. Kusumakar, D., Pancholi, D., & Tiwari, V. (2019). GFDM 5G cellular network using MIMO system. *International Journal of Computer Applications, 178*(21), 1–3.
6. Cheng, H., Xia, Y., Huang, Y., Yang, L., & Mandic, D. P. (2019). Joint channel estimation and Tx/Rx I/Q imbalance compensation for GFDM systems. *IEEE Transactions on Wireless Communications, 18*(2), 1304–1317.
7. Danneberg, M., et al. (2015). Implementation of a 2 by 2 MIMO-GFDM transceiver for robust 5G networks. In *2015 International symposium on wireless communication systems (ISWCS)* (pp. 236–240). Brussels.
8. Öztürk, E., et al. (2019). Multiple-input multiple-output generalized frequency division multiplexing with index modulation. *Physical Communication, 34*, 27–37.
9. Nimr, A., Zhang, D., Martinez, A., & Fettweis, G. (2017). A study on the physical layer performance of GFDM for high throughput wireless communication. In *25th European signal processing conference (EUSIPCO)* (pp. 638–642), Kos, Greece.
10. Lim, B., & Ko, Y. (2017). SIR analysis of OFDM and GFDM waveforms with timing offset, CFO, and phase noise. *IEEE Transactions on Wireless Communications, 16*(10), 6979–6990.
11. Borges, R., et al. (2018). Implementation of a Multi-Gbit/s and GFDM-based optical-wireless 5G network. *Journal of Microwaves, Optoelectronics and Electromagnetic Applications, 17*, 579–589.
12. Datta, J., Lin, H., & Lin, D. (2015). A method to implement interference avoidance based MIMO-GFDM using spatial modulation. In *2015 International conference on advanced technologies for communications (ATC)* (pp. 572–577), Ho Chi Minh City.
13. Gaspar, I. S., Mendes, L. L., Michailow, N., et al. (2014). A synchronization technique for generalized frequency division multiplexing. *EURASIP Journal of Advances Signal Processing, 2014*, 67.
14. Hu, S., Liu, Z., Guan, Y. L., Jin, C., Huang, Y., & Wu, J.-M. (2017). Training sequence design for efficient channel estimation in MIMO-FBMC systems. *IEEE Access, 5*, 4747–4758.
15. Zhong, J., Chen, G., Mao, J., Dang, S., & Xiao, P. (2018). Iterative frequency domain equalization for MIMO-GFDM systems. *IEEE Access, 6*, 19386–19395.
16. Udayakumar, E., & Krishnaveni, V. (2019). A review on interference management in millimeter-wave MIMO systems for future 5G networks. In *Proceedings of international conference on innovations in electrical and electronics engineering (ICIEEE 2019)* (pp.715–721), Hyderabad, 26–27 July 2019.
17. Zewail, I. (2018). Maximizing the total throughput for GFDM system using hybrid PSO–PS algorithm. *Journal of Electrical Systems and Information Technology, 5*, 594–607.
18. Udayakumar, E., Krishnaveni, V. (2020). Performance evaluation of phase noise reduction for MIMO FBMC-OQAM systems. *Solid State Technology, Extension Media, 63*(5), 5668–5675.
19. Tai, C., Su, B., & Jia, C. (2019). Frequency-domain decoupling for MIMO-GFDM spatial multiplexing. In *ICASSP 2019—IEEE international conference on acoustics, speech and signal processing (ICASSP)* (pp. 4799–4803). Brighton, United Kingdom.
20. Öztürk, E., Basar, E., & Çirpan, H. A. (2017). Generalized frequency division multiplexing with flexible index modulation. *IEEE Access, 5*, 24727–24746.
21. Dias, W. D., Mendes, L. L., & Rodrigues, J. J. P. C. (2019). Low complexity GFDM receiver for frequency-selective channels. *IEEE Communications Letters, 23*(7), 1166–1169.
22. Öztürk, E., Basar, E., & Çırpan, H. A. (2016). Spatial modulation GFDM: A low complexity MIMO-GFDM system for 5G wireless networks. In *2016 IEEE international black sea conference on communications and networking (BlackSeaCom)* (pp. 1–5), Varna.
23. Al-Juboori, G., Doufexi, A., & Nix, A. R. (2017). System level 5G evaluation of MIMO-GFDM in an LTE-A platform. In *2017 24th International conference on telecommunications (ICT)* (pp. 1–5). Limassol.
24. Udayakumar, E., & Krishnaveni, V. (2022). Phase noise effect on millimeter wave MIMO-OFDM and FBMC-OQAM systems. In *AIP conference proceedings of international conference on applied data science and smart systems*, Punjab, 4–5 Nov 2022.
25. Wang, Y., & Fortier, P. (2022). Union bound on the bit error rate for MIMO-GFDM systems. *Wireless Personal Communications, 123*, 1825–1839.

26. Shimodaira, H., et al. (2016). Enhanced next generation millimeter-wave multicarrier system with generalized frequency division multiplexing. *International Journal of Antennas and Propagation.*
27. Udayakumar, E., & Krishnaveni, V. (2019). Analysis of various interference in millimeter-wave communication systems: a survey. In *Proceedings of 10th IEEE international conference on computing, communication and networking technologies (ICCCNT 2019)* (pp. 1–5), Uttar Pradesh, 6–8 July 2019.
28. Kumar, A., & Santosh, S. (2018). GFDM for next generation wireless communication system. *International Journal of Students Research in Technology & Management, 6*(2), 60–69.
29. Vetrivelan, P., et al. (2015). PAPR redcton for OQAM/OFDM signal using optimized iterative clipping & filtering technique. In: *IEEE international conference on soft-computing & net security 2015* (ICSNS'15) (pp. 1–6).
30. Matthé, M., Mendes, L., Gaspar, I., et al. (2016). Precoded GFDM transceiver with low complexity time domain processing. *Journal on Wireless Communication and Networking, 2016*, 138.
31. Mohammadian, A., & Tellambura, C. (2019). Full-duplex GFDM radio transceivers in the presence of phase noise, CFO and IQ imbalance. In *ICC 2019 IEEE international conference on communications (ICC)* (pp. 1–6). Shanghai, China.
32. Udayakumar, E., & Krishnaveni, V. (2022). Analysis of phase noise issues in millimeter wave systems for 5G communications. *Wireless Personal Communications, Springer, 126*, 1601–1619.
33. Mohammadian, A., Tellambura, C., & Valkama, M. (2020). Analysis of self-interference cancellation under phase noise, CFO, and IQ imbalance in GFDM full-duplex transceivers. *IEEE Transactions on Vehicular Technology, 69*(1), 700–713.
34. Vetrivelan, P., et al. (2015). PAPR reduction for OQAM/OFDM signal by using neural network. *International Journal of Applied Engineering Research, 10*, 30292–30297.
35. Michailow, N., et al. (2014). Generalized frequency division multiplexing for 5th generation cellular networks. *IEEE Transactions on Communications, 62*(9), 3045–3061.
36. Chang, Y.-K., et al. (2018). A novel MIMO-GFDM receiver for next generation communication. *Emerging Telecommunication Technologies, 29*(6), 1–15.
37. Udayakumar, E., & Krishnaveni, V. (2023). Interference and phase noise in millimeter wave MIMO-NOMA and OFDM systems for beyond 5G networks. In *Security and privacy schemes for dense 6G wireless communication networks.* IET book chapter (pp. 349–370).
38. Tahir, S. (2019). Pilot placement schemes for channel estimation of proposed 5G-GFDM system. In *ITM Web of Conferences,* vol. 27, p. 01001
39. Zhong, Z., & Guo, J. (2016). Bit error rate analysis of a MIMO-generalized frequency division multiplexing scheme for 5th generation cellular systems. In *2016 IEEE international conference on electronic information and communication technology (ICEICT)* (pp. 62–68), Harbin.
40. Rajappa, A. C. J., Ramadhas, S. D., Anjaneyulu, B. M., et al. (2020). Golden Coded GFDM for 5G Communication. *Wireless Personal Communication,* 1–14.

Credit Card Fraud Detection Using Deep Learning for Internet-of-Things Enabled Smart Digital Financial System

Anchal Chand, Sachin Sharma, Piyush Agarwal, and Seshadri Mohan

Abstract It is crucial to adhere to established protocols and effective techniques to prevent the theft of any sensitive information when paying with a credit card as banking and finance activities through the Internet-of-Things enabled smart digital financial system increase. Due to many factors, including the lack of financial data and the dynamic nature of transactions and the accompanying frauds, detecting such frauds is a significant challenge. This indicates the need for efficient strategies to combat credit card theft. In this study, a few machine learning-based methods and the associated difficulties have been briefly described. This chapter focuses on a deep learning method for Internet-of-Things enabled smart digital financial system. We discussed a case study of HSBC that serves nearly 40 million customers in over 63 countries. The relevance and benefits of the proposed methodology for the financial institutions has tremendous utilization of Internet-of-Things and advanced technologies, whether at local level or internationally, are highlighted by the instance of HSBC like financial institutions.

Keywords Fraud detection · Digital financial system · Internet-of-things · Deep learning · HSBC

A. Chand
Graphic Era Hill University, Dehradun, Uttarakhand, India
e-mail: anchalchand93@gmail.com

S. Sharma (✉) · P. Agarwal
Graphic Era Deemed to be University, Dehradun, Uttarakhand, India
e-mail: sachin.cse@geu.ac.in

P. Agarwal
e-mail: piyush221292@gmail.com

S. Mohan
University of Arkansas at Little Rock, Little Rock, AR, USA
e-mail: sxmohan@ualr.edu

A. Prasad et al. (eds.), *Communication Technologies and Security Challenges in IoT*, Internet of Things, https://doi.org/10.1007/978-981-97-0052-3_23

Introduction

Banks provide their customers with credit cards as a form of payment to enable them to make cashless purchases of goods and services. It differentiates itself from an ATM card or a debit card by its pre-set credit limit. The bank or the card issuer determines this credit limit depending on the income, credit history and credit score of a customer. The total amount spent to purchase a commodity, or a service can be replaced with low-cost EMIs by using a credit card. This amount can be repaid within a grace period without any interest. An interest or a penalty charge is included in the repayment amount only after the grace period. This prevents the need to carry money and is indeed helpful in a financial emergency. In this way, the 'short-term loan' support by the credit cards provides convenience to the cardholders. The banks issues two types of credit cards to their customers (i) personal or consumer credit cards, and (ii) business credit cards. The business credit cards have higher limits as compared to personal or consumer credit cards. The credit card transaction can be done through two methods: physical method and virtual method. In the physical method, the physical possession of credit card is required to swipe the card for the transaction. While the virtual method, or generally called as Card not Present (CNP), requires the name of the cardholder, CVV number, password and few other details for online transaction. Credit card frauds are the illegal act committed by the scammers to make unauthorized use of the account or any financial transaction from the account of the cardholder without his/her prior knowledge. Credit card frauds can be internal and external. The consent between the cardholder and the bank due to some false identity is a reason of the internal credit card fraud. A stolen credit card or its sensitive information is used unethically for the external fraud. The proposed system is described in Fig. 1.

- *Application Fraud*: It includes the unethical approach to use fake documents or synthetic identity of a cardholder to open an account in another person's name.

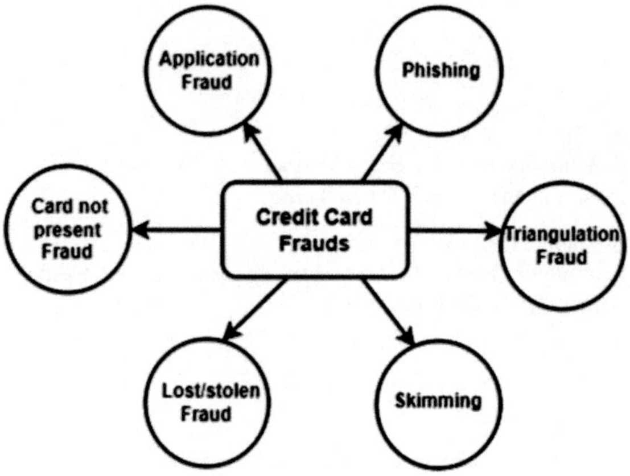

Fig. 1 Types of credit card frauds in Internet-of-Things enabled smart digital financial system

- *Card Not Present Fraud*: This is done through online transactions, mobile phone or mails when the card is not present for a visual check by the merchant. This fraud is done by tracing customers' information like device fingerprint, social media lookup, IP address analysis, address details, or CVV (Card Verification Value) code.
- *Lost/stolen Fraud*: The physical possession of the credit card by an unauthorized person can allow him/her to withdraw all the amount from a cardholder's account in case the card is lost/stolen.
- *Phishing*: Phishing refers to an attempt to steal sensitive information from a cardholder by tricking them that can legitimate a scammer to access the bank account of a customer. It is one of the common credit card frauds.
- *Skimming*: It is done using a device called card skimmer which is kept on the credit card swiping machine to steal the credit card information through the magnetic strip of the card at an ATM or POS when the card is swiped. The fraudsters can steal all the money from the account or can sell the card to third parties to achieve a similar purpose by replicating the financial information.
- *Triangulation*: In this fraud, the fraudster may hijack the e-commerce buying process and can operate as the merchant. The cardholder details are then stolen on the pretext of completing the payment on that ecommerce site.

With emerging state-of-art techniques, digitalization has made possible the automation of tedious manual tasks and is helping to ease the human lifestyle. Online purchasing of groceries (Bigbasket, Grofers), medicines and medical assistance (Apollo 247, PharmEasy), e-education (Udemy, Byju's, Coursera, YouTube), food delivery services (Swiggy, Zomato), housekeeping services like appliance repairing, plumbing (Urban Company) and a lot more adds to the list of facilities possible due to digitalization. Online payment services are another such asset. The lockdown period resulted in a hike of these online services. This made most of the payment through online mode. The online purchases and payment services provide users multiple advantages, may it be the ease and convenient of usage, time and resource saving facilities or better financial analytics. The financial sector deals with the most crucial information and thus it is essential to implement effective measures to prevent any theft or misuse of such sensitive information. As the security methodologies are being enhanced in the financial sector to tackle the issues of security breaches and theft of sensitive information by unauthorized people, the modes of frauds are also being modified with the advancement of time and technology. This adds up to the need for a robust and highly efficient system to detect such security breaches and fraud to prevent huge losses.

Literature Review

The supervised learning technique will treat a transition as legitimate even if it does not match the database records. If used, this technique can be a disadvantage. In order to identify credit card fraud, Pumsirirat and Yan [1] concentrated on scenarios that are difficult to identify using supervised algorithms. They employed Restricted Boltzmann Machine (RBM) and Auto-Encoder (AE). By compressing and encoding the input data, followed by its reconstruction as an output, encoders' primary goal is to reduce data noise. Through this, the encoders reduce dimensionality to use a lower-dimensional representation for a higher-dimensional data. These two deep learning algorithms were implemented for German, Australian, and European datasets, the later one with the most data as compared to prior two datasets. AE is efficient for large datasets. As a result, the AUC score was more for the European dataset and least for the German dataset using AE algorithm. The concept drift problems result in the high data imbalance as they change over time in an unpredictive manner. Authors in [2], focused on dealing with this problem by using Machine Learning based approach. The initial step is to segregate the cardholders into various clusters depending on the amount of transaction. Then, a sliding-window concept is used to extract some features from the window in order to get the behavioral patterns of the transactions by the cardholders. The Matthews Correlation Coefficient (MCC) and on-class classifiers are the other two ways used for handling the imbalanced dataset. Support Vector Machine (SVM), Deep learning, Artificial Immune System, Logistic Regression, and other machine learning based techniques are analyzed in [3]. Artificial Immune System (AIS) has detector cells that are used to detect foreign bodies and are generated in random numbers. This algorithm is evaluated using multiple classification strategies. AIS is divided into negative selection and clone selection. In the clone selection ASI, the generated detector cells have a short lifespan and extended to fight with the intruder in case any antigen is detected. The cells best suited for detecting the antigens are termed as the survival cells. The cell creation is arbitrary in the negative selection ASI. The Bayesian Belief Network works on the principle of probability and Logistic Regression is used for binary classification for data with unambiguous attributes to predict patterns. Both are statistical classification techniques. The Neural Network is a computational approximation of the human brain [3]. Support Vector Machine uses polynomial function to remodel dataset by mapping the higher dimensional space point and the input space point. To determine the best method for identifying credit card fraud, four distinct Deep Learning topologies have been assessed, and several model-building parameters have been examined in [4]. The size of the network has the greatest impact on performance when comparing the parameters taken into account; larger networks perform better than smaller networks [5]. The dataset is acquired from Kaggle and contains the dates, amounts, and transaction classes (0 for legitimate transactions and 1 for fraudulent transactions) for the last two days. The local values of the sample are compared to their neighbours in the Local Outlier Factor algorithm, and the value that is lower than its neighbours' is taken as the outlier. The split value between a feature's maximum and minimum values is chosen at random using the Isolation Forest method. Shorter path length samples are regarded as abnormalities [6]. The technique's flaw is that the results are static regardless of the dataset's kind or size. The most accurate architecture was an ANN. Data that has been sampled and pre-processed improves the decision tree's accuracy. Contrarily, Logistic Regression performs better when given unfiltered, unsampled data. SVM and Random Forest both do well with small datasets and categorical data, respectively. A genetic algorithm is used to configure an artificial neural network as a method of identifying credit card fraud in [7]. This configuration addresses the difficulties of training ANN by reducing the dataset through mutation and crossover using genetic algorithms in light of the fact that the transactions differ from one another. An advantage of using ANN is that it eliminates the requirement to relate models before grouping. Consumer spending patterns are taken into account in [8] because they are just as important as other transactional characteristics like the sum, the timing, and the place. It can be used to determine the dependence between subsequent transactions. As a sequence learner and a dynamic pattern recognition classifier for this, LSTM networks have been employed. As neural networks only deal with numerical data, the Min-Max normalization approach is employed for data normalization, where each input data is normalized to the range value [0,1]. The long-term sequence dependencies can be learned via the feedback connections between the hidden units of the LSTM, a unique sort of ANN. The dataset created by BankSim and obtained from Kaggle contains 594,643 transactions that span a six-month period. The network's AUC value over 10 epochs was 0.9955 [9]. All these methods exhibit their pros and cons. The Fuzzy Darwinian system showed accuracy of 100% true positive, but with a low speed of processing. The Hidden Markov Model signified fast processing speed but with low accuracy. AIS presented fast as well as accurate results. KNN showed good results if taken a large value of k. Bayesian Network and Fuzzy Neural Network possessed higher speed as compared to other techniques, the latter being highly expensive in terms of computational cost. The outcome in [9] stresses a hybrid approach to construct a fraud detection system.

Overview of Fraud Detection Techniques

The financial sector is the most sensitive fields that possess risk in almost each of its related activities. The increase in volume of data also increased the expense to analyze and compute it. The manual methods to do so are time consuming too. As a result, the rule-based fraud prevention system is inefficient because the same pre-defined rules cannot work equally effective for different scenarios of fraud. Artificial Intelligence (AI) based Fraud Detection algorithms provide the advantage of fast computation and extract new patterns from the data. The capability of an AI based model to process the redundant task of data analysis makes it more approachable in terms of accuracy of the fraud detection in Internet-of-Things enabled smart digital financial system. Professionals also examine the outcome reports to confirm the validity of the card-holders' transactions before providing input to the algorithm's training mechanism.

Artificial intelligence and statistical data analysis are two of the primary methods used to identify credit card fraud. Either supervised or unsupervised statistical fraud detection methods are available. The prior transactions are employed in the supervised learning process to distinguish between the new transactions as legal and fraudulent. Unsupervised approaches use unlabeled data samples from previous transactions to identify patterns and dependencies in the dataset and to organise the data samples into groups based on shared characteristics. As a result, the samples of data don't need to be manually labelled. K-Nearest Neighbour, Hidden Markov Model, Decision Tree, Support Vector Machines, Bayesian Networks, Logistic Regression, etc. are some examples of approaches used to detect credit card fraud. Decision trees are a two-phase statistical data mining technique focused on the identification of credit card fraud. The first stage involves creating the decision tree from the training data, and the second stage involves putting the decision rules into practise to determine the type of incoming transaction. Naïve Bayes forecasts the class value of the incoming instance using a labelled training dataset. The term "naïve" refers to a probabilistic supervised machine learning approach that treats class variables as stand-alone attributes. To determine the likelihood of the correct class, classification uses the Bayes theorem of probability. For detecting fraud, ML algorithms outperform the rule-based approach by a wide margin. Even more sophisticated algorithms are used in deep learning. Automatic feature extraction and modelling techniques in deep learning take the role of the need to extract and construct features in machine learning. Deep Learning approaches, which are able to train without any rules, can be used to reduce the expense and effort associated with data labelling for ML models. As a result, Deep Learning algorithms are superior to ML techniques. All of the many methods and algorithms used for detecting credit card fraud have their distinct benefits and drawbacks. In terms of speed of fraud detection, some strategies may be more practical, while others may outperform owing to the accuracy they offer. To stay up with emerging fraud patterns and to maintain the effectiveness of fraud detection with precise results, the system for fraud detection must undergo ongoing training. When receiving new data entries, this training must be completed.

Importance of Internet-of-Things enabled smart digital financial system: Internet-of-Things enabled smart digital financial system provides financial services to the customers through the digital medium via internet. There are numerous advantages of digital financial system over the traditional mode of finance and conventional banking for both customers and business organizations such as easy accessibility, time and resource saving, real-time updating of transactions, technical support, faster and automatic execution, low operational cost and better compliance. The emerging technical domains like machine learning, deep learning and big data analytics are utilized to perform multiple tasks like preparing annual financial statements, real-time data processing and reporting, tracking assets, forecasting overall gain percentage and much more. Internet-of-Things enabled smart digital financial system derives the analytical understanding of the customer's behavior. In the banking sector, data is a crucial resource, and the digital banking system works for the efficient management of this data. ML algorithms are used to analyze a large volume of customers data, both structured and unstructured, and discover patterns and trends for better decision making, investment prediction and assessing financial threats. The integration of machine learning and big data makes these systems capable of handling complex and large amount of customers data. The process of financial monitoring helps to identify transaction anomalies, compliance issues and to ensure that the transactions are accurate. The intelligent process automation contributes to the business productivity and improved customer service experience at a low operational cost. Amy from HSBC, Ally Assist from Ally bank, Electronic Virtual Assistant (EVA, 2017) from HDFC, SBI Intelligent Assistant (SIA, 2017) from SBI, Keya (2019) from Kotak Bank are few examples of virtual banking assistants developed for personalized customer service. ML-powered banking applications have also been developed like YONO SBI, which allow customers to analyze their daily transactions and to perform other financial tasks. For example, the SBI YONO application used a Response Surface Methodology (RSM) model for pre-approved personal loans (PAPL) and in the year 2019–20, the bank disbursed a total of Rs. two billion in the form of PAPL.

Challenges in Credit Card Fraud Detection for Internet-of-Things enabled smart digital financial system: The dynamic nature of normal and fraudulent behavior profiles signifies the extent to which a fraudulent transaction seems almost a legitimate one. The credit card transaction datasets required to develop an efficient credit card detection system are hardly available and even if available, are highly imbalance and include both labelled and unlabeled samples [10].

Proposed Methodology

The proposed methodology depicts the deep learning technique to detect the fraudulent transactions related to credit card for Internet-of-Things enabled smart digital financial system. It is based on a sequential model with 1D Convolution Layer. A Sequential model has exactly one input and one output associated to each layer. The associated sequential data may include streams of text, time-series data, audio clips etc. It forms a sequential neural network where the layers are stacked together, one after another. Often, a Convolution Neural Network (CNN) denotes a two-dimensional CNN, used for image data, but basically there are two more types—one-dimensional CNN in which the kernel slides along only one direction example time-series data and three-dimensional CNNs.

A linear function ReLu is used. Rectified Linear Unit or ReLu produces output same as the input if the input is positive, else it will produce zero as the output. It prevents the activation of all the neurons at once and thus helps in avoiding the vanishing gradient problem and the exponential computation growth in the network. Batch normalization is done along mini batches instead of the complete data set between the layers of a Neural Network. It boosts training and stabilizes the network by normalizing the output from the previous layer before transferring it to the next layer. Flatten layer converts the multidimensional data of a layer into a 1-dimensional array for the next layer. It is used to transmit input from the convolution layer to the full connected layer. At last, sigmoid activation function is used that produces output

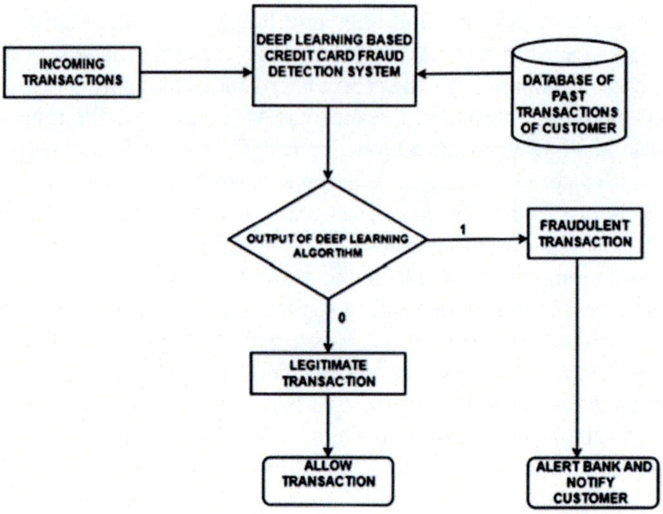

Fig. 2 Workflow of the proposed methodology

value with range of 0 to 1. The dataset consists of transactions by a cardholder from Europe with a positive class percentage of 0.172%, or 492 frauds out of a total of 2.84 lakhs transactions. Figure 2 shows how a technique for detecting credit card fraud operates.

Result Analysis

For experiment, the dataset was collected from the Kaggle data repository with 2.84 lakh transactions and 25+ parameters. The accuracy of the proposed method when implemented with 50 epochs is 97.71% (Fig. 3). While with 80 epochs it was 99.87% (Fig. 4). Table 1 shows the comparative analysis of the different method. The proposed method outperformed with highest accuracy of 99.87%.

Case Study: HSBC

The enhancement in computing power, storage and generation of large volume of data have advanced Artificial Intelligence (AI). Well-trained AI algorithms replace methods to address manually intensive queries. The computational expense is thus reduced. The increased quality of customer service is another feature. AI also speeds up the processing of manual tasks with better accuracy than humans and increases the overall productivity. The abundant features of AI have made its existence in almost

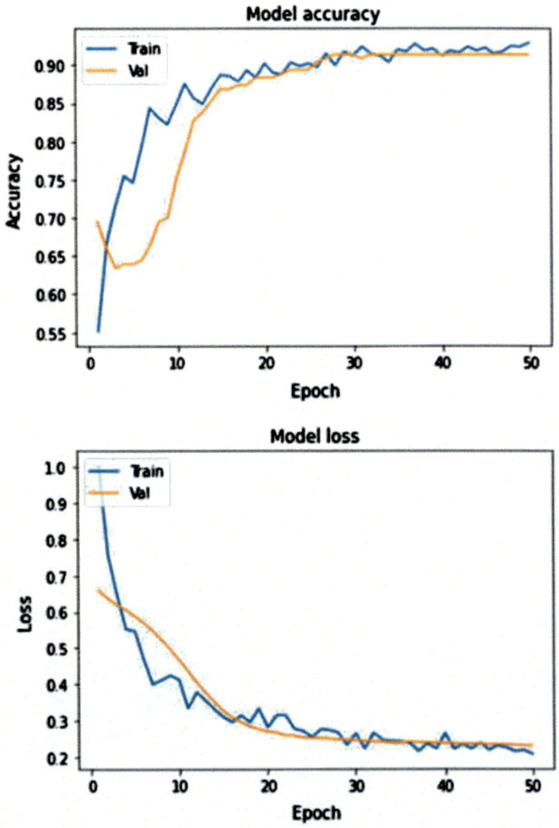

Fig. 3 Result with 50 epochs

every sector. The financial sector is one such. HSBC in India has been contributing to the development of the Indian banking industry since 1853. The first ATM in India was set up by HSBC in 1987. With a history of 157 long years, increased competition, global recession, technological enhancement were some challenges faced by HSBC. HSBC have worked to keep pace with the emerging technologies and to provide productive facilities to its customers.

HSBC has also utilized AI tools to enhance asset allocation and to analyze and monitor investments. Screening the unstructured data from mobile devices, web traffic and social media is one step to do that. With an aim to develop an AI-enabled anti-money laundering solution, HSBC partnered with Ayasdi, an AI startup, in 2018. Ayasdi's solutions work on anomaly detection technology. To segregate the customer into categories using an unsupervised learning method by analyzing their past and current activities that helps in understanding the probability of the customers to be the money launderers. The NLP recognizes the contextual patterns from the data. The IT team of HSBC and the data scientists and developers of Ayasdi work collaboratively for the smooth functioning of the solution. The IT team works to allow Ayasdi in accessing and parsing the Anti Money Laundering (AML) data of

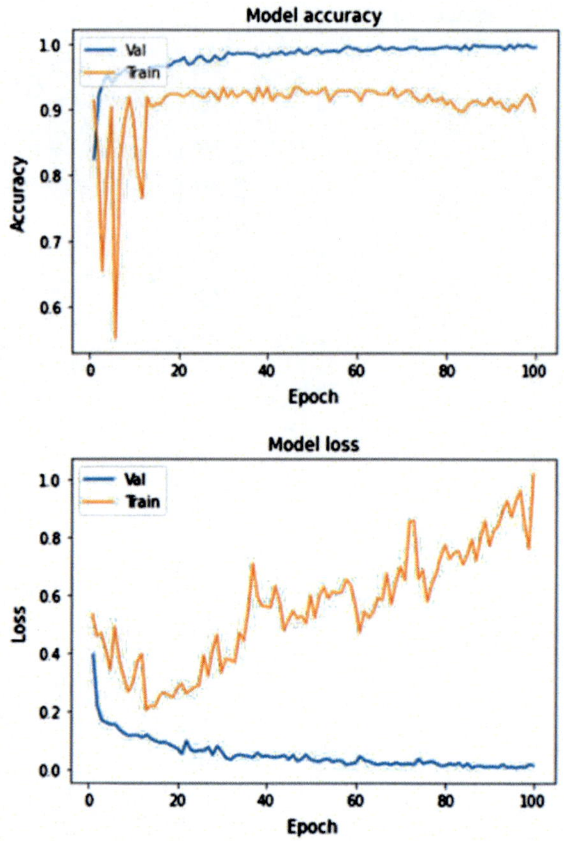

Fig. 4 Result with 80 epochs

HSBC to create more accurate models for customer behavior and to analyze the source and destination of customer payments. This ensures the legitimate sources of funds. According to [16], the false positive rate in HSBC is reduced by 20%. The AI-driven robotic technologies have eased the customer services at HSBC. In the online banking environment, the virtual assistants support customers for online transactions to resolve their queries. In the physical branch, a human-like robot with a tablet screen can be a suitable option for assisting customers. In 2018, HSBC started using Pepper, a humanoid robot, in the USA. Manufactured by SoftBank Robotics, Pepper uses AI to provide self-services that include greetings, navigation assistance, answer customer queries and ask relevant questions for analyzing customer's needs. Another application of robotics is a virtual assistant of HSBC, Amy that can be accessed through a smartphone or a personal computer. The 24/7 availability of chatbots to address customer queries helps to boost the customer service. Just a year after Amy's debut, the chatbot was able to address 97,000 questions with a 90% success rate and hold volume of interactions of 38,000 customer

Table 1 Comparative analysis of existing methods

Technology used	Algorithm	Accuracy (%)
Machine learning [11]	Logistic Regression Classifiers K-Nearest Neighbor Naïve Bayes	54.86 97.69 97.92
Machine learning [12]	Naïve Bayes Decision Tree Random Forest SVM	80 83 88 86
Machine learning [13]	Optimized Light Gradient Boosting Machine (OLightGBM)	92.88
Machine learning and deep learning [14]	Bayesian Learning Hidden Markov Model BLAST-SSAHA Hybridization SVM K-Nearest Neighbour Neural Networks Naïve Bayes	98 70 86 70 80 70 66
Deep learning [15]	CNN	99.76
Deep learning (proposed)	CNN	99.87

conversations. HSBC Connect is a single interface that connects nearly 40 countries to access multiple banking services provided by HSBC. It supports automation through easy to follow payment instructions. In the Middle East, the processing time and number of resources required for issuing payments to students decreased by 30% through HSBC Connect. With a decrease in the transaction costs too, the tuition and maintenance payments of over 40,000 students locally and 8,000 overseas for higher education students is automated. The number of incoming enquiry calls has been replaced by the payment confirmation notifications. It enabled the clients with a quick payment process upon authorization and an automated process to send daily statements to students. In North America, the HSBC solution of Global Disbursements and a Foreign Exchange rate service supported the central government with an automated cross border payments in multiple currencies. As a result, the currency transaction costs of clients have been reduced, with an increase in control of their funds and efficiency in the related business and the transactional capability of the cheque issuance of over 200. The government-run organizations that aim to support the research projects financially in the field of food security, technology, health, employment and research and development often send funds across the globe. Many of these recipients are developing countries. The HSBC Global Disbursements and HSBC Connect promote the centralization of cross border payment. Thus, HSBC serves as an ideal bank for effective management of international multicurrency payments and the incoming funds, through which the client can opt for a single standard for both domestic and international payments. HSBC partnered with Silent Eight in 2021 that uses AI models and Robotic Process Automation to automate manual tasks performed by human operators like customer screening, transaction monitoring and alert adjudication. For over 20 years, Celent has been an advisory firm and the leading research that focuses on technology for financial institutions globally. A Featurespace's platform, ARIC uses a real-time AI algorithm to monitor transactions in retail and insurance and is the first AML technology to use industry-leading rules and ML models, deployed within the bank. It generates automatic and improved alert quality, reduces false positives and the risks of missing the genuine cases of money laundering and identifies new risk scenarios and typologies. HSBC is awarded Asia's best bank for sustainable finance 2022, consequently fifth time.

One another example is the work done by Visa, USA one of the biggest payment networks in the world. Since 2013, Visa has used deep learning to enhance its fraud detection capabilities. Each day, Visa's deep learning models examine more than 500 million transactions in search of trends that might point to fraud. The system is made to identify new and developing fraud patterns, which can be difficult for conventional systems that use rules-based algorithms for fraud detection. To provide a complete picture of each transaction, Visa's deep learning models draw on a range of data sources, including transactional data, device data, and behavioural data. The models' real-time anomaly detection capabilities allow Visa to take prompt action in the event of probable fraud. The approach has been incredibly effective at lowering fraud losses. The company's fraud rate is currently at its lowest point in more than 30 years, according to Visa. False positives, which may be a significant problem for conventional fraud detection systems, have been reduced thanks to the deep learning models. The industry has acknowledged Visa's effectiveness with deep learning in fraud detection. Visa received the SC Awards' Best Mobile Security or Anti-Fraud Solution honour in 2017, which honours the best in cybersecurity. Deep learning models can enable more accurate and efficient fraud detection by combining data sources and real-time analysis, which can reduce losses and boost security for customers and financial institutions.

The work done by the Commonwealth Bank of Australia (CBA), one of the biggest banks in the nation, is one case study of how credit card fraud is detected in Australia using deep learning. Since 2017, CBA has used deep learning to enhance its fraud detection abilities. The bank has created a deep learning model that can quickly analyse a vast volume of transaction data and spot irregularities that might be signs of fraud. A variety of data sources, including transactional data, device data, and behavioural data, are used by the deep learning model. Traditional fraud detection methods struggle to spot unusual patterns, but the model can. The deep learning model is combined with CBA's current method for detecting fraud, which is based on regulations. The deep learning model adds extra levels of security, allowing the bank to see suspected fraud in real-time and take immediate action. The deep learning model used by CBA has had great success in lowering fraud losses. The method, according to the bank, has cut false positives by as much as 75%, which has significantly decreased the cost of managing fraud. The industry has acknowledged CBA's accomplishments with deep learning in fraud detection. At the Australian Banking and Finance Awards, which honour achievement in the financial services sector, the bank earned the Best Security Initiative prize in 2018. Deep learning models can enable more accurate and efficient fraud detection by merging a variety of data sources and applying real-time analysis, resulting in decreased losses and increased security for customers and financial institutions.

Royal Bank of Canada (RBC), Canada is one of the biggest banks in the nation. Since 2016, RBC has been enhancing its fraud detection skills by utilizing deep learning. The bank has created a deep learning model that can quickly analyze a vast volume of transaction data and spot irregularities that might be signs of fraud. To provide a complete picture of each transaction, the deep learning model draws on a range of data sources, including transactional data, device data, and behavioural data. The programme can identify aberrant patterns that are difficult for conventional fraud detection systems to spot. The deep learning model is combined with RBC's current rules-based algorithm-based fraud detection system. The deep learning model adds extra levels of security, allowing the bank to see suspected fraud in real-time and take immediate action. The deep learning model used by RBC has had great success in lowering fraud losses. The method, according to the bank, has cut false positives by as much as 50%, which has significantly decreased the cost of managing fraud. The industry has acknowledged RBC's accomplishments with deep learning in fraud detection. At the Canadian Mortgage Awards, which honour achievement in the mortgage sector, the bank won the Best Security Initiative prize in 2019. Deep learning models can enable more accurate and efficient fraud detection by combining data sources and real-time analysis, which can reduce losses and boost security for customers and financial institutions.

HDFC Bank is one of the biggest private sector banks in the nation. Since 2017, HDFC Bank has used deep learning to enhance its fraud detection abilities. The bank has created a deep learning model that can quickly analyse a vast volume of transaction data and spot irregularities that might be signs of fraud. A variety of data sources, including transactional data, device data, and behavioural data, are used by the deep learning model. Traditional fraud detection methods struggle to spot unusual patterns, but the model can. The existing fraud detection system at HDFC Bank, which is based on a rules-based algorithm, has been linked with the deep learning model. The deep learning model adds extra levels of security, allowing the bank to see suspected fraud in real-time and take immediate action. The deep learning model used by HDFC Bank has had great success in lowering fraud losses. The method, according to the bank, has cut false positives by as much as 95%, which has significantly decreased the cost of managing fraud. The industry has acknowledged HDFC Bank's performance with deep learning in fraud detection. The bank received the Indian Banks' Association Technology Awards' Best Use of Artificial Intelligence and Machine Learning Award in 2019, which honours technological excellence in the banking sector. Deep learning models can enable more accurate and efficient fraud detection by combining data sources and real-time analysis, which can reduce losses and boost security for customers and financial institutions.

Challenges

Deep learning for fraud detection in an IoT enabled smart digital financial system comes with a number of difficulties. The key obstacles are as follows (Fig. 5):

- *Data security and privacy*: Because credit card transactions contain significant amounts of financial and personal data, there is a high danger of data breaches and privacy violations. Therefore, it is crucial to guarantee that the data is in a safe manner.
- *Data diversity and quality*: For deep learning models to be accurate and useful, high-quality, diverse data must be used for training. However, credit card data is frequently noisy, unbalanced, and incomplete, which might impair the model's effectiveness.
- *Real-time processing*: To detect credit card fraud, massive volumes of data must be processed in real-time. A high-performance computer environment that can handle the data volume and processing speed is needed for this.
- *Adaptability*: To get beyond fraud detection systems, con artists are continually improving their methods. Deep learning models must therefore be flexible and ready to pick up on new fraud trends.
- *Explainability*: Understanding how the model makes decisions and seeing potential biases may be difficult as a result.
- *Integration with current systems*: It can be difficult to integrate deep learning models into current credit card fraud detection systems. The models might need to be altered to fit the unique system architecture if there are compatibility problems.
- *Cost*: The implementation of deep learning techniques can be expensive, and the price may rise dramatically for large-scale systems. Therefore, it's crucial to strike a balance between the advantages of adopting deep learning and their implementation costs.

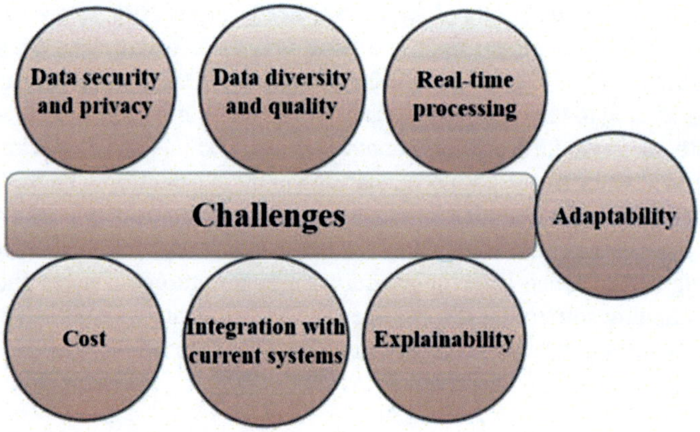

Fig. 5 Challenges

Future Perspective

Deep learning for an IoT enabled smart digital financial system has promising future prospects. The following possible developments could have an impact on how this field develops in the future (Fig. 6):

- *Greater accuracy*: Deep learning models have demonstrated good results in identifying credit card fraud, and further study may provide even more precise models. This could be accomplished by including more data sources and more advanced algorithms and methodologies.
- *Real-time detection*: Preventing losses from credit card fraud requires real-time detection. A big advancement over current techniques could come from deep learning models that can process data in real-time and swiftly find anomalies.
- *Advanced explainability*: Deep learning models' explainability is a key component, and there is continuing research into how to make these models easier to understand. Future advancements might make it possible to provide stronger justifications for the judgements the models make, making it simpler to see and correct such biases.
- *Better integration*: Integrating deep learning models for fraud detection with current systems can be challenging. The development of models that are simpler to integrate and more compatible with current systems could be the main goal of future study.
- *Increased security*: A vital component of any digital financial system is the security of data and transactions. Future studies could concentrate on creating deep learning models with improved fraud detection and prevention capabilities as well as increased attack resistance.

Fig. 6 Future perspectives

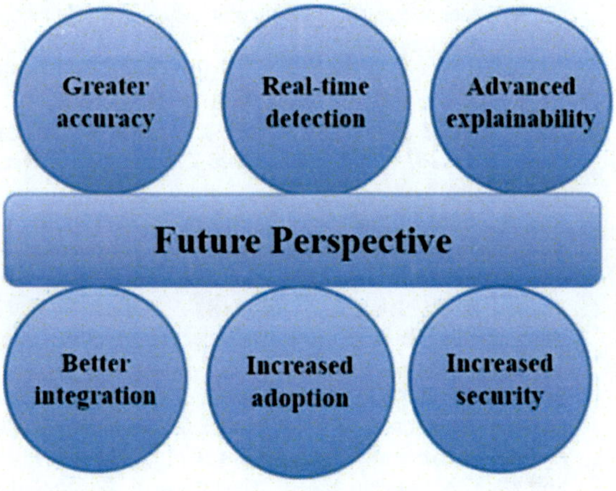

- *Increased adoption*: As the advantages of deep learning are more widely understood, there may be an increase in the use of these methods for detecting credit card fraud. This might result in less loss for both customers and financial institutions and more efficient and effective fraud detection.

Conclusion

The suggested solution for detecting credit card fraud for an Internet-of-Things enabled smart digital financial system is based on deep learning theory. To determine whether a particular transaction should be labelled as fraudulent or not, historical transaction data is used. Customer transactions are classified into levels ranging from level 1 to level 5 depending on the accuracy rate. Since levels of authentic transactions decrease with increasing rates, level 5 will consist solely of fraudulent transactions. It will make it simple for the system to notify the connected bank about any fraudulent transactions and promptly notify the client. Implementing a deep learning neural network with 86 epochs results in an accuracy of 99.87%. The relevance and benefits of financial institutions implementing cutting-edge security measures and giving clients ease in related financial transactions on Internet-of-Things enabled smart digital financial system, whether at local level or internationally, are highlighted by the instance of HSBC.

References

1. Pumsirirat, A., & Yan, L. (2018). Credit card fraud detection using deep learning based on auto-encoder and restricted Boltzmann machine. *International Journal of Advanced Computer Science and Applications, 9*(1).
2. Dornadula, V. N., & Geetha, S. (2019). Credit card fraud detection using machine learning algorithms. *Procedia Computer Science, 165*, 631–641.
3. Popat, R. R., & Chaudhary, J. (2018). A survey on credit card fraud detection using machine learning. In *2018 2nd International Conference on Trends in Electronics and Informatics (ICOEI)* (pp. 1120–1125). IEEE.
4. Roy, A., Sun, J., Mahoney, R., Alonzi, L., Adams, S., & Beling, P. (2018). Deep learning detecting fraud in credit card transactions. In *2018 Systems and Information Engineering Design Symposium (SIEDS)* (pp. 129–134). IEEE.
5. Maniraj, S. P., Saini, A., Ahmed, S., & Sarkar, S. D. (2019). Credit card fraud detection using machine learning and data science. *International Journal of Engineering Research, 8*(9), 110–115.
6. Sadineni, P. K. (2020). Detection of fraudulent transactions in credit card using machine learning algorithms. In *2020 Fourth International Conference on I-SMAC (IoT in Social, Mobile, Analytics and Cloud) (I-SMAC)* (pp. 659–660). IEEE.
7. Shirodkar, N., Mandrekar, P., Mandrekar, R. S., Sakhalkar, R., Chaman Kumar, K. M., & Aswale, S. (2020). Credit card fraud detection techniques—A survey. In *2020 International Conference on Emerging Trends in Information Technology and Engineering (ic-ETITE)* (pp. 1–7). IEEE.

8. Benchaji, I., Douzi, S., & El Ouahidi, B. (2021). Credit card fraud detection model based on LSTM recurrent neural networks. *Journal of Advances in Information Technology, 12*(2).
9. Zareapoor, M., Seeja, K. R., & Alam, A. (2012). Analysis on credit card fraud detection techniques: Based on certain design criteria. *International Journal of Computer Applications, 52*(3).
10. Gupta, S., Dangi, S., & Sharma, S. (2022). A perspective on blockchain-based cryptocurrency to boost futuristic digital economy. In *The data-driven blockchain ecosystem: Fundamentals, applications, and emerging technologies* (pp. 81–100). CRC Press. https://doi.org/10.1201/9781003269281-6
11. Awoyemi, J. O., Adetunmbi, A. O., & Oluwadare, S. A. (2017). Credit card fraud detection using machine learning techniques: A comparative analysis. In *2017 International Conference on Computing Networking and Informatics (ICCNI)* (pp. 1–9). IEEE.
12. Borah, L., Saleena, B., & Prakash, B. (2020). Credit card fraud detection using data mining techniques. *Seybold Report, 15*, 2431–2436.
13. Taha, A. A., & Malebary, S. J. (2020). An intelligent approach to credit card fraud detection using an optimized light gradient boosting machine. *IEEE Access, 8*, 25579–25587.
14. Bhatia, S., Bajaj, R., & Hazari, S. (2016). Analysis of credit card fraud detection techniques. *International Journal of Science and Research, 5*(3), 1302–1307.
15. Voican, O. (2021). Credit card fraud detection using deep learning techniques. *Informatica Economica, 25*(1).
16. Qi, Z., & Qi, C. (2019). Evaluate the managerial effectiveness of artificial intelligence—Case studies from HSBC and iQiYi. In *Proceedings of the 2019 International Conference on Artificial Intelligence and Computer Science* (pp. 542–547).

Development of a Secure and Transparent Blockchain for Electricity Bill Management in Smart Cities Using Enhanced Proof of Energy Consumption

Narendra Kumar Dewangan and Preeti Chandrakar

Abstract Blockchain technology offers transparency, immutability, security, privacy, and trust, but it faces challenges such as inability to obtain real-time data on past electricity usage, lack of customer privacy, and opaque payment mechanisms. A blockchain-based model for electricity use in smart cities aims to address these issues while maintaining transparency and protecting consumers' privacy. Smart meters, IoT-based devices, collect electricity consumption data at regular intervals and send it to energy providers and consumers. These devices are connected to cloud servers, which function as blockchain nodes. The proposed smart meter-based electricity consumption reading and bill management system uses the blockchain as a decentralized and distributed data storage system for electricity consumption. The system analyzes data generated by smart meters for particulate areas and sends reminders for exceeding the bill amount set by the consumer. The system is implemented on a Python-based blockchain using a customized consensus algorithm called enhanced proof-of-consumption (EPoCon) for energy management. The system considers scenarios of owner changing and meter defects, using Ed25519 and consumer information to generate meter IDs. The system also provides security analysis for the connection between smart meters and the cloud, demonstrating faster transaction rates and security against known attacks.

Keywords Blockchain · Energy management · Privacy · Security · Smart cities · Enhanced proof of consumption

N. K. Dewangan (✉) · P. Chandrakar
National Institute of Technology Raipur, G.E. Road, Raipur, India
e-mail: narendra.dewangan89@gmail.com

P. Chandrakar
e-mail: preeti.pubnitrr@gmail.com

A. Prasad et al. (eds.), *Communication Technologies and Security Challenges in IoT*, Internet of Things, https://doi.org/10.1007/978-981-97-0052-3_24

Introduction

In a smart city, various services are online and controlled through IoT devices and smart networks. These networks must be secure and fast for communication. Electricity service is an essential service in a smart city. Because of the electricity service, many other services like industry, transportation, and health services are running. Because of our dependency on electricity, the requirement and production of electric energy face a lot of pressure. The generation, transmission, and distribution of electricity are very time-consuming and tedious processes. Many factors are involved in the hurdles surrounding distribution and supply to the end user, including natural and human-made problems. Wrong billing, no real-time monitoring, and no proper storage of real-time metering data also influence the electricity generation financial system.

Blockchain is a distributed, decentralized, immutable, transparent, secure, and fast system to store data. Devices like the IoT (Internet of Things) are applied as smart devices in a real-time scenario. Blockchain can store real-time data in distributed storage. This storage is free from single points of failure because of the availability of data at any point. Blockchain data, once placed in the block, cannot be changed or deleted. Altering data is not possible on the blockchain. In blockchain, each block is connected to its previous block by the cryptographic hash pointer. This hash is known as the previous hash in the blockchain's structure. This part defines the structure and working of the blockchain, cloud storage, and connectivity of the IoT devices with the cloud. At the end of the introduction part, the goals of the chapter are defined along with the organization of the chapter.

Blocks in the blockchain comprise the data structure known as the Markle tree. This tree holds the list of transactions in binary tree format. The transaction is converted into cryptographic hash form before compiling on the blockchain. This transaction is headed by the root known as the Markle root in the block. Other block structures included the timestamp for the recording of the block creation time and verification process in future transactions. Block headers consist of the previous hash (block header of the previous block), nonce, and block index. Nonce is defined as the number only used once in any blockchain for identification of the block's uniqueness. All these data are compressed with one hash known as the block hash. These block hashes are used to point to the next and previous blocks. One block is connected to another, making a chain. This structure is known as the blockchain.

In this smart city environment, IoT and cloud computing are playing a major role in setting up the data collection and processing unit. IoT devices have small storage and low computation power. So, for making a blockchain node, sometimes it is not useful. Blockchain required large storage in some cases and high computations. Smart city environments connect smart devices to the cloud for data storage and computation. These gadgets periodically transmit data to the cloud servers. The blockchain environments can make use of these cloud servers as nodes. Large data sets can be stored in clouds, where they can then be processed as block-chain transactions. In smart cities, energy conservation and metering can be done using smart meters [1, 2]. The loss and quality of services related to the smart meters must be accurate because once the data is committed to a blockchain, it cannot be deleted or changed.

The organization of the chapter consists of the literation review in section "Literature Review". The proposed scheme and theoretical concepts are given in section "Proposed System". Experimental setup, results and comparisons are given in section "System Implementations". Section "Conclusion and Future Scope" concludes the chapter with future scope.

Literature Review

In their thorough study of the literature, the development of blockchain technology in Singapore, as well as its current and possible future applications, were highlighted by Hasankhani et al. [3]. This study aims to describe SG's organizational structure in the context of blockchain applications and discuss the benefits and drawbacks of blockchain across all SG disciplines. The explanation of the blockchain-based multiple energy trading (B-MET) system for carrying out a smart contract by Guo and others [4]. Ante et al. [5] have completed yet another analysis of the energy industry. In this study, they use co-citation analysis to examine the relationship between blockchain and energy using the underlying references from 166 publications. Six different study streams are being used in exploratory factor analysis. Stekli et al. [6] discussed the various ways DERs can participate in the electricity market, the architectures and methodologies used to model their participation, the past, present, and future development of communications technologies that make this participation possible, and commercial applications of these approaches. In their publication [7], Cantillo-Luna et al. present a comprehensive examination of blockchain applications for DER administration and integration. With the aforementioned frameworks, Dinesha and Balachandra [8] investigate the potential for creating blockchain-enabled smart microgrids (BSMG). It intends to create the transaction protocols and process flows for the BSMG conceptual framework. Wu et al. [9] examine the socio-technical interaction and mechanism for sustainable P2P energy trading from the social perspective of multi-level market players cooperating, the technical perspective of cutting-edge flexibility exchange, and the economic perspective of cooperative decentralized or regulated marketplaces. Using Hyperledger technology, Yan et al. [10] present a blockchain-based trustworthy collecting system architecture for smart meter quality data that incorporates smart meters, edge computing nodes, blockchain, and data storage platforms. Second, in order to realize automation of data collection and enhance data dependability, the fundamental contract, identity contract, and data contract must be meticulously established. Mukherjee et al.'s [11] proposal of an e-chain architecture secured by blockchain technology aims to accomplish smart energy consumption. The confidentiality and privacy of the destruction data kept and assessed for the same are all guaranteed by this architecture, along with efficient generation, uniform distribution, and shielded monitoring. Peer-to-peer energy trading amongst all prosumers (as consumers and sellers) and retailers is made possible by Mehdinejad et al.'s [12] design of a prosumer and retailer competitive market. Without the help of outside parties or the disclosure of player privacy information, the intended market is cleared using the novel decentralized primal–dual sub-gradient algorithm (PDSGA). They evaluate current cloud computing models' use of blockchain-related research trends. A thorough examination of how blockchain is used in the cloud computing concept to offer security services was provided by Zou et al. [13]. Sharma et al.'s [14] probabilistic approach simplifies data routing between nodes and local clouds in an IoT network utilizing a multitier trust and encryption technique for secure data delivery. The suggested solution outperforms industry-standard protocols like PRoPHET, MaxProp, ProWait, and TCAFE in attack resilience and routing efficiency in simulations across two data sets, emphasizing Sec-CIoT operations. The Hewa et al. [15] service makes it simpler to protect the confidentiality of equipment-cloud communications and to authenticate CM equipment while upholding anonymity and unlikability over the blockchain. They developed the suggested architecture using Hyperledger Fabric and evaluated its performance advantages over leading-edge alternatives. Chaganti et al. [16] offer a cloud-enabled smart-farm security monitoring system to effectively monitor device status and sensor anomalies and to avoid security assaults by using behavioral patterns. The security-monitoring-framework prototype for smart farms was developed, and network latency was evaluated, using the Arduino Sensor Kit, ESP32, AWS cloud, and the smart contract on the Ethereum Rinkeby Test Network. Cha et al. [17] used a distributed approach to increase data integrity and security in external cloud services. CSPs use blockchains to verify user data and simplify data access. CSP uses the blockchain and Secret Sharing algorithm to securely store user data in a distributed manner that is more secure than centralized systems. The proposed solution outperforms earlier studies in transaction speed and data storage efficiency and improves security and privacy through security analysis. Rezaeimozafar et al. [18] examined BTM ESSs placed in end users' premises, related technologies, billing and pricing systems, and their prospective capabilities from system operators' and end users' viewpoints. Several BTM case studies and a brief but detailed examination of optimization strategies for BTM energy management challenges are included. Finally, BTM systems implementation issues in the power system are investigated, and future study paths and new solutions are offered.

Zhao et al. [19] present reviews of smart meter operating systems and associated programming methodologies. Future research might be guided by the analysis of the problem of energy theft and the available detection methods by Xia et al. [20]. Olivares-Rojas et al. [21] proposed a novel cybersecurity architecture for smart metering systems based on blockchains in order to significantly increase data security. In order to improve performance, the proposed architecture is divided into various tiers and uses edge, cloud, and fog computing models. The authors also offer smart meter proofs of effectiveness and an enhanced consensus method. Their results show that the suggested architecture may greatly improve data security and cybersecurity for smart metering systems when put into practice. By using the Proof-of-Efficacy algorithm, consensus is achieved. A Survey of Blockchain Applications in the Energy Sector is provided by Bao et al. [22]. In this survey, they go over the numerous blockchain applications in the energy industry. Tian et al. [23] state that four examples are looked at in order to forecast energy on various scales in the energy industry. Aslam et al. [24] provide reviews of smart meter operating systems and associated programming methodologies. The approach proposed by Gonçalves et al. [25] provides data privacy protection for the covariance matrix and coefficients in addition to supporting distributed model estimation. Additionally, during model fitting, asynchronous peer communication is addressed, and both centralized and peer-to-peer collaborative models are taken into consideration. Utilizing K-means or socioeconomic affinities, the architecture can be used to classify consumers based on similarities in their consumption patterns. Sav and Olivadese [26] developed a local training technique that uses calendar variables (hour, weekday, and average consumption of past days) and weather-related data (clustering apparent temperature). Uniejewski and Weron [27] show that its predictive performance is superior, particularly when the r regularization parameter is selected ex-ante using the Bayesian Information Criterion (BIC). The Bayesian Information Criterion (BIC) is used to pre-select the regulation parameter.

Proposed System

The chapter proposed an energy management architecture based on blockchain technology. In this chapter, we propose a smart energy unit recording in blockchain and their billing system. The units are recorded by the smart meters, which send them to the cloud servers. The entire process of the proposed system is divided into three parts. The first part deals with the smart meters, the cloud connection, and data collection. Second, it deals with cloud-to-blockchain data transactions and approvals. The third part is the bill generation part. The consensus algorithm is explained in the second part, and based on the consensus algorithm, a blockchain data bill is generated for the third part. The overall architecture of the proposed system is shown in Fig. 1.

Smart Meters and Clouds

In this section, the smart meters are acting as IoT devices and monitoring the unit of electricity burned per hour. In each hour, the meters send data to the cloud servers allocated to them. This data collection is done through the open channel and public network. So, the security and privacy of the meter identity, data attached to the meter, and cloud servers are important at this point. The smart meter is denoted as the IoTSM, and cloud servers are denoted as the CN. The symbols and abbreviations used in this chapter are detailed in Tables 1 and 2. To connect the smart meters with the cloud nodes, secure channel creation is required. This channel is created using authentication and message passing between the smart meters and cloud nodes. Message passing and verification include key generation and login phases for smart meters and cloud nodes. Key generation and verification of the identities are done using the central key generation and verification authority (CKGVA). This CKGVA provides keys to the smart meters and cloud nodes and controls the registration and

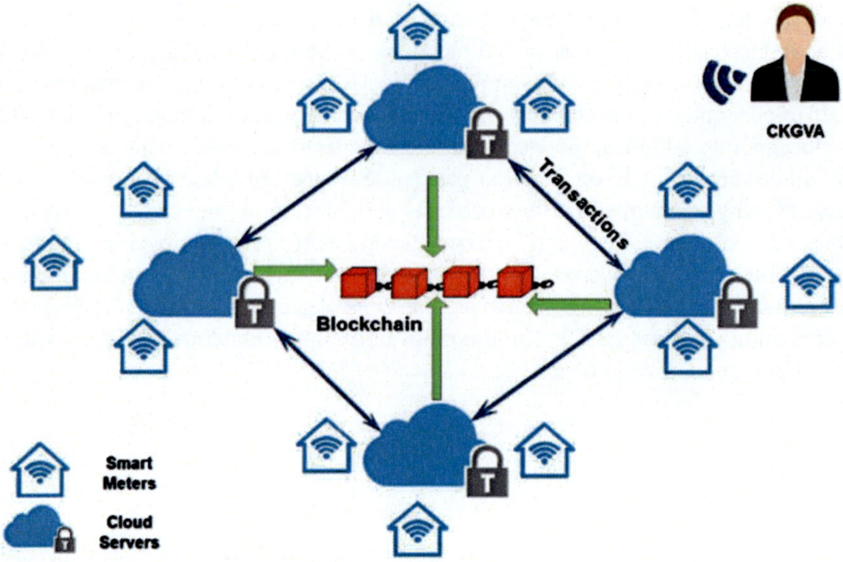

Fig. 1 Overall system architecture of proposed model

login processes. Figure 2 illustrates the key generation and registration processes of the smart meters and cloud servers with the CKGVA.

Smart Meter Registration

Let the identity of the smart meter (SM) be IDSM. SM sends the request for the pseudo-identity generation of the smart meters and the computation of the keys for registration and login. Let the random key generated by the CKGVA be X. So, the SM sends the request for the pseudo-identity generation in the following form:

$$SM\,(ID_{SM})\xrightarrow{(ID_{SM}\,,TSP_1)}CKGVA$$

CKGVA computes

Table 1 Symbols used in this chapter

Symbol	Description	Symbol	Description
IoT_{SM}	Smart meter	$\|\|$	Concatenation
CN	Cloud server	R_{num}	Random number
ΔT	Time delay	PID_{SM}	Pseudo identity of smart meter
ID_{SM}	Identity of smart meter	$hash256(x)$	SHA-256 of x
X	Random key generated by X	$PvtKey(x)$	Private key of x
TSP_x	Time stamp x	$PubKey(x)$	Public key of x

Table 2 List of abbreviations in this chapter

Abbreviation	Details	Abbreviation	Details
EPoCon	Enhanced proof of energy consumption	BCM	Byzantine-based consensus mechanism
Ed25519	Edward curve 25519	DES	Distributed energy system
IoT	Internet of Things	DER	Distributed energy resources
SM	Smart meters	BSMG	Blockchain enabled smart microgrids
CN	Cloud server	PDSGA	Primal–dual sub-gradient algorithm
B-MET	Blockchain-based multiple energies trading	BTM	Behind-the-meter
ESS	Energy storage system	CKGVA	Central key generation and verification authority

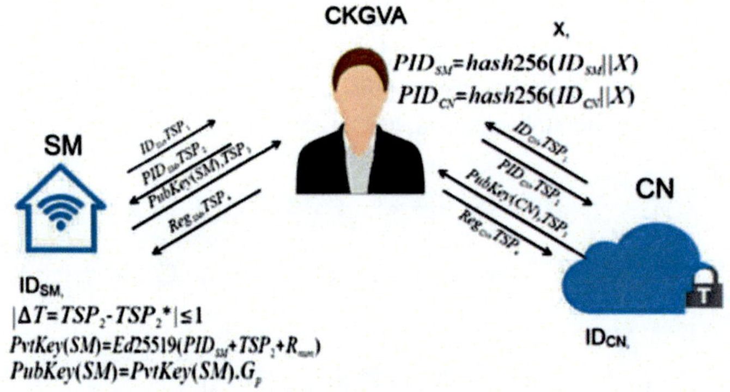

Fig. 2 Registration phase of cloud and smart meter with CKGVA

$$PID_{SM} = hash256(ID_{SM}||X) \quad (1)$$

$$SM \xleftarrow{(PID_{SM}, TSP_2)} CKGVA$$

At SM the time stamp difference is calculated as the $|\Delta T = TSP_2 - TSP_2^*| \leq 1$. If this is true, then the key generation for SM:

$$PvtKey(SM) = Ed25519(PID_{SM} + TSP_2 + R_{num}) \quad (2)$$

$$PubKey(SM) = PvtKey(SM).G_p \quad (3)$$

Now the SM sends the public key with the timestamp to the CKGVA

$$SM \xrightarrow{(PubKey(SM),TSP_3)} CKGAV$$

CKGAV calculates the time delay at the receiving end as $|\Delta T = TSP_3 - TSP_3^*| \leq 1$. If the value is true then the calculation of registration identity of the SM is done as follows:

$$Reg_{SM} = hash256(PubKey(SM) + TSP_3 + PvtKey(CKGAV)) \quad (4)$$

Now the CKGAV sends the registration identity of the smart meter and the all the authority of the blockchain with available cloud nodes too.

$$SM \xleftarrow{(Reg_{SM},TSP_4)} \text{CKGAV}$$

$$CN_i \xleftarrow{(Reg_{SM},TSP_4)} \text{CKGAV}$$

Cloud Registration

For the cloud node registration, the identity of the node is ID_{CN}. This CN sends this to the CKGVA to generate pseudo-identities of the CN.

$$CN \xrightarrow{(ID_{CN},TSP_1)} CKGVA$$

Now the CKGVA calculates the pseudo identity of the CN by using X, and send it to the CN.

$$PID_{CN} = hash256(ID_{CN}||TSP_1||X) \quad (5)$$

Now CKGVA sends this PID_{CN} to the CN.

$$CN \xleftarrow{(PID_{CN},TSP_2)} CKGVA$$

CN verifies the received message by using timestamp check $|\Delta T = TSP_2 - TSP_2^*| \leq 1$. If verification is true then the CN accepts the $PID_{CN} and$ computes the key pair.

$$PvtKey(CN) = Ed25519(PID_{CN}||TSP_2||X) \quad (6)$$

$$PubKey(CN) = PvtKey.G_p \quad (7)$$

Now the CN sends the public key to the CKGVA for generation of registration identity of the CN.

$$CN \xrightarrow{(PubKey(CN),TSP_3)} CKGVA \tag{8}$$

Now verifies the received message by using timestamp check $|\Delta T = TSP_3 - TSP_3^*| \leq 1$. If verification is true then the CKGVA computes the registration identity of the CN as follows:

$$Reg_{CN} = hash256(PubKey(SM) + TSP_3 + PvtKey(CKGAV))$$

This information is sends to the all-online cloud nodes by CKGVA.

$$CN_i \xleftarrow{(Reg_{SM},TSP_4)} CKGVA$$

Login of Cloud and Smart Meters

For the login of cloud and smart meters, the credential created and shared with the cloud are used. For the smart meter login with cloud node, SM sends the login request to the cloud node and cloud node verifies the request by signature and timestamp on the request.

$$CN \xleftarrow{(PID_{SM},TSP_L,Sign_{Req})} SM$$

Algorithm 1 Registration of Smart meters and Cloud

Input: Identity of Smart Meter (ID_{SM}) and Identity of Cloud Node ID_{CN}
Output: Registration of SM to CN
Step-1: SM and CN sends their identity to the CKGVA at TSP_1
Step-2: at CKGVA if ($|\Delta T = TSP_1 - TSP_1^*| \leq 1$) then
Pseudo identity of SM and CN is calculated as
$PID_{SM} = hash256(ID_{SM} \| X)$ & $PID_{CN} = hash256(ID_{CN} \| X)$
And send to the CN and SM at TSP_2
Step-3: at CN and SM if ($|\Delta T = TSP_2 - TSP_2^*| \leq 1$) then
Pseudo key pair are generated for the CN and SM
$PvtKey(SM) = Ed25519(PID_{SM} + TSP_2 + R_{num})$, $PubKey(SM) = PvtKey(SM).G_p$ and
$PvtKey(CN) = Ed25519(PID_{CN} \| TSP_2 \| X)$, $PubKey(CN) = PvtKey.G_p$

And Public key is end to the CKGVA TSP_3
Step-4: at CKGVA if ($|\Delta T = TSP_3 - TSP_3^*| \leq 1$) then
Registration Number of SM and CN is calculated as
$Reg_{SM} = hash256(PubKey(SM) + TSP_3 + PvtKey(CKGAV))$ and
$Reg_{CN} = hash256(PubKey(SM) + TSP_3 + PvtKey(CKGAV))$

This registration identity is shared with the blockchain and all node in the blockchain.

CN computes the time delay as $|\Delta T = TSP_L - TSP_L^*| \leq 1$. If the verification is true, then the CN granted for the login. The collection of data is started from this point till device get disconnected or getting off. Pseudo code of the registration and login process is detailed in the Algorithm 1.

Cloud and Blockchain Level

The cloud nodes are now sending data to the authority nodes. Since this is a private blockchain, the rights to transparency are maintained by the authority nodes in the blockchain. The consensus algorithm used here is Enhanced Proof of Energy Consumption. The pseudocode of the algorithm is given in Algorithm 2. The cloud sends the transition with the meter registration ID and the energy consumption unit in the transaction for each meter. At the authority node, the consumption is verified by all the authority nodes, but at least (n − 1)/2 nodes. The transaction approval is based on the verification of the power consumed by the customer at the smart meter level. The transaction is signed by the cloud node, and at the authority node it is verified for authentication. The approved transaction is added to the current unmined block. The blocks are mined after each minute. If the Markle tree is unbalanced because of an insufficient number of transactions, then the extra transactions are added to the next block and the current block is mined.

Algorithm 2 Enhanced Proof-of-Energy Consumption (EPoECon) Algorithm

Input: Energy units from CN

Output: Approved Transactions and Block
Step-1: Energy units are received from the cloud server
Step-2: CN send the T_x with the meter id to the authority node
Step-3: Authority node verifies if ($|\Delta T = TSP_x - TSP_x^*| \leq 1$) and verified by (n-1)/2 members then
The consumption unit is regulated and verified by the authority and this $Verified(T_x)$ message send to all nodes
Step-4: Transaction is added into the current block.
$BL = (T_x^1 + T_x^2 \ldots..)$
Step-5: The block is formed in every ten minutes by adding all available transactions
$BL = BL(Previous) + 1$

Bill Generation Process

From the collected data on the blockchain, the bill is calculated on a weekly and monthly basis. The collected data are selected by the meter ID without regard to the cloud server node. The amount of the bill is calculated using the tariff set by the authority node.

System Implementations

This part comprises the first part of the implementation of the IoT cloud security protocol verification and data collection from the IoT devices to the cloud servers. The second part of implementation comprises the creation of the blockchain, transactions, and their verification. The last and third part of implementation discusses bill generation from the blockchain. Implementation parameters are described in Table 3.

IoT Cloud Security Verification and Data Collection from Smart Meters

We are using scyther tool for the security protocol verification between the smart meters and cloud servers. Input of the protocol is shown in Figs. 3 and 4. The security protocols mentioned above are verified and results are discussed in the experimental results section of this chapter.

Blockchain Implementation

The blockchain is implemented with custom python-based blockchain with Couchdb. Couchdb is used for the distributed ledger and the consensus algorithm is written in the python. The energy consumption is verified at the python codes and the bill is generated form the blockchain. Blockchain consist of the wallet generation process, transaction process and verification process. To generate wallet for the nodes, Ed25519 is used in the. Example of the key generation is shown in Fig. 5. For implementation of proposed system, [28–30] are taken as reference.

Transaction is done between one cloud node, which is collecting data from the smart meters. These cloud nodes are sending the transaction to the authority nodes. Since this is a private blockchain architecture, the authority nodes rea known. The transaction format is shown in Fig. 6.

Bill Generation from the Blockchain

The bill generated from the approved transaction is done in a time interval. We set the time interval for the 72 h for unit monitoring and 15 days for the bill generation process. The user can check the updated unit of the energy consumption in each 72 h and pay the generated bill in 15 days of interval.

Table 3 System implementation parameters

System part	Details
OS	Windows-10, 64-Bit
Processor	Intel i7, 10 Generation
Blockchain	Python-based custom blockchain
Distributed database	CouchDB V 3.2.2
Security verification tool	Scyther

```
const exp: Function; const hash: Function; hashfunction h; const XOR:Function;
const hl:Function; const plus:Function;const mod:Function;
protocol devicelogin(SM,CKGVA)
{
role SM {
const RegIDi, Pi, Bi, SIDk,k,s,b,g,p;
fresh N1: Nonce;
var N2: Nonce;
macro b = h(b);
macro k = mod(exp(g,b),p); macro Pij = h(XOR(XOR(h(RegIDi), h(h(RegIDi),
h(s))), plus(h(N2),1)),h(RegIDi,k)); macro Lij = XOR(h(N2) , h(RegIDi,k));
send_1(SM,CKGVA, XOR(h(SIDk,N1), h(RegIDi)),N1);//C1
recv_2(CKGVA, SM, XOR(h(N2), h(N1,1)));//C2
send_3(SM,CKGVA,h(XOR(XOR(h(RegIDi),
h(h(RegIDi),
h(s))),
plus(h(N2),1)),h(RegIDi,k)));//Pij
send_4(SM,CKGVA,Lij);
claim_i1(SM, Secret, XOR(h(SIDk,N1), h(RegIDi)));//C1
claim_i2(SM,Secret,XOR(h(N2), h(N1,1)));//C2
claim_i3(SM,Secret,h(XOR(XOR(h(RegIDi),
h(h(RegIDi),
h(s))),
plus(h(N2),1)),h(RegIDi,k)));//Pij
claim_i4(SM, Secret, h(s)); claim_i10(SM,Secret,k); claim_i11(SM,Secret,h(RegIDi));
claim_i12(SM,
Secret,N1);
claim_i5(SM,
Secret,
h(N2));
claim_i13(SM,Secret,Lij);
claim_i6(SM,
Niagree);
claim_i7(SM,Nisynch);claim_i8(SM,
Alive);claim_i9(SM,Weakagree);
claim_i10(SM, Commit, CKGVA,N1,N2);
}
role CKGVA{
const RegIDi,Pi,N2,Bi, SIDk,k,s,b,g,p;
var N1:Nonce; fresh N2: Nonce;
recv_1(SM,CKGVA, XOR(h(SIDk,N1), h(RegIDi)),N1);//C1
send_2(CKGVA, SM, XOR(h(N2), h(N1,1)));//C2
recv_3(SM,CKGVA,
h(XOR(XOR(h(RegIDi),
h(h(RegIDi),
h(s))),
plus(h(N2),1)),h(RegIDi,k)));//Pij
recv_4(SM,CKGVA,Lij);
claim_r13(CKGVA, Secret,Lij); //Lij
claim_r1(CKGVA,  Secret, XOR(h(SIDk,N1), h(RegIDi)));//C1
claim_r2(CKGVA,  Secret, XOR(h(N2), h(N1,1)));//C2
claim_r2(CKGVA,
Secret,
h(XOR(XOR(h(RegIDi),
h(h(RegIDi),
h(s))),
plus(h(N2),1)),h(RegIDi,k)));//Pij
claim_r3(CKGVA,
Secret,h(s));
claim_r10(CKGVA, Secret,k);
claim_r4(CKGVA, Secret,h(RegIDi));
claim_r5(CKGVA,  Secret, h(N2)); claim_r6(CKGVA , Alive); claim_r7(CKGVA, Niagree);
claim_i10(CKGVA,  Running, SM,N1,N2);claim_r8(CKGVA, Nisynch);
claim_r9(CKGVA,  Weakagree);
}}
```

Fig. 3 Scyther input for security protocol verification between CKGVA and SM

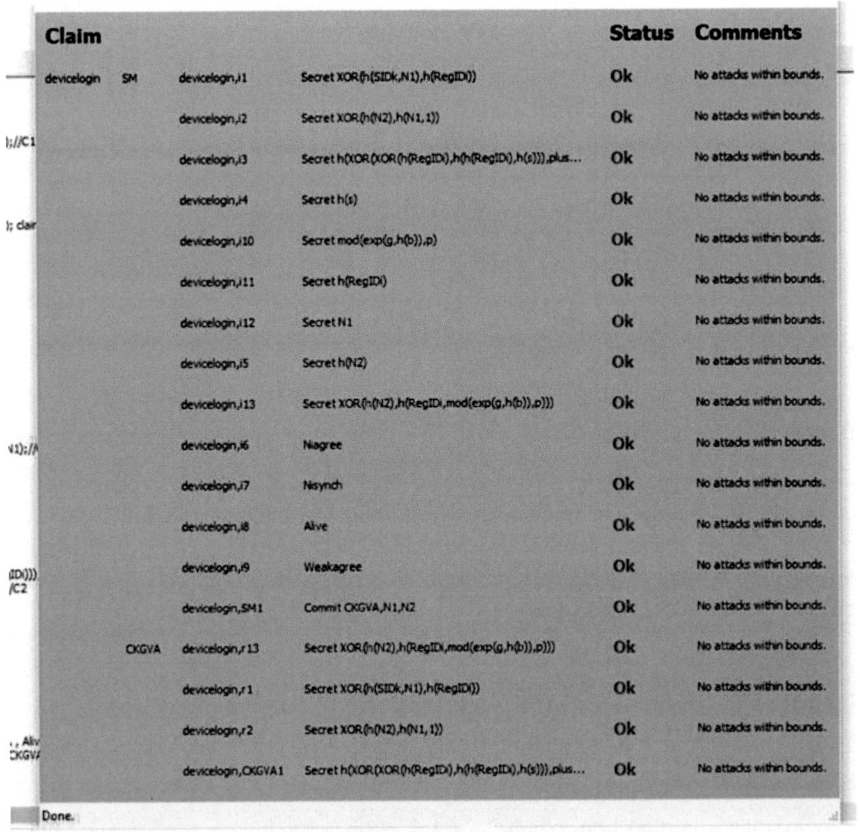

Fig. 4 Scyther verification result for the smart meter and CKGVA

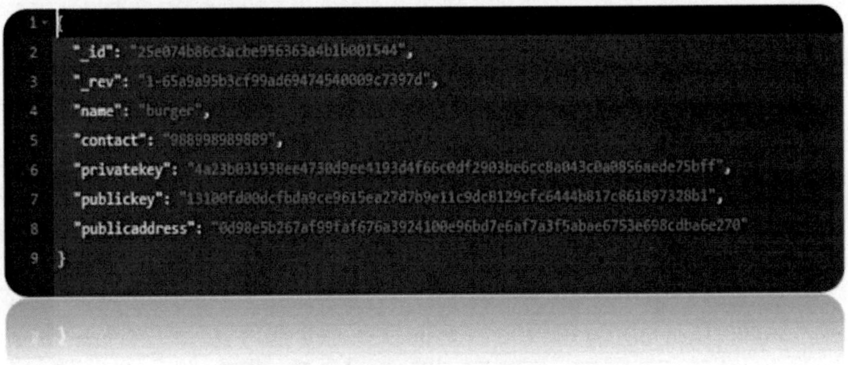

Fig. 5 Wallet generation for the smart meter in the blockchain

Fig. 6 Transaction example in CouchDB

Results and Comparisons

The experimental results are divided into three parts. First part is security simulation results. Second is blockchain results and the third one is bill generation results. The blockchain results are comprises the transaction speed, transactions per second, signature speed and verification speed, number of messages in secure channel establishment and transaction verification speed.

Security Simulation Results

From the Fig. 4, we can see that there are no security attacks presented in our proposed scheme. Security analysis of the proposed scheme is detailed in the next section.

Blockchain Results

This part is divided into the following subsections:

- Transaction Speed: The transaction speed of the proposed system is 0.071 s. The proposed scheme taking less time with compare to the other developed systems. This comparison is given in the Fig. 7.
- Transaction Per Second (TPS): There are approximately 14 transactions per seconds are generated. Comparison of the transactions per seconds are given in the Fig. 8.
- Signature and Verification Speed: Signing speed of the node on the transactions is 0.014 s and verification speed is 0.037 s. The comparison of the signing and verification speed of the proposed system with the previously developed system are given in the Fig. 9.

Message Counts: Total messages required for the communication are five. For the registration phase, device and CKGVA, message is 4 and 1 message is required for the login process. These messages are illustrated in Fig. 10.

Time versus Transaction graph is shown in Fig. 11, where we can see that the number of transactions load is increasing as the nodes and time are increasing.

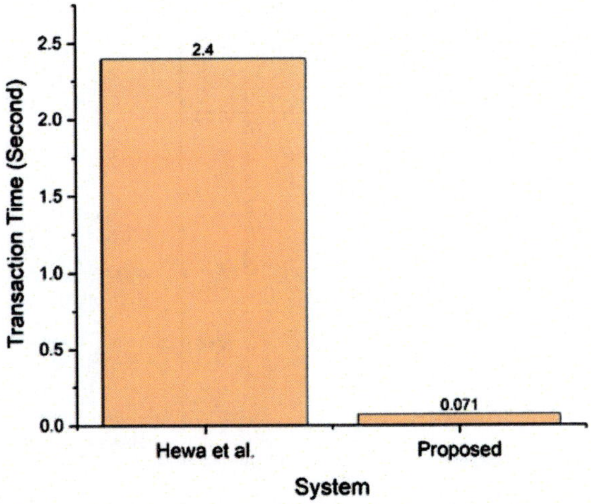

Fig. 7 Transaction time for first transaction

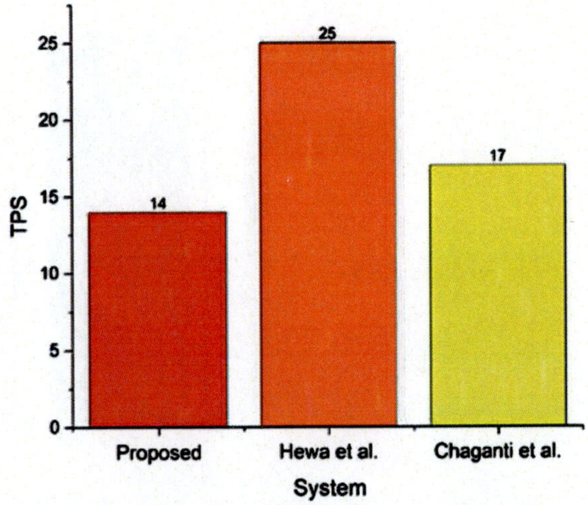

Fig. 8 Transactions per second comparison of chart

From Table 4, we can see that our proposed system is more scalable, flexible, reliable, secure and user friendly. It has property like security and privacy of user data. This system is specially designed for the energy meters in the smart cities.

Conclusion and Future Scope

Smart city needs smart services facilities. In order to maintain the smart facilities in the developing smart cities. We need to provide ground breaking technology solutions. Blockchain is an innovative technology, which is famous for its features

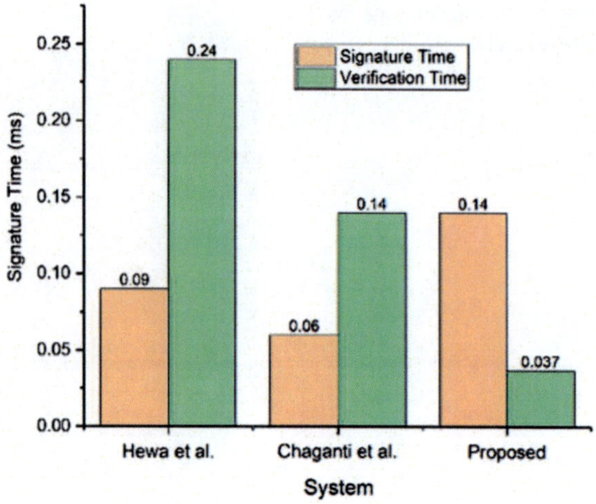

Fig. 9 Signature and verification time compare for proposed system with state-of-the-art systems

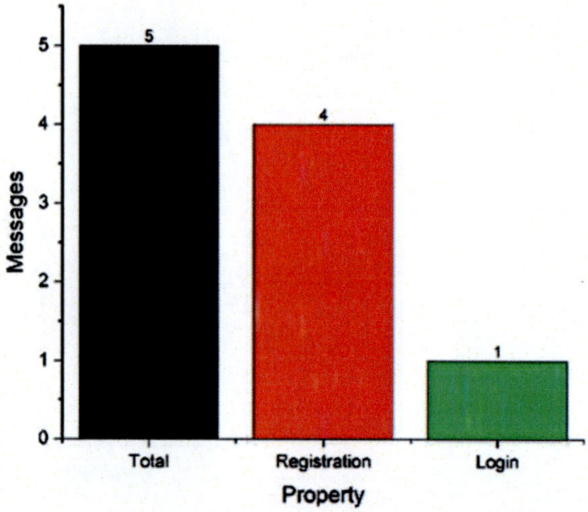

Fig. 10 Number of messages during communication

like immutability and transparency. Our Proposed system maintains the security and privacy during each transaction and message passing process. The key generation process, device registration process, authentication of messages and transactions, and verification of signatures are done using the Ed25519 algorithm. Since our proposed system is novel and based on EPoECon and blockchain, we have only a few papers for the comparison section. But according to the comparisons, our proposed system is better in terms of device privacy, security, and security of the messages. Our proposed scheme also provides security analysis and verification of the security protocol proposed in this chapter.

In future goals, these schemes can be used in the E-governance schemes to control the different services and manage data in the blockchain-based e-governance system.

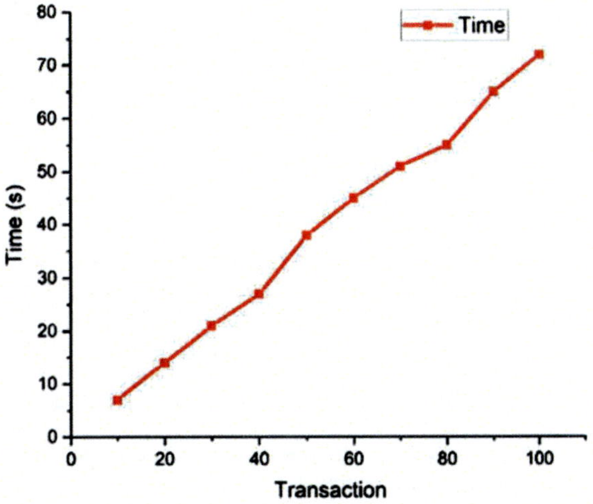

Fig. 11 Transaction versus time graph for proposed system

Table 4 Comparison of proposed system with state-of-the-art systems

Property	Proposed	Guo et al. [4]	Mehdinejad et al. [12]	Hewa et al. [15]	Chaganti et al. [16]
Blockchain	Yes	Yes	Yes	Yes	Yes
Platform	Python-based custom blockchain	NA	NA	Hyperledger fabric	Ethereum
Consensus	Enhanced proof of energy consumption (EPoCon)	Byzantine based consensus mechanism	Markle-based algorithm	Raft	PoS
Type	Private	Private	Private	Private	Public
Security analysis	Yes	No	No	Yes	Yes
Privacy analysis	Yes	No	No	No	Yes
Device verification	Yes	No	No	Yes	Yes

The real time implementation with smart city infrastructure can be done with the proposed scheme. The readers and researchers can use the proposed scheme with the integration in the data science and IoT to for smart city governance. Data analysis and rectification in the present system can make more easy for the citizen of smart cities to use these technology and scheme.

References

1. Yang, Q., & Wang, H. (2021). Privacy-preserving transactive energy management for IoT-aided smart homes via blockchain. *IEEE Internet of Things Journal, 8*(14), 11463–11475. https://doi.org/10.1109/JIOT.2021.3051323
2. Li, D., Deng, L., Cai, Z., & Souri, A. (2022). Blockchain as a service models in the Internet of Things management: Systematic review. *Transactions on Emerging Telecommunications Technologies, 33*, e4139. https://doi.org/10.1002/ett.4139
3. Hasankhani, A., Hakimi, S. M., Bisheh-Niasar, M., Shafie-khah, M., & Asadolahi, H. (2021). Blockchain technology in the future smart grids: A comprehensive review and frameworks. *International Journal of Electrical Power & Energy Systems, 129*, 106811.
4. Guo, J., Ding, X., & Wu, W. (2022). An architecture for distributed energies trading in Byzantine-based blockchains. *IEEE Transactions on Green Communications and Networking, 6*(2), 1216–1230. https://doi.org/10.1109/TGCN.2022.3142438
5. Ante, L., Steinmetz, F., & Fiedler, I. (2021). Blockchain and energy: A bibliometric analysis and review. *Renewable and Sustainable Energy Reviews, 137*, 110597. ISSN 1364-0321. https://doi.org/10.1016/j.rser.2020.110597
6. Stekli, J., Bai, L., Cali, U., Halden, U., & Dynge, M. F. (2022). Distributed energy resource participation in electricity markets: A review of approaches, modeling, and enabling information and communication technologies. *Energy Strategy Reviews, 43*, 100940. ISSN 2211-467X. https://doi.org/10.1016/j.esr.2022.100940
7. Cantillo-Luna, S., Moreno-Chuquen, R., Chamorro, H. R., Sood, V. K., Badsha, S., & Konstantinou, C. (2022). Blockchain for distributed energy resources management and integration. *IEEE Access, 10*, 68598–68617. https://doi.org/10.1109/ACCESS.2022.3184704
8. Dinesha, D. L., & Balachandra, P. (2022). Conceptualization of blockchain enabled interconnected smart microgrids. *Renewable and Sustainable Energy Reviews, 168*, 112848. ISSN 1364-0321. https://doi.org/10.1016/j.rser.2022.112848
9. Wu, Y., Wu, Y., Cimen, H., Vasquez, J. C., & Guerrero, J. M. (2022). P2P energy trading: Blockchain-enabled P2P energy society with multi-scale flexibility services. *Energy Reports, 8*, 3614–3628. ISSN 2352-4847. https://doi.org/10.1016/j.egyr.2022.02.074
10. Yan, L., Angang, Z., Huaiying, S., Lingda, K., Guang, S., & Shuming, S. (2021). Blockchain-based reliable collection mechanism for smart meter quality data. In X. Sun, X. Zhang, Z. Xia, & E. Bertino (Eds.), *Artificial intelligence and security. ICAIS 2021. Lecture notes in computer science* (Vol. 12737). Springer. https://doi.org/10.1007/978-3-030-78612-0_38
11. Mukherjee, P., Barik, R. K., & Pradhan, C. (2021). eChain: Leveraging toward blockchain technology for smart energy utilization. In R. Kumar, R. K. Dohare, H. Dubey, & V. P. Singh (Eds.), *Applications of advanced computing in systems. Algorithms for intelligent systems*. Springer. https://doi.org/10.1007/978-981-33-4862-2_8
12. Mehdinejad, M., Shayanfar, H., & Mohammadi-Ivatloo, B. (2022). Peer-to-peer decentralized energy trading framework for retailers and prosumers. *Applied Energy, 308*, 118310. ISSN 0306-2619. https://doi.org/10.1016/j.apenergy.2021.118310
13. Zou, J., He, D., Zeadally, S., Kumar, N., Wang, H., & Choo, K. R. (2021). Integrated blockchain and cloud computing systems: A systematic survey, solutions, and challenges. *ACM Computing Surveys, 54*(8), Article 160. https://doi.org/10.1145/3456628
14. Sharma, D. K., Bhardwaj, K. K., Banyal, S., Gupta, R., Gupta, N., & Nkenyereye, L. (2022). An opportunistic approach for cloud service-based IoT routing framework administering data, transaction, and identity security. *IEEE Internet of Things Journal, 9*(4), 2505–2512. https://doi.org/10.1109/JIOT.2021.3078810
15. Hewa, T., Braeken, A., Liyanage, M., & Ylianttila, M. (2022). Fog computing and blockchain-based security service architecture for 5G industrial IoT-enabled cloud manufacturing. *IEEE Transactions on Industrial Informatics, 18*(10), 7174–7185. https://doi.org/10.1109/TII.2022.3140792

16. Chaganti, R., Varadarajan, V., Gorantla, V. S., Gadekallu, T. R., & Ravi, V. (2022). Blockchain-based cloud-enabled security monitoring using Internet of Things in smart agriculture. *Future Internet, 14*(9), 250. https://doi.org/10.3390/fi14090250
17. Cha, J., Singh, S. K., Kim, T. W., & Park, J. H. (2021). Blockchain-empowered cloud architecture based on secret sharing for smart city. *Journal of Information Security and Applications, 57*, 102686. ISSN 2214-2126. https://doi.org/10.1016/j.jisa.2020.102686
18. Rezaeimozafar, M., Monaghan, R. F.D., Barrett, E., & Duffy, M. (2022). A review of behind-the-meter energy storage systems in smart grids. *Renewable and Sustainable Energy Reviews, 164*, 112573. ISSN 1364-0321. https://doi.org/10.1016/j.rser.2022.112573
19. Zhao, X., Zhang, B., Yin, S., Xie, Q., Zhang, L., & Wu, H. (2022). Discussion on application of embedded operating system in dual-core smart electric meter. *Scientific Programming, 2022*, 1880811. https://doi.org/10.1155/2022/1880811
20. Xia, X., Xiao, Y., Liang, W., & Cui, J. (2022). Detection methods in smart meters for electricity thefts: A survey. *Proceedings of the IEEE, 110*(2), 273–319. https://doi.org/10.1109/JPROC.2021.3139754
21. Olivares-Rojas, J. C., Reyes-Archundia, E., Gutiérrez-Gnecchi, J. A., Cerda-Jacobo, J., & González-Murueta, J. W. (2020). A novel multitier blockchain architecture to protect data in smart metering systems. *IEEE Transactions on Engineering Management, 67*(4), 1271–1284. https://doi.org/10.1109/TEM.2019.2950410
22. Bao, J., He, D., Luo, M., & Choo, K.-K.R. (2021). A survey of blockchain applications in the energy sector. *IEEE Systems Journal, 15*(3), 3370–3381. https://doi.org/10.1109/JSYST.2020.2998791
23. Tian, J., Li, K., & Xue, W. (2021). An adaptive ensemble predictive strategy for multiple scale electrical energy usages forecasting. *Sustainable Cities and Society, 66*, 102654. ISSN 2210-6707. https://doi.org/10.1016/j.scs.2020.102654
24. Aslam, S., Herodotou, H., Mohsin, S. M., Javaid, N., Ashraf, N., & Aslam, S. (2021). A survey on deep learning methods for power load and renewable energy forecasting in smart microgrids. *Renewable and Sustainable Energy Reviews, 144*, 110992. ISSN 1364-0321. https://doi.org/10.1016/j.rser.2021.110992
25. Gonçalves, C., Bessa, R. J., & Pinson, P. (2021). Privacy-preserving distributed learning for renewable energy forecasting. *IEEE Transactions on Sustainable Energy, 12*(3), 1777–1787. https://doi.org/10.1109/TSTE.2021.3065117
26. Savi, M., & Olivadese, F. (2021). Short-term energy consumption forecasting at the edge: A federated learning approach. *IEEE Access, 9*, 95949–95969. https://doi.org/10.1109/ACCESS.2021.3094089
27. Uniejewski, B., & Weron, R. (2021). Regularized quantile regression averaging for probabilistic electricity price forecasting. *Energy Economics, 95*, 105121. ISSN 0140-9883. https://doi.org/10.1016/j.eneco.2021.105121
28. Dewangan, N. K., & Chandrakar, P. (2022). Patient-centric token-based healthcare blockchain implementation using secure Internet of Medical Things. *IEEE Transactions on Computational Social Systems*. https://doi.org/10.1109/TCSS.2022.3194872
29. Dewangan, N. K., Chandrakar, P., Kumari, S., et al. (2022). Enhanced privacy-preserving in student certificate management in blockchain and interplanetary file system. *Multimedia Tools and Applications*. https://doi.org/10.1007/s11042-022-13915-8
30. Dewangan, N. K., & Chandrakar, P. (2021). Enhanced privacy and security of voters' identity in an interplanetary file system-based e-voting process. In Blockchain for information security and privacy (pp. 113–132). Auerbach Publications.

A Comparative Study of Threat Detection for IoT Devices Using Machine Learning Techniques

Gowri Priya and K. V. Greeshma

Abstract "Cybersecurity threats" comprise computer viruses, data breaches, DoS, and other attack methods. These threats try to steal data, corrupt data, or otherwise disrupt digital life. IoT devices lack the required built-in security to combat these types of security threats. As a result of its unsatisfactory hardware and software design and also numerous other factors, IoT devices are at risk of cybersecurity threats. Machine learning techniques are used to handle global computer security issues like malware detection, ransomware recognition, fraud detection, and spoofing identification. This study examines how machine learning is applied to computer defense. Here, the primary focus lies on the prevalent threat of malware (A common threat) and its detection. The fundamental concept used here is that malware is discovered by a machine using the training set after it has been programmed with new data using various algorithms and classification approaches. The most commonly used algorithms for this include supervised neural networks, support vector machine, K-means clustering, KNN (k-nearest neighbors), etc. The results are obtained based on accuracy. Various analyses showed that the highest accuracy was achieved by the SVM algorithm, Decision Tree, and LSTM model. The challenges of malware detection are likewise analyzed.

Keywords Threat detection · IoT · Machine learning · Neural network · Malware

G. Priya (✉)
Student, M.Sc. Forensic Science, Department of Forensic Science, University of Calicut, Kerala Police Academy, Thrissur 680631, Kerala, India
e-mail: sarovaram001@gmail.com

K. V. Greeshma
Assistant Professor, Department of Forensic Science, University of Calicut, Kerala Police Academy, Thrissur 680631, Kerala, India
e-mail: greeshmakv@gmail.com

A. Prasad et al. (eds.), *Communication Technologies and Security Challenges in IoT*, Internet of Things, https://doi.org/10.1007/978-981-97-0052-3_25

Introduction

IoT (Internet of Things) devices are physical objects or devices that have embedded sensors, software, and some other mechanisms that enable internet connectivity, communicate with other devices, and share data. By collecting and transmitting vast amounts of data to other devices, cloud platforms, and backend servers, IoT devices have the potential to improve processes, optimize performance, and enhance decision-making across many industries. The growing need for IoT devices is driven by a desire to enhance efficiency, reduce costs, improve decision-making, increase convenience, improve safety and security, and provide personalized experiences. However, IoT devices can be vulnerable to various security threats due to their interconnected nature and their ability to collect and transmit sensitive data. Common security issues in this field include Malware, Phishing, DDoS, Data breaches, Insider threats, zero-day attacks, etc. Malware attacks are one of the most prevalent types of cybersecurity threats.

This is a type of harmful software intentionally designed to exploit computer systems, networks, or devices. These attacks are used by cybercriminals to obtain unauthorized access to sensitive information, steal intellectual property, or cause damage to critical infrastructure. To achieve these goals, different kinds of malicious software, including viruses, worms, and Trojans, can be employed [6]. Static and Dynamic are the two main analysis techniques used for the detection of this type of threat. Static analysis is performed by examining the programming code without actually running it. There are several sub-techniques within static analysis including signature or pattern-oriented detection [15], heuristic detection which uses rules to identify suspicious behavior, and static analysis tools which automate the analysis process. "On the contrary, dynamic analysis entails executing a program within a controlled environment and carefully observing its behavior" [7].

The presence of harmful programs in a user's system often goes unnoticed until the system's performance begins to degrade, leading to suspicion of malicious activity. The use of ML techniques can automate the process of malware detection [7]. In ML, the main objective considered is to develop an advanced model using a specific algorithm that can accomplish a certain task, such as classification, clustering, or regression. This model is created by training it on a set of input data, and once trained, it can be used to make predictions. The type of output produced by the model depends on the task it was designed to perform and how it was implemented [10].

ML-based threat detection systems have the ability to detect unknown or emerging threats that traditional rule-based systems may miss. These algorithms can learn from new data and update their models to detect new and previously unseen threats and Optimize the accuracy of detecting security risks. However, there are also challenges and limitations to using ML techniques. An example of a notable challenge for these algorithms is the quality and accessibility of data. Additionally, the lack of interpretability of ML models can make it challenging to understand how the models arrive at their decisions. Nonetheless, ML algorithms can operate in real time, enabling them to respond immediately to potential threats.

As our reliance on digital technologies continues to grow, the need for efficient and reliable threat detection systems has become increasingly vital. Apart from the cybersecurity sector, ML algorithms have found their applications in other fields like social media, healthcare, and physical security. For instance, in physical security, ML techniques are employed to identify potential threats and anomalies through video surveillance systems. In healthcare, ML algorithms are utilized for detecting diseases, predicting outcomes, and assisting in medical diagnosis. The application of ML algorithms for threat detection demands constant monitoring and updating to keep up with the ever-evolving threat landscape. The models have to be frequently updated with new data and retrained to ensure their accuracy and effectiveness.

The intention of this study focuses on the presentation of an overview of the latest developments in using ML methods for detecting threats to IoT devices. We will examine different ML algorithms that have been utilized in pursuing this precise objective, including "Support Vector Machines (SVM), Random Forest (RF), Logistic Regression, and Decision Trees", among others, and compare their effectiveness in threat detection.

Literature Review

The relevance of security in today's world has resulted in a growing interest in the application of ML techniques for threat detection, making it an area of study that is gaining prominence. Researchers [25] identified IoT devices that were not authorized for use on an organization's premises using ML techniques. They used Random Forest and the trained classifiers achieved an average accuracy of 96%. A comprehensive examination was conducted [11] on a set of 41 IoT devices to classify them within the context of smart homes using the logistic regression method, combined with Logitboost, and developed a classification model.

A thorough comparison of various methods for detecting malware has been conducted [3]. Among the methods evaluated, the "J48 algorithm and Hybrid analysis" demonstrated superior performance for detecting malware in Windows systems. "The Decision Tree algorithm using Dynamic analysis" also achieved the same level for malware detection in Android systems. A recent study [30] demonstrates the effectiveness of combining multiple ML methods in a "hybrid learning approach" to identify malicious software where it merges together "Random Forest and K-Nearest Neighbor Classifier" to create a model for detecting newer types of malware using an updated dataset comprising 10,000 samples. The introduction of a flexible framework utilizes various ML algorithms to distinguish malware files, also clean them, to minimize false positives [13]. The framework is presented using two types of cascade one-sided perceptrons, namely the standard version and the kernelized version. The study also identifies the "cascade one-sided perceptron and its explicitly mapped variant" as the most reliable algorithms among those presented, especially given the limitations of speed and memory that commercial anti-virus products face.

A study shows that Memory analysis, along with data mining and ML techniques, holds great promise for improving malware detection in the future [33]. They also compare different malware detection methods. The use of artificial intelligence and ML techniques can be an effective way to reduce the risks associated with malware hidden within images, while also helping to identify and prevent malicious users from exploiting these vulnerabilities [4]. The authors recommend a specific model for use in detecting malware. Image file types that are most vulnerable to malware include BMP, JPEG, PNG, and GIF, and malware can be embedded in these files during the creation process. To check for malware, users can upload images to a website like VirusTotal for scanning.

The literature [26] emphasizes the potential of ML, AI, and Blockchain to address IoT security issues and highlight key research obstacles. A novel technique for detecting malware using ML Classifiers was discovered [16]. This study involved building and testing methods for detecting malware using a dataset of 19,612 records. Employing "independent sample t-tests", the researchers found a statistically meaningful difference between the accuracy of the two models, with the "Random Forest Algorithm" outperforming the "Decision Tree Algorithm".

Introduction of a model for detecting obfuscated and swiftly progressing malicious software targeting particular IoT devices utilized five in-demand models, including "SVM, RF, LR, Bagging, and MLP" [35]. The best results were achieved using "Random Forest (RF)". The previous studies [32] have proposed a model, utilizing two levels of classification: Macro and Micro. The Weka Framework was employed for the development of ML models using training datasets, with the "J48 Decision tree". A lightweight system for identifying malware on Android devices using ML was developed [37]. The proposed system utilizes a "support vector machine (SVM)" model for classification. It is designed to detect unknown Android applications based on ML, setting it apart from traditional detection methods. Additionally, a new feature selection algorithm called PCA-RELIEF is introduced to process raw features.

Introduction of the "Intelligent Malware Detection System (IMDS)", utilized "objective-oriented Association (OOA) mining" [39]. The system's effectiveness was evaluated through experimentation, with promising results indicating that the "IMDS" system outperforms popular anti-virus software as well as previous data "mining-based detection systems". To detect "polymorphic malware", Akhtar and Feng [2] utilized various ML techniques. The study found that three classifiers, namely "Decision Tree (DT), Convolutional Neural Network (CNN), and Support Vector Machine (SVM)", achieved high levels of accuracy in detecting these types of malicious threats. Specifically, DT achieved an accuracy rate of 99%, followed by CNN with 98.76%, and SVM with 96.41%.

To address issues like Zero-day attacks, Researchers [12] employed "Two-Layer CNN, Four-Layer CNN, VGG16, and under the category of CNN classifiers, and Logistic Regression, Support Vector Machine, and K-Nearest Neighbours under Ordinary Classifiers". Findings demonstrated SVM classifier with MobileNetv2 features achieved the highest performance, reaching an accuracy rate of 94%. The proposed solution by Kamboj et al. [20] involves utilizing various ML approaches. To develop and test the ML models, the researchers trained them on a dataset and used various balancing techniques to ensure that the dataset was representative. The balancing techniques included Oversampling, Undersampling, SMOTE, and Balanced Bagging Classifier.

IoT Device Security: Protecting Against Cyber Threats

IoT devices are frequently exploited by attackers to infiltrate larger networks. Even a single compromised IoT device can provide an attacker with a foothold to launch a significant cyber attack leading to substantial harm to people and organizations. Additionally, IoT devices collect and transmit confidential data such as personal, financial, and healthcare, etc. These data can be accessed by unauthorized parties if proper security measures are not in place resulting in identity thefts, financial fraud, and other cyber crimes. Therefore safeguarding these devices is crucial for protecting personal and confidential details, preventing cyber threats, ensuring safety and reliability, and maintaining trust and confidence in IoT technologies.

Typically security is implemented in three distinct layers: "the Perception layer, network layer, and application layer" which can safeguard IoT devices, thereby reducing the vulnerability to cyber threats. This method also poses potential security-related issues that must be addressed [40]. Each layer has its own set of security measures that are designed to prevent potential threats. This passage delves into further details about these three layers of data processing. The perceptual Layer uses sensors for gathering and interpreting facts regarding the real world, while the Network Layer is responsible for transmitting this data reliably over various networks [10]. At last, the final layer is accountable for providing personalized applications to users through a variety of devices [25]. Essentially, the Perceptual Layer captures data, Network Layer transmits it, and Application Layer delivers it to users. Researchers have already addressed security issues related to these layers [34]. The First layer can be vulnerable to security breaches, such as tampering, DoS, and data transit attacks (in wireless sensor networks or radio sensor networks). The network layer is vulnerable to routing attacks, data transit attacks, spoofed routing information, and eavesdropping. On the other hand, the application layer offers services to customers, and its main security concerns include data leakage and denial of service attacks.

Security measures are crucial for ensuring the safety of IoT devices. Among the various security mechanisms employed, some common techniques include "Encryption and Hashed Based Security", where Encryption converts a message into an unknown form. This method uses a key that is double the length of the message, making it difficult to break. A "hash function" is for detecting and recovering any alterations made by attackers to the message [22]. "Public Key Infrastructure (PKI)" is another method for encrypting and decrypting. A real PKI protocol incorporates additional security measures like digital signatures and revocation mechanisms to guarantee the authenticity and integrity of public keys and digital certificates [23]. Embedded Security Framework, OpenHab Technology, IoTOne Technology, Identity Management, and Virtual Identity (VID) Framework are the other security mechanisms explained by researchers to assist security-related issues in the security architecture layer of IoT [11].

ML for IoT Security

The integration of ML techniques in IoT devices and networks enables the development of intelligent security systems, which provide effective solutions to address a wide range of security challenges [3]. They can enable smart devices to derive meaningful insights from device-generated or human-generated data, facilitating automated behavior or situational variations. For an effective IoT solution, this ability to learn and adapt based on knowledge is considered essential [18]. However, with the development of smart attacks, IoT devices must determine key security parameters and choose defense policies in heterogeneous and dynamic networks [38].

Many studies have identified several major security challenges faced by IoT devices, such as weak, default, or no passwords, and storing data online without protection. Many devices with default or easily remembered passwords, or without any password at all, making them highly vulnerable to exploitation by hackers [30]. This vulnerability exposes users' privacy [1].

ML-based security presents various challenges including limited observation of the state of the system, high computation and communication overhead, and the need for backup security solutions [38]. Detecting malware poses several challenges, including the difficulty of verifying classification results in real-world scenarios, the need for more advanced techniques to improve active learning, and the development of methods to detect specific types of attacks and handle large datasets while also reducing dimensionality.

ML is an increasingly important tool in IoT security that can help address various security problems such as malware detection and vulnerability exploitation. Its ability to learn and adapt based on knowledge is crucial in enabling effective IoT solutions.

Malware—A Common Threat

"Malicious software, commonly known as malware, is a type of computer program that is designed to cause harm, or disruption to computers, networks, and other interconnected assets" [24]. Malware can infect computers without the knowledge of their owners. Malware creators constantly come up with new techniques to evade detection, and therefore, they design malware to morph and evolve to make detection difficult [36]. To overcome the limitations of current malware identification techniques, ML as well as data mining approaches integrated with current techniques to improve their performance [10].

Types of Malware

The following list outlines the most frequent and hazardous types of malicious software.

Virus

It is a type of program disguised within another program, and it has the capability of producing duplicates of itself while inserting them into other programs or files. Typically, a computer virus is designed to cause harm, such as destroying data or other malicious actions. It is important to note that computer viruses are not naturally occurring, but are instead created by humans. Once a virus is released, it can spread without direct human control.

Worm

Worms and viruses are both types of computer programs that can make copies of themselves and spread to other computers through networks. They can also disrupt the regular functioning of the system, but there is one key difference between them. Worms present as a standalone entity [33]. However, people often refer to worms as viruses due to their similarities.

Ransomware

This affects the system and makes the files in an unreadable format. For decryption, they ask for a payment, most of the time as Bitcoin [4]. "Crypto-ransomware, and locker ransomware" are the two common forms of ransomware [29].

Trojan Horse

It performs actions that are not disclosed to the user, but which were intended by the programmer. If users were aware of these actions, they would not accept the program. Some definitions categorize viruses as a specific type of Trojan horse that can infect other programs, effectively turning them into Trojan horses as well. Other definitions state that a virus that only replicates without causing intentional harm is not considered a Trojan horse.

Backdoors

A backdoor attack is an intrusive method employed by attackers to obtain unauthorized remote entry to a target's computer system. Successful execution of the attack may enable the attacker to extract sensitive information from the system, corrupt, delete or transfer data, without the victim's knowledge. The malware used for this purpose is typically hidden within the code of programs, operating systems, firmware, routers, or other hardware. It is often utilized by network administrators who need to reset employee passwords or anyone who wants to bypass the authentication process without being detected [17].

Rootkits

A rootkit can infiltrate a computer by exploiting vulnerabilities in security systems, using phishing techniques, or cracking passwords. Once the rootkit gains access, it searches for another program to conceal its code, making it difficult to detect. To obtain unauthorized access to the computer is the objective behind this. It can be achieved by giving the hacker administrator privileges during installation [17].

Keyloggers

Keyloggers can be surreptitiously placed on a variety of devices. They work by recording the specific keys that are being pressed by a user, to capture sensitive information. The primary focus of keyloggers is to collect data such as login credentials and debit card details, which are then transmitted back to the hacker [17].

Malware Analysis Techniques

Malware analysis involves the examination and comprehension of malicious software. Its goal is to recognize how the malware behaves and operates and to determine the most effective means of defending against it and reducing its impact. By understanding the motivations behind the creation of the malware, it is possible to develop appropriate defensive measures. There are three categories of malware analysis techniques, categorized according to their timing and methodology.

Static Analysis

The act of inspecting the programming instructions of malware to infer its behavioral characteristics. The primary benefit of this analysis is its ability to uncover all feasible behavioral scenarios. By scrutinizing the code, the analyst can identify all possible ways the malicious software could execute, beyond its present situation. Furthermore, this technique is less risky than dynamic analysis since the malware is not run, thereby eliminating the possibility of harmful consequences for the system. However, static analysis is generally more time-consuming [10].

Dynamic Analysis

It entails running a file, such as a software program or application, to observe and analyze its behavior and characteristics, that is in a virtual environment to determine whether this contains any malicious code or behavior that could potentially harm a system or network [3, 10]. By executing the file inside the virtual environment, analysts can observe its behavior without putting their system or network at risk [26]. This approach allows for the safe and thorough analysis of potentially harmful files and can help to prevent the spread of malware and other malicious code.

Hybrid

This approach combines both code-level examination and runtime behavior analysis, allowing it to leverage the advantages of each. The software is initially analyzed using code analysis techniques to identify any malware signatures, and subsequently executed within a virtual environment to observe its true behavior [36].

Exploring Malware Detection

Techniques for Detecting Malware are utilized to identify and prevent malicious software, or malware, from causing harm to computer systems, networks, or users.

Signature Based

Malware creators embed a unique set of instructions, known as a signature, in their code to identify the malware's family. Antivirus programs typically use signature-based detection techniques [31]. Examining the disassembled binary code of malware to extract features that can be used to construct a signature [21]. This process is time-consuming and laborious, as different signatures must be extracted for various types of malware [3].

Heuristic Based

Heuristics relies on rules that can be determined either by experts or ML classifiers to identify suspicious activities and detect unknown malicious code [14]. It typically involves two stages: the "training phase and the detection phase". During the first step, the operation of the system is monitored using ML techniques, which are then used in the detection phase to flag potential attacks by comparing the current behavior to the established profile [21].

Specification Based

Specification-based detection techniques examine applications accordant with their specifications and look for usual and unusual behaviors [31]. This technique is similar to anomaly detection, but it differs in that it relies on manually developed specifications that define legitimate system behavior rather than ML techniques [31]. Positive and negative aspects of Malware detection techniques can be seen in Table 1 [5, 31].

Table 1 Benefits and drawbacks of malware detection techniques

Malware detection techniques	Pros	Cons
Signature based	• Widely available for access • Simple and straightforward to execute • Rapidly identifies threats • Provides a thorough analysis of malware • This method consumes fewer resources when compared to other techniques	• Unable to identify polymorphic malware • Quickly identifying threats through replication of data from a large database • Malware that is unfamiliar or unknown cannot be identified
Heuristic based	• This method can identify both known and unknown/new types of malware	• Regular updating of data is necessary to account for new and unfamiliar malware • This process requires significant time and storage resources • This approach has a high incidence of false positives
Specification based	• Malware, whether familiar or unfamiliar, can be identified • The occurrence of false positives is minimal	• The detection method is not efficient in identifying new types of malware • There is a high chance of false negatives • Developing specifications for detection is a time-consuming process

Detecting Malware Using ML

Artificial intelligence (AI) is the basis of ML. Computers are trained using previously made data sets and based on this training, computers can take decisions when introducing new datasets. Major ML detection techniques include:

- **Supervised learning**

It requires the machine to receive external guidance to learn. This guidance is provided by using a labeled dataset to train supervised learning models. These models can be categorized mainly into: "classification and regression" [16]. Two of the most commonly utilized models within this approach are simple linear regression and decision trees.

- **Unsupervised learning**

This approach does not require external guidance for learning from data. The reason this learning approach is referred to as unsupervised is due to the fact that the models can be trained using a dataset that lacks labels, classifications, or categorizations. As a result, the algorithms must operate on this data without any guidance or supervision. There are two types of unsupervised learning: clustering and association [35]. One example of unsupervised learning is K-means clustering.

- **Semi-supervised learning**

This approach can be considered as a middle ground between supervised (which does not use labeled training data) and supervised learning (which uses fully labeled training data). In semi-supervised learning, some of the training examples may not have labels, but research has shown that there is a significant enhancement in learning accuracy when incorporating unlabeled data in conjunction with a limited quantity of labeled data [32].

- **Reinforcement learning**

This approach is distinct from other ML approaches in that it does not rely on labeled data. Rather, machines learn through experience and must employ a trial-and-error method to improve their decision-making abilities. Reinforcement learning is built on a feedback-based process.

ML Algorithms

Here are some of the widely employed ML algorithms.

Linear Regression

This involves modeling continuous variables and making predictions. It is limited to linear relationships and handling complex patterns can be challenging, and incorporating appropriate polynomials into the model can be a difficult task [28].

Logistic Regression

It involves assigning weights to the features in the input data, taking the logarithm of these weights, and then combining them in a linear fashion. This involves multiplying each feature by its corresponding weight and adding them up. This regression method is specifically designed for predicting the probability of an event or outcome. It accomplishes this by fitting the available data to a logistic function [27], resulting in a binomial outcome. Some benefits of logistic regression include simplicity in execution and Processing speed. However, it has limitations such as it is not able to fix non-linear challenges, susceptibility to overfitting, and the requirement to identify all independent variables for it to work effectively [28].

Decision Tree

It is a graphical representation resembling a tree structure that effectively showcases choices and their corresponding outcomes. In this graph, nodes are utilized to symbolize events or choices, while the edges serve to depict decision rules [8]. Decision trees offer advantages such as visual interpretability, applicability to both classification and regression tasks, and handling of categorical and quantitative data, however, they are prone to sampling errors and offer only locally optimal solutions, not globally optimal ones [37]. Possibility for individual classification trees to exhibit overfitting tendencies is an additional limitation [19, 28].

Random Forest

The technique of utilizing multiple randomized decision trees and averaging their predictions has demonstrated exceptional performance in scenarios where the value of variables surpasses the value of data points. It has the capability to handle large-scale problems, is highly adaptable for diverse learning tasks, and produces evaluations of variable significance [9].

Support Vector Machines (SVM)

These are supervised learning models in ML that come with learning algorithms to examine data used for regression analysis and classification [8]. Some advantages include their ability to perform well with highly intricate data [39]. However, they are more susceptible to overfitting compared to other algorithms and can be viewed as a "black box" approach, as they do not provide metrics on how predictors are merged to optimize hyperplane [19].

K-Nearest Neighbors (KNN)

This technique uses a collection of data points that have already been grouped into specific classes. The algorithm then attempts to categorize new data points by treating them as a problem of classification. Some advantages of this algorithm include its simplicity, ease of implementation, flexibility, and suitability for multi-modal classes with multiple class labels [2]. However, it can be relatively expensive to classify unknown records since it requires computing the Separation of KNN. Whereas, the K-Means Clustering Algorithm has drawbacks including the challenge of determining the optimal value of K, reduced performance when clusters are globular, and sensitivity to different initial partitions leading to different final clusters [28].

K Means Clustering

It is commonly employed for addressing clustering problems. One of its advantages is that it is computationally more efficient than hierarchical clustering when dealing with large variables [28].

Supervised Neural Network

Supervised neural networks are used in feed-forward networks, where the predicted output is compared to the known output for a given input, and if there is an error, the network's parameters are adjusted and the input is fed through the network again [8].

Steps for Performing Malware Detection

Various steps for performing malware detection using MLare following [30].

- **Gathering malware samples**

It is the first step in which malware samples are collected from different online platforms. It can be private or public platforms.

- **Creating a dataset**

It involves structuring the collected data into a format such as CSV, JSON, or XML.

- **Analyzing the features of the dataset**

In ML, feature analysis is a crucial step that involves identifying the most important features or variables for Training. This is necessary because some features may not be relevant or useful in predicting the target variable, and including irrelevant or redundant features can harm the model's performance. The feature analysis process involves selecting, engineering, and reducing features to obtain the most relevant ones. This helps to enhance the efficiency and accuracy of ML models, which leads to better predictions and decision-making.

- **Processing the data to prepare it for training**

This structured data needs to be prepared for use by removing errors, missing values, and irrelevant information. Additionally, the data may require transformations, scaling, or normalization to ensure it is appropriate for ML algorithms. For evaluating a ML technique effectively, To ensure a reliable and accurate assessment of a model's performance, it is crucial to partition the dataset into distinct subsets for training, validation, and testing purposes. The training set serves as the foundation for teaching the model. While the training set is essential for model development, the validation set serves a distinct purpose. It is used to fine-tune the model's hyperparameters [12].

- **Training models using the prepared dataset**

Different algorithms are used to train the models. Different ML methods come with their own unique advantages and disadvantages, which is why it is essential to use multiple models to improve metrics analysis. By training models using a variety of methods, we can obtain a more accurate understanding of how each model performs on different sets of data.

- **Evaluating the efficiency of the trained models**

This step is necessary for verifying accuracy and effectiveness. This can be done using various metrics, such as "accuracy, precision, recall, and F1 score" [20]. This is employed for determining if a technique is acceptable for use, and models with the highest accuracy are generally more effective in identifying malware during real-time cybersecurity threats.

- **Exporting the trained models for use in detection or analysis**

After evaluating and testing ML models, they can be exported and stored for future use. This enables easy access to the models without requiring them to be retrained from the beginning. These exported models can also be applied to further research and development, including the creation of antimalware programs for identifying and preventing cybersecurity threats. Utilizing the models as a foundation for antimalware programs can lead to the development of more effective and efficient solutions for safeguarding against malware attacks (Table 2).

Table 2 ML techniques for malware detection

Author	Machine learning technique/s	Dataset	Tools	Most reliable algorithms	Accuracy (%)	Pros	Cons
Meidan et al.	Random Forest	Network traffic data from 17 distinct IoT devices represent nine types of IoT devices	n/a	n/a	99.49	• Good classifiers' transportability • Resilience to cyber attacks • High classification accuracy • Detection speed is high • More traffic data was collected	n/a
Cvitić et al.	Logitboost (Logistic regression method enhanced by the concept of supervised ML)	Primary: 103 files in.pcap Secondary: 41 files of the same format	WEKA software tool with the support of MS Excel 2016	n/a	99.79	• It is possible to accurately classify new and unseen IoT devices and their traffic flow into predefined classes Sufficient to identify which class the device belongs to	• Only usable as a support process for various other activities • It is not possible to know the traffic profiles for each device
Sethi et al.	J48 Decision Tree, Random Forest, and SMO	220 samples of data containing malware from OpenMalware and clean files from Softonic	Cuckoo Sandbox and WEKA	J48	100	High accuracy	Limited samples are used and they are made from scratch due to unavailability of samples

(continued)

Table 2 (continued)

Author	Machine learning technique/s	Dataset	Tools	Most reliable algorithms	Accuracy (%)	Pros	Cons
Rimon and Haque	Hybrid ML Algorithm (Random Forest and K-Nearest Neighbor Classifier have been combined)	Collected from VirusShare in Windows Portable Executable (PE) format	The training of the model was done in Google Colaboratory	n/a	98	• High accuracy • The proposed system provides better results for both binary classification and multi-classification tasks, making it more effective for system development	This paper did not mention the influence of this approach on Android malware, macOS malware, etc.
Haritha and Rajendran	Random Forest Algorithm and Decision Tree Algorithm	Collected from the open forum domain Kaggle with 19,612 rows and 79 columns	SPSS instrument	Random Forest Algorithm	99	High accuracy	• The dataset collection and preprocessing of the data takes too much time: high complexity and difficulty handling multiple features
Tahir and Qadir	Support Vector Machine (SVM), Logistic Regression (LR), Bagging/Bootstrap Aggregating, Random Forest (RF), Multi-Layer Perceptron (MLP)	524 malware samples and 524 benign samples from different sources	Windows 11 machine with VMWare Work station 16.2.4 Pro	Random Forest (RF)	99.04	• High accuracy • Effective for the detection of rapidly evolving and obfuscated malware	• It uses only 69 system calls extracted

(continued)

Table 2 (continued)

Author	Machine learning technique/s	Dataset	Tools	Most reliable algorithms	Accuracy (%)	Pros	Cons
Ye et al.	IMDS, NB, SVM, and J48	29,580 files, 17,366 malware, and 12,214 benign	W32Dasm WEKA and C++	IMDS	93.07	Efficiency, accuracy, and scalability are high compared to other methods	Only provides binary predictions
Wen and Yu	SVM	2,000 Android applications, 1,000 malware (from the Drebin Project) and 1,000 benign (from Google Play)	Androguard and DroidBox	n/a	95.2	It can detect the unknown Android application based on the ML	n/a
Akhtar and Feng	DT, CNN, and SVM	51 different groups of malware (from the Canadian Institute for Cybersecurity)	Virtual machine (VM), PEiD software	DT	99	The highest accuracy for detecting and accurately characterizing malicious software (we can identify without executing)	n/a

(continued)

Table 2 (continued)

Author	Machine learning technique/s	Dataset	Tools	Most reliable algorithms	Accuracy (%)	Pros	Cons
Kamboj et al.	Logistic Regression, Decision Tree, Random Forest, XGBoost Model, Gradient Boosting Model, AdaBoost Classifier, Principal Component Analysis, and K-Means Clustering	130,000 rows (data points) and 57 columns (features) collected from different sources	Jupyter Notebook Or Google Colab	Random Forest Classifier	99.99	• High accuracy • Helps to identify malware	• Issues related to datasets • Very time-consuming
Meidan et al.	Random Forest	Network traffic data from 17 distinct IoT devices represent nine types of IoT devices	n/a	n/a	99.49	• Good classifiers' transportability • Resilience to cyber attacks • High classification accuracy • Detection speed is high • More traffic data was collected	n/a

*n/a: not applicable

Conclusion

In this paper, various ML techniques and analyzed their performance in the context of detecting malware. We also covered fundamental concepts related to malware analysis and detection and emphasized the importance of using ML techniques for securing IoT devices and protecting personal and confidential information. We surveyed and compared several representative studies based on their accuracy and other factors, and compiled the results in a comparison table. Based on our analysis, Sethi et al. yielded highly promising results (about 100%) in the domain of malware detection on Windows systems using the J48 Decision tree, followed by Kamboj et al. with a Random Forest model achieving an accuracy of 99.99%. The study conducted by, Cvitić et al. presents a significant contribution to the field of IoT data analysis. They achieved an accuracy rate of 99.79% using the Logitboost method, which is a supervised ML approach based on logistic regression. However, each method has its own set of challenges as outlined in the comparison table, and hybrid ML techniques appear to be more reliable. There is still a lot of room for future work and several areas for improvement in this field.

References

1. Ahmad, R., & Alsmadi, I. (2021). Machine learning approaches to IoT security: A systematic literature review [formula presented]. *Internet of Things (Netherlands), 14*. https://doi.org/10.1016/j.iot.2021.100365
2. Akhtar, M. S., & Feng, T. (2022). Malware analysis and detection using machine learning algorithms. *Symmetry, 14*(11). https://doi.org/10.3390/sym14112304
3. Al-Janabi, M., & Arab, A. A. (2020). A comparative analysis of machine learning techniques for classification and detection of malware. *ieeexplore.ieee.org* [Preprint]. Retrieved March 10, 2023, from https://ieeexplore.ieee.org/abstract/document/9300081/
4. Alodat, I. (2023). Malware: Detection and defense. *Malware - Detection and Defense* [Preprint]. https://doi.org/10.5772/INTECHOPEN.108434
5. Alsmadi, T., & Alqudah, N. (2021). A survey on malware detection techniques. In *2021 International Conference on Information Technology, ICIT 2021 - Proceedings* (pp. 371–376). https://doi.org/10.1109/ICIT52682.2021.9491765
6. Alzubaidi, A. (2021). Recent advances in Android mobile malware detection: A systematic literature review. *IEEE Access, 9*, 146318–146349. https://doi.org/10.1109/ACCESS.2021.3123187
7. Awodiji, T. O. (2022). Malicious malware detection using machine learning perspectives. *Journal of Information Engineering and Applications*, (December), 9–17. https://doi.org/10.7176/jiea/12-2-02
8. Batta, M. (2018). Machine learning algorithms—A review. *International Journal of Science and Research (IJSR), 18*(8), 381–386. https://doi.org/10.21275/ART20203995
9. Biau, G., & Scornet, E. (2016). A random forest guided tour. *Test, 25*(2), 197–227. https://doi.org/10.1007/s11749-016-0481-7
10. Chumachenko, K. (2017). Machine learning methods for malware detection and classification. In *Proceedings of the 21st Pan-Hellenic Conference on Informatics—PCI 2017* (p. 93).
11. Cvitić, I., Peraković, D., Periša, M., & Gupta, B. (2021). Ensemble machine learning approach for classification of IoT devices in the smart home. *International Journal of Machine Learning and Cybernetics, 12*(11), 3179–3202. https://doi.org/10.1007/s13042-020-01241-0

12. El-Sayed, R., El-Ghamry, A., Gaber, T., & Hassanien, A. E. (2021). Zero-day malware classification using deep features with support vector machines. In *Proceedings—2021 IEEE 10th International Conference on Intelligent Computing and Information Systems, ICICIS 2021*, April 2022 (pp. 311–317). https://doi.org/10.1109/ICICIS52592.2021.9694256
13. Gavriluţ, D., Cimpoeşu, M., Anton, D., & Ciortuz, L. (2009). Malware detection using machine learning. In *Proceedings of the International Multiconference on Computer Science and Information Technology, IMCSIT '09*, 4 May 2014 (pp. 735–741). https://doi.org/10.1109/IMCSIT.2009.5352759
14. Gupta, A., Maurya, S., Kapil, D., Mehra, N., & Negi, H. S. (2020). Android malware detection using machine learning. *International Journal of Recent Technology and Engineering, 8*(2S12), 65–70. https://doi.org/10.35940/ijrte.b1011.0982s1219
15. Gyamfi, N. K., & Owusu, E. (2019). Survey of mobile malware analysis, detection techniques and tool. In *2018 IEEE 9th Annual Information Technology, Electronics, and Mobile Communication Conference, IEMCON 2018*, November (pp. 1101–1107). https://doi.org/10.1109/IEMCON.2018.8614895
16. Haritha, B., & Rajendran, T. (2022). Improved malware detection for IoT devices using random forest algorithm comparing with decision tree algorithm. *Advances in Parallel Computing, 0*(41), 597–603. https://doi.org/10.3233/APC220085
17. Harjeevan, B., & Gill, S. (2022). Malware: Types, analysis, and classifications.
18. Hussain, F., Hussain, R., Hassan, S. A., & Hossain, E. (2020). Machine learning in IoT security: Current solutions and future challenges. *IEEE Communications Surveys and Tutorials, 22*(3), 1686–1721. https://doi.org/10.1109/COMST.2020.2986444
19. Jiang, T., Gradus, J. L., & Rosellini, A. J. (2020). Supervised machine learning: A brief primer. *Behavior Therapy, 51*(5), 675–687. https://doi.org/10.1016/j.beth.2020.05.002
20. Kamboj, A., Kumar, P., Bairwa, A. K., & Joshi, S. (2022). Detection of malware in downloaded files using various machine learning models. *Egyptian Informatics Journal, 24*(1), 81–94. https://doi.org/10.1016/j.eij.2022.12.002
21. Landage, J., & Wankhade, M. (2013). *Malware and malware detection techniques: A survey. International Journal of Engineering Research & Technology, 2*(12), 61–68. http://www.ijert.org/browse/volume-2-2013/december-2013-edition?download=6744%3Amalware-and-malware-detection-techniques--a-survey&start=10
22. Li, F., & Xiong, P. (2013). Practical secure communication for integrating wireless sensor networks into the Internet of Things. *IEEE Sensors Journal, 13*(10), 3677–3684. https://doi.org/10.1109/JSEN.2013.2262271
23. Li, Z., Yin, X., Geng, Z., Zhang, H., Li, P., Sun, Y., Zhang, H., & Li, L. (2013). Research on PKI-like protocol for the Internet of Things. *ieeexplore.ieee.org* [Preprint]. Retrieved March 10, 2023, from https://ieeexplore.ieee.org/abstract/document/6493878/
24. Maniriho, P., Mahmood, A. N., & Chowdhur, M. J. M. (2022). A study on malicious software behaviour analysis and detection techniques: Taxonomy, current trends and challenges. *Elsevier* [Preprint]. Retrieved April 26, 2023, from https://www.sciencedirect.com/science/article/pii/S0167739X21004751?casa_token=t_Gh0GvuDmIAAAAA:19J2kSuKqMdHMhkPyUaLaqXRUi7iP9cQECE_LrW-ViAX8p6WnlhLkkBHSfdmHs7p3-jlI77liNE
25. Meidan, Y., Bohadana, M., Shabtai, A., Ochoa, M., Tippenhauer, N. O., Guarnizo, J. D., & Elovici, Y. (2017). Detection of unauthorized IoT devices using machine learning techniques. September. http://arxiv.org/abs/1709.04647
26. Mohanta, B. K., Jagadeesh, B. N., & Narasimhamurthy, M. S. (2020). Survey on IoT security: Challenges and solution using machine learning, artificial intelligence and blockchain technology. *Internet of Things (Netherlands), 11*(May). https://doi.org/10.1016/j.iot.2020.100227
27. Nasteski, V. (2017). An overview of the supervised machine learning methods. *Horizons.B, 4*(December 2017), 51–62. https://doi.org/10.20544/horizons.b.04.1.17.p05
28. Ray, S. (2019). Introduction to machine learning and different types of machine learning algorithms. In *Proceedings of the International Conference on Machine Learning, Big Data, Cloud and Parallel Computing: Trends, Prespectives and Prospects, COMITCon 2019* (pp. 35–39).

29. Richardson, R., & North, M. (2017). Ransomware: Evolution, mitigation and prevention. *International Management Review, 13*(1), 10–21.
30. Rimon, S. I., & Haque, M. M. (2023). Malware detection and classification using hybrid machine learning algorithm. In *Lecture notes in networks and systems* (Vol. 569 LNNS, pp. 419–428). https://doi.org/10.1007/978-3-031-19958-5_39
31. Sehrawat, S., & Singh, D. D. (2022). Malware and malware detection techniques: A survey. *International Journal for Research in Applied Science and Engineering Technology, 10*(5), 3947–3953. https://doi.org/10.22214/ijraset.2022.43287
32. Sethi, K., Chaudhary, S. K., Tripathy, B. K., & Bera, P. (2017). A novel malware analysis for malware detection and classification using machine learning algorithms. In *ACM International Conference Proceeding Series*, October (pp. 107–116). https://doi.org/10.1145/3136825.3136883
33. Sihwail, R., Omar, K., & Ariffin, K. A. Z. (2018). A survey on malware analysis techniques: Static, dynamic, hybrid and memory analysis. *International Journal on Advanced Science, Engineering and Information Technology, 8*(4–2), 1662–1671. https://doi.org/10.18517/ijaseit.8.4-2.6827
34. Singh, D., Mishra, M. K., & Lamba, A. (2020). Security issues in different layers of IoT and their possible mitigation. *International Journal of Scientific and Technology Research, 9*(4), 2762–2771
35. Tahir, I., & Qadir, S. (2022). Machine learning-based detection of IoT malware using system call data. 0–10. https://doi.org/10.21203/rs.3.rs-2384013/v1
36. Tahir, R. (2018). A study on malware and malware detection techniques. *International Journal of Education and Management Engineering, 8*(2), 20–30. https://doi.org/10.5815/ijeme.2018.02.03
37. Wen, L., & Yu, H. (2017). An Android malware detection system based on machine learning. *AIP Conference Proceedings, 1864*. https://doi.org/10.1063/1.4992953
38. Xiao, L., Wan, X., Lu, X., Zhang, Y., & Wu, D. (2018). IoT security techniques based on machine learning. 1–20. http://arxiv.org/abs/1801.06275
39. Ye, Y., Wang, D., Tao, L., Ye, D., & Jiang, Q. (2008). An intelligent PE-malware detection system based on association mining. *Journal in Computer Virology, 4*(4), 323–334. https://doi.org/10.1007/s11416-008-0082-4
40. Zhang, J., Jin, H., Gong, L., Cao, J., & Gu, Z. (2019). Overview of IoT security architecture. In *Proceedings—2019 IEEE 4th International Conference on Data Science in Cberspace, DSC 2019* (pp. 338–345). https://doi.org/10.1109/DSC.2019.00058

Exhaustive Theoretical Study of Practical Free Space Optical Cooperative Relaying Technology: New Trends in IoT Communication

Anu Goel and Richa Bhatia

Abstract The Internet of Things (IoT) technology is vigorously evolving, necessitating cutting-edge technologies and solutions to effectively realize global wireless communication needs. In response to the explosive growth of data-hungry applications arising from the enormous connectedness of the IoT devices, 6G wireless network, the current radio frequency (RF) range of operation is inadequate to accommodate the growing needs of massive throughput due to congestion and scarcity of bandwidth. Optical Wireless Communication (OWC) technology has been among the emerging communication trends that may help with IoT implementation. If the OWC is operated in infrared wavelength ranges, then it is said to be Free Space Optical Communication (FSO). Future smart infrastructures and systems could tremendously benefit from the use of FSO technology to facilitate device-to-device connectivity. Hybrid access network technologies will be required to support the growing number of IoT devices. Therefore, in this chapter, a review of research works reported in the field of RF/OWC has been presented. The literature on wireless optical systems has been presented to highlight the various factors that impact the operating conditions of FSO systems. Afterward, the combination of radio and Optical Wireless systems has been described and various published research works have been exhaustively reviewed. The objective of this chapter is to bring insights into the key components of Optical Wireless transmission and to highlight various factors that need to be included in mixed radio/Optical Wireless relaying systems to quantify the impact of its practical problems.

Keywords FSO · Optical wireless communication with IoT · Cooperative relaying · Interference · Hardware impairment · Mixed RF/FSO

A. Goel (✉)
Department of Electronics and Communication Engineering, NSUT East Campus (Formerly AIACTR), Affiliated to GGSIPU, New Delhi, India
e-mail: anu.goel@ymail.com

R. Bhatia
Department of Electronics and Communication Engineering, NSUT East Campus (Formerly AIACTR), New Delhi, India
e-mail: richatimes@gmail.com

A. Prasad et al. (eds.), *Communication Technologies and Security Challenges in IoT*, Internet of Things, https://doi.org/10.1007/978-981-97-0052-3_26

Background

Wireless network deployments have increased at an exponential rate in recent years. Web surfing, music and video downloads, video telephony and the Internet of Things (IoT) are all examples of current and future multimedia applications that require higher data rates and, as a result, increased transmission capacity. The majority of the IoT networks in use today use radio frequency (RF) communication technology. With the growth of IoT-based services, the need for faster data rate communication lines is rapidly rising. But the current electromagnetic (EM) spectrum under radio range has been pronounced as a major limited resource due to the fast development in the number of customers who want multimedia services via wireless communications. Congestion and overcrowding in the available spectrum present a substantial barrier to the development of greater data-rate radio communication lines. Researchers must either boost the effectiveness of the existing congested RF spectrum or find a suitable alternative of RF systems to meet these demands. Free space optical (FSO) communication is envisaged as a practicable contender to augment traditional radio systems due to the overpopulation of the radio spectrum. The transmission of the optical signal in free space is feasible only when a line-of-sight (LoS) is applicable which sends data-bearing signals wirelessly in free space using the optical spectrum. The wavelength ranges used in FSO communication include infrared (IR) and visible light [1]. The spectrum above 300 GHz, i.e., the optical spectrum, must be adequately exploited because the whole electromagnetic frequency ranges of up to 300 GHz is practically fully allotted [1–10]. Because it encompasses the range of frequency from 300 GHz to 800 THz, the optical domain has a far higher usable bandwidth than the microwave frequency range. The optical spectrum, in particular, is largely under-utilized for wireless data transmission.

Free Space Optical Communication (Overview)

To enable various broadband services, next-generation wireless networks should be able to provide increased data capacity. Low data speeds, expensive licensing, and a crowded spectrum are limitations of copper and coaxial cables, as well as modern broadband radio/microwave technologies, whereas optical fiber deployment is a difficult challenge for network engineers. Furthermore, the installation of optical fiber by excavating the ground in rural locations is not an economically viable idea. Many government, military, and commercial entities have presented Optical Wireless communication systems as possible candidates for meeting their data transmission demands. High bandwidth, physical security, and the ability to reuse wavelengths are all hindered by Optical Wireless systems [2–12]. Furthermore, Optical Wireless systems operate in an unlicensed spectrum, allowing the operator to sidestep any licensing issues. Wireless optical systems communicate with the help of light-emitting diodes (LED) and LASERs as low-power optical sources to transport data using an optical beam in the visible spectrum. The Optical Wireless system's fast deployment is one of its main advantages with context to new generation IoT systems. An Optical Wireless channel can be quickly implemented if there is a LoS link between the telescope and the aperture, as opposed to the time-consuming process of digging up sidewalks to build fiber cables. This property of the Optical Wireless communication system makes it a viable option for deploying a network in places prone to natural disasters [7]. Moreover, it is relatively simple to connect distant sites on an ad hoc basis, such as on borders and surveillance sites, in geographically challenging areas, and in deserts with massive data transmission requirements. This is a major advantage for IoT-based communication set-up. Optical Wireless is especially useful for creating satellite linkages, and its compact antennas/telescopes make it an attractive contender for deep space missions and inter-satellite links [2, 5, 13]. Optical Wireless systems have also proven to be a good fit for military applications because of their high bandwidth.

Despite the huge number of options provided by Optical Wireless systems, air losses and weather conditions along the propagation path have a negative influence on the wireless optical channel. Rain, fog, dust particles, snow, path loss, signal loss due to turbulence, and pointing error are all major degrading variables in the free space for optical signals [8–12]. Because of changes in the profile of the refractive index of the atmosphere, turbulence is generated by fast modifications in the intensity and phase of the FSO signal [1]. Temperature changes in the atmosphere give rise to optical signal fluctuations, which is a chaotic condition of air movements. Eddies or spherical areas constitute the turbulent medium in the atmosphere, with randomly variable sizes and refraction indices. In this randomly varying-refractive-index medium, the optical beam traveling in the turbulence channel undergoes random temporal and spatial variations, resulting in severe fading.

Dynamic wind loads, building movement, and/or weak seismic tremors can contribute to the misalignment of transmitter and receiver apertures, resulting in pointing inaccuracy [3–5] in Optical Wireless systems. The Beers-Lambert Law, on the other hand, determines the path loss of a wireless optical network [4]. Various mitigation strategies have been presented in the literature to compensate for the aforementioned obstacles. Forward error correction (FEC) has generally been considered in digital systems for improving reliability. Automated repeat request (ARQ) and dynamic thresholding have all been proven to be viable options [5, 14–19]. Aperture averaging has been used at the physical layer to circumvent the Optical Wireless systems' bottlenecks. The receiver aperture is raised in this case, which averages out the quick changes generated by minor eddies. In certain Optical Wireless systems, a feedback control technique called adaptive optics (AO) is utilized to minimize turbulence by applying a pre-corrected beam to the transmitter before delivering the signal into the turbulent channel. The conjugate of the atmospheric turbulent channel is applied to the beam for pre-correction. Certain Optical Wireless systems benefit from modulation scheme selection based on bandwidth and/or power efficiency. On–off keying (OOK) [20] and pulse position modulation (PPM) are commonly used optical modulation methods [21]. Jitter, coding, rejection, isolation, and background noise cancellation are some of the additional mitigating strategies used in Optical

Wireless communications. The relevance of the cooperative relaying approach, which was used as a mitigation mechanism in this study, is described in the next section.

Optical Wireless Transmission Systems

The information-bearing signal is converted into an electrical domain that modulates one or the other parameter of the optical signal while implementing an Optical Wireless system. The modification of the parameters of the laser beam gives a direct conversion from the electrical to the optical domain. The transmitter consists of a driver, laser/LED, modulator, and transmitting telescope. The nano-technology is being used to manufacture this equipment nowadays to enhance system characteristics [22] making it viable for IoT systems. The light intensity of the laser is directly modulated using a data stream, converting it into an optical signal. The property of the optical signal that is generally changed with respect to the information signal is the intensity. The information is converted into digital bit streams which is used to govern the signal fluctuations of the optical signal.

To control the current flowing through the light source, modulators and a driver circuit can be used. The sources generally employed are LASER or LEDs. The properties of high coherence optical beam always make the LASER as a preferred option among the two. Fabry–Perot (FP) which uses mirrors with separation as per linewidth required, Vertical Cavity Surface Emitting Laser (VCSEL) designed with heterojunction concept and Distributed-Feedback Laser (DF) designed based on Bragg's diffraction are a few types of LASERs that are designed and the choice of selection generally depends on the linewidth, bandwidth and power requirements of the systems. Generally, in industrial applications, the VCSELs are more affordable and perform better [23, 24].

The signal is received and processed using an optical receiver after it has traveled across the Optical Wireless channel. An optical receiver's five fundamental parts are the receiving telescope, filter, amplifier, detector, and demodulator. The photodetector (PD) receives the focused optical light from the channel through a telescope. A large detector telescope's aperture is crucial for gathering different uncorrelated radiations, averaging them, and focusing them onto the PD. The purpose of PD is to act as a converter of an optical signal into an electrical signal. The signal generated from the PD circuitry is applied at the input of optical filters, which considerably minimizes sun irradiation by removing the unwanted wavelength components. Traditionally, the avalanche photodiode (APD) and PIN photodiode are the two main types of PD that are most suitable for optical systems.

The design of the wireless optical systems depends on the available link budget. The choice of selection of the optical source and the sink is generally controlled by the several losses that take place in the medium. Generally, the link design is the most critical part of the implementation. Generally, the amplification of optical signal can be done in the optical domain using amplifiers that are used to enhance the quality of service. The optical signal received at the input of the optical receiver has a very small average signal power and the same needs to be improved for which preamplifier is used to boost the optical signal's intensity after being degraded by various atmospheric conditions. It suppresses the thermal noise created in the electrical amplifier that has a limiting influence on the receiver, hence considerably increasing receiver sensitivity. The purpose of the receiver section is to obtain an estimate of the phase of the received signal. The phase of the received signal contains the content of the information and is generally detected at the output of the PD.

Types of FSO Systems

Based on the number of optical beams, single-beam and multiple-beam systems are the two types of Optical Wireless systems that are available for IoT systems. A multi-beam beam Optical Wireless system aims to transmit data by using only one beam rather than numerous beams [23]. The air turbulence caused by large raindrops or haze impacts the characteristics of single beam Optical Wireless systems. This makes it exceedingly unlikely for a beam to reach the receiver in these circumstances. Since at least one of the numerous beams must strike the target, using several beams increases the chance of striking the target. The cost-effectiveness of the design is generally improved with single-beam system because of its simplicity. In a multi-beam Optical Wireless system, a laser beam combination method improves characteristics by minimizing the degradation of Optical Wireless power brought on by air turbulence. In addition, it is possible to minimize the effect of scintillation in the environment by using the laser beam merging technique, which is particularly helpful in Optical Wireless systems. It may be utilized, among other things, to resolve detector beam dispersion and power loss issues. It has been observed that a multi-beam method may be utilized to increase connection availability in coastal locations where there is restricted eyesight and coastal fog is assumed to be a confined phenomenon.

Transmission Schemes

The main transmission/reception scheme that is used in optical wireless systems is called the intensity modulation direct detection (IM/DD) system. The decoding approach is frequently employed to identify square-low levels of light, even though the information is sent through variations in light intensity. The optical technique used by IM/DD systems most frequently is called OOK. The OOK modulation has constraint spectrum efficiency due to its ease of usage and tolerance to nonlinearity. To increase spectrum efficiency, the OFDM technology simultaneously delivers data on orthogonal sub-carriers.

The resilience of multipath propagation is also improved by the OFDM modulation. The Optical Wireless system employs the multiple-input multiple-output

(MIMO) technology for increasing bandwidth and throughput. The technology of optical code division multiple access (OCDMA) also offers increased network capacity and privacy enhancement in addition to flexible channel allocation and asynchronous operation. OCDMA technology, on the other hand, is a bad choice for many applications because of its complexity and expensiveness. The ODM method is considered for enhancing the system throughput, number of user count, reliability in rain fading and link budget in the Optical Wireless system.

The combination of PPM and minimum shift keying (MSK) leads to a new scheme called the subcarrier intensity modulation (PPM-MSK-SIM) in Optical Wireless signals which is gaining prominence. The PPM-MSK-SIM modulation approach may make use of MSK's potent anti-interference qualities as well as the high power utilization ratio of PPM. The error rate characteristics across lognormal optical signal fluctuations channels may be thoroughly examined using the hybrid modulation approach proposed by several researchers. To get the best results, using an APD detector based on simulation is an option. In optical systems, studies have confirmed that PPM and BPSK-SIM are inferior compared to PPM-MSK-SIM in terms of reliability characteristics. The order of modulation can be changed as per the available channel conditions and amount of fading.

The above-mentioned techniques are modulated into the carrier of an optical system, where they will be conveyed, based on their properties such as amplitude, frequency, phase, and polarization. The phase modulation and IM/DD in combination with a coherent receiver system are generally suited for the best form of communication. The IM/DD scheme's great appeal is a result of how easily it can be put into practice [1, 2].

An additional relay-equipped method utilized in Optical Wireless systems is the multi-hop transmission mechanism. Using this method, the source and the destination nodes communicate using a series of intermediate relay nodes. Receiving and decoding the signal is the responsibility of every relay device. The decoded signal is sent to the following link in the chain, and the process is continued until the signal reaches the receiver. For reducing the impact of turbulence on the Optical Wireless network and increase its dependability, several relay nodes are utilized. The relaying procedure increases the coverage and enhances the error rate characteristics of the system.

Challenges in FSO Systems

Wireless optical technology offers several advantages such as less effect of electromagnetic interference, a large bandwidth leading to high data rates and no spectrum licensing. Instead, Optical Wireless is severely impacted by poor weather conditions. The enormous theoretical bandwidth that is available in the optical domain is limited by the losses that occur in the medium. Some sources of losses that exist in Optical Wireless system [2] are indicated in Fig. 1. Fog, rain, cloud cover and visibility factors always deteriorate the signal quality of the optical signal when transmitted in free space [1, 2]. A free space path loss factor is also available in FSO IoT services. The alignment between the telescope and aperture also induces fading called the pointing error or the misalignment fading. Atmospheric turbulence and interference also impact the reliability of optical wireless networks.

In this circumstance, external characteristics are connected to atmospheric conditions, and these factors are the source of losses. The physical and climatic conditions of the area where an Optical Wireless link is placed have a significant influence on how well it performs. Weather and installation factors such as atmospheric attenuation, visibility, alignment, interference from solar radiation, and obstructions in the LoS, all affect the signal quality and characteristics of the Optical Wireless link.

Following are the main challenges in FSO system:

Atmospheric Turbulence

The FSO system uses the free space environment as its transmission medium. A variety of concentric layers emerge in the atmosphere. The three primary layers that make up the hemisphere are the troposphere, stratosphere, and mesosphere. The temperature variations between these levels can be distinguished by their heights. The Optical Wireless communication route connects various strata inside the troposphere. Most of the atmospheric phenomena take place in the troposphere, including physical abstractions and scintillation [1, 2]. Although the theoretical method for detecting air turbulence has been around for a while, the experimental findings defy

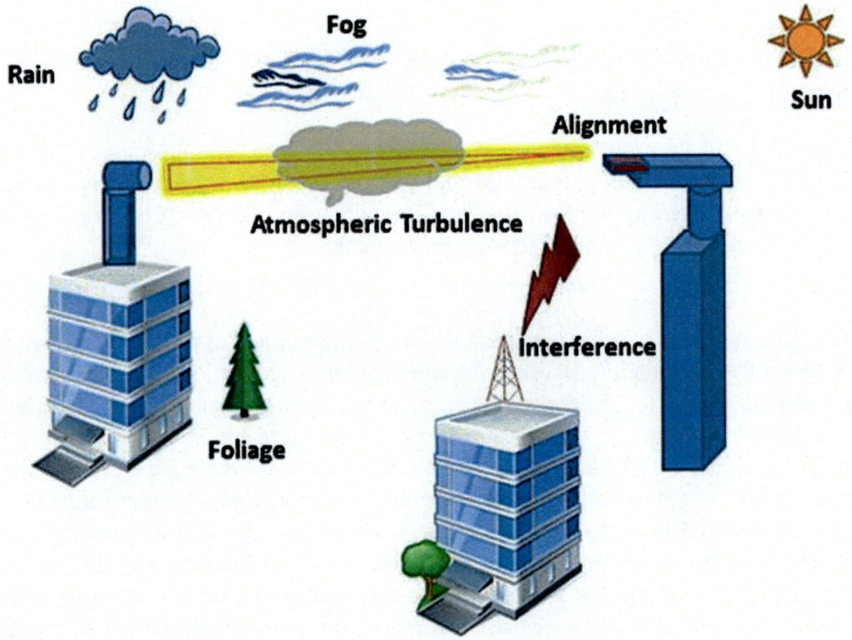

Fig. 1 Schematic representation of challenges of optical wireless system

the theory. Whatever the level of air turbulence in the area, operating the optical wireless system with an appropriate error rate is preferred. To improve Optical Wireless communication systems, horizontal propagation has received more attention. Several statistical models have been used to represent channels of varying intensities of optical signal fluctuations. For severe turbulence channels, the K-fading model has been selected as a helpful model because it offers strong congruence between theoretical and practical data. For modeling the strong fading conditions, the negative exponential model has been implemented. Log-normal distribution models the weak turbulence conditions while a wide range is modeled by Gamma-Gamma distribution. Some latest distributions facilitating the modeling of the fading conditions in the FSO IoT systems are Malaga and double generalized Gamma distributions.

Pointing Error

FSO system uses a narrow beam (Gaussian distribution) emitter and a wireless medium for data transmission, any inadequate transmitter and receiver positioning can cause pointing loss very frequently. When the Gaussian power distribution center is located close to the center of the detectors, it is feasible to establish excellent alignment conditions in an Optical Wireless system. But the characteristics are compromised when the detector and transmitter are not well aligned. The optical detector needs complete LoS track of the received signal and misalignment leads to signal fading [4]. The base motion of the structure is the most frequent source of misalignment in a practical Optical Wireless system. Skyscrapers themselves are subject to a significant amount of sway. An FSO system can be linked with an automated pointing and tracking system to reduce misalignment loss, with the tracking system continuously modifying the Optical Wireless system to get the best alignment.

Absorption and Scattering Loss

Scattering and absorption are the primary causes of loss in the atmosphere medium. Laser beams pass through particles and gas molecules as they pass through the atmosphere. Like any wireless transmission condition, the optical signal channel suffers from absorption and scattering phenomena.

Fog

Fog is defined as a weather condition with a visibility range of 0 to 2,000 m, which is generally the situation. Because fog produces both absorption and dispersion, it creates air attenuation. There can be a strong attenuation of more than 350 dB/km when there is severe fog. When water droplets have a size similar to the wavelength of IR radiation, the fog attenuation is greatest, and this is where the worst attenuation occurs. Very deep fog causes extremely high attenuation, with attenuation rates exceeding 100 dB/km. To put this in perspective, a practical FSO system may have a fading margin of 20–30 dB when deployed, implying a deployment distance of less than 500 m in fog. Fog is the most essential meteorological phenomenon in the research of Optical Wireless because it contains tiny water droplets of the same size as infrared wavelengths. The distribution of particle size changes depending on how much fog there is at any one time which can influence the IoT communication.

Rain

Rain has a different effect than fog. Raindrops have a larger diameter than the FSO communication wavelength. There is a 2.5 mm/h attenuation loss in light rain and a 25 mm/h attenuation loss in severe rain. When Optical Wireless modulated light travels through free space, atmospheric properties like scattering, non-selective scattering and absorption all have a substantial influence on its behavior. The wavelength characteristic is highly significant in absorption-based attenuation, with UV wavelengths experiencing the most severe absorption (below 200 nm). In addition, the wavelength has a considerable influence on how light scatters, and rain introduces non-selective scattering because of the size of the raindrops in the atmosphere. The climate in the European region, which is vulnerable to fog and severe snow, has a direct influence on how well an FSO link performs. Convective rainfall is caused by the production of convective clouds. These clouds occur when a moist atmosphere becomes overheated in comparison to the surroundings. This results in a strong upward movement as a result. True, this type of rain occurs as a result of a very unstable atmosphere, and it is differentiated by the presence of heavier and larger rain particles. The occurrence of this condition over a brief period and its characteristic rainfall rates of more than 10 mm/h has also been identified. These types of rainfall are caused by weather systems that exist around the equator of the planet. When warm air and cold air clash in low-pressure zones, it most usually results in heavy rain. Even though fusiform rain is the most prevalent type of rainfall, it cannot be discriminated. It occurs in clouds that form horizontally rather than vertically, such as nimbostratus clouds.

Snow

The dimension of snow lies in between that of fog and raindrops. Snow attenuates sound more than rain but less than fog. The attenuation is between 30 and 350 dB/km. Snowflakes are made up of ice crystals that come in a variety of sizes and shapes. Because the size of snowflakes is huge in comparison to the operational wavelength of the device, whiteout conditions may cause the beam to be attenuated. This is not a serious worry for Optical Wireless devices. The degree of attenuation in snowy situations ranges from 3 to 30 dB/km.

Attenuation

In the absence of fog, which is particularly frequent in tropical climates, rain attenuation is a major contributor to air scattering. The three most prevalent types are non-selective scattering, Mie scattering, and Rayleigh scattering. Mie scattering takes place when the size of particles is equivalent to wavelength. Rayleigh scattering effects if the particle's size is lower than the wavelength; however, at 1550 nm, the wavelength is too short for the particle to disperse, hence Rayleigh scattering is not present. Non-selective scattering happens when a particle's drop size is larger than the wavelength of light; this type of scattering is not wavelength dependent. The fact that the size of the raindrop spans from 200 to 2000 microns is what defines it as non-selective scattering. Given that this factor is far larger than the wavelength of the

Optical Wireless source light, it can be presumed that scattering has no appreciable influence on the Optical Wireless system. As a result of scattering effects, rainfall in Malaysia's tropical environment is the main cause of signal attenuation. This condition might result in signal degradation or even full signal loss because rain falls at a rate of more than 145 mm/h in some areas throughout the year, degrading or even eradicating the signal.

Mitigation Techniques

For FSO communications to be dependable and smooth, they must be able to provide large data rates at the required level of quality to communication service endpoints while operating in a variety of atmospheric conditions. This implies that Optical Wireless communications need to be more trustworthy to maintain operations when working in bad weather. In the end, atmospheric conditions have an influence on the functionality of Optical Wireless communications systems, both for terrestrial applications and for uplink-downlink.

For successful construction of Optical Wireless link, there are several significant issues that must be addressed. These issues are thoroughly explained in this section. The environment channel has a variety of drawbacks that can result in significant signal channel disturbance and, in extreme cases, complete signal loss. Weather-related problems like air absorption, atmospheric dispersion, and optical signal fluctuations might decrease the effectiveness of the Optical Wireless link. Scintillation or intensity changes can drastically reduce system characteristics, which is often measured by the error rate. Factors like beam wander and scintillations generally impact the signal-to-noise ratio (SNR) at the receiver front end. This degrades the received optical signal and generally, it is caused by air turbulence. Random SNR loss, therefore, influences the error rate, which measures system characteristics. The possibility of an outage is another criterion provided by channel fading for assessing the characteristics of communications systems. Long-distance broadcasts and communication with a moving platform can both benefit from the capacity to fade in and out. In the late afternoon, the system's reliability is negatively influenced by the weather. The optical signal sent through the environment is significantly reduced because of particle matter in the atmosphere's ability to both absorb and scatter light. In an effort to optimize operation in a variety of scintillation, the Optical Wireless equipment must be able to dynamically adapt to shifting atmospheric conditions. To accomplish dynamic characteristics optimization under shifting atmospheric circumstances, the free-space communication system must be created effectively. Understanding the communication consequences brought on by the atmospheric channel, which were covered in depth in the previous part, is necessary before deploying complex mitigation measures and improving overall characteristics. The following section is supported by a number of Optical Wireless mitigation techniques.

Aperture Averaging

One of the most straightforward methods to lessen turbulence is to increase the effective receiver aperture. If the receiving aperture is greater than the spatial dimension (due to turbulence) that produces the irradiance changes, it is possible for the aperture to average the received signal. The point receiver will be more sensitive in this situation, resulting in fewer irradiance variations or scintillations during detection. Aperture averaging causes the spectrum of the optical signal to move towards a lower spectrum because it averages out the turbulence's rapid small-scale size fluctuations. This is brought on by aperture averaging. Scale sizes larger than the aperture consequently cause the observed scintillation generated by a receiving aperture. Of course, the best receiver size will depend on the specifications of the system in question as well as the architecture of the communication link. A severe decline in scintillation occurs when the aperture size is increased up to the coherence time in the presence of significant turbulence, and this dip in scintillation lasts forever. There is a substantial increase in average power SNR due to averaging of signal power. This enhancement in SNR improves Optical Wireless system characteristics in the presence of air turbulence by bringing BER down to a manageable level.

Space Diversity Techniques

In Optical Wireless communications, two different types of diversity mechanisms can be used for reducing the impact of irradiance fluctuations and scattering, i.e., temporal and geographical diversity. Optical communication in free space has the advantages of being a license-free, low-cost, and high-bandwidth access mechanism for high-data-rate applications that do not call for a specific license. By reducing the influence of irradiance on Optical Wireless systems, diversity reception makes it possible to significantly improve characteristics. Data is sent from sources to receivers using diversity techniques using several identical broadcasting channels. To lessen the effective degree of scintillation and the likelihood of channel disturbance, uncorrelated channels are utilized in combination with multiple transmitters and receivers. This method establishes at least two pathways between the transmitter and the receiver using spatial transmission and reception variety to reduce turbulence. The transmitter and receiver must be set farther away than the link's lateral correlation distance in order to achieve the variation. Each road suffers individual channel disturbance when these roads are uncorrelated, which improves system efficiency because it is unlikely that all pathways would experience a profound fade at the same time. But, if a wide spatial separation is required, it will be difficult to execute the system since every source and sink should line up with one another.

Time Diversity Techniques

It is possible to use temporal diversity as an alternative to spatial diversity. Channel coherence time is used to send multiple copies of the same signal in different instances. Data transfers that include just one transmitter and one receiver are referred to as serial transmissions. Scintillation in the Optical Wireless system influences the signal quality with disturbances in the IM/DD system. Due to scintillation, the optical signal oscillates at the receiver, causing a signal disturbance in the system. The receiver output may fall below a predetermined threshold, and this estimate is used to assess the likelihood of this happening. The incoming signal in the turbulence channel will fade in and out at random intervals, based on the channel's probability distributions. Therefore, a precise probability density distribution (PDF) is required to forecast the probability of fade for an active Optical Wireless communication system.

Wavelength Diversity

The wavelength diversity approach has several applications. Mechanisms for delivering the same signal over many wavelengths have been recently revealed. Different wavelengths of channel disturbance qualities, such as intensity variations at the receiver, fluctuate due to random refractive index oscillations. By transmitting optical signals at different wavelengths, it is possible to minimize the impact of irradiance which affect a specific wavelength.

Coding Techniques

Additionally, it is feasible to increase Optical Wireless communications systems' dependability while maintaining the predefined error rate by applying coding strategies that have been proven effective in radio communication technologies and are applicable to optical communications as well. Coding methods can be used in terrestrial Optical Wireless communications systems for both SISO and MIMO mode transmission. However, because of practical restrictions like power, weight, and system size, the diversity technique for allowing redundant transmissions in a communication scenario like airborne terminals is not feasible. Forward Error Correction (FEC) is a simple method of reducing the influence of air turbulence that can be used in a variety of situations. Many research works have been done and are still being done on this subject. Each programmed Optical Wireless communication has a unique set of guidelines and augments the architecture with new components. Coding is employed to weaken the turbulence at the expense of cost, system complexity and applicability. For instance, FEC codes might introduce faults by adding check bits to the data stream to increase power and bandwidth. The Optical Wireless mechanism in FEC has a substantial overhead. The communication setup, which takes range, throughput, and reliability into account, frequently affects the choice of the coding scheme. The coding is decided using the appropriate modulation methods, the quantity of available laser transmitter power, and the detector sensitivity, if applicable. The only way forward is to choose the finest possible coding system. The characteristics of a communication system in a channel disturbance condition are improved via error-correcting, a fade mitigation approach. The delivered data stream now has redundancy, which requires a higher rate and results in an increase in bandwidth use. An Optical Wireless system with data speeds of 10 Gbps or greater should take into account the computational complexity and power requirements of FEC implementation. Power efficiency and spectral efficiency are trade-offs to consider while selecting a particular FEC.

Cooperative Relaying Technology

Multiple transmitter and receiver structure implementations have gained attention in the literature [25–30]. These multiple input multiple output (MIMO) systems may be divided into three groups. The first is the spatial multiplexing system, in which the same data signal is sent by numerous transmitters and receivers to obtain multiplexing advantages. The second kind is a spatial diversity system, in which redundant data is sent from many transmitters to better the optical link's condition. Different combining schemes are used with great success in wireless communications, such as selection combining, only processing signals with the maximum SNR, and maximal ratio combining (MRC), where each signal is weighted according to its received signal strength [31]. The third type of multiple-transceiver system uses a beam-forming approach to direct the transmitter beam to the intended target. It's worth noting that the multiplexing and/or diversity advantage in each of these three approaches comes at the expense of increased hardware. To circumvent this, cooperative relaying systems are utilized to provide variety to the communication system by re-purposing existing equipment. Communication between two nodes is accomplished using the hardware of intermediary nodes in a cooperative relaying scheme. Because the intermediary nodes are often existing mobile users in the communication system, this approach can provide spatial diversity/multiplexing gain based on channel conditions without raising hardware requirements [32]. Cooperative relaying strategies are used to describe virtual MIMO systems in which numerous transmitters and receivers are dispersed among several devices. Apart from delivering diversity benefits, such relaying systems may also expand the network's coverage area [10].

In cooperative relaying strategy, the communication between two nodes takes place with the help of the hardware of one or more intermediate nodes called as a relay node. As indicated in Fig. 2, in dual-hop cooperative relaying technology transmission of information is done in two hopes; the first hop is called as a source to relay node (S-R) and the second hop is called a relay to the destination node (R-D). This mechanism increases the capacity and coverage of the network. It can also bring spatial diversity/multiplexing gains as per channel condition, but without an increase in hardware requirement as the intermediate nodes are generally the existing mobile users in the communication system. The following main types of relying systems can be found in the open literature:

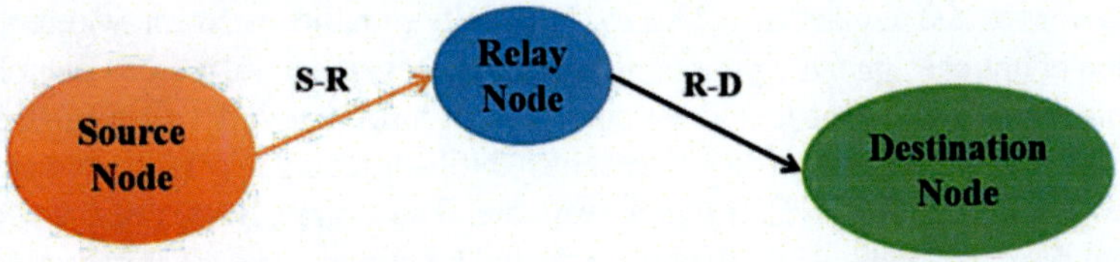

Fig. 2 Schematic representation of dual hop cooperative relaying technology

Amplify and Forward Relay

After collecting the information, the relays first amplify the analog signal and then forward the amplified version to the intended device in amplify and forward (AF) relaying. These are also known as non-regenerative relays since no decoding procedure is done. AF relaying systems may also be divided into fixed and variable gain networks. At the start of data transmission, the relay gain of fixed gain relays is set [6, 33]. Relay gains in a variable gain system, on the other hand, are governed by the condition of the channel of the source-to-relay link [9, 34, 35].

Decode and Forward Relay

The message provided by the source is first decoded by the relay node. The decoded message is retransmitted as a newly encoded version to the destination node in decode-and-forward (DF) relaying systems [36, 37]. Regenerative relaying is the name given to this type of relaying architecture. The DF relay technique usually comes in three flavors. The first is the basic DF scheme, in which the message in Phase-I is decoded by the relay and, if the decoding in Phase-I is successful, passes the new coded message in Phase-II. The second form is the selection DF, in which if it is not possible to decrypt the message in Phase-I, permits the source node to retransmit the original message in Phase-II [32]. Demodulate-and-forward (DeF) relaying is the third type of DF relaying, in which if it is not possible to decode the original version of the signal, it can demodulate the message symbol-by-symbol.

Coded Cooperation

The DF relay technique may be thought of as a broader variant of coded cooperative relaying methods. Stronger channel codes are used in such cooperative relaying schemes to boost the overall system's spectral efficiency [38–40]. Different pieces of the same message are broadcast in two phases in this system. The same version of the message is not sent twice by the source and relay nodes. In essence, in Phase-I, the source delivers the coded form of the message without the incremental redundancy that is sent in Phase-II by either the source or the relay node.

Asymmetric Mixed RF/FSO Relaying Schemes

Because of its advantages, FSO technology has piqued the interest of the research arena over the last few decades. Due to the paucity of radio spectrum, wireless transmission of optical signal is a promising wireless access potential that can complement or replace radio solutions to address current "last mile" network issues [6–9, 33–37, 41–50]. RF systems, on the other hand, have built-in fading resistance and are often low-cost solutions for an LoS is not available. The characteristics of both RF and Optical Wireless channels is adversely affected by climatic and other natural factors. When compared to Optical Wireless connections, radio links perform better in poor channel circumstances. Furthermore, when it comes to mobility, radio technology

Fig. 3 Schematic representation of mixed RF/optical wireless asymmetric dual hop relaying technology

will always be the favored mode of communication. Optical Wireless systems have been offered as an option for backhauling for multiplexing many radio channels due to properties such as enormous accessible capacity in the optical domain, minimal latency, and easy implementation. Due to this reason, mixed radio/Optical Wireless relaying systems as shown in Fig. 3 have found enormous attention of researchers.

The authors of [6] proposed asymmetric radio/Optical Wireless relaying strategy, where the source-to-relay node channel is designed to be a radio connection, and the link between the relay and destination nodes is an Optical Wireless connection. As both hops of communication fade at different rates, this relaying is said to be asymmetric relaying. Rayleigh and Nakagami-m, Rician, η-μ, Generalized-K fading models channel models are used in RF systems. The turbulence (or irradiance variations) on the Optical Wireless connection is captured according to the degree of fading, which ranges from weak to moderate to strong. In the literature, the negative exponential (NE) model for irradiance has been used for strong turbulence whereas the log-normal (LN) fading model represents weak turbulence conditions. The Gamma-Gamma turbulence model can simulate environments with moderate to high scintillation. The relay approach used in [6] is AF relaying, which results in a low-cost, low-complexity system, whereas multiple carriers are superimposed using the subcarrier intensity modulation (SIM) technology.

In the realm of mixed communications with Optical Wireless technology, Refs. [51–55] were recently reported. While the authors of [51] offered an analysis of a hybrid Optical Wireless/radio selective DF relaying method with switching control, the authors of [52] presented an analysis of a deep space-terrestrial Optical Wireless/RF paradigm for both types of AF relaying systems. The research paper [53] evaluates the secrecy characteristics of mixed RF/Optical Wireless relaying systems by determining the key parameters to be considered for eavesdropper assaults. Furthermore, the work [54] demonstrates that the hybrid Optical Wireless/radio mode of communication outperforms ground-to-satellite communications. The authors of [55] provided an analysis of deep space relaying communication using mixed radio/Optical Wireless relaying to connect with a geostationary satellite and obtain the expressions of probability of outage, error rate, and ergodic capacity.

Four transmit antenna selection strategies for mixed radio/Optical Wireless systems in the presence of numerous eavesdroppers have been presented in [56] to display the statistical modeling of the system thus devising the PDF and cumulative distribution function (CDF) of a mixed RF/Optical Wireless MIMO system. In [56], the fading on the radio link has been assumed to follow Nakagami-m and Optical

Wireless link experiences Malaga distributed turbulence-induced fading. For this framework, an approximate bound for the effective secrecy capacity was derived. The authors of [57] provided a cognitive network model with mixed RF/Optical Wireless relays and deduced an equation for the user's probability of outage. On the other hand, in [58], the secrecy capacity and probability of outage were studied in the face of a friendly jammer, with the radio connections undergoing Nakagami-m and the Optical Wireless link undergoing exponential Weibull fading models. Maximization of the average signal-to-interference-plus-noise ratio (SINR) of the users has been done in [59] using the beamforming technique, for mixed radio/FSO relaying networks for constructing satellite-aerial-terrestrial systems. The outage performance has been determined using closed-form formulas where Nakagami-m and Malaga distributed channel models have been used. The likelihood of positive secrecy capacity and secrecy outage probability has been analyzed in [53], demonstrating the relevance of mixed RF/Optical Wireless communications.

Recent efforts in the realm of Optical Wireless technology, however, are not restricted to mixed radio/Optical Wireless systems. The application of deep learning (DL) detection algorithms to overcome scintillation constraints is demonstrated in [60]. A method for detecting OOK signals was given that did not require prior information of the channel and used multiple models of DL for varying strength turbulent channels. The results of [60] show that DL decoders improve Optical Wireless system operation while reducing power consumption. The authors of [61] created a restricted stochastic optimization framework that covers numerous Optical Wireless case studies such as relay selection and power adaption. The authors presented two alternate resource allocation techniques based on machine learning. The authors of [62] implemented orbital angular momentum-shift keying for Optical Wireless communication, emphasizing that modulation strategy is important for delivering videos over Optical Wireless channels. In [62], a 3D chaotic interleaved multi-coded system was used to transmit video frames, with the 3D convolution neural network (3DCNN) and convolution recurrent neural network (CRNN) being exploited to improve the efficiency of systems. When compared to existing modulation methods, the findings of this study showed that CRNN and 3DCNN performed better. In the study [63], the influence of irradiance on phase fronts of orbital angular momentum (OAM) beams was investigated using deep learning algorithms. The OAM beams were modeled in machine learning with enhanced cognitive accuracy. In [64], a deep learning model was recently developed to investigate adaptive demodulation techniques for following OAM on Optical Wireless lines. The published model may demodulate many types of higher-order modulation schemes by generalizing the influence of irradiance variations.

Mixed RF/Optical Wireless Relaying as Backhaul Networks

In 2020, the data traffic demand on cellular networks is estimated to surpass 500 hexabytes, according to [65]. Because of the increased capacity required, a lot of small cell base stations have been installed to deliver higher data rates with reduced latency [7]. The technology used as backhaul to connect these base stations determines the overall characteristics of the communication system. Combined RF/Optical Wireless relaying systems have been offered as a key possibility as a backhaul solution in this area [7]. Backhauling options have traditionally been copper, microwave, or optical fiber communications. With data speeds of 1.544 Mbps for T1 and 2.048 Mbps for E1 carrier systems, leased T1/E1 copper wire backhaul lines have been used in second (2G) and third-generation (3G) mobile systems [66]. Multiple parallel connections of copper lines can be used to enhance data rates of future mobile communications, but this increases cost. Optical fiber connection is an alternative for obtaining higher data speeds, but it comes with a higher initial cost and setup time. Furthermore, if wired lines are not possible to construct, wireless setups such as RF and FSO are preferable. Although RF backhaul networks even function when a LoS is not present, they have a greater latency. In contrast to RF channels, LoS Optical Wireless links may function with very low latency, but they are badly influenced by weather conditions. Due to the availability of several intermediate base stations, next-generation wireless networks provide instances where the backhaul connections between the two base stations are a mix of NLoS and LoS links. A joint venture of RF and Optical Wireless communication technologies can improve the overall user experience in terms of service quality in such instances. Therefore, combined RF/Optical Wireless cooperative relaying systems have the potential to provide backhaul solutions for next-generation mobile networks.

Related Work

FSO Channel Models

Turbulence is created in the Optical Wireless channel change of homogeneity of the refractive index of the free space as a cause of temperature and pressure changes [1]. Even when the sky is clear, optical wireless networks may experience signal channel disturbance due to the turbulent environment. In the open literature, several statistical distributions have been presented to characterize the amplitude and phase variations occurring in the wireless channel with regard to scintillation. In [67], K distribution was proposed, which was used to estimate the irradiance on the Optical Wireless connection for a high turbulence environment. The LN distribution, on the other hand, has been used to mimic weak turbulence [68]. Inverse-K (IK) and log-normal Rician scintillation models have been derived from K and LN irradiance models in [69, 70], respectively. Furthermore, the NE distribution has been used to mimic high turbulence conditions (i.e., saturation regime and beyond) [71]. Andrews et al. introduced the Gamma-Gamma turbulence model which is the most important optical power irradiance model in [72]. Researchers have focused their attention on this distribution because different combinations of model parameters may replicate turbulence situations ranging from mild to moderate to high. Using the α-μ channel disturbance model, a Gamma-Gamma turbulence model has been effectively created in [73] as per the generalization capability of the model. Various combinations of α and μ can be utilized to represent the complex envelope of the wireless signal. The authors of [74] have also created a new PDF of the Malaga distribution [74], whereas the generic exponential Weibull distribution has been presented by the authors of [75]. In [74], the authors have also successfully derived most of the previously discussed models as special cases of the proposed Malaga distribution. Yet another significant irradiance fluctuation model, introduced by Kashani et al. in [76] is called the D-GG turbulence model. The product of two independent Generalized Gamma distributed random variables yields this model. This distribution can closely resemble the optical wireless channel disturbance channel's tiny and large-scale irradiance changes. The D-GG turbulence model may be viewed as a particular instance of the previously mentioned models.

Double Generalized Gamma Model (D-GG)

The D-GG channel disturbance irradiance I can be described as the multiplication of two Gamma distributed random variates. Specifically, it is defined as $I = I_1 I_2$, where $I_1 \sim GG(\alpha_1, \beta_1, \Omega_1)$ and $I_2 \sim GG(\alpha_2, \beta_2, \Omega_2)$. Shaping parameters are denoted as β_1 and β_2, whereas α1, Ω1, α2 and Ω2 depend on the variances of small and large scale field variations. It is to be noted that the D-GG model reduces to most of the current distributions when different combinations of model parameters are used as descriptive in Table 1. The PDF for the SNR irradiance fluctuation over D-GG distribution is formulated as [76]

$$f_\gamma(\gamma) = \frac{D_1}{r\gamma} G^{\lambda+\sigma+1,0}_{1,\lambda+\sigma+1}\left[D_2 z^y \left(\frac{\gamma}{\mu_r}\right)^{\frac{y}{r}} \middle| \begin{matrix} \tau_2 \\ \tau_1 \end{matrix} \right] \tag{1}$$

where $D_1 = \frac{\left(\xi^2 \sigma^{\beta_1 - \frac{1}{2}} \lambda^{\beta_1 - \frac{1}{2}} (2\pi)^{1-\frac{\sigma+\lambda}{2}}\right)}{\Gamma(\beta_1)\Gamma(\beta_2)}$

$D_2 = \frac{\beta_1^\sigma \beta_2^\lambda}{\lambda^\lambda \sigma^\sigma \Omega_1^\sigma \Omega_2^\lambda}$

$\tau_1 = \left[\frac{\xi^2}{y}, \Delta(\sigma : \beta_1)\Delta(\lambda : \beta_2)\right]$, $\tau_2 = \left[1 + \frac{\xi^2}{y}\right]$, $\mu_r = \frac{(\eta E^r[I])}{N_0}$.

The D-GG model is generalized model and can represent existing channel models as special cases for particular combination of the model parameters as given below.

Table 1 Generalization capability of D-GG model [76]

S.No.	Special cases	Models
1	$\alpha_i \to 0, \beta_i \to \infty$	Log-normal
2	$\alpha_i = 1, \Omega_i = 1$	Gamma-Gamma
3	$\beta_i = 1$	Double Weibull
4	$\alpha_i = 1, \Omega_i = 1, \beta_i = 1$	K-channel

Table 2 Generalization capability of Malaga model

S.No.	Distribution	Special cases
1	Rice-Nakagami	$\rho = 0, \|Var[U_L]\| = 0$
2	Gamma	$\rho = 0, \gamma = 0$
3	HK distribution	$\|Var[G]\| = 0, \rho = 0, X = \gamma$
4	Gamma-Gamma	$\rho = 1$, then $\gamma = 0, \Omega' = 1$
5	Log-normal	$\rho = 0, \|Var[U_L]\| = 0, \gamma \to 0$
6	K-distribution	$\Omega = 0, \rho = 0$ or $\beta = 1$
7	Exponential	$\Omega = 0, \rho = 0, \alpha \to 0$

MALAGA Model

The Malaga distribution was developed in [74] based on the fact that as a wave travels through a medium, some of the power is dispersed along with the type of scattering determines the shape of the PDF of the statistics of the channel. Three terms make up the observable field at the receiver: first term indicates LOS, U_L, next is quasi-forward scattered by the eddies represented by U_C and the remaining term is the energy scattered due to eddies represented as U_G. Utilizing the mentioned terminology, the PDF has been derived for the generalized case in research work [74]. The Malaga distribution PDF is as given below [74]:

$$f_\gamma(x) = \frac{\xi^2 A_2}{2^r x} \sum\nolimits_{i=1}^{\beta} x_i G_{1,3}^{3,0}\left[A_3 x^{\frac{1}{r}} \middle| \begin{matrix} \xi^2 + 1 \\ \xi^2, \alpha, i \end{matrix} \right] \tag{2}$$

where, β is a natural number depending on the channel disturbance, ξ is the pointing error parameter and, $A_3 = \frac{2\alpha^{\frac{\alpha}{2}}}{g^{\left(1+\frac{\alpha}{2}\right)}\Gamma(\alpha)} \left(\frac{g\beta}{g\beta+\Omega} \right)^{\beta+\frac{\alpha}{2}}$.

The details of the remaining parameters can be obtained from [74]. Importantly, the Malaga distribution can represent the various statistical irradiance fading models that have been used for representing the Optical Wireless channels. Table 2 illustrates the capability of the model for representing the various existing models.

Optical Wireless Relaying Systems

Because of its benefits, such as energy efficiency and enhanced coverage, as well as the increased capacity afforded by overall communication networks, relaying techniques have sparked a lot of attention. In [6], a dual-hop AF relaying technique has been introduced and investigated, where the relay to destination optical wireless connection is following the Gamma-Gamma irradiance distribution while the RF link is following the Rayleigh channel disturbance model. Hybrid RF/Optical Wireless relaying is discussed in research work [7] as a major backhauling solution for succeeding wireless systems. Authors of [8] extend the study of [6] to investigate asymmetric relaying systems with pointing problems on the Optical Wireless connection. In particular, the RF connection in [8] is expected to follow Rayleigh distribution, but the optical wireless link is influenced by air turbulence represented as Gamma-Gamma distribution. In work [9], signal fading on the RF channel is modeled as the Nakagami-m model, which also represents Rayleigh statistical model, while turbulence on the Optical Wireless connection is captured as Gamma-Gamma distribution. Here, two types of AF relaying strategies were considered: fixed gain and variable gain relaying. In research studies, combined RF/Optical Wireless relaying has been further investigated [33–37]. The influence of hardware impairments was explored in [33], whereas the authors of [34] published an analysis of outage probability for variable gain combined RF/Optical Wireless AF relays. A similar study, [35], the existence of obsolete channel state information was observed in a variable gain combined RF/Optical Wireless AF relay network (CSI). Research studies [36], combined RF/Optical Wireless DF have demonstrated the characteristics of mixed DF relaying networks, which is a significant addition. Likewise, the authors of [41] examined the performance of Optical Wireless systems exposed to Malaga distribution. In [41], the generalized Malaga distribution is exploited to represent fading, capturing Malaga distribution and the Gamma-Gamma model as specific examples to examine the characteristics of fixed gain AF cooperative relays in the face of alignment loss. Mathematical analytical formulas for the total system's outage probability, ergodic capacity, error rate, and moment generating function (MGF). Furthermore, the authors have provided the partial relay selection (PRS) strategy for combined radio/Optical Wireless transmission scheme in [42]. Furthermore, the influence of an eavesdropper on a combined radio/Optical Wireless AF system has been studied by the authors of [43] where a jamming model has been proposed that allow for transmission of information in a secured manner. As the order of modulation is increased, the quality of service can be enhanced in optical wireless links subject to slight enhancement in error rate performance as declared in the research work presented by the authors of [44]. Furthermore, asymmetric RF/Optical Wireless relaying, comprising of millimeter-wave (mm-Wave) link has been discussed in [45]. The Rician distribution has been considered for modeling the signal fading on the mm-Wave link. The Optical Wireless channel is considered on the second hop for analyzing the effects of combining the capabilities of mm-Wave and optical communication systems. Researchers in [46] obtained accurate and asymptotic approximate formulas for the outage probability and throughput that can be obtained from Optical Wireless/mm-Wave relaying systems in a similar study. The authors of [47] looked into asymmetric RF/Optical Wireless relaying systems from a cognitive radio viewpoint, where the signal intensity of users on the radio channel is captured as a Nakagami-m statistical distribution and scintillation on the FSO channel was considered to undergo D-GG statistical. The authors of [48] investigated MIMO RF/Optical Wireless DF relay networks for cognitive radio, whereas the authors in [49] studied the effect of aggregated hardware impairment on the behavior of combined radio/Optical Wireless relaying systems. In [50], the authors study the Optical Wireless systems with misalignment fading over D-GG statistical channels. The authors have developed accurate and high SNR approximate analytical formulas for the BER, outage probability, and ergodic capacity of the combined radio/Optical Wireless cooperative network. Table 3 provides a quick summary of published important research works in this field.

Multi-hop and bidirectional relaying, as well as user diversity strategies, have been promoted with varying success for boosting the overall characteristics of terrestrial Optical Wireless communication systems. In research efforts [77–80] the relaying idea has been expanded to a multi-hop relaying scheme which enhances the energy efficiency and coverage area of the cooperative relays. The average error rate and outage characteristics of a multiple hop DF parallel relaying Optical Wireless transmission system operating in joint impacts of path loss, misalignment fading, and irradiance fading captured as Malaga distribution have been presented in Ref. [77]. The characteristics analysis of serial relays is presented in [78]. In addition to this, Ref. [79] investigates the characteristics of a system where data transmission is done using multi-hops with pointing error. In [80], the background in Ref. [79] is extended to deduce the analytical formulas for the error rate and ergodic capacity for the multiple-hop Optical Wireless system. In order to execute relay selection using PRS, the authors constructed formulas for the probability of outage, error rate, and ergodic capacity with obsolete CSI. The bidirectional relaying technology known as two-way relaying (TWR) can increase the spectral efficiency of wireless channel disturbance channels [96, 97]. As a result, research papers [81–84] have been published on the application of TWR method for Optical Wireless systems. The authors of [81] looked at the capacity of an all-optical TWR system utilizing the relay selection protocol and an MGF-based technique, whereas the authors of [82] looked at the outage analysis of TWR systems. The authors recently published a characteristics investigation of TWR Optical Wireless systems across D-GG turbulence channels in [83] with non-zero boresight pointing error.

Several research studies have also used diversity methods to improve the dependability of wireless channel disturbance channels. In a wireless setting, user diversity advantages may be acquired by observing that assigning the resource to various users optimally can boost the overall efficiency in terms of dependability and data rates among numerous users with independently fluctuating channels. Opportunistic scheduling is a scheduling technique in which resources are assigned to the wireless user operating in favoring conditions. It has garnered the attention of researchers. Particularly in the downlink, where, despite the fact that all users share the same communication channel, only the best user is assigned to send data.

In Optical Wireless systems, multiuser diversity (MUD) schemes have been incorporated in research works [84–89] for Optical Wireless transmission systems in order to bring about improvement with the aid of user diversity. The authors of [84] looked at the characteristics of DF mixed RF/Optical Wireless TWR systems that used user diversity to enhance the RF connection. Furthermore, the authors of [84] have provided a power allocation technique reduction in probability of outage for DF TWR system under consideration. While the authors of [85] proposed a user scheduling method, the authors of [86] offered a study of a hybrid Optical Wireless/RF system. The idea of effective capacity in terms of service quality, on the other hand, has been

Table 3 Summary of published important research works

Reference	First hop	Second hop	Remarks
[6]	Rayleigh	Gamma-Gamma	Outage probability
[8]	Rayleigh	Gamma-Gamma	Outage probability, capacity, error rate
[9]	Nakagami-m	Gamma-Gamma	Outage probability, capacity, error rate
[33]	Rayleigh	Gamma-Gamma	Outage probability, capacity, bit error probability
[34]	Rayleigh	Malaga	Outage probability (AF, variable gain)
[35]	Rayleigh	Gamma-Gamma	Outage probability (AF, CSI assisted relaying)
[36]	Gamma-Gamma	Gamma-Gamma	Outage probability, DF relaying
[37]	Malaga	Malaga	OP, and ASER, DF
[41]	Rayleigh	Malaga	Outage probability, BER, capacity
[42]	Rayleigh	Gamma-Gamma	Outage probability and the average BER
[43]	Rayleigh	Gamma-Gamma	Secrecy outage probability
[44]	NA	Gamma-Gamma	Outage probability, BER
[45]	Rician	Malaga	Outage probability, BER, capacity, mm-Wave
[52]	Rayleigh	Gamma-Gamma	Outage
[46]	κ–μ	Malaga	Outage probability, mm-Wave
[47]	Nakagami-m	Optical losses	OP for cognitive AF fixed and variable gain
[48]	Nakagami-m	Optical losses	Effect of mobility for DF cognitive relays
[49]	Rayleigh	Malaga	OP, capacity, SEP, hardware impairment
[50]	Rayleigh	D-GG	Outage probability, BER, capacity
[77]	Malaga	Malaga	Outage probability, BER for DF optical wireless relaying
[78]	Gamma-Gamma	Gamma-Gamma	Outage probability, BER for AF and DF
[79]	Gamma-Gamma	Gamma-Gamma	Outage probability, BER
[80]	Gamma-Gamma	Gamma-Gamma	BER, capacity for AF and DF relaying
[81]	Gamma-Gamma	Gamma-Gamma	ASR for two way relays using MGF approach
[82]	Gamma-Gamma	Gamma-Gamma	Outage probability for two way relaying
[84]	Rayleigh	Gamma-Gamma	OP, BER, capacity for TWR with MUD
[85]	NA	NA	Weak-turbulence conditions, MUD
[86]	Nakagami-m	Gamma-Gamma	OP, BER, MUD
[87]	Gamma	NA	MIMO, effective capacity
[88]	Rayleigh	Gamma-Gamma	OP, MIMO
[89]	η–μ	Malaga	OP, BER, capacity, MUD
[90]	Rayleigh	D-GG	OP, BER, CCI

(continued)

Table 3 (continued)

Reference	First hop	Second hop	Remarks
[91]	Nakagami-*m*	Gamma-Gamma	OP, CCI
[92]	Malaga	*K*-model	OP, BER, capacity, CCI, MUD
[93]	Malaga	*K*-model	OP, BER, CCI
[94]	Nakagami-*m*	Gamma-Gamma	OP, intercept probability, CCI
[95]	Nakagami-*m*	D-GG	Outage

examined in [87]. Similarly, the authors of [88] showed the outage characteristics for fixed and variable gain cooperative relay networks with diversity combining and selection. Yang et al. [89] studied the characteristics of the MUD-RF/Optical Wireless AF relays, where the η-μ distributed fading model has been used to represent user diversity on the RF link and irradiance on the Optical Wireless link was captured using the Malaga fading model. The analytical expressions for the error rate, ergodic capacity and probability of outage for the relaying network have been published by the authors.

Interference in Combined RF/FSO Systems

Though the interference does not bother FSO systems, using radio systems in a combined RF/FSO relaying approach exposes the total system to interference from nearby communication systems. Furthermore, largely the erstwhile studies on the characteristics of combined RF/FSO relay network have assumed interference-free information signal transmission. Incorporating interference into asymmetric relaying can improve the system model's practicality. The influence of interference on the functioning of combined radio/Optical Wireless relays was previously discussed in [90–95, 98–100]. The authors of [90] investigated the characteristics of asymmetric AF cooperative relaying systems for understanding the influence of numerous co-channel interferers (CCIs). Investigation of a relaying approach with interference where a direct link also exists between the two communicating nodes has been considered in this paper. After combining the signal at the destination node using MRC approach, the total network's outage characteristics were examined. The optical channel in [90] has been captured with a D-GG channel, whereas the radio link is anticipated to go through Nakagami-m statistical model.

The probability of outage for a combined radio/Optical Wireless AF relaying method influenced by numerous CCIs near the intermediate node when the Optical Wireless link is vulnerable to pointing error has been investigated by Petkovic et al. in [91]. The influence of unwanted signaling on AF relays has been examined in [91] by modeling the interferers and the information-carrying signal on the RF link following Nakagami-m model channel disturbance, whilst the FSO link suffers from Gamma-Gamma signal variations. Outage probability equations for the scenario where the number of pointing mistakes approaches zero have also been included in this work. The authors have also obtained the analytical formula for the capacity of the studied system for multi-aperture Optical Wireless/multi-user RF cooperative systems in [92]. The influence of CCI on the characteristics of combined radio/FSO systems is addressed in [93], where interference has been addressed over generalized channel disturbance channels. Furthermore, the authors of [94] performed a characteristics analysis of a multiuser combined RF/FSO cooperative strategy in the vicinity of eavesdroppers and presented a power allocation technique to improve the system's outage characteristics. For CCI-influenced combined radio/FSO system with PRS scheme, the authors have derived accurate and asymptotic closed-form formulas for the probability of outage, error rate, and ergodic capacity in [95]. In [95], the interfering signals are captured as Nakagami-m distributed, while the FSO link is captured as D-GG turbulence. The authors of [98], examined multi-user diversity for a combined RF/FSO DF interference restricted relay, while [99], implemented the idea of user diversity and assumed that the RF and FSO links fade according to the Nakagami-m and D-GG models to examine the effect of interference. Another significant contribution was made by the authors of [100], who examined combined radio/FSO AF semi-blind relaying systems to determine how interference and non-zero boresight errors affected the overall system characteristics.

Impact of Hardware Impairment

Non-ideal restrictions occur from carrier component imbalances, high power amplifier (HPA) non-linearities, and phase noise, to name a few physical flaws that are pervasive in relaying systems using low-cost hardware [101]. Furthermore, relaying strategies with FSO link need to function on a high frequency. Due to this, the hardware switching rates are very high. Hardware is more susceptible to switching flaws. In research studies on radio transmission systems, the influence of hardware impairments was studied in research works [101–104]. The authors of [101], on the other hand, studied the aggregate influence of hardware irregularities on TWR networks on a macro level. The authors of [104] have made an important addition by analyzing the characteristics of combined radio/FSO AF and DF relaying with practical hardware degraded operation of the three nodes involved. Studies [105–109] investigate the influence on the characteristics of combined RF/FSO relays. Authors have published an investigation of PRS for a hybrid RF/FSO AF relaying system with non-ideal hardware in [105]. Although some results existed for a network with similar conditions, no results for non-ideal characteristics of hardware was published before. This makes the contribution important. In [106], a PRS protocol using outdated CSI for selecting the best functioning relay for combined radio/FSO relaying with non-ideal hardware operation has been presented. Importantly, recent research works [110, 111] have analyzed the influence of CCIs and hardware irregularities on combined RF/FSO relaying systems.

Future Scope and Relevance to IoT

The survey of the literature presents several key opportunities for the upcoming researchers to study the functionalities related to combined radio/FSO relaying on topics like IoT still not attended. It is clear that there is no research work that has attempted key-hole communication channel with FSO systems. Analysis of full duplex two-way relaying systems has been avoided by researchers till date due to the involved mathematical complexity. The full duplex systems have the influence of self-interference involved in their functioning which can be minimized using cancellation techniques, which are still to be evaluated for combined RF/FSO relays.

It is very interesting to note that there is no work that has investigated coverage probability for combined radio/FSO relaying systems. Although the mathematical computation required will remain to be seen, it may be taken up to advocate the influence of the addition of RF with FSO to enhance the coverage. The investigation of underwater FSO relaying strategy with various related interference has been missing in the literature. Another important that is to be attempted is to develop a relatively simpler PDF for FSO systems where the impact of turbulence is considered and can be utilized to analyze more complex scenarios such as MUD and MIMO to find coding gain and diversity order on the FSO link.

It is important to visualize that when IoT systems are built, the scarcity of spectrum needs to be solved. For ad-hoc basis implementations, which is generally the key requirement for any IoT system, FSO technology will be a viable contender. However, due to limitations in the range of communication with FSO technology, it is necessary to include some other form of communication, such as RF. Therefore, for quick deployment of IoT systems, the combined RF/FSO systems are very convenient and can prove to be fruitful. The future generation of IoT systems that will connect millions of devices also need the protection of the security aspect of signaling which is inbuilt in the FSO communication. Once the practical constraints such as interference and hardware impairment are checked and mitigated, the combined RF/FSO cooperative relaying systems have the capability to introduce scalability in the integration of IoT devices and maintain seamless connection among them. The research on combined RF/FSO relaying for IoT devices needs proper attention as the involvement of RF systems also brings some security concerns and a proper power allocation scheme are indispensable to save the battery of the connected devices in IoT framework. Therefore, the researchers in the area of combined radio/FSO cooperative communication need to propose scheme that maintain the required security and power levels in device-to-device communication.

Conclusion

The review presented in this chapter provides an overview of the possibility of integrating FSO technology with IoT. FSO is a quickly growing technology and provides unlicensed parts of the EM spectrum. The chapter discussed how FSO technology fits in the IoT architecture due to various favorable features such as quick deployment and high security. These features make the integration between the FSO and IoT systems seamless. However, several key restricting features of the FSO systems also limit the extent of usage in practical scenarios. Also, the already installed hardware which operates in RF frequency also needs to be utilized. Therefore, the vision that can drive the IoT systems will be a combination of FSO and RF systems.

Moreover, in this chapter, the existing literature on the combined RF/FSO relaying systems has been presented. More specifically, the works that have considered one or more kinds of practical design issues have been attended. Initial trends in the joint venture of RF/FSO technology were mostly considered by researchers as an aid to boost the capacity of next-generation wireless networks. However, recent studies in this field have suggested even though FSO provides a very large bandwidth to operate with, the impact of important system parameters needs to be addressed. The effect of interference on combined RF/FSO relaying system will be a determining factor while implementing such systems in principle. Moreover, the high switching requirements of FSO networks have the capability to provoke hardware impairment phenomenon on the overall characteristics of the combined form of relaying.

References

1. Ghassemlooy, Z., Popoola, W., & Rajbhandari, W. (2012). *Optical wireless communications: System and channel modeling with MATLAB*. CRC Press.
2. Arnon, S., Barry, J., Karagiannidis, G., Schober, R., & Uysal, M. (2012). *Advanced optical wireless communication*. Cambridge University Press.
3. Farid, A. A., & Hranilovic, S. (2007). Outage capacity optimization for free-space optical links with pointing errors. *Journal of Lightwave Technology, 25*, 1702–1710.
4. Sandalidis, H. G., Tsiftsis, T. A., Karagiannidis, G. K., & Uysal, M. (2008). BER performance of FSO links over strong atmospheric turbulence channels with pointing errors. *IEEE Communications Letters, 12*, 44–46.
5. Kaushal, H., & Kaddoum, G. (2017). Optical communication in space: Challenges and mitigation techniques. *IEEE Communications Surveys Tutorials, 19*(2), 57–96.
6. Lee, E., Park, J., Han, D., & Yoon, G. (2011). Performance analysis of the asymmetric dual-hop relay transmission with mixed RF/FSO links. *IEEE Photonics Technology Letters, 23*(21), 1642–1644.
7. Dahrouj, H., Douik, A., Rayal, F., Al-Naffouri, T. Y., & Alouini, M. S. (2015). Cost-effective hybrid RF/FSO backhaul solution for next generation wireless systems. *IEEE Wireless Communications, 22*(5), 98–104.
8. Ansari, I. S., Yilmaz, F., & Alouini, M. S. (2013). Impact of pointing errors on the performance of mixed RF/FSO dual-hop transmission systems. *IEEE Wireless Communications Letters, 2*(3), 351–354.

9. Zedini, E., Ansari, I. S., & Alouini, M. S. (2015). Performance analysis of mixed Nakagami-m and Gamma-Gamma dual-hop FSO transmission systems. *IEEE Photonics Journal, 7*(1), 1–20.
10. Safari, M., & Uysal, M. (2008). Relay-assisted free-space optical communication. *IEEE Transactions on Wireless Communication, 7*(12), 5441–5449.
11. Jaiswal, A., Bhatnagar, M. R., & Jain, V. K. (2017). Performance evaluation of space shift keying in free-space optical communication. *IEEE/OSA Journal of Optical Communication and Networking, 9*(2), 149–160.
12. Jaiswal, A., Bhatnagar, M. R., & Jain, V. K. (2017). Performance of optical space shift keying over Gamma-Gamma fading with pointing error. *IEEE Photonics Journal, 9*(2), 1–16.
13. Hamza, A. S., Deogun, J. S., & Alexander, D. R. (2019). Classification framework for free space optical communication links and systems. *IEEE Communications Surveys Tutorials, 21*(2), 1346–1382.
14. Leoraj, R., & Selvi, J. A. V. (2016). Comparative performance analysis of forward error correcting codes for free space optical communication. In *2016 International Conference on Emerging Trends in Engineering, Technology and Science (ICETETS)* (pp. 1–6).
15. Abu Almaalie, Z., Ghassemlooy, Z., Al-Rubaie, A. A. S., Lee, I. E., & Le-Minh, H. (2016). Forward error correction with physical layer network coding in two-way relay free space optical links. In *2016 8th Computer Science and Electronic Engineering (CEEC).*
16. Liu, J., Zhang, X., Blow, K., & Fowler, S. (2017). Performance analysis of packet layer FEC codes and interleaving in FSO channels. *IET Communications, 11*(13), 2042–2048.
17. Kiasaleh, K. (2010). Hybrid ARQ for FSO communications through turbulent atmosphere. *IEEE Communications Letters, 14*(9), 866–868.
18. Mai, V. V., & Pham, A. T. (2016). Cross-layer designs and analysis of adaptive-rate transmission and ARQ for free-space optical communications. *IEEE Photonics Journal, 8*(1), 1–15.
19. Le, H. D., Mai, V. V., Nguyen, C. T., & Pham, A. T. (2019). Design and analysis of sliding window ARQ protocols with rate adaptation for burst transmission over FSO turbulence channels. *IEEE/OSA Journal of Optical Communications and Networking, 11*(5), 151–163.
20. Joseph, T., Kaushal, H., Jain, V. K., & Kar, S. (2014). Performance analysis of OOK modulation scheme with spatial diversity in atmospheric turbulence. In *2014 23rd Wireless and Optical Communication Conference (WOCC)* (pp. 1–3).
21. Djordjevic, I. B., Vasic, B., & Neifeld, M. A. (2006). Multilevel coding in free-space optical MIMO transmission with Q-ary PPM over the atmospheric turbulence channel. *IEEE Photonics Technology Letters, 18*(14), 1491–1493.
22. Jain, S., Swarup, A., & Garg, M. (2013). Design and simulation of robust controller for flexure stage based piezo-actuated nano positioning device. *International Journal of Control and Automation, 6*(2), 17–28.
23. Singhal, M., Chauhan, R., & Sharma, G. (2011). A new approach for finding the various optimal variable to generate the BDD of a computer communication networks. *International Journal of Computer Applications, 31*(3), 1702–1710.
24. Sukavanam, N., Balasubramanian, R., & Kumar, S. (2007). Error estimation in reconstruction of quadratic curves in 3-D space. *International Journal of Computer Mathematics, 84*(1), 121–132.
25. Kumar, P., Pattanayak, P., & Roy, K. M. (2014). Analysis of a new MIMO broadcast channel limited feedback scheduling algorithm with user grouping. *Springer Wireless Personal Communication, 80*(3), 1079–1094.
26. Wang, Z., Zhong, W., Fu, S., & Lin, C. (2009). Performance comparison of different modulation formats over free-space optical (FSO) turbulence links with space diversity reception technique. *IEEE Photonics Journal, 1*(6), 277–285.
27. Jiang, S., Yang, G., Wei, Y., Bi, M., Lu, Y., Zhou, X., Hu, M., & Li, Q. (2015). Performance analysis of space-diversity free-space optical links over exponentiated Weibull channels. *IEEE Photonics Technology Letters, 27*(21), 2250–2252.

28. Zhu, B., Cheng, J., & Wu, L. (2015). A distance-dependent free-space optical cooperative communication system. *IEEE Communications Letters, 19*(6), 969–972.
29. Dabiri, M. T., Saber, M. J., & Sadough, S. M. S. (2017). On the performance of multiplexing FSO MIMO links in log-normal fading with pointing errors. *IEEE/OSA Journal of Optical Communications and Networking, 9*(11), 974–983.
30. AlQuwaiee, H., Ansari, I. S., & Alouini, M. (2016). On the maximum and minimum of double generalized Gamma variates with applications to the performance of free-space optical communication systems. *IEEE Transactions on Vehicular Technology, 65*(11), 8822–8831.
31. Simon, M. K., & Alouini, M. S. (2005). *Digital communication over fading channels* (2nd ed.). Wiley.
32. Hong, Y.-W.P., Huang, W.-J., & Kuo, C.-C.J. (2010). *Cooperative communications and networking: Technologies and system design.* Springer.
33. Balti, E., Guizani, M., Hamdaoui, B., & Khalfi, B. (2017). Mixed RF/FSO relaying systems with hardware impairments. In *2017 IEEE Global Communications Conference* (pp. 1–6).
34. Petkovic, M. I., & Trpovski, Z. (2018). Exact outage probability analysis of the mixed RF/FSO system with variable-gain relays. *IEEE Photonics Journal, 10*(6), 1–14.
35. Petkovic, M. I., Djordjevic, G. T., & Ivanis, P. N. (2016). Partial relay selection with variable gain relays and outdated CSI in mixed RF/FSO system. In *2016 24th Telecommunications Forum (TELFOR)* (pp. 1–4).
36. Bhatnagar, M. R. (2012). Performance analysis of decode-and-forward relaying in Gamma-Gamma fading channels. *IEEE Photonics Technology Letters, 24*(7), 545–547.
37. Sharma, P. K., & Garg, P. (2014). Bi-directional decode-XOR-forward relaying over M-distributed free space optical links. *IEEE Photonics Technology Letters, 26*(19), 1916–1919.
38. Yue, G., Wang, X., Yang, Z., & Host-Madsen, A. (2008). Coding schemes for user cooperation in low-power regimes. *IEEE Transactions on Signal Processing, 56*(5), 2035–2049.
39. Elfituri, M., Hamouda, W., & Ghrayeb, A. (2009). A convolutional-based distributed coded cooperation scheme for relay channels. *IEEE Transactions on Vehicular Technology, 58*(2), 655–669.
40. Bao, X., & Li, J. T. (2011). Generalized adaptive network coded cooperation (GANCC): A unified framework for network coding and channel coding. *IEEE Transactions on Communications, 59*(11), 2934–2938.
41. Ansari, I. S., Yilmaz, F., & Alouini, M.-S. (2016). Performance analysis of free-space optical links over Malaga (M) turbulence channels with pointing errors. *IEEE Transactions on Wireless Communications, 15*(1), 91–102.
42. Djordjevic, G., Petkovic, M., Cvetkovic, A., & Karagiannidis, G. (2015). Mixed RF/FSO relaying with outdated channel state information. *IEEE Journal on Selected Areas in Communications, 33*(9), 1935–1948.
43. El-Malek, A. H. A., Salhab, A. M., Zummo, S. A., & Alouini, M. S. (2016). Security-reliability trade-off analysis for multiuser SIMO mixed RF/FSO relay networks with opportunistic user scheduling. *IEEE Transactions on Wireless Communications, 15*(9), 5904–5918.
44. Touati, A., Abdaoui, A., Touati, F., Uysal, M., & Bouallegue, A. (2016). On the effects of combined atmospheric fading and misalignment on the hybrid FSO/RF transmission. *OSA Journal of Optical Communications and Networking, 8*(10), 715–725.
45. Trinh, P. V., Thang, T. C., & Pham, A. T. (2017). Mixed mmWave RF/FSO relaying systems over generalized fading channels with pointing errors. *IEEE Photonics Journal, 9*(1), 1–14.
46. Trigui, I., Cherif, N., & Affes, S. (2017). Relay-assisted mixed FSO-RF systems over Malaga and shadowed fading channels κ-μ. *IEEE Wireless Communications Letters, 6*(5), 682–685.
47. Arezumand, H., Zamiri-Jafarian, H., & Soleimani-Nasa, E. (2017). Outage and diversity analysis of underlay cognitive mixed RF-FSO cooperative systems. *IEEE/OSA Journal of Optical Communications and Networking, 9*(10), 909–920.
48. Varshney, N., & Jagannatham, A. K. (2017). Cognitive decode-and-forward MIMO-RF/FSO cooperative relay networks. *IEEE Communication Letters, 21*(4), 893–896.
49. Balti, E., Guizani, M., Hamdaoui, B., & Khalfi, B. (2018). Aggregate hardware impairments over mixed RF/FSO relaying systems with outdated CSI. *IEEE Transactions on Communication, 66*(3), 1110–1123.

50. AlQuwaiee, H., Ansari, I. S., & Alouini, M. S. (2015). On the performance of free-space optical communication systems over double generalized Gamma channel. *IEEE Journal on Selected Areas in Communications, 33*(9), 1829–1840.
51. Sharma, S., Madhu Kumar, A. S., & Swaminathan, R. (2019). Switching-based cooperative decode-and-forward relaying for hybrid FSO/RF networks. *Journal of Optical Communications and Networking, 11*(6), 267–281.
52. Xu, G., Zheng, Z., & Wang, W. (2020). Dual-hop deep space-terrestrial FSO/RF communication under solar scintillation: Performance analysis and challenges. *China Communications, 17*(7), 27–37.
53. Erdogan, E., Altunbas, I., Kurt, G. K., & Yanikomeroglu, H. (2022). The secrecy comparison of RF and FSO eavesdropping attacks in mixed RF-FSO relay networks. *IEEE Photonics Journal, 14*(1), 1–8.
54. Swaminathan, R., Sharma, S., Vishwakarma, N., & Madhu Kumar, A. S. (2021). HAPS-based relaying for integrated space/air/ground networks with hybrid FSO/RF communication: A performance analysis. *IEEE Transactions on Aerospace and Electronic Systems, 57*(3), 1581–1599.
55. Xu, G., & Zhang, Q. (2021). Mixed RF/FSO deep space communication system under solar scintillation effect. *IEEE Transactions on Aerospace and Electronic Systems, 57*(5), 3237–3251.
56. Lei, H., Luo, H., Park, K.-H., Ansari, I. S., Lei, W., Pan, G., & Alouini, M. S. (2020). On secure mixed RF-FSO systems with TAS and imperfect CSI. *IEEE Transactions on Communications, 68*(7), 4461–4475.
57. Kandelusy, O. M., & Kirsch, N. J. (2021). Cognitive buffer-aided mixed RF/FSO backhauling network with switch-based rate adaptation. *IEEE Communications Letters, 25*(8), 2644–2648.
58. Wang, Y., Tong, Y., & Zhan, Z. (2022). On secrecy performance of mixed RF-FSO systems with a wireless-powered friendly jammer. *IEEE Photonics Journal*, 1–1.
59. Liu, X., Gu, C., Guo, K., Cheng, M., Lin, M., & Zhu, W.-P. (2021). Robust beamforming and outage performance of uplink multiuser satellite-aerial-terrestrial networks with mixed RF-FSO channels. *IEEE Photonics Journal, 13*(4), 1–8.
60. Darwesh, L., & Kopeika, N. S. (2020). Deep learning for improving performance of OOK modulation over FSO turbulent channels. *IEEE Access, 8*, 155275–155284.
61. Gao, Z., Eisen, M., & Ribeiro, A. (2021). Resource allocation via model-free deep learning in free space optical communications. *IEEE Transactions on Communications*, 1–1.
62. El-Meadawy, S. A., Shalaby, H. M. H., Ismail, N. A., Farghal, A. E. A., El-Samie, F. E. A., Abd-Elnaby, M., & El-Shafai, W. (2021). Performance analysis of 3D video transmission over deep-learning-based multi-coded N-ary orbital angular momentum FSO system. *IEEE Access, 9*, 110116–110136.
63. Hao, Y., Zhao, L., Huang, T., Wu, Y., Jiang, T., Wei, Z., Deng, D., Luo, A.-P., & Liu, H. (2020). High-accuracy recognition of orbital angular momentum modes propagated in atmospheric turbulences based on deep learning. *IEEE Access, 8*, 159542–159551.
64. Wang, Z., & Guo, Z. (2019). Adaptive demodulation technique for efficiently detecting orbital angular momentum (OAM) modes based on the improved convolutional neural network. *IEEE Access, 7*, 163633–163643.
65. Andrews, J. G. (2014). What will 5G be? *IEEE JSAC, 32*(6), 1065–1082.
66. Li, Y., Pioro, M., & Angelakisi, V. (2013). Design of cellular backhaul topology using the FSO technology. In *Proc. 2nd Intl. Wksp. Optical Wireless Commun. Newcastle Upon Tyne, U.K* (pp. 6–10).
67. Jakeman, E. (1980). On the statistics of K-distributed noise. *Journal of Physics America, 13*, 31–48.
68. Phillips, R. L., & Andrews, L. C. (1981). Measured statistics of laser light scattering in atmospheric turbulence. *Journal of the Optical Society of America, 71*(12), 1440–2144.
69. Andrews, L. C., & Phillips, R. L. (1985). I-K distribution as a universal propagation model of laser beams in atmospheric turbulence. *Journal of Optical Society America A2, 2*, 160–163.

70. Churnside, J. H., & Clifford, S. F. (1987). Log-normal Rician probability density function of optical scintillations in the turbulent atmosphere. *Journal of Optical Society America, 4*, 1923–1930.
71. Andrews, L., & Phillips, R. L. (2005). *Laser beam propagation through random media* (2nd ed.). SPIE Press.
72. Al-Habash, M., Andrews, L. C., & Phillips, R. L. (2001). Mathematical model for the irradiance probability density function of a laser beam propagating through turbulent media. *Optical Engineering, 40*, 1554–1562.
73. Yacoub, M. D. (2007). The α-μ distribution: A physical fading model for the Stacy distribution. *IEEE Transactions on Vehicular Technology, 56*(1), 27–34.
74. Navas, A. J., Balsells, J. M. G., Paris, J. F., & Notario, A. P. (2011). A unifying statistical model for atmospheric optical scintillation. *Numerical Simulations of Physical and Engineering Processes* (Chap. 8).
75. Barrios, R., & Dios, F. (2013). Exponentiated Weibull model for the irradiance probability density function of a laser beam propagating through atmospheric turbulence. *Optical Laser Technology, 45*, 13–20.
76. Kashani, M. A., Uysal, M., & Kavehrad, M. (2015). A novel statistical channel model for turbulence-induced fading in free-space optical systems. *IEEE/OSA Journal of Lightwave Technology, 33*(11), 2303–2312.
77. Wang, P., Wang, R., Guo, L., Cao, T., & Yang, Y. (2016). On the performances of relay-aided FSO system over distribution with pointing errors in presence of various weather conditions. *Optics Communications, 367*, 59–67.
78. Tsiftsis, T. A., Sandalidis, H. G., Karagiannidis, G. K., & Sagias, N. C. (2006). Multihop free-space optical communications over strong turbulence channels. *IEEE International Conference on Communications, 6*, 2755–2759.
79. Tang, X., Wang, Z., Xu, Z., & Ghassemlooy, Z. (2014). Multihop free-space optical communications over turbulence channels with pointing errors using heterodyne detection. *Journal of Lightwave Technology, 32*(15), 2597–2604.
80. Zedini, E., & Alouini, M. S. (2015). On the performance of multihop heterodyne FSO systems with pointing errors. *IEEE Photonics Journal, 7*(2), 1–10.
81. Puri, P., Chatzidiamantis, N. D., Garg, P., Aggarwal, M., & Karagiannidis, G. K. (2015). Two-way relay selection in multiple relayed FSO networks. *IEEE Wireless Communications Letters, 4*(5), 485–488.
82. Park, J., Chae, C., & Yoon, G. (2017). Amplify-and-forward two-way relaying system over free-space optics channels. *Journal of Communications and Networks, 19*(5), 481–492.
83. Yang, L., Alouini, M. S., & Ansari, I. S. (2018). Asymptotic performance analysis of two-way relaying FSO networks with nonzero boresight pointing errors over double-generalized Gamma fading channels. *IEEE Transactions on Vehicular Technology, 67*(8), 7800–7805.
84. Al-Eryani, Y. F., Salhab, A. M., Zummo, S. A., & Alouini, M. S. (2018). Two-way multiuser mixed RF/FSO relaying: Performance analysis and power allocation. *IEEE/OSA Journal of Optical Communications and Networking, 10*(4), 396–408.
85. Abouei, J., & Plataniotis, K. N. (2012). Multiuser diversity scheduling in free-space optical communications. *Journal of Lightwave Technology, 30*, 1351–1358.
86. Chen, L., Wang, W., & Zhang, C. (2016). Multiuser diversity over parallel and hybrid FSO/RF links and its performance analysis. *IEEE Photonics Journal, 8*(3), 1–9.
87. Chen, L., & Wang, W. (2017). Effective capacity of MIMO free-space optical systems over Gamma-Gamma turbulence channels. *Optics Communications, Elsevier, 382*, 450–454.
88. Chen, L., & Wang, W. (2017). Multi-diversity combining and selection for relay-assisted mixed RF/FSO system. *Optics Communications, 405*, 1–7.
89. Yang, L., Hasna, M. O., & Ansari, I. S. (2017). Unified performance analysis for multiuser mixed η - μ and Malaga distribution dual-hop RF/FSO systems. *IEEE Transactions on Communications, 65*(8), 3601–3613.
90. Soleimani-Nasab, E., & Uysal, M. (2016). Generalized performance analysis of mixed RF/FSO cooperative systems. *IEEE Transactions on Wireless Communications, 15*(1), 714–727.

91. Petkovic, M. I., Cvetkovic, A. M., Djordjevic, G. T., & Karagiannidis, G. K. (2017). Outage performance of the mixed RF/FSO relaying channel in the presence of interference. *Wireless Personal Communication, 96*(2), 2999–3014.
92. Cherif, N., Trigui, I., & Affes, S. (2017). On the performance analysis of mixed multi-aperture FSO/multiuser RF relay systems with interference. In *18th IEEE International Workshop on Signal Processing Advances in Wireless Communications (SPAWC)* (pp. 1–5).
93. Trigui, I., Cherif, N., Affes, S., Wang, X., Leung, V., & Stphenne, A. (2017). Interference-limited mixed Malaga- and generalized-K dual-hop FSO-RF systems. In *Proceedings IEEE Annual International Symposium Personnel, Indoor, and Mobile Radio Communication, Montreal, QC, Canada* (pp. 1–6).
94. El-Malek, A. H. A., Salhab, A. M., Zummo, S. A., & Alouini, M. (2017). Effect of RF interference on the security-reliability trade-off analysis of multiuser mixed RF/FSO relay networks with power allocation. *Journal of Lightwave Technology, 35*(9), 1490–1505.
95. Balti, E., & Guizani, M. (2018). Mixed RF/FSO cooperative relaying systems with co-channel interference. *IEEE Transactions on Communications, 66*(9), 4014–4027.
96. Gurjar, D. S., Upadhyay, P. K., da Costa, D. B., & de Sousa, R. T., Jr. (2017). Beamforming in traffic-aware two-way relay systems withchannel estimation error and feedback delay. *IEEE Transactions on Vehicular Technology, 66*(10), 8807–8820.
97. Sharma, P. K., & Upadhyay, P. K. (2017). Performance analysis of cooperative spectrum sharing with multiuser two-way relaying over fading channels. *IEEE Transactions on Vehicular Technology, 66*(2), 1324–1333.
98. Upadhya, A., Dwivedi, V. K., & Singh, G. (2019). Multiuser diversity for mixed RF/FSO cooperative relaying in the presence of interference. *Optics Communications, 442*, 77–83.
99. Upadhya, A., Dwivedi, V. K., & Alouini, M. S. (2019). Interference-limited mixed MUD-RF/FSO two-way cooperative networks over double generalized Gamma turbulence channels. *IEEE Communications Letters, 442*, 1551–1555.
100. Upadhya, A., Dwivedi, V. K., & Karagiannidis, G. K. (2020). On the effect of interference and misalignment error in mixed RF/FSO systems over generalized fading channels. *IEEE Transactions on Communications, 68*(6), 3681–3695.
101. Bjornson, E., Matthaiou, M., & Debbah, M. (2013). A new look at dual-hop relaying: Performance limits with hardware impairments. *IEEE Transactions on Communications, 61*(11), 4512–4525.
102. Dardari, D., Tralli, V., & Vaccari, A. (2000). A theoretical characterization of nonlinear distortion effects in OFDM systems. *IEEE Transactions on Communications, 48*(10), 1755–1764.
103. Bjornson, E., Zetterberg, P., Bengtsson, M., & Ottersten, B. (2013). Capacity limits and multiplexing gains of MIMO channels with transceiver impairments. *IEEE Communications Letters, 17*(1), 91–94.
104. Matthaiou, M., Papadogiannis, A., Bjornson, E., & Debbah, M. (2013). Two-way relaying under the presence of relay transceiver hardware impairments. *IEEE Communications Letters, 17*(6), 1136–1139.
105. Balti, E., Guizani, M., Hamdaoui, B., & Maalej, Y. (2016). Partial relay selection for hybrid RF/FSO systems with hardware impairments. In *2016 IEEE Global Communications Conference (GLOBECOM)*.
106. Balti, E., Guizani, M., Hamdaoui, B., & Khalfi, B. (2018). Aggregate hardware impairments over mixed RF/FSO relaying systems with outdated CSI. *IEEE Transactions on Communications, 66*(3), 1110–1123.
107. Goel, A., & Bhatia, R. (2020). Hybrid RF/MIMO-FSO relaying systems over Gamma-Gamma fading channels. In *Proceedings of International Conference on Innovative Computing and Communication (ICICC-2020)*. Springer
108. Goel, A., & Bhatia, R. (2020). On the performance of mixed user diversity-RF/spatial diversity-FSO cooperative relaying AF systems. *Optics Communications, 477*.
109. Goel, A., & Bhatia, R. (2022). Performance analysis of multi-user diversity schemes on interference-limited D-GG atmospheric turbulence channels. In *Innovations in Electrical and Electronic Engineering: Proceedings of ICEEE* (Vol. 1). Springer.

110. Goel, A., & Bhatia, R. (2021). Joint impact of interference and hardware impairments on the performance of mixed RF/FSO cooperative relay networks. *Optical & Quantum Electronics, 442.*
111. Goel, A., & Bhatia, R. (2022). Double relay selection for combating the impact of interference and hardware impairments in mixed RF/FSO two-way relay networks. *Optical and Quantum Electronics.*